Managing Forest Ecosystems

Volume 40

Series Editors
Margarida Tomé, Instituto Superior de Agronomía, Lisboa, Portugal
Thomas Seifert, Faculty of Environment and Natural Resources,
University of Freiburg, Freiburg, Germany
Mikko Kurttila, Natural Resources Institute, Helsinki, Finland

The aim of the book series *Managing Forest Ecosystems* is to present state-of-the-art research results relating to the practice of forest management. Contributions are solicited from prominent authors. Each reference book, monograph or proceedings volume will be focused to deal with a specific context. Typical issues of the series are: resource assessment techniques, evaluating sustainability for even-aged and uneven-aged forests, multi-objective management, predicting forest development, optimizing forest management, biodiversity management and monitoring, risk assessment and economic analysis.

More information about this series at https://link.springer.com/bookseries/6247

Roberto Tognetti • Melanie Smith
Pietro Panzacchi

Editors

Climate-Smart Forestry in Mountain Regions

 Springer

Editors
Roberto Tognetti
Agricoltura, Ambiente e Alimenti
Universita del Molise
Campobasso, Campobasso, Italy

Melanie Smith
Inverness College
University of Highlands and Islands
Inverness, UK

Pietro Panzacchi
Faculty of Science and Technology
Free University of Bozen-Bolzano
Bolzano, Italy

ISSN 1568-1319   ISSN 2352-3956   (electronic)
Managing Forest Ecosystems
ISBN 978-3-030-80769-6   ISBN 978-3-030-80767-2   (eBook)
https://doi.org/10.1007/978-3-030-80767-2

This Springer imprint is published by the registered company Springer Nature Switzerland AG
The registered company address is: Gewerbestrasse 11, 6330 Cham, Switzerland

This book is dedicated to the memory of Professor Giustino Tonon: our friend, colleague, and a co-author of many of these chapters. Giustino conveyed great passion and commitment for forests and conservation; it showed in his words, his actions, and the intensity of his listening. Yet his calm and steadfast manner allowed him to work with everyone. His work brought a rigorous scientific approach to understanding the challenges faced by humans and nature in a time of rapid change. He was unfailingly caring and considerate of others, and his positive outlook on life invariably lifted the mood of those in his company. When he passed away on 7 July 2021, he was in a place he loved, the mountain forests of the Dolomites.

Preface

This book is one of the outputs of the COST Action CA15226, Climate-Smart Forestry in Mountain Regions (CLIMO). Funded by the EU's Horizon 2020 COST Action programme, CLIMO has been developing in the last few years as a new concept, which is central to the changes in the way forestry resources are used by the European community. While climate change is increasingly filling the policy agenda at global level, mountain regions are extremely vulnerable to its effects. Because climate change increases the frequency and intensity of ecosystem imbalances, the economic value and adaptive capacity of these regions is jeopardised, which has led to a call for changes in forestry polices and management.

Initiated in October of 2016 and finished in April of 2021, CLIMO addressed the complex issue of forest management, which plans for the long term, while dealing with uncertainties related to the productivity and health of forest ecosystems, and their adaptation to short-term environmental changes. The establishment of this network, which mobilized more than two hundred researchers from 28 countries, focused on mountain environments, considered as a climate change hotspot. Given the growing pressure on mountain regions by climate change, there is a need to emphasise forest production systems that are resilient to climate-driven disturbances. These climate targets can be mainstreamed through multidisciplinary Climate-Smart Forestry, paying attention to regional circumstances, opportunities and challenges. This multidisciplinary approach, from tree to landscape and with a variety of tools, was made possible through networking and stakeholders' engagement.

A three-dimensional approach was presented to enhance adaptation and resilience to climate change within forest ecosystems, optimising the provision of ecosystem services. A new definition of Climate-Smart Forestry advocated by this COST Action, supported by the development of indicators, allowed a balanced understanding of adaptation and mitigation potentials of mountain forests facing climate change. CLIMO also contributed to the debate concerning the resilience of forest and the provision of ecosystem services. This way, the seeds of progress in forest practices were sown, which hopefully will germinate into results that will allow for a more sustainable future.

The multiscale and multidisciplinary approach allowed evaluating social-ecological resilience of tree individuals and forest stands to climate change. In particular, the Action advanced our understanding on how to assess adaptation and mitigation trade-offs and synergies over time, in forest systems of mountain regions. In addition, a network of about 200 temporary and long-term experimental plots was established, including major forest tree species (beech, spruce, fir). Monitoring tools and advances of forest processes and ecosystem services were also addressed thanks to the collaboration between scientists from different disciplines, also involving colleagues from Brazil and Canada. There is now potential to build process-based monitoring networks and invest in schemes for payment for ecosystem services.

All of the above converged in several academic articles and in this book which has been planned as the way to put together concretely many authors from many countries on the same topic, while promoting interdisciplinarity. This book appeals to academics and researchers in forestry and related areas, also providing practical support to forest managers and decision makers.

Campobasso, Italy   Roberto Tognetti

Inverness, UK   Melanie Smith

Bolzano, Italy   Pietro Panzacchi

Acknowledgements

This book is based upon work from COST Action CA15226, Climate-Smart Forestry in Mountain Regions, supported by COST (European Cooperation in Science and Technology). COST (European Cooperation in Science and Technology) is a funding agency for research and innovation networks. COST Actions help connect research initiatives across Europe and enable scientists to grow their ideas by sharing them with their peers. This boosts their research, career and innovation. www.cost.eu

Contents

About the Editors

Roberto Tognetti completed undergraduate and master's degrees, majoring in forestry, at the University of Firenze (Italy). He got a PhD in botany from the Trinity College of Dublin (Ireland). He is full professor of forest ecology and management in the Department of Agricultural, Environmental and Food Sciences at the University of Molise (Italy), acting as chairman of the second level degree courses in forestry and environmental sciences. He studies the ecophysiological mechanisms underlying plant responses to environmental conditions and the influence these responses have on ecological patterns and processes. Plants live in a wide range of environments, and the conditions in these environments fluctuate over the time scale of seconds to years and beyond. He uses the combined potential of biometeorology and ecology to study the effects of disturbances on tree productivity and plant development. He focuses on the basic environmental physiology of carbon, water and nutrient cycling and strives to integrate these physiological processes to gain an understanding of plant functions and ecosystem processes, in a changing global environmental setting. Experimental observations are made at a range of spatial scales and a modelling framework is used in an effort to relate mechanistic responses to ecosystem functions and services.

Pietro Panzacchi got his master's degree in forestry from the University of Firenze (Italy) and a PhD in forest ecology from the University of Bologna (Italy). He currently works in the Center for Inland Areas and Apennines (ARIA) at the University of Molise (Italy) where he acted as project manager of the Cost Action CLIMO. His research focusses on the effect of climate change in forests biogeochemical cycles, with special interests for carbon and nitrogen cycling. Atmospheric nitrogen deposition, climate change and carbon stock potential of forests are strictly interwoven and their study at field level is challenging. In the last 10 years, his collaboration between University of Bologna (Italy), Free University of Bolzano/Bozen (Italy) and University of Molise (Italy) put him in the privilege position to study the effect of different drivers on different environments, from fruit orchards and poplar plantations in the Po plain (Italy) to miscanthus plantations in the UK to Alpine forest in South Tyrol. He started his collaboration with Roberto Tognetti while working for the EFI's Project Centre on Mountain Forests (MOUNTFOR) when he firstly approached the concept of Climate-Smart Forestry.

Melanie Smith graduated with her undergraduate joint honours degree in biology and geography from Royal Holloway, University of London. She completed a PhD in palaeoecology and woodland history with London University and Historic Scotland, investigating the interactions through the Holocene between people and their environment in northern Scotland. Currently she is assistant principal academic and research with Inverness College, University of the Highlands and Islands (Scotland), where she has worked since 2003, primarily leading the development and delivery of research and innovation. Throughout her career of nearly 30 years, she has led research projects and teaching in ecology and conservation, landscape ecology, forest history and catchment management. Her research in application of palaeoecological data to forest and conservation management led her to investigate further how an identification and understanding of forest functional traits over long time frames can inform the management of forests as complex adaptive systems for Climate-Smart Forestry.

Contributors

Iciar Alberdi Instituto Nacional de Investigación y Tecnología Agraria y Alimentaria, Madrid, Spain

A. Avdagić Faculty of Forestry, Department of Forest Management Planning and Urban Greenery, University of Sarajevo, Sarajevo, Bosnia and Herzegovina

João C. Azevedo Instituto Politécnico de Bragança, Escola Superior Agrária, Campus de Santa Apolónia, Bragança, Portugal

Ignacio Barbeito Southern Swedish Forest Research Center, Swedish University of Agricultural Sciences, Alnarp, Sweden
Université de Lorraine, AgroParisTech, INRA, UMR Silva, Nancy, France

Johan Barstad University College for Green Development, Bryne, Norway

Viera Baštáková SlovakGlobe: Slovak Academy of Sciences and Slovak University of Technology, Bratislava, Slovakia

Nicolas Bélanger Centre d'étude de la forêt, Université du Québec à MontréalMontréal, QC, Canada
Département Science et Technologie, Téluq, Université du QuébecMontréal, QC, Canada

Luca Belelli Marchesini Department of Sustainable Agro-Ecosystems and Bioresources, Fondazione Edmund Mach, San Michele all'Adige, Italy

Yves Bergeron Centre d'étude de la forêt, Université du Québec à Montréal, Montréal, QC, Canada
Forest Research Institute, Université du Québec en Abitibi-Témiscamingue, Montréal, QC, Canada

Kamil Bielak Department of Silviculture, Institute of Forest Sciences, Warsaw University of Life Sciences, Warsaw, Poland

Franz Binder Sachgebiet Schutzwald und Naturgefahren, Bayerische Landes-anstalt für Wald und Forstwirtschaft, Freising, Germany

Olivier Blarquez Département de Géographie, Université de Montréal, Montréal, QC, Canada

Andrej Bončina Department of Forestry and Renewable Forest Resources, Biotechnical Faculty, University of Ljubljana, Ljubljana, Slovenia

Michal Bosela Faculty of Forestry, Technical University in Zvolen, Zvolen, Slovakia
National Forest Centre, Zvolen, Slovakia

Giorgia Bottaro Land Environment Agriculture and Forestry Department (TeSAF), University of Padua, Padua, Italy

Alessandra Bottero Swiss Federal Institute for Forest, Snow and Landscape Research WSL, Birmensdorf, Switzerland

Euan Bowditch Inverness College UHI, University of the Highlands and Islands, Inverness, UK

Stanislava Brnkaľáková SlovakGlobe: Slovak Academy of Sciences and Slovak University of Technology, Bratislava, Slovakia

Jader Nunes Cachoeira Environmental Monitoring and Fire Management Center, University of Tocantins, Gurupi, Brazil

Han Y. H. Chen Faculty of Natural Resources Management, Lakehead University, Thunder Bay, ON, Canada

Paolo Cherubini Swiss Federal Research Institute WSL, Birmensdorf, Switzerland
Department of Forest and Conservation Sciences, Faculty of Forestry, University of British Columbia, Vancouver, BC, Canada

Luis Coll Department of Agriculture and Forest Engineering, School of Agrifood and Forestry Science and Engineering, University of Lleida, Lleida, Spain
Joint Research Unit CTFC-AGROTECNIO, Solsona, Spain

Philip G. Comeau Department of Renewable Resources, University of Alberta, Edmonton, AB, Canada

Sonia Condés Department of Natural Systems and Resources, School of Forest Engineering and Natural Resources, Universidad Politécnica de Madrid, Madrid, Spain

Damiana Beatriz da Silva Environmental Monitoring and Fire Management Center, University of Tocantins, Gurupi, Brazil

Louis De Grandpré Natural Resources Canada, Canadian Forest Service, Laurentian Forestry Centre, Quebec City, QC, Canada

Sylvain Delagrange Centre d'étude de la forêt, Université du Québec à Montréal, Montréal, QC, Canada
Institut des Sciences de la Forêt Tempérée (ISFORT), Université du Québec en Outaouais (UQO), Ripon, QC, Canada

Miren del Río INIA, Forest Research Centre, Madrid, Spain
iuFOR, Sustainable Forest Management Research Institute, University of Valladolid & INIA, Valladolid, Spain

Annie DesRochers Centre d'étude de la forêt, Université du Québec à Montréal, Montréal, QC, Canada
Forest Research Institute, Université du Québec en Abitibi-Témiscamingue, Montréal, QC, Canada

Amanda Diochon Department of Geology, Lakehead University, Thunder Bay, ON, Canada

L'ubica Ditmarová Institute of Forest Ecology, Slovak Academy of Sciences, Zvolen, Slovakia

Loïc D'Orangeville Faculty of Forestry and Environmental Management, University of New Brunswick, Fredericton, NB, Canada

Pierre Drapeau Centre d'étude de la forêt, Université du Québec à Montréal, Montréal, QC, Canada
Département des sciences biologiques, Université du Québec à Montréal, Montréal, QC, Canada

Lenka Dubova Faculty of Social and Economic Studies, Institute for Economic and Environmental Policy, J. E. Purkyne University, Usti nad Labem, Czech Republic

Louis Duchesne Direction de la Recherche Forestière, Ministère des Forêts, de la Faune et des Parcs du Québec, Quebec City, QC, Canada

Marek Fabrika Faculty of Forestry, Technical University in Zvolen, Zvolen, Slovakia

Gianluca Filippa ARPA Valle d'Aosta, Saint-Christophe, AO, Italy

Elise Filotas Centre d'étude de la forêt, Université du Québec à Montréal, Montréal, QC, Canada
Département Science et Technologie, Téluq, Université du Québec, Montréal, QC, Canada

Christoph Fischer Swiss Federal Institute for Forest, Snow and Landscape Research WSL, Birmensdorf, Switzerland

Ludovico Frate Freelance, Ispra, Varese, Italy

Paola Gatto Dipartimento Territorio e Sistemi Agroforestali, Università degli Studi di Padova, Legnaro, Padova, Italy

Rachel Gaulton School of Natural and Environmental Sciences, Newcastle University, Newcastle upon Tyne, UK

Fabio Gennaretti Forest Research Institute, Université du Québec en Abitibi-Témiscamingue, Montréal, QC, Canada

Veronika Gežík Institute for Forest Ecology, Slovak Academy of Sciences, Bratislava, Slovakia

Francesco Giammarchi Faculty of Science and Technology, Free University of Bolzano-Bozen, Bolzano, Italy

Damiano Gianelle Department of Sustainable Agro-Ecosystems and Bioresources, Fondazione Edmund Mach, San Michele all'Adige, Italy

Marcos Giongo Environmental Monitoring and Fire Management Center, University of Tocantins, Gurupi, Brazil

Giacomo Grassi European Commission, Joint Research Centre, Ispra, Varese, Italy

Stefano Grigolato Department of Land, Environment, Agriculture and Forestry, Università degli Studi di Padova, Legnaro, PD, Italy

Thomas Gschwantner Federal Research and Training Centre for Forests, Natural Hazards and Landscape BFW, Vienna, Austria

Berthold Heinze Department of Forest Genetics, BFW Austrian Federal Research Centre for Forests, Vienna, Austria

Torben Hilmers School of Life Sciences Weihenstephan, Technical University of Munich, Freising, Germany

Maria Höhn Faculty of Horticultural Science, Department of Botany, Szent István University, Budapest, Hungary

Daniel Houle Science and Technology Branch, Environment and Climate Change Canada, Montreal, QC, Canada

Mohammad Imangholiloo Department of Forest Sciences, University of Helsinki, Helsinki, Finland

Gabriela Jamnická Institute of Forest Ecology, Slovak Academy of Sciences, Zvolen, Slovakia

Milica Kašanin-Grubin Institute for Chemistry, Technology and Metallurgy, University of Belgrade, Belgrade, Serbia

Matija Klopčič Biotechnical Faculty, Department of Forestry and Renewable Forest Resources, University of Ljubljana, Ljubljana, Slovenia

Tatiana Kluvánková SlovakGlobe: Slovak Academy of Sciences and Slovak University of Technology, Bratislava, Slovakia

Daniel Kneeshaw Centre d'étude de la forêt, Université du Québec à Montréal, Montréal, QC, Canada
Département des sciences biologiques, Université du Québec à Montréal, Montréal, QC, Canada

Anu Korosuo European Commission, Joint Research Centre, Ispra, Varese, Italy

Benoit Lafleur Forest Research Institute, Université du Québec en Abitibi-Témiscamingue, Montréal, QC, Canada

David Langor Natural Resources Canada, Canadian Forest Service, Edmonton, AB, Canada

Nicola La Porta Department of Sustainable Agroecosystems and Bioresources, IASMA Research and Innovation Centre, Fondazione Edmund, Mach, Trento, Italy

Simon Lebel Desrosiers Centre d'étude de la forêt, Université du Québec à Montréal, Montréal, QC, Canada
Département Science et Technologie, Téluq, Université du Québec, Montréal, QC, Canada

Jerzy Lesiński Department of Forest Biodiversity, Faculty of Forestry, University of Agriculture in Krakow, Krakow, Poland

Francois Lorenzetti Centre d'étude de la forêt, Université du Québec à Montréal, Montréal, QC, Canada
Institut des Sciences de la Forêt Tempérée (ISFORT), Université du Québec en Outaouais (UQO), Ripon, QC, Canada

Sebastiaan Luyssaert Department of Ecological Sciences, Faculty of Sciences, University of Amsterdam, Amsterdam, The Netherlands

Rongzhou Man Ontario Forest Research Institute, Ontario Ministry of Natural Resources and Forestry, Sault Ste. Marie, ON, Canada

John D. Marshall Department of Forest Ecology and Management, SLU, Umea, Sweden

Katarína Merganičová Faculty of Forestry and Wood Sciences, Czech University of Life Sciences PraguePraha, Suchdol, Czech Republic
Department of Biodiversity of Ecosystems and Landscape, Slovak Academy of Sciences, Institute of Landscape Ecology, Nitra, Slovak Republic

Christian Messier Centre d'étude de la forêt, Université du Québec à Montréal, Montréal, QC, Canada
Institut des Sciences de la Forêt Tempérée (ISFORT), Université du Québec en Outaouais (UQO), Ripon, QC, Canada

Ilona Mészáros Faculty of Science and Technology, Department of Botany, University of Debrecen, Debrecen, Hungary

Miguel Montoro Girona Forest Research Institute, Université du Québec en Abitibi-Témiscamingue, Montréal, QC, Canada
Restoration Ecology Group, Department of Wildlife, Fish and Environmental Studies, Swedish University of Agricultural Sciences, Umeå, Sweden

Micael Moreira Santos Environmental Monitoring and Fire Management Center, University of Tocantins, Gurupi, Brazil

Gert-Jan Nabuurs Wageningen Univ & Res, Wageningen, Netherlands

Bożydar Neroj Bureau for Forest Management and Geodesy, Sekocin Stary, Poland

Charles Nock Department of Renewable Resources, University of Alberta, Edmonton, AB, Canada

Maciej Pach Department of Forest Ecology and Silviculture, Faculty of Forestry, University of Agriculture in Krakow, Kraków, Poland

Pietro Panzacchi Centro di Ricerca per le Aree Interne e gli Appennini (ArIA), Università degli Studi del Molise, Campobasso, Italy
Department of Biosciences and Territory, Università degli Studi del Molise, Pesche, Italy
Faculty of Science and Technology, Free University of Bolzano-Bozen, Bolzano, Italy

Christoforos Pappas Centre d'étude de la forêt, Université du Québec à Montréal, Montréal, QC, Canada
Département Science et Technologie, Téluq, Université du Québec, Montréal, QC, Canada

Davide Pettenella Land Environment Agriculture and Forestry Department (TeSAF), University of Padua, Padua, Italy

Michael Pfatrisch Bureau for Forest Management and Geodesy, Sekocin Stary, Poland

Gianni Picchi Institute of Bioeconomy of the National Research Council (CNR-IBE), Sesto Fiorentino, FI, Italy

Roberto Pilli European Commission, Joint Research Centre, Ispra, Varese, Italy

Hans Pretzsch School of Life Sciences Weihenstephan, Technical University of Munich, Freising, Germany

Jakub Sandak InnoRenew CoE, Izola, Slovenia

Giovanni Santopuoli Dipartimento Agricoltura, Ambiente e Alimenti, Università degli Studi del Molise, Campobasso, Italy
Centro di Ricerca per le Aree Interne e gli Appennini (ArIA), Università degli Studi del Molise, Campobasso, Italy

Murat Sarginci Faculty of Forestry, Duzce University, Duzce, Turkey

Yusuf Serengil Faculty of Forestry, Department of Watershed Management, Istanbul University, Istanbul, Turkey

Jerzy Skrzyszewski Department of Forest Ecology and Silviculture, Faculty of Forestry, University of Agriculture in Krakow, Krakow, Poland

Lenka Slavikova Institute for Economic and Environmental Policy, Faculty of Social and Economic Studies, J. E. Purkyne University, Usti nad Labem, Czech Republic

Melanie Smith Inverness College, University of Highlands and Islands, Inverness, UK

Peter Spathelf Faculty of Forest and Environment, Eberswalde University for Sustainable Development, Eberswalde, Germany

Radoslaw Sroga Bureau for Forest Management and Geodesy, Sekocin Stary, Poland

Branko Stajić Faculty of Forestry, University of Belgrade, Belgrade, Serbia

Kilian Stimm School of Life Sciences Weihenstephan, Technical University of Munich, Freising, Germany
Bavarian State Institute of Forestry (LWF), Freising, Germany

Roar Stokken Department for Social Sciences and History, Volda University College, Volda, Norway

Katarina Strelcova Faculty of Forestry, Technical University in Zvolen, Zvolen, Slovakia

Christian Temperli Swiss Federal Institute for Forest, Snow and Landscape Research WSL – Forest Resources and Management, Birmensdorf, Switzerland

Barb R. Thomas Department of Renewable Resources, University of Alberta, Edmonton, AB, Canada

Roberto Tognetti Dipartimento di Agricoltura, Ambiente e Alimenti, Università degli Studi del Molise, Campobasso, Italy
Centro di Ricerca per le Aree Interne e gli Appennini (ArIA), Università degli Studi del Molise, Campobasso, Italy

Giustino Tonon Faculty of Science and Technology, Free University of Bolzano/Bozen, Bolzano, Italy

Chiara Torresan Institute of BioEconomy (IBE) – National Research Council of Italy, San Michele all'Adige, TN, Italy

Enno Uhl School of Life Sciences Weihenstephan, Technical University of Munich, Freising, Germany
Bavarian State Institute of Forestry (LWF), Freising, Germany

Riccardo Valentini DIBAF – Department for Innovation in Biological, Agri-Food and Forest Systems, University of Tuscia, Viterbo, Italy
Smart Urban Nature Laboratory, RUDN University, Moscow, Russia

Violeta Velikova Institute of Plant Physiology and Genetics – Bulgarian Academy of Sciences, Sofia, Bulgaria

Matteo Vizzarri European Commission, Joint Research Centre, Ispra, Varese, Italy

Andrew Weatherall National School of Forestry, University of Cumbria, Ambleside, UK

Timothy Work Centre d'étude de la forêt, Université du Québec à Montréal, Montréal, QC, Canada
Département des sciences biologiques, Université du Québec à Montréal, Montréal, QC, Canada

Tzvetan Zlatanov Institute of Biodiversity and Ecosystem Research, Bulgarian Academy of Sciences, Sofia, Bulgaria

Chapter 1
An Introduction to Climate-Smart Forestry in Mountain Regions

Roberto Tognetti, Melanie Smith, and Pietro Panzacchi

Abstract The goal to limit the increase in global temperature below 2 °C requires reaching a balance between anthropogenic emissions and reductions (sinks) in the second half of this century. As carbon sinks, forests can potentially play an important role in carbon capture. The Paris Agreement (2015) requires signatory countries to reduce deforestation, while conserving and enhancing carbon sinks. Innovative approaches may help foresters take up climate-smart management methods and identify measures for scaling purposes. The EU's funding instrument COST has supported the Action CLIMO (Climate-Smart Forestry in Mountain Regions – CA15226), with the aim of reorienting forestry in mountain areas to challenge the adverse impacts of climate change.

Funded by the EU's Horizon 2020, CLIMO has brought together scientists and experts in continental and regional focus assessments through a cross-sectoral approach, facilitating the implementation of climate objectives. CLIMO has provided scientific analysis on issues including criteria and indicators, growth dynamics, management prescriptions, long-term perspectives, monitoring technologies, economic impacts, and governance tools. This book addresses different combinations of CLIMO's driving/primary objectives and discusses smarter ways to develop forestry and monitor forests under current environmental changes, affecting forest ecosystems.

R. Tognetti (✉)
Dipartimento di Agricoltura, Ambiente e Alimenti, Università degli Studi del Molise, Campobasso, Italy
e-mail: tognetti@unimol.it

M. Smith
Inverness College, University of Highlands and Islands, Inverness, UK
e-mail: melanie.smith.ic@uhi.ac.uk

P. Panzacchi
Department of Biosciences and Territory, Università degli Studi del Molise, Pesche, Italy

Faculty of Science and Technology, Free University of Bolzano-Bozen, Bolzano, Italy

R. Tognetti et al. (eds.), *Climate-Smart Forestry in Mountain Regions*, Managing Forest Ecosystems 40, https://doi.org/10.1007/978-3-030-80767-2_1

1.1 Forests and Climate Change

The recent report of IPCC on climate change, desertification, land degradation, sustainable land management, food security, and greenhouse gas (GHG) fluxes in terrestrial ecosystems (IPCC 2019) highlights a considerable increase in mean land surface air temperature, since the preindustrial period, even in comparison with the global mean surface temperature. In parallel, an estimated 23% of total anthropogenic GHG emissions has occurred in the decade 2007–2016, due to agriculture, forestry, and other land uses, which affects more than 70% of the land surface (22% for managed and plantation forests). Curtis et al. (2018) calculated that 27% of global forest loss in the 2001–2015 period was due to deforestation, through permanent land use change for commodity production. In the remaining areas, which maintained the same land use, forest loss was attributed to forestry operations (26%), shifting agriculture (24%), forest fires (23%), and urban expansion (0.6%). Forest management strategies may interact with climate forces other than GHG (e.g., surface albedo, canopy roughness, biogenic emissions, stand evapotranspiration). Indeed, changing forest species composition may affect the emissions of biogenic volatile organic compounds, and the formation of secondary organic aerosols, which, in turn, have warming and cooling effects, depending on time, location, and type of emission (Šimpraga et al. 2019). Generally, deciduous trees have higher reflectivity (albedo), with consequent reduced warming effect, than conifer species. This is nuanced by different forest types and management intensities having different albedos, depending on the degree of exposure of the forest floor toward the atmosphere, particularly when covered by snow.

Climate change has already caused many changes in forest ecosystems and negative effects prevail, including warming-induced shifts in species distribution (Lindner et al. 2010; Boisvert-Marsh et al. 2014) and drought-related increases in tree mortality (Allen et al. 2010; Cailleret et al. 2017). Impacts of climate change magnify local disturbances (Fig. 1.1), such as environmental pollution, nitrogen deposition, habitat fragmentation, forest fire, pest outbreak, and alien species, altering forest development trajectories and decreasing capacity for resistance (Millar and Stephenson 2015; Johnstone et al. 2016). The climate emergency challenges traditional silvicultural strategies that have evolved through hundreds of years of relative climate stability. This indicates the necessity of developing and adopting the adaptable management framework of climate-smart forestry (CSF).

Climate-smart forestry is forestry that sustainably raises timber productivity (production), increases resilience (adaptation), stores carbon (mitigation), and enhances the achievement of development goals. Climate-smart forestry involves monitoring forest functions and anticipating disturbance effects, while undertaking resilient actions to avoid the negative consequences on the provision of ecosystem services and forest productivity (Bowditch et al. 2020). Climate-smart forest structure and functions require flexible silvicultural approaches, which may address mitigation needs and adaptation circumstances at different spatial scales. Climate-smart forestry, as defined in Chap. 2 of this book (Weatherall et al. 2021), can be

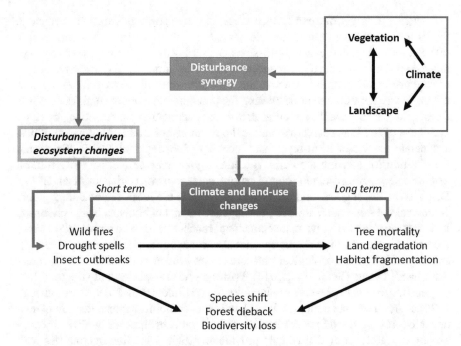

Fig. 1.1 Conceptual and empirical models that address the impacts of climate and land-use changes on forest vulnerability need to focus on landscape dynamics and vegetation processes, integrating disturbance strategies

considered a toolbox for sustainable management of forests, where the goal is to increase the resilience to environmental disturbance, reduce the deviation from natural structure, and use the wood for carbon storage.

All the climate drivers and relevant sectors (including forestry, agriculture, energy, transportation, industry, construction) should be considered, when studying the contribution of land use and biomass systems in reducing the radiative forcing in the atmosphere over varying time scales. Yet, the interaction among climate change, human activities, and ecosystem processes (e.g., forest aging and natural disturbances), as well as the capacity of forests to sequester and store carbon over time, needs to be addressed for all biogeographic regions. In doing this, the climate impact of CSF, considering CO_2 and other GHG, the effects of albedo and evapotranspiration, biogenic emissions, and the various aspects related to biodiversity and resilience, needs to be assessed and monitored (Fig. 1.1). Engaging forest managers, policy makers, and relevant stakeholders is central to ensure applicability of CSF at different levels (local to European), and to pursue the integration with forest inventory data and decision support systems.

The Paris Agreement sets the global warming target to be "well below +2 °C compared to the preindustrial levels." In order to achieve this, the Parties have given voluntary pledges to decrease their net emissions of GHG (the so-called Intended Nationally Determined Contributions – INDCs). The latter are clearly not enough to reach the +2 °C target, and extra efforts are required as soon as possible. Nevertheless, the Paris Agreement endorsed the role of forests as the most important carbon sink that can be managed to balance emissions and removals. Therefore, forest management should aim to reduce CO_2 concentration in the atmosphere (Articles 4 and 5), as well as the radiative imbalance at the top of the atmosphere (Article 2), without raising the air temperature or decreasing the precipitation amount (Article 7). While forests subject to prevailing natural drivers are regarded as representing the baseline, most of the European forestland is the result of millennia of human activities, a dynamic mosaic of disturbance and recovery (Sabatini et al. 2018). Significant contribution for meeting the Paris Agreement goals may derive from silvicultural activities (e.g., reforestation, conservation, management) (Griscom et al. 2017). Demand for forest goods and services is growing in the local, regional, and global context; at the same time, however, the negative impact of climate change on forests is increasing and alternative management approaches need to be employed to avoid the risk of exceeding thresholds for global chronic stress (Hartmann et al. 2018), stability of long-term ecological processes (Hanewinkel et al. 2013), and discrete disturbance events (Seidl et al. 2014). Fostering mixed-species and structural heterogeneous forests to preserve productivity (e.g., Hilmers et al. 2019; Torresan et al. 2020) and/or a balanced mosaic forest landscape, combining intensification of sustainable forest production with retention of high value habitats for biodiversity conservation (Ceddia et al. 2014), may provide an efficient solution in countries with high quality of governance, and a template exemplar for others.

1.2   A Climate-Smart Perspective: Becoming Climate Smart

Trees and forests are passively subject to climate, but are also able to dynamically modify it, providing important ecosystem services. Indeed, although 5–10 GtCO$_2$e per year come from deforestation and forest degradation, forest protection and the reduction of forest degradation represent effective mitigation options, in terms of social and environmental benefits (0.4–5.8 Gt CO_2e per year). Therefore, to maximize the climate benefits of forests, more forest landscapes should be maintained, and objectives set to restore those lost (Fig. 1.2). Overall, forest should be managed more sustainably. Managing forests to more closely recall the structure, function, and composition of natural forests, compared with traditional approaches, has been the goal of several silvicultural frameworks aimed to sustain the productivity of healthy forests and to maintain the provision of ecosystem services across successional stages (e.g., Seymour and Hunter 1999; Franklin et al. 2007; Puettmann et al. 2015).

Traditionally, the main objective of forest management has been to harvest timber Puettmann et al. 2009), often converting uneven-aged mixed-species stands to even-aged homogeneous monocultures, mimicking an agricultural approach, increasing the standing volume, volume increment, and allowable cut. The transformation of ecological functions (e.g., watershed conservation, soil retention,

Fig. 1.2 Lago di Carezza (1534 m a.s.l.) and the Latemar forest, a typical multifunctional landscape of the Dolomites (Italy)

biodiversity protection, carbon sequestration) into environmental services occurs when people take benefits. Nevertheless, society has historically focused on timber commodity chains and nonwood forest products, neglecting most environmental services. Recently, a growing interest has emerged in achieving multiple Sustainable Development Goals (SDGs) through effective interactions or synergies among measures and policies (Rio + 20 conference in 2012). In particular, the urgent need of mitigating climate change has led to promote forest protection (SDG 15) in order to accumulate carbon in trees and soils, and, therefore, mitigate climate change (SDG 13). Diversification and flexibility of production systems are considered crucial approaches for increasing resilience of forests to natural disturbances.

More forest area than ever since the Middle Ages, although tree cover cannot accurately assess the surface of forest ecosystems, poses high expectation on the European forest sector (Nabuurs et al. 2019). Large mitigation effects from forest in the European Union (EU) can be expected by increasing sequestration potentials and through substitution effects (Grassi et al. 2019). However, reducing temperature increase through forestry (mitigation) may have a limited effect on climate (Fig. 1.3), since trade-offs exist between options to meet climate objectives (Luyssaert et al. 2018). Therefore, a question arises on whether shifting forest management from timber-oriented approaches to CSF should privilege adaptation (climate prospect) rather than mitigation (carbon view), particularly in the vulnerable mountain landscape (Fig. 1.3). Forest ecosystems and local circumstances of European mountains

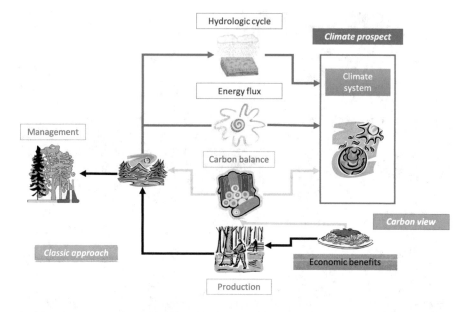

Fig. 1.3 Forests store large amounts of carbon in tree biomass, understory vegetation, and soil compartments, but mitigation may have poor effect on climate

are well suited to address questions associated with climate-smart potentials of the forest sector (European Environment Agency. 2016). Locally tailored CSF measures need to minimize trade-offs among bioeconomy, adaptation (biodiversity, disturbance), and mitigation (carbon, substitution) options. In Europe, 41% of mountain areas are covered by forests providing diverse ecosystem services, with an impact both at local communities' scale and at regional scale (Price et al. 2011).

Nabuurs et al. (2018), following the Climate-Smart Agriculture concept developed by FAO, proposed a CSF approach for European forests, suggesting safeguarding the mitigation potential of forests against climate change through an array of regionally tailored measures. Kauppi et al. (2018), focusing on bioeconomy issues and policy instruments, stressed the need to tackle multiple policy goals with CSF, by reducing or removing GHG emissions, adapting and building forest resilience, and sustainably increasing forest productivity and incomes. Verkerk et al. (2020), trying to indicate possible synergies between the mitigation and adaptation pillars of CSF with other forest functions, considered the sustainable climate mitigation potential of the whole forest and wood product chain, including material and energy substitution and accounting for local circumstances. However, the implementation of CSF approaches and the establishment of disturbance-resilient forest systems have not yet become rooted in silvicultural practices and with forest operators that promote sustainable forest management, consequently some significant open questions remain:

- How to implement adaptation, mitigation, and production options in forestry systems, in synergy with the development of bioeconomy and the preservation of biodiversity?

- How to achieve multifunctional landscapes, through addressing the sustainable provision of environmental services to accomplish the tenure security of owner-ship rights?
- How to interconnect local circumstances, opportunities, and challenges with governance goals to reap the potential of the forest sector in meeting the climate targets?

Forest managers that want to meet climate change issues through sustainable forest management practices will need to plan at multiple spatial and temporal scales and operate with multidisciplinary approaches and design flexibility (Innes 2009). Threats to forest ecosystems can be reduced by increasing the adaptive capacity, resilience strength, and resource-use efficiency of forestry systems. Although forest managers have always dealt with patchiness in environmental conditions, particularly in mountain areas, the escalation in climatic variation at the global level and the increase in uncertainty at the local scale call for more flexible and rapid response capacity.

A comprehensive definition of CSF and the process for selecting suitable indicators of climate-smartness were proposed by Bowditch et al. (2020). Adaptation and mitigation issues, as well as the social dimension, were the core focus of this definition, which recognizes the need to integrate and avoid development of these aspects in isolation. The need for integration derives from the complementarity of the resilience concepts in forestry, quantified as the recovery time after a disturbance (Nikinmaa et al. 2020): engineering, ecological, and social-ecological resilience. The complexity of social-ecological and practical challenges, associated with changing climate, disturbance, and governance, results in difficulties for forest managers in applying resilience concepts and CSF. The COST Action CLIMO (Climate-Smart Forestry in Mountain Regions) started responding to these multifaceted needs by laying the foundations of long-term experimental forest sites in mountain areas, which aim to facilitate the exchange of knowledge and practice and build forest manager–scientist partnerships.

Climate-smart forestry fosters the integration of scientific knowledge, technical skill, and policy action toward smart prospects through learning, monitoring, planning, coordinating, supporting, and financing the road to resilient forest systems and flexible management strategies (Fig. 1.4). Involving local communities and administrations in participatory design of specific solutions has a key role in fostering successful policies and actions. Operating forestry systems to accomplish climate targets follows the three pillars of CSF:

1. The development of mitigation opportunities for carbon sequestration in forests
2. The advancement of adaptation strategies for resilience to climate change
3. The intensification of socio-ecological sustainability for natural resource management

Although these three pillars of CSF require common consideration, the relative importance of each objective may vary locally, as much as trade-offs and synergies do (Lipper et al. 2014).

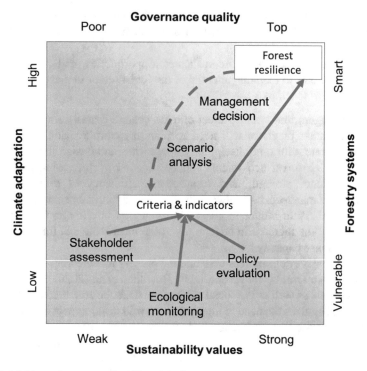

Fig. 1.4 Linking values to sustainability with climate smartness of the forestry system

Through effective knowledge exchange with forest managers and the entire sector, forestry research has the potential to reorient management practices and utilization approaches, as well as to develop evidence-based policy for CSF. In this context, CSF builds on well-established sustainable forest management approaches, for example, close-to-nature forestry (Bauhus et al. 2013), continuous-cover forestry (Pommerening and Murphy 2004), ecological forestry (Franklin et al. 2007), adaptive silviculture (Nagel et al. 2017), and systemic silviculture (Ciancio and Nocentini 2011). Combinative approaches have been developed to coordinate diverse management objectives on the same land, aimed at satisfying multiple societal demands (Aggestam et al. 2020); conversely, segregative approaches to land use target the maximization of specific objectives in separate spatial contexts (Phalan et al. 2011). To meet the increasing demand for ecosystem services, forestry needs to consider territorial-specific potentialities and limitations, and land-allocation procedures (Messier et al. 2019). Climate-smart forestry may also encompass sustainable intensification on abandoned agricultural land, for mitigation purposes (e.g., short-rotation forestry), and the establishment or strengthening of protected forest areas to compensate for the loss or degradation of biodiversity. Indeed, rapidly changing climatic conditions and land uses call for continuous updating of forestry practice through innovation, knowledge exchange, and learning. Whether land sharing or land sparing may better limit environmental impacts, ensuring the supply of goods, is still a matter of debate. Some agricultural studies argue that more intensive and efficient farming would increase farm productivity on reduced farmed hectarage, this would reduce the total land required for agriculture, and release land for other ecosystem services, for example, biodiversity, flood mitigation, and carbon capture. (Phalan et al. 2016). In contrast, other studies claim that low-intensity agriculture can satisfy the increased demand for food, while promoting biodiversity and other ecosystem services (Tscharntke et al. 2012). The latter approach has been traditionally implemented in the EU's agricultural schemes to compensate potential loss of income by farmers who mitigate detrimental effects of intensification on biodiversity. As forests are planted on agricultural land for carbon capture and nature conservation, there is a need to understand and plan how these forests integrate into emerging dynamic landscapes. Recently, Messier et al. (2019) have proposed to integrate functional diversity and redundancy concepts into complex spatial network approaches, as an adaptable strategy to manage forest systems at multiple scales. This strategy can be further informed through reference to true long-term ecological data.

1.3   Referencing True Long-Term Ecological Data for CSF

Key to achieving the goals of climate-smart forestry is an understanding of how species assemblages respond to environmental change. Accessing comprehensive long-term ecological data is an important part of building this understanding (Pretzsch et al. 2017; del Río et al. 2021), but until recently, "long term" generally referred to data sets of decadal records. However, *true* long-term ecological (tLTE) data encompasses millennial scale information (Willis et al. 2010; Rull 2014), drawing upon transdisciplinary sources including paleoecology, archeology, and history (Fig. 1.5). The epochal scale of the climate change now underway demands consideration of the last time such changes occurred, for example, at the end of the

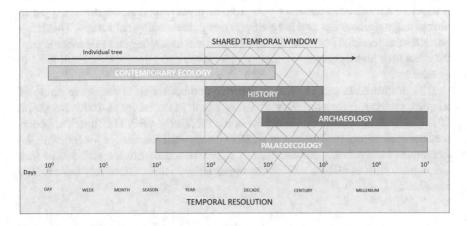

Fig. 1.5 Shared window of temporal evidence sources. (Adapted from Rull 2014)

last ice age. It also demands conceptual tools for analyzing how plant species assemble into communities.

Messier et al. (2019) describe one set of tools based on the notions of functional diversity and functional redundancy. High diversity in functional traits (e.g., between gymnosperms and angiosperms) can be achieved through the presence of just two species within a forest stand, but in such a stand, the functional redundancy is potentially low, because the functional traits are only represented by each of the two species. If one species disappears, the entire representation of those functional traits also disappears. High functional diversity and redundancy can be achieved where several different species in the same stand display the same functional traits. By viewing tLTE evidence through the lens of functional trait analyses, it may be possible to infer how species and populations responded to past climatically induced disturbances. Hamilton et al. (2018) discuss and identify some of the challenges to furthering the field of functional paleoecology and resilience, primarily the need to identify traits that are ecologically meaningful and quantifiable in the sediment record. However, considerable advances in techniques to improve taxonomic resolution are now ready to address some of these challenges. For example, Bálint et al. (2018) describe the application of environmental DNA, (eDNA) from sediments to create continuous time series to resolve ecological dynamics and move toward the harmonization of multidisciplinary data (Fig. 1.5).

Davies et al. (2017) show the response of vegetation over long-time series to microclimatic variations between upland and lowland sites, with local site conditions variously influencing the degree and trajectory of vegetation response to climate shifts. Analysis of the palynological record shows that where high functional diversity and redundancy are evident, the forest has greater capacity to change composition, maintenance of continuous canopy cover, and thus resilience. This suggests that tLTE can indicate the temporal and spatial scales at which complex adaptive forest systems can most effectively deal with changing conditions and unpredicted disturbances. Bonsall et al. (2020) attempt to assess ecosystem resilience measured by the role of mycorrhizal association in a plant-nitrogen dynamic model in a long-time series palaeoecological record. This demonstrates the potential of synthesizing plant nutrient modelling and palaeoecological data, but also highlights the significant challenges and questions that still need to be addressed.

tLTE information can be used to enhance understanding of forests as complex adaptive systems, informing close to nature forestry (Bauhus et al. 2013; Nocentini 2015), as well as becoming a standard in the toolkit for CSF. The functional complex network approach (Messier et al. 2019) is a useful framework for focusing the techniques for synthesis and analysis of tLTE data, and with the addition of new techniques, such as eDNA, there is the opportunity to more effectively inform CSF practice.

1.4 Integrating Forest Disturbance and Ecological Stability

Forests are increasingly challenged by changing environmental conditions and growing societal demands. Forest managers and authorities, responsible for forestry operation and decision making, are required to address multiple, sometimes conflicting, objectives, such as the provision of forest products, the conservation of biodiversity, and the sequestration of carbon. To tackle these issues, novel management strategies, planning guidance, and policy recommendations are needed, at stand and regional scales. However, stakeholders and decision makers lack simple indicators to assess the suitability of forest ecosystems to achieve the vast array of objectives facing global change. This shortcoming hinders the adoption of appropriate strategies and may cause forest policy and management to be inadequate for addressing multiple future challenges. The strength of CLIMO is the network of long-term experimental plots, with detailed structural data, which generate operational examples of management options for merging silvicultural strategies with environmental changes and the social dimension (Fig. 1.6). In this context, engaging stakeholders will be central in developing, testing, and refining local-level indicators of "smartness," and producing a practical toolkit to help implement CSF.

Forest dynamics are changing globally, because of the interplay between the increasing anthropogenic impact on the environment and the changing course of natural disturbances. In Europe, during the twentieth century, tree growth and forest area have been enhanced by the fertilization effect of CO_2 and the cessation of deforestation, though there are signs of carbon sink saturation in forest biomass (Nabuurs et al. 2013). Legacies of past silvicultural practices have resulted in altered forest composition, often favoring species more vulnerable to drought (e.g.,

On scope of resistance	On scope of resilience	On scope of sustainability
• Maintain the species richness and ecosystem complexity in mountain landscapes	• Promote the deployment of protection forests in mountain landscapes to help prevent disasters and protect soils	• Propose innovative schemes of PES useful to develop policies supporting the delivery of ecosystem services
• Encourage forest natural regeneration and tree species mixture to support existing food web interactions	• Foster mountain forests that feature a good mix of species, ages and structures	• Encourage the incorporation of climate impact in mountain forestry investment projects
• Retain soil fertility, preserve water storage capacity, and maintain structural and biotic integrity	• Establish an integrated technological platform to monitor environmental changes and test adaptive strategies	• Endorse the inclusion of climate-smart mountain forests in REDD+ strategies and finance

Fig. 1.6 Important recommendations for decision makers, who need to consider the major contribution of emission reductions per produced unit and the consequences of carbon sink saturation of mountain forests, when sustaining the mitigation potential of forests in mountain landscapes

European beech, Norway spruce), which adds to increase in warming-induced evaporative demand. Accelerating processes of increased tree mortality, indeed, are apparently forcing forests to become both younger and shorter (McDowell et al. 2020). Climate-smart forestry warrants to address the consequences of the shifts in forest dynamics related to these changes, considering multiple plant traits associated with drought resilience and additional stress agents (Coble et al. 2017).

Uncertain climatic conditions and erratic extreme events warrant careful consideration of synergies and trade-offs among mitigation, adaptation, biodiversity conservation, and provisioning of ecosystem services. Accordingly, CSF management strategies need to be customized to match the different ecological and social contexts, assessing the interconnected implications. The complex dynamics of forest ecosystems and rapidly accelerating environmental changes increase the uncertainty of historical knowledge and require novel management approaches and silvicultural operations to maintain forest integrity and halt the loss of resilience (Pretzsch et al. 2020). Yet, the shift in societal expectations and growing socioeconomic disparity, on one hand, add to the simplification of forest structures and increasing forest fragmentation, on the other, making long-term landscape planning difficult (Messier et al. 2019). Managing forests for uncertainty calls for approaches able to adapt promptly and flexibly, and that change in accordance with unexpected (even extreme) events (Millar et al. 2007).

Flexible forest management approaches have been proposed to address accelerating environmental and societal challenges, including disturbance-based management (Franklin et al. 2007), complexity-related management (Puettmann et al. 2009), and mixed-forest management (Bravo-Oviedo et al. 2018). These approaches try to incorporate principles of natural forest dynamics, including the role of disturbances in generating the landscape mosaics, applying silvicultural practices and operations that limit the impact of forest management on environmental functions and services. Managing forests as natural infrastructures, within a matrix of diverse land cover types, may provide appropriate "nature-based solutions" for balancing integration and segregation processes in the landscape mosaics, with respect to different ecological and social contexts. Facing climate change requires, however, a rethinking of current forest management challenges that promote adaptive (resistance, resilience, and response options) and mitigation strategies, while maintaining current goals (e.g., wood production, biodiversity conservation, carbon sequestration, water provision, social value) under uncertain climate scenarios (Nagel et al. 2017; Kauppi et al. 2018; Bowditch et al. 2020). In a few words, management should assure the stability and sustainability of forest ecosystems (Larsen 1995) (Fig. 1.7). Mimicking disturbance regimes with forestry practices needs trend analysis to extract an underlying pattern or time series analysis to deal with data collected over time.

Trend detection helps determine whether there has been a departure from the background or historical conditions (see Chap. 6: Pretzsch et al. 2021b), whereas trend estimation is useful to quantify the nature of the change and investigate models that provide the interpretation of the triggering process. To assess short- and long-term impacts of climate change on forest resilience and implement indicators

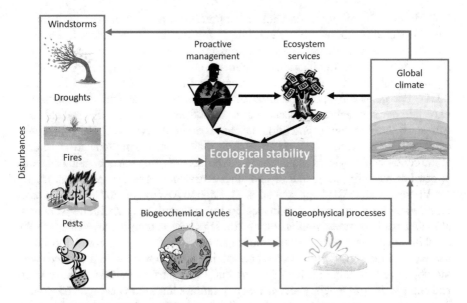

Fig. 1.7 A climate-smart perspective considers the ecological stability of forests, which comprehends resistance and resilience, and therefore, synergies between climate and other benefits, and between adaptation and mitigation. The interlinkages between the three dimensions of CSF (stability, productivity, flexibility) need to be taken into consideration for a holistic approach to any management transformation

of climate-smart forest management and utilization, as well as decision support tools, for adapting forest systems to changing environmental conditions (Santopuoli et al. 2020), quality monitoring of permanent plots is mandatory (see Chaps. 3 and 4: del Río et al. 2021; Temperli et al. 2021). Yet, to identify major drivers of forest change and improve the quality of data, time series and forest inventory analyses can be integrated with enhanced long-term controlled experiments (see Chap. 10: Tognetti et al. 2021) and further scaled geographically (see Chap. 11: Torresan et al. 2021). Comparing historical versus current performance, a network of plots for major forest species (monocultures vs. admixtures) established along adaptation gradients may allow accommodate a range of potential tree growth trajectories, across a variety of ecosystem types and geographic regions, for being used at an operational spatial scale (see Chaps. 5 and 6: Pretzsch et al. 2021a, b). Data collected in permanent plots, designed as long-term studies, together with tLTE data (see Sect. 1.3), can be used to parametrize growth models to facilitate testing species responses to climate change and CSF patterns (see Chap. 7: Bosela et al. 2021). Forest managers and landscape planners facing climate-related uncertainty and unknown forest dynamics require operational examples to inform decision-making processes and prepare forest ecosystems and silvicultural recommendations for climate-driven and socially driven challenges (see Chaps. 8 and 9: Pach et al. 2021; Picchi et al. 2021).

To inform climate-smart decision making in forest management and utilization, valuable demonstrations are needed, integrating multiscale indicators to compare and shape proactive strategies, testing alternative management approaches, conveying the silvicultural implications of climate change, and identifying locally to regionally appropriate planning processes (see Chap. 13: Bottaro and Pettenella 2021). For generating policy instruments that foster climate-smart responses to uncertain environmental prospects, interaction between scientists and stakeholders (forest owners, managers, practitioners, etc.) is compulsory (see Chap. 14: Dubova et al. 2021). The science-policy partnerships may advance communication on CSF at the continental to global scale, across countries, weighing the cobenefits and side effects between climate regulation and other ecosystem services (see Chaps. 15, 16, 17: Vizzarri et al. 2021; Pappas et al. 2021; Giongo Alves et al. 2021). Examples of climate-smart measures include managing forest disturbances and extreme events, selecting resilient trees, implementing forest reserves (high-nature value, HNV), combining carbon storage, sequestration, and substitution, using forest bioenergy and wood in the construction sector, and valuing ecosystems and their services with the objective that help halt land degradation. Payment for environmental services and other forms of monetary incentives may support long-term ecologic, economic, and social perspectives of CSF, particularly when involving local communities, and aligning social norms with personal values (see Chaps. 12 and 13: Gežík et al. 2021; Bottaro and Pettenella 2021).

1.5   The Climate-Smart Forestry Framework

Climate is a dynamic system that is in equilibrium, driven by the amount of energy. However, if the amount of energy kept by the atmosphere from sunlight increases, then climate changes. The exchange of energy, water, and CO_2 within forests influences climate, interacting with the atmosphere. Anthropogenic impact on forests (land cover) alters biogeochemical (carbon cycle) and biogeophysical (albedo and evapotranspiration) cycles, in turn changing the global climate (Bonan 2008). Forest–atmosphere–human interactions cause complex climate forcing, feedback, and response (IPCC 2019). Since rapid changes in external factors (e.g., land cover, GHG concentration) may push the climate system into a new mode, the resulting augmented level of uncertainty requires urgent actions to increase the resilience of forests (Vose et al. 2018). Increasing frequency and magnitude of forest disturbances, as well as long-term changes in vegetation dynamics, add concern for forest productivity, health, and biodiversity and, consequently, for the provision of ecosystem services. Mostly, the knowledge of forest–climate interactions comes from models that strive to simulate complex physical, chemical, and biological processes. Nevertheless, understanding of forest functioning needs to be gained from site-specific permanent sample plots, and management solutions for resilience tested across biogeographic regions (Fig. 1.8).

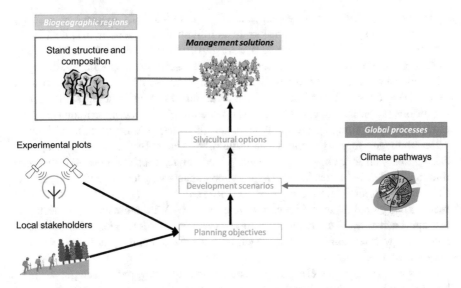

Fig. 1.8 Stakeholder engagement and experimental input to develop adaptable CSF solutions in different biogeographic regions

Technological advances in manufacturing miniaturized machine tools and Internet network connecting systems have prompted the development of cyberinfrastructures to address ecological questions (Rundel et al. 2009; Torresan et al. 2021). The use of satellite imagery and statistical modelling has recently allowed the generation of a spatially continuous map of forest tree density at a global scale (Crowther et al. 2015). Being able to count the global sum of trees, a question emerges: can we monitor a representative number of individual trees as biogeo-physical-chemical units of major forest types and/or biogeographic areas?

To cope with increasing societal demands on forests, a holistic view of the options is essential, including the provision of wood and nonwood products, as well as biodiversity and other ecosystem services. Therefore, suitable climate-smart indicators, for evaluating and monitoring forests and forestry in relation to their adaptation to changes in disturbance regime and intensity, require to be implemented to assure a stable provision of forest products and ecosystem services (Bowditch et al. 2020; Santopuoli et al. 2020; see also Chap. 2 of this book, Weatherall et al. 2021). Indicators can be conveniently used to develop management tools and design novel forest systems, and continuously monitored to measure progress. European mountain systems are experiencing forest colonization and densification due to land abandonment and, as such, represent good standards to identify restoration targets and segregative instruments, considering a range of opportunities, risks, and constraints of the forest value chain in the long run. In this context, the view that mountain forests provide services for lowland uses should be replaced with a perspective focused on forest health and resilience per se, and on strategies to reduce forest vulnerability to changing environment, in order to strengthen the protective functions of these forests.

Promoting climate-smart governance of mountain forests requires testing: (i) whether local experiences on forest systems influence manager–environment interactions to meet real goals (expert partnership), (ii) how research methods help understand the relationships between forests and stakeholders and the feasibility of silvicultural options (stakeholder dialogue), (iii) whether the monitoring of forests helps the understanding of past–present relations while envisioning future forest conditions (scientific contribution), (iv) how the learning process facilitates the generation of novel silvicultural approaches and the interpretation of existing knowledge (manager training), and what trade-offs exist between mitigation of and adaptation to climate change, and between biodiversity conservation and forest production (Ogden and Innes 2007). Disturbance impacts vary locally and need adaptation and communication strategies at different levels (Lindner et al. 2014), reconciling global issues and regional objectives, which requires an interdisciplinary approach toward understanding forest systems, dissemination of quantitative evaluation to forest stakeholders, as well as planning and managing forests sustainably for the long term.

Long-term effects and efficacy of designed smart-management strategies need to be addressed, considering current policy and future consequences on forests. A participatory approach with forest managers, landowners, decision makers, local communities, and the forest industry is, therefore, crucial to implement CSF practices at the stand and landscape scales. To this aim, simple indicators to assess the suitability of forest ecosystems to achieve the multiple demands made on these systems and appropriate tools to quantify these challenges are needed to tailor local policies and incentives. A multiscale indicator approach may help integrate the research strands (long-term experimental plots, region-specific case studies), management options, silvicultural guidelines, and advanced modelling with stakeholder panels.

A CSF network needs to be integrated with that of protected areas (Natura 2000, national schemes), particularly with segregated old-growth forests. Remaining primary forests are the richest terrestrial ecosystems in biodiversity, and keep removing carbon from the atmosphere, while storing significant carbon stocks in biomass and soils (Luyssaert et al. 2008). A trans-biogeographic network of climate-smart and old-growth forest nodes connected by ecological corridors may generate a green infrastructure of healthy terrestrial ecosystems at the European scale. This strategy requires an effort in the restoration of damaged landscapes (including afforestation and reforestation in support of ecosystem restoration), ensuring the long-term productivity of forest capital and the ecological value of natural habitats. Promoting the health and diversity of forest ecosystems increases the sector's resilience to climate change, disturbance risk and economic crisis, while creating green job opportunities, for example, in small fruit farming or in forest therapy activities.

In parallel with segregation of high nature value landscapes, sustainable intensification through scientific and technological innovations is important in landscape zoning that maximizes the efficiency of production and reduces the competition for land. In this sense, the sustainable use of forest biomass for energy production and the shift toward advanced biofuels based on residues, as well as the efficient use of wood-based products for substitution purposes (the use of wood fuel in place of fossil fuels, for energy, and the use of wood fiber in place of cement, etc., whose production emits large amounts of CO_2), have great potential for the long-term reduction of carbon emissions. An important challenge is represented by the need of integrating biomass production and nature conservation into forest management, as well as implementing European forest policy (Aggestam et al. 2020).

Changes in disturbance regimes (Seidl et al. 2016), in a cascade effect, may affect tree growth, stand structure, species composition, and regeneration processes, that is, forest dynamics, potentially impairing the purposes of forest management. Consequently, habitat conservation, timber supply, carbon stock, nontimber products, recreation, infrastructure safety, and cultural values can also suffer the consequences of multiple simultaneous changes. An integrated approach is, therefore, needed to reconcile critical trade-offs between the multiple goals of forest management, and direct adaptation measures in regions with a long legacy of land uses and cover changes, such as Europe. Because of the uncertain scenarios, disturbance risk monitoring and early warning recommendations are an important means of improving the efficiency of decision-making process and climate-smart support (Millar et al. 2007).

1.6 A European Way to Climate-Smart Forestry

Achieving food security and avoiding climate change are two major goals of our time (FAO 2013). Since deforestation and forest degradation account for about 12% of global anthropogenic carbon emissions, forests, forestry, and the whole forest-based sector play an important role in providing sustainable solutions to halting climate change. Forests and other wooded land cover at least 43% of the surface of the EU (28 countries; 182 million hectares), and the forest sector employs at least half a million European citizens directly and 2.6 million indirectly. Forests, the forest-based sector, and the bioeconomy have a crucial role in meeting the goals of the European Green Deal and EU's climate, energy, and environmental objectives. Yet, the EU's international commitments, such as the UN SDGs, Kyoto Protocol, and Paris Agreement, are impossible to achieve without climate benefits of multifunctional resilient forests, CSF, and ecosystem services provided by forests. Indeed, the EU Forest Strategy needs to address the continuous evolving policy options (e.g., referring to Bioeconomy Strategy, Deforestation Action Plan, 2030 Climate and Energy Framework) (Fig. 1.9).

The EU Forest Strategy needs to deploy innovation in support of rural development and scale up the role of the forest industry, while protecting the environment and ensuring circularity. Therefore, forest owners and local administrations, including regional governments, have a key part to play in strengthening CSF. Through sustainable development plans and local bioeconomy strategies, regional governments may support the deployment of renewable energy and promote entrepreneurship in the forest sector (Fig. 1.10). The EU regulation emphasizes the role of forests

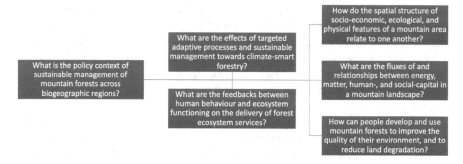

Fig. 1.9 Driving questions and principles toward a green economy in European mountain regions

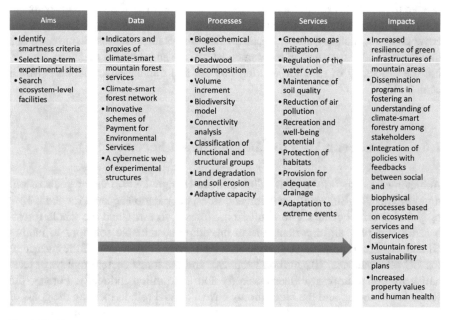

Fig. 1.10 Conceptual structure of smartness-related actions and flow chart of ecosystem service quantification

and wood in reducing emissions and in carbon sequestration, further underpinning CSF. Yet, Member States have national or sectoral adaptation strategies to climate change. The European Forest Strategy also emphasizes the potential of forest management plans, or equivalent instruments, for a balanced delivery of ecosystem services and forest products.

Long-term data sets are critical for detecting patterns in environmental disturbance and trends in forest resilience. Forest health and tree productivity in permanent plots serve as benchmarks against which change in the response to natural and anthropic impacts on the landscape can be gauged. A network of biogeographically distributed permanent plots has the potential to serve as a long-term multisite platform for detection of environmental change over time, for forests of Europe. A strategy for implementing a multisensor platform, or modernizing existing infrastructures in forest ecosystems, requires developing low-cost wireless technologies for the collection and transmittal of data at the local to continental scales. Permanent plots are also useful for monitoring biodiversity. The pandemic of 2020 (COVID-19) highlighted how forest degradation (biodiversity loss) and landscape fragmentation (habitat loss) might increase the risk of infectious diseases. Curbing interference with natural ecosystems and restoring functionality in resilient forests are factors in preventing the spread of infectious disease outbreaks. Changes in the opportunity space for pathogens may facilitate mixing of infectious agents and expand grouping of potential hosts.

The new growth strategy of the EU (The European Green Deal) has the ambition to transform the Union into a resource-efficient, green, and competitive low-carbon society. The bioeconomy approach, based on renewable biological resources and sustainable bio-based solutions, offers a unique opportunity to address these complex and interconnected challenges. Appropriate indicators may help address possibilities and uncertainties related to the potential capacity of forest resources in contributing to the sustainable development of bioeconomy (Wolfslehner et al. 2016). Key indicators and monitoring tools need to consider differences in forest resources and management approaches among biogeographic regions and Member States. A continuous update of monitoring instruments and the adaptation of general principles to local circumstances are necessary to integrate decision-making processes. Autonomous wireless sensor networks, connected and communicating with each other, may measure relevant parameters within an area at many sites simultaneously, for a detailed picture about the environmental conditions on site and the early stages of tree response (Bayne et al. 2017). Forest managers, in general, do not have access to low-cost and high-frequency measurements of productivity-related parameters and, therefore, they can hardly be able to monitor the changeable conditions and complex interrelationships in forest stands. Combination of automated sensor networks of monitored trees, collecting key management data in real time, with UAV (Unmanned Aerial Vehicle) and multi-/hyperspectral remote sensing, will benefit transfer information wireless with LPWAN (low-power wide-area network) – LoRa (Long Range) and satellite communication, as well as allow forest inventory based on single tree measurements.

Different regions in Europe can be classified for the purpose of drawing down structural incentives to establish and play CSF actions. This analysis can be based on detailed criteria specific to each site and defined to identify thresholds for future silvicultural actions (Table 1.1). In this sense, the establishment of qualified partnerships is essential to effectively design and implement permanent sample plots, connecting research staff, forest managers, policy makers to develop, test, and monitor climate-smart management approaches. Information on the impacts of climate change on forestry systems is also required to build on the present status of sustainable forest management and develop climate-smart measures. Partnership needs to be adaptable to local circumstances and able to share costs and benefits associated with research activities and development plans.

Table 1.1 Challenges and opportunities for CSF (and the wood industry) to align research activities with development plans at the local to European scales

Silvicultural traditions	Complement advanced tradition with local specificity, balancing supply and demand
Wood resources	Consider forest structure, tree species, size distribution, forest fragmentation.
Certification bodies	Develop the implementation of forest certification schemes.
Forest ownership	Reduce fragmentation of forest ownership and incentivize entrepreneurship.
Data availability	Harmonize forest inventories and monitor the results of alternative treatment.
Forest roads	Improve the quality of road network and mobilization of forest products.
Forest industries	Distribute costs and benefits between private and public bodies, and rise competitiveness.
Service networks	Expand the utilization of innovations and the accessibility to digital platforms.
Support systems	Adopt decision support systems for forestry performance.
Data digitization	Increase the quality of data and their accessibility to small and medium enterprises.
Local communities	Dialogue with associations, cooperatives, consortia, and other stakeholders.
Governance quality	Ponder management agencies, institutional incentives, public policies.
Value-chain development	Increase the degree of mechanization, capacity of investment, and multifunctional material.
Nongovernmental organizations	Balance landscape intensification vs. ecosystem services, nature conservation.
Technical skills	Progress the formation of forest practitioners and education programs.
Communication tools	Interpret public opinion and tackle administrative constraints.
Production value	Challenge market requirements and support new and high-quality products.
Cross-sectoral cooperation	Integration with agriculture and environment, and expansion of bioeconomy.

Urban areas host already complex networks of digital infrastructures. The social-political focus on emerging digital technologies and smart cities risks leaving rural territories behind, preventing forest managers from using e-services, particularly if operating in mountain environments. Environmental discrimination is an increasingly growing phenomenon, poorly addressed in forestry research, posing urban dwellers and rural communities as dichotomous categories. Nevertheless, the proliferation of technology-driven initiatives requires balancing the interests of elite players with the concerns of local stakeholders (Gabrys 2020). Many efforts, from local to global initiatives (e.g., Bonn Challenge, Initiative 20x20, AFR100 African Forest Landscape Restoration Initiative, and The United Nations General Assembly declared 2021–2030 the United Nations Decade on Ecosystem Restoration) are underway to support the restoration of degraded forest landscapes and present Internet of Things (IoT) as necessary to meet global change targets. Therefore, companies that provide technology relevant for precision forestry may try to get benefits from ecological restoration initiatives, though criticism raises owing to the possible disconnection between people and nature, as well as the requirements of intellectual, financial, and material resources necessary for collecting, processing, and interpreting data.

1.7 Pilot Forests

Long-term experiments and permanent plots, addressing consequences of changing environmental conditions and disturbance regimes to ensure sustainable forestry, have been generally planned for local targets and with specific designs (tree growth, stand productivity). Networks of complex research infrastructures, such as the ESFRI (European Strategy Forum on Research Infrastructures) infrastructure ICOS—Integrated Carbon Observing System (www.icos-ri.eu), may provide the basis for a deeper understanding of forest ecosystem functioning (Rebmann et al. 2018). These infrastructures, designed for long-term monitoring of sources and sinks of GHG, have several disadvantages, including the relatively high personnel costs, installation difficulties, power requirements, managing skills, and maintenance needs. At these experimental stations, the principal technique for measuring ecosystem–atmosphere exchange of CO_2 and H_2O at the stand scale (forest level) is the eddy covariance, which is an evolution of physiological measurements of gas exchange in individual leaves (organ level). Nevertheless, integrated measurements of CO_2 and H_2O fluxes at the tree level (stem enlargement, tree transpiration) are gaining new perspectives (Steppe et al. 2015; Valentini et al. 2019). Applying a modular approach would be extremely useful to combine various monitoring tools and contribute toward CSF.

Advances in sensor technology, wireless communications, and software applications have enabled the development of low-cost, low-power multifunctional environmental sensors and sensor networks that can communicate forest conditions to researchers, managers, and the public in real time. These sensors generate information at unprecedented temporal and spatial scales, and offer transformational opportunities to better understand the physical, chemical, and biological "pulse" of forest ecosystems. Ground data can be coupled with remote sensing. For example, multi-temporal monitoring of forest growth patterns on a plot scale across phenological stages using light detection and ranging (LiDAR) data and the use of remotely sensed sun-induced chlorophyll fluorescence (SIF) for tracking forest photosynthesis offer great potential to follow changes in gross primary productivity (GPP) of forests (Brocks et al. 2016). Wireless sensor nodes designed to provide interoperability with space observations, inventory data, meteorological records, and forest operations might reduce uncertainties and increase reliability of CSF indicators (see Chaps. 9, 10 and 11 of this book: Picchi et al. 2021; Tognetti et al. 2021; Torresan et al. 2021). These real-time "windows on mountain forests" also provide compelling new ways to engage the public and provide novel tools for resource managers. They may influence the forestry sector to meet the scientific and technological challenges emerging in mountain environments, providing solutions for proactive silviculture, while bringing trees closer to people (*serba me, servabo te*).

Traditional systems to collect basic information of forest functioning are labor intensive and the delivery of data to end users is slow, although reliable, data collection is of difficult expansion in time and space. In addition, traditional systems do not respond to environmental changes in real time. In this sense, IoT (Internet of Things) tools for data collection, processing, visualization, and device management provide an integrated technological platform for advanced monitoring of forests. A vision for a comprehensive redesign and standardization of environmental data collection and delivery at experimental sites is critically needed. Deployment of cyber technology can be envisaged in selected experimental forests, in order to develop a network of sites in which trees are monitored intensively and over time. Collecting long-term data sets is critical for detecting patterns and trends in climate–productivity relationships in health forests and responding to tree mortality issues.

The establishment of a permanent network of coordinated and distributed climate-smart forests, in which to test pre- and postdisturbance data collection, requires a substantial organization capacity and a preliminary analysis of environmental conditions and forest settings (Halbritter et al. 2020). This helps develop future scenarios and hypothesizes alternative management options, which can be assessed in local circumstances, following cost–benefit analysis. Continuously monitoring of forest functions and ecosystem services in long-term sampling plots is recommended. Climate-smart forestry and its placement into forest management decision processes need support for addressing trade-offs and synergies (Fig. 1.11).

Technological advances do not come without risks. Climate-smart sensor networks with wireless communication links are keys to monitor tree functions, forest conditions, and forestry operations. A CSF network should include a common suite of low-cost sensors for data collection, real-time data delivery to single web access points, and interactive data visualizations. However, wireless sensors deployed in forest stands still need improvements for data collection, transmission, control, and processing. Processors enabling continuous measurement and real-time storage of data are not always low-cost solutions for forest monitoring and forestry applications. Coordination and standardization of measuring methods and sampling protocols across different sites for establishing CSF networks is also a difficult task. Large amounts of data generated in long-term monitoring studies and their accessibility require property right analysis (Clarke et al. 2011), in order to make the best use of information related to the forestry sector. Ground-level data obtained from sensor nodes need to be related to stand structural complexity and scaled to forest management units by means of statistical modeling and remote sensing (e.g., UAV imagery), for being operative. However, environmental conditions near the ground (due to vegetation and topography) affect tree function and stand structure, and hardly represent a realistic picture of the tree–environment or stand–environment interface (Zellweger et al. 2019).

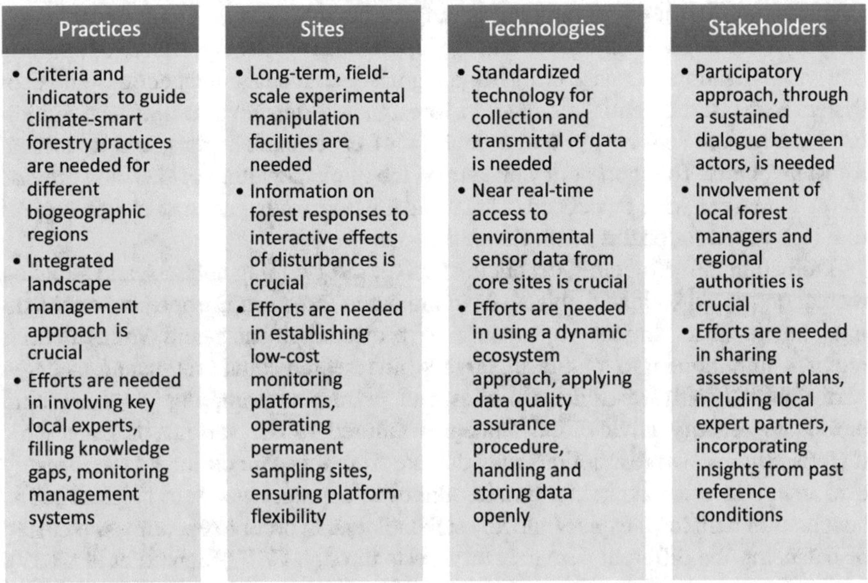

Practices	Sites	Technologies	Stakeholders
• Criteria and indicators to guide climate-smart forestry practices are needed for different biogeographic regions • Integrated landscape management approach is crucial • Efforts are needed in involving key local experts, filling knowledge gaps, monitoring management systems	• Long-term, field-scale experimental manipulation facilities are needed • Information on forest responses to interactive effects of disturbances is crucial • Efforts are needed in establishing low-cost monitoring platforms, operating permanent sampling sites, ensuring platform flexibility	• Standardized technology for collection and transmittal of data is needed • Near real-time access to environmental sensor data from core sites is crucial • Efforts are needed in using a dynamic ecosystem approach, applying data quality assurance procedure, handling and storing data openly	• Participatory approach, through a sustained dialogue between actors, is needed • Involvement of local forest managers and regional authorities is crucial • Efforts are needed in sharing assessment plans, including local expert partners, incorporating insights from past reference conditions

Fig. 1.11 Key messages in support of CSF

Electronic data collection requires a continuous data flow from the sensor network. Gaps may derive from instrument failures, power interruptions and bad weather, or instrument problems and maintenance stops (Rundel et al. 2009). Therefore, gap-filling techniques are required to produce continuous data time series (Moffat et al. 2007). Sensors and devices expose themselves to aspects of environmental conditions and material features that may degrade data integrity, making calibration an essential process. Poor calibration may also cause damage to hardware and the general infrastructure. Therefore, achieving good-quality data and maintaining error-free data collection in wireless sensor networks is challenging and calibration is essential to limit environmental noise and hardware failure (Barcelo-Ordinas et al. 2019). Energy consumption and sensor connectivity in wireless sensor networks are also crucial issues for the network effectiveness and efficiency in terms of lifetime, cost, and operation. Finally, the deployment of devices in CSF sites will require experienced and well-trained forestry personnel, which may cause digital inequalities due to the shift of control from traditional actors to elite services.

Forest monitoring detects the impacts and trends of climate change, natural disturbances, and human activities and is an essential element in CSF schemes. Legacies of natural disturbance and land use direct stand development and monitoring and understanding their role are keys to implement management techniques that contribute to maintaining forest stability (Franklin et al. 2007). Disturbance legacies and ecosystem conditions comprise ecological memory (Johnstone et al. 2016), which contributes to draw the trajectories of forest reorganization after disturbance

(Jõgiste et al. 2018). Forest monitoring must be flexible, adapting tools and methodologies to the targeted objectives and biogeographic regions, for effective reporting and harmonization. Because of the time lag between management actions and ecosystem responses, essential elements of forest monitoring are a common set of indicators, a remote sensing system, a network of experimental plots, and a national forest inventory. To report policy makers on the implementation of CSF and provide forest managers with evidence on smart-adaptation strategies, a decision support tool is required for dealing with the challenges.

Protecting infrastructures, producing timber, safeguarding habitats, and allowing recreation are consolidated objectives in mountain forests in Europe, and establishing climate-smart targets in response to projected impacts and vulnerabilities requires commitment to long-term, large-scale research collaboration and manager partnership. Highly instrumented sites and permanent sampling plots are both needed to identify silvicultural strategies tailored to the various biogeographic regions. Since new prescriptions may deviate from those practitioners traditionally use, an expert and stakeholder training dialogue approach needs to be searched in cooperation with local experts and key stakeholders, in order to encompass the need of balancing the different demands on forests through CSF (Tognetti et al. 2017). Discussion panels may seek to answer specific questions or build consensus about management objectives.

Although experimental sites (permanent plots, instrumented sites) exist at the national level, and can be sampled to gain data and test models, as well as provide insights on tree responses over time, there can be limitations, due to the original study target and design (Fig. 1.12).

Building on previous experience and knowledge, new studies may develop local approaches further, specifically addressing the response of trees and the resilience of forests to disturbance in the context of climate change. Indeed, while HNV forests can be used as a benchmark for restoring managed forests, to some extent (Jandl et al. 2018), climate change may push ecosystems beyond historical limits (Millar et al. 2014; Dumroese et al. 2015). Novel information may contribute to and deliver a range of prescriptions according to experience and expertise in sustainable forest management, though durable commitments and consistent resources are required to ensure long-term operativity and development of management options.

Certainly, flexible forest management strategies are highly desirable to tackle spatially and temporally variable environmental conditions, with the aim of maintaining species mixtures, complex structures, and multiple functions of forests, as a way of enhancing the resilience to natural disturbances (Franklin et al. 2007; Puettmann et al. 2009; O'Hara and Ramage 2013), on a local basis over time. As an example, in locations where there are recurrent extreme events, for example, windstorms, a decrease in the growing stock can also be envisaged to reduce the vulnerability of the forest stands and support the potential of the forestry industries. Conversely, remote, HNV forests can be set aside to protect biodiversity. Yet, regeneration of drought-tolerant species may gradually replace that of co-occurring, more vulnerable species (e.g., Hilmers et al. 2019; Torresan et al. 2020). Genetic materials more adapted to environmental modifications can also be tested and used.

Fig. 1.12 "Aldo Pavari" permanent experimental plot of *Pinus nigra* ssp. *laricio* established in 1919; Foresta Demaniale di Fiorentini, Sardinia, Italy)

Permanent sampling forest plots are the basic unit of functional and structural indicator measurements. Their size may range, though experimental plots covering 1 ha (100 × 100 m) can be considered suitable for simulating various forestry scenarios and their impacts on the provision of ecosystem services and/or the preservation of habitat diversity (Santopuoli et al. 2019). Within these plots, such as the Marteloscope (Bruciamacchie et al. 2005), all trees are counted, numbered, marked, and mapped in the census. Circular sampling subplots, with radii ranging from 4 m up to 20 m, can then be considered to quantify specific indicators (Lombardi et al. 2015). Repeated measurements of differential forest traits, resulting from changes in governance or environment compared to a defined business-as-usual scenario or normality mode, are needed to predict future forest ecosystem trajectories and new safe operating spaces (Johnstone et al. 2016). These measurements require that modular instruments (soil–plant–atmosphere continuum devices and ancillary sensors) are in place and monitoring methods are well established (Chave et al. 2019). Therefore, changes in patterns of tree growth and health status, as well as successional development and anthropogenic impact, can be recorded. The information is useful for planning observational networks or storing model data at different temporal and spatial scales. Since disturbances may shape the structure of forest stands, boundaries and dimensions of ecological or operational units vary over time, depending on environmental circumstances or management objectives to follow the natural stand dynamics (O'Hara and Nagel 2013).

1.8 Putting Climate-Smart Forestry into Practice

Climate-smart forestry integrates the overlaps, synergies, and trade-offs of global objectives and local needs into forest governance. Indeed, the vulnerability to climate change in the forest sector affects both rural communities and the international arena. Therefore, climate-smart forest strategies need to flow into policies on rural development, climate mitigation, and the bioeconomy. Climate-smart forestry may play an important role in strengthening the resilience of rural communities, by coordinating risk assessment, planning efforts, management activities, support policies, and incentive strategies. Understanding the role of CSF requires targeting specific areas of vulnerability, while maintaining a broad vision of forest-related options and their potential impacts on adaptation to and mitigation of climate change.

While climate change, disturbance regime, and unsustainable exploitation jeopardize natural ecosystems, nature-based solutions, including sustainable forest management, can be fundamental in combating climate and land-use changes (Seddon et al. 2019). Landowners, forest enterprises, and end users (in addition to decision makers, forest managers, and research scientists) are key actors in halting biodiversity loss, while benefiting from healthy forests. Reconciling the sustainable supply of economic goods with a wise demand of ecosystem services is crucial to build forest resilience and prevent landscape degradation. Climate-smart forestry approach focuses directly on forest functions and silvicultural systems by assessing the vulnerability of forest sustainable management objectives, recognizing the need to balance the three dimensions of development goals, that is, the economic, social, and environmental. This requires appropriate technological capacity for applying the conceptual framework in practice, which may form the basis of decision support

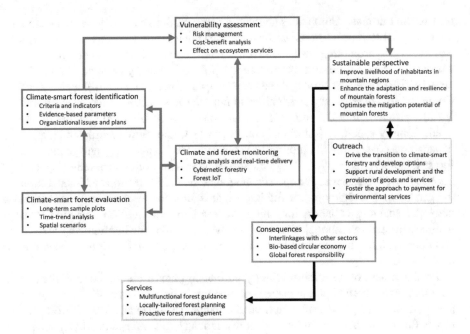

Fig. 1.13 Feedback diagram of the CSF approach

tools that assist in the development and implementation of climate-smart options in response to system-related vulnerabilities (Fig. 1.13).

Because forests provide essential ecosystem services, the inclusion of payment schemes for ecosystem services in adaptation plans needs to be strengthened. These schemes increase the adaptive capacity of vulnerable communities (like those of mountain areas) and provide an incentive mechanism for long-term sustainability (adoption of specific support measures). Climate-smart forestry needs to incorporate emerging technology, identifying gaps in capacity building of the forest sector and changes in development objectives of modern society. Assessing the impact of climate change on the forest sector is prerequisite to implement management options that increase the resilience of forest ecosystems and decrease the vulnerability of rural communities. As an example, forest roads are required to transport timber and other forest products to markets, generating income to local enterprises, but also to fight forest fires, reducing risk of extreme events. These efforts need to be cross-sectoral and require multilevel actions to prepare effective responses of the forest sector to climate change.

Forest coverage varies considerably among Member States of the EU. In Finland and Sweden, forests cover more than 60% of the country, while only 11% in the Netherlands and the United Kingdom. Although the forest landscape has been largely shaped by man, in the EU, the area of land covered by forests is growing, as a result of both natural growth and afforestation work. Only 4% of the forested area can be considered pristine. However, as the Treaties make no specific reference to forests, the European Union does not have a common forestry policy. With the aims to ensure that Europe's forests are managed sustainably and to strengthen the EU's contribution to promoting sustainable forest management and tackling deforestation worldwide, the Commission set out a new EU Forest Strategy (beyond 2020). The new 2030 Biodiversity Strategy and the European Green Deal, developed by the Commission, are relevant opportunities to put the forest sector on the international agenda, provided that a common communication and collaboration plan is developed. Country-specific institutional context, in fact, may slow the implementation of new concepts, like CSF, aiming to assess climate change risks and implement forest adaptation strategies. A systematic approach to common monitoring and knowledge sharing across Europe is encouraged by COST (European Cooperation in Science and Technology), a funding organization for research and innovation networks enabling collaboration among researchers. Stronger coordination with institutional and operational levels is, however, essential for harmonizing assessment protocols, managing field trials, and transferring research results at the continental scale.

Establishment of a multidisciplinary partnership approach and commitment to long-term experimental plot management are keys to the success of CSF that aim to develop and test large-scale silvicultural alternatives (Nagel et al. 2017). Differences across European countries in terms of forest industry, governance quality, manager experience, investment capacity, and training system are challenges that need to be harmonized by a shared strategy. Cross-site analysis of standardized monitoring plot networks and perceived ecosystem service benefits in multifunctional forest landscapes, with varying spatial and temporal resolution, may improve our understanding of forest responses to chronic stresses, extreme events, and successional processes. A European forest network approach may provide examples of adaptation, mitigation, and production strategies, spanning a range of management and governance options. Evidence-informed management approach of threatened forests is key to help data-driven gradual transition to more stable ecosystems, as well as to fine-tune the provision of ecosystem services.

The application of CSF strategies and the bottom-up approach of the Nationally Determined Contributions (NDCs) in the Paris Agreement may help comply with the target of reducing climate risks. Nevertheless, the benefits from enhancing carbon sequestration through forest management can be counterbalanced or amplified by concurrent management-induced changes in complex forest–atmosphere interactions (Bravo et al. 2017). Therefore, managing forests for halting climate warming through mitigation may result only in compensating CO_2 emissions. Carbon extraction and sequestration through bioenergy plantations, forest restoration, and urban forestry will not suffice alone to thin the atmospheric CO_2 blanket and reduce the risk of CO_2-induced global warming. Although these cost-effective land-based sink options may complement negative emission technologies, they are also vulnerable to both natural and anthropic disturbances. Should climate benefits from CSF remain important only on a local scale or in specific geographic circumstances (e.g., mountain environments), forest adaptation to changing climate will still be essential to sustain the provision of goods and services, while avoiding positive climate feedback from disturbances.

References

Aggestam F, Konczal A, Sotirov M et al (2020) Can nature conservation and wood production be reconciled in managed forests? A review of driving factors for integrated forest management in Europe. J Environ Manag 268:110670. https://doi.org/10.1016/j.jenvman.2020.110670

Allen CD, Macalady AK, Chenchouni H et al (2010) A global overview of drought and heat-induced tree mortality reveals emerging climate change risks for forests. For Ecol Manag 259:660–684. https://doi.org/10.1016/j.foreco.2009.09.001

Bálint M, Pfenninger M, Grossart H-P et al (2018) Environmental DNA time series in ecology. Trends Ecol Evol 33:945–957. https://doi.org/10.1016/j.tree.2018.09.003

Barcelo-Ordinas JM, Doudou M, Garcia-Vidal J, Badache N (2019) Self-calibration methods for uncontrolled environments in sensor networks: a reference survey. Ad Hoc Netw 88:142–159. https://doi.org/10.1016/j.adhoc.2019.01.008

Bauhus J, Puettmann KJ, Kühne C (2013) Close-to-nature forest management in Europe: does it support complexity and adaptability of forest ecosystems? In: Managing forests as complex adaptive systems. Routledge, pp 201–227

Bayne K, Damesin S, Evans M (2017) The internet of things – wireless sensor networks and their application to forestry. 61:5

Boisvert-Marsh L, Périé C, de Blois S (2014) Shifting with climate? Evidence for recent changes in tree species distribution at high latitudes. Ecosphere 5:art83. https://doi.org/10.1890/ES14-00111.1

Bonan GB (2008) Forests and climate change: forcings, feedbacks, and the climate benefits of forests. Science 320:1444–1449

Bonsall MB, Froyd CA, Jeffers ES (2020) Resilience: nitrogen limitation, mycorrhiza and long-term palaeoecological plant–nutrient dynamics. Biol Lett 16:20190441. https://doi.org/10.1098/rsbl.2019.0441

Bowditch E, Santopuoli G, Binder F et al (2020) What is climate-smart forestry? A definition from a multinational collaborative process focused on mountain regions of Europe. Ecosyst Serv 43:101113. https://doi.org/10.1016/j.ecoser.2020.101113

Bravo F, Jandl R, LeMay V, von Gadow K (2017) Introduction. In: Bravo F, LeMay V, Jandl R (eds) Managing forest ecosystems: the challenge of climate change. Springer International Publishing, Cham, pp 3–12

Bravo-Oviedo A, Condés S, del Río M et al (2018) Maximum stand density strongly depends on species-specific wood stability, shade and drought tolerance. Forest Int J Forest Res 91:459–469. https://doi.org/10.1093/forestry/cpy006

Brocks S, Bendig J, Bareth G (2016) Toward an automated low-cost three-dimensional crop surface monitoring system using oblique stereo imagery from consumer-grade smart cameras. J Appl Remote Sens 10:046021. https://doi.org/10.1117/1.JRS.10.046021

Bruciamacchie M, Pierrat J-C, Tomasini J (2005) Modèles explicatif et marginal de la stratégie de martelage d'une parcelle irrégulière. Ann For Sci 62:727–736. https://doi.org/10.1051/forest:2005070

Cailleret M, Jansen S, Robert EMR et al (2017) A synthesis of radial growth patterns preceding tree mortality. Glob Change Biol 23:1675–1690. https://doi.org/10.1111/gcb.13535

Ceddia MG, Bardsley NO, Gomez-y-Paloma S, Sedlacek S (2014) Governance, agricultural intensification, and land sparing in tropical South America. Proc Natl Acad Sci 111:7242. https://doi.org/10.1073/pnas.1317967111

Chave J, Davies SJ, Phillips OL et al (2019) Ground data are essential for biomass remote sensing missions. Surv Geophys 40:863–880. https://doi.org/10.1007/s10712-019-09528-w

Ciancio O, Nocentini S (2011) Biodiversity conservation and systemic silviculture: concepts and applications. Plant Biosyst – Int J Deal Asp Plant Biol 145:411–418. https://doi.org/10.108 0/11263504.2011.558705

Clarke N, Fischer R, de Vries W et al (2011) Availability, accessibility, quality and comparability of monitoring data for European forests for use in air pollution and climate change science. IForest – Biogeosci Forestry 4:162–166. https://doi.org/10.3832/ifor0582-004

Coble AP, Vadeboncoeur MA, Berry ZC et al (2017) Are northeastern U.S. forests vulnerable to extreme drought? Ecol Process 6:34. https://doi.org/10.1186/s13717-017-0100-x

Crowther TW, Glick HB, Covey KR et al (2015) Mapping tree density at a global scale. Nature 525:201–205. https://doi.org/10.1038/nature14967

Curtis PG, Slay CM, Harris NL et al (2018) Classifying drivers of global forest loss. Science 361:1108–1111. https://doi.org/10.1126/science.aau3445

Davies AL, Smith MA, Froyd CA, McCulloch RD (2017) Microclimate variability and long-term persistence of fragmented woodland. Biol Conserv 213:95–105. https://doi.org/10.1016/j. biocon.2017.06.006

del Río M, Vergarechea M, Hilmers T et al (2021) Effects of elevation-dependent climate warming on intra- and inter-specific growth synchrony in mixed mountain forests. For Ecol Manag 479:118587. https://doi.org/10.1016/j.foreco.2020.118587

Dumroese RK, Williams MI, Stanturf JA, Clair JBS (2015) Considerations for restoring temperate forests of tomorrow: forest restoration, assisted migration, and bioengineering. New For 46:947–964. https://doi.org/10.1007/s11056-015-9504-6

European Environment Agency (2016) European forest ecosystems: state and trends. Publications Office, LU

FAO (2013) The state of food insecurity in the world. FAO, Rome

Franklin JF, Mitchell RJ, Palik BJ (2007) Natural disturbance and stand development principles for ecological forestry. U.S. Department of Agriculture, Forest Service, Northern Research Station, Newtown Square

Gabrys J (2020) Smart forests and data practices: from the internet of trees to planetary governance. Big Data Soc 7:2053951720904871. https://doi.org/10.1177/2053951720904871

Grassi G, Cescatti A, Matthews R et al (2019) On the realistic contribution of European forests to reach climate objectives. Carbon Balance Manag 14:8. https://doi.org/10.1186/ s13021-019-0123-y

Griscom BW, Adams J, Ellis PW et al (2017) Natural climate solutions. Proc Natl Acad Sci 114:11645–11650. https://doi.org/10.1073/pnas.1710465114

Halbritter AH, Boeck HJD, Eycott AE et al (2020) The handbook for standardized field and laboratory measurements in terrestrial climate change experiments and observational studies (ClimEx). Methods Ecol Evol 11:22–37. https://doi.org/10.1111/2041-210X.13331

Hamilton R, Brussel T, Asena Q et al (2018) Assessing the links between resilience, disturbance and functional traits in paleoecological datasets. Past Glob Chang Mag 26:87–87. https://doi. org/10.22498/pages.26.2.87

Hanewinkel M, Cullmann DA, Schelhaas M-J et al (2013) Climate change may cause severe loss in the economic value of European forest land. Nat Clim Chang 3:203–207. https://doi. org/10.1038/nclimate1687

Hartmann H, Moura CF, Anderegg WRL et al (2018) Research frontiers for improving our understanding of drought-induced tree and forest mortality. New Phytol 218:15–28. https://doi. org/10.1111/nph.15048

Hilmers T, Avdagić A, Bartkowicz L et al (2019) The productivity of mixed mountain forests comprised of Fagus sylvatica, Picea abies, and Abies alba across Europe. Forestry Int J Forest Res 92:512–522. https://doi.org/10.1093/forestry/cpz035

Innes J (2009) Forest fragments need fencing and pest control. Open Space 75:15–16

IPCC (2019) Climate change and land: an IPCC special report on climate change, desertification, land degradation, sustainable land management, food security, and greenhouse gas fluxes in terrestrial ecosystems, Geneva

Jandl R, Ledermann T, Kindermann G et al (2018) Strategies for climate-smart Forest Management in Austria. Forests 9:592. https://doi.org/10.3390/f9100592

Jõgiste K, Frelich LE, Laarmann D et al (2018) Imprints of management history on hemiboreal forest ecosystems in the Baltic States. Ecosphere 9:e02503. https://doi.org/10.1002/ecs2.2503

Johnstone JF, Allen CD, Franklin JF et al (2016) Changing disturbance regimes, ecological memory, and forest resilience. Front Ecol Environ 14:369–378. https://doi.org/10.1002/fee.1311

Kauppi P, Hanewinkel M, Lundmark T et al (2018) Climate smart forestry in Europe. European Forest Institute, Joensuu

Larsen JB (1995) Ecological stability of forests and sustainable silviculture. For Ecol Manag 73:85–96. https://doi.org/10.1016/0378-1127(94)03501-M

Lindner M, Maroschek M, Netherer S et al (2010) Climate change impacts, adaptive capacity, and vulnerability of European forest ecosystems. For Ecol Manag 259:698–709. https://doi.org/10.1016/j.foreco.2009.09.023

Lindner M, Fitzgerald JB, Zimmermann NE et al (2014) Climate change and European forests: what do we know, what are the uncertainties, and what are the implications for forest management? J Environ Manag 146:69–83. https://doi.org/10.1016/j.jenvman.2014.07.030

Lipper L, Thornton P, Campbell BM et al (2014) Climate-smart agriculture for food security. Nat Clim Chang 4:1068–1072. https://doi.org/10.1038/nclimate2437

Lombardi F, Marchetti M, Corona P et al (2015) Quantifying the effect of sampling plot size on the estimation of structural indicators in old-growth forest stands. For Ecol Manag 346:89–97. https://doi.org/10.1016/j.foreco.2015.02.011

Luyssaert S, Schulze E-D, Börner A et al (2008) Old-growth forests as global carbon sinks. Nature 455:213–215. https://doi.org/10.1038/nature07276

Luyssaert S, Marie G, Valade A et al (2018) Trade-offs in using European forests to meet climate objectives. Nature 562:259–262. https://doi.org/10.1038/s41586-018-0577-1

McDowell NG, Allen CD, Anderson-Teixeira K et al (2020) Pervasive shifts in forest dynamics in a changing world. Science 368:eaaz9463. https://doi.org/10.1126/science.aaz9463

Messier C, Bauhus J, Doyon F et al (2019) The functional complex network approach to foster forest resilience to global changes. Forest Ecosyst 6:21. https://doi.org/10.1186/s40663-019-0166-2

Millar CI, Stephenson NL (2015) Temperate forest health in an era of emerging megadisturbance. Science 349:823–826

Millar CI, Stephenson NL, Stephens SL (2007) Climate change and forests of the future: managing in the face of uncertainty. Ecol Appl 17:2145–2151. https://doi.org/10.1890/06-1715.1

Millar CI, Swanston CW, Peterson DL (2014) Adapting to climate change. In: Peterson DL, Vose JM, Patel-Weynand T (eds) Climate change and United States forests. Springer Netherlands, Dordrecht, pp 183–222

Moffat AM, Papale D, Reichstein M et al (2007) Comprehensive comparison of gap-filling techniques for eddy covariance net carbon fluxes. Agric For Meteorol 147:209–232. https://doi.org/10.1016/j.agrformet.2007.08.011

Nabuurs G-J, Lindner M, Verkerk PJ et al (2013) First signs of carbon sink saturation in European forest biomass. Nat Clim Chang 3:792–796. https://doi.org/10.1038/nclimate1853

Nabuurs G-J, Verkerk PJ, Schelhaas M-J et al (2018) Climate-smart forestry: mitigation impacts in three European regions. European Forest Institute

Nabuurs G-J, Verweij P, Van Eupen M et al (2019) Next-generation information to support a sustainable course for European forests. Nat Sustain 2:815–818. https://doi.org/10.1038/s41893-019-0374-3

Nagel LM, Palik BJ, Battaglia MA et al (2017) Adaptive Silviculture for climate change: a National Experiment in manager-scientist partnerships to apply an adaptation framework. J For 115:167–178. https://doi.org/10.5849/jof.16-039

Nikinmaa L, Lindner M, Cantarello E et al (2020) Reviewing the use of resilience concepts in Forest sciences. Curr For Rep 6:61–80. https://doi.org/10.1007/s40725-020-00110-x

Nocentini S (2015) Managing forests as complex adaptive systems: an issue of theory and method. In: Proceedings of the second international congress of Silviculture. Accademia Italiana di Scienze Forestali, Florence, pp 913–918

O'Hara KL, Nagel LM (2013) The stand: revisiting a central concept in forestry. J For 111:335–340. https://doi.org/10.5849/jof.12-114

O'Hara KL, Ramage BS (2013) Silviculture in an uncertain world: utilizing multi-aged management systems to integrate disturbance†. For Int J For Res 86:401–410. https://doi.org/10.1093/forestry/cpt012

Ogden AE, Innes J (2007) Incorporating climate change adaptation considerations into forest management planning in the boreal forest. Int For Rev 9:713–733

Phalan B, Onial M, Balmford A, Green RE (2011) Reconciling food production and biodiversity conservation: land sharing and land sparing compared. Science 333:1289–1291. https://doi.org/10.1126/science.1208742

Phalan B, Green RE, Dicks LV et al (2016) How can higher-yield farming help to spare nature? Science 351:450–451. https://doi.org/10.1126/science.aad0055

Pommerening A, Murphy ST (2004) A review of the history, definitions and methods of continuous cover forestry with special attention to afforestation and restocking. For Int J For Res 77:27–44. https://doi.org/10.1093/forestry/77.1.27

Pretzsch H, Rötzer T, Forrester DI (2017) Modelling mixed-species Forest stands. In: Pretzsch H, Forrester DI, Bauhus J (eds) Mixed-species forests: ecology and management. Springer, Berlin/Heidelberg, pp 383–431

Pretzsch H, Hilmers T, Biber P et al (2020) Evidence of elevation-specific growth changes of spruce, fir, and beech in European mixed mountain forests during the last three centuries. Can J For Res. https://doi.org/10.1139/cjfr-2019-0368

Price M, Gratzer G, Alemayehu Duguma L, et al (2011) Mountain Forests in a Changing World: realizing values, adressing challenges. Food and Agriculture Organization of the United Nations (FAO) and Centre of …

Puettmann KJ, Coates D, Messier C (2009) A critique of Silviculture: managing for complexity. Island press, Washington DC

Puettmann KJ, Wilson SM, Baker SC et al (2015) Silvicultural alternatives to conventional even-aged forest management – what limits global adoption? For Ecosyst 2:8. https://doi.org/10.1186/s40663-015-0031-x

Rebmann C, Aubinet M, Schmid H et al (2018) ICOS eddy covariance flux-station site setup: a review. Int Agrophysics 32:471–494. https://doi.org/10.1515/intag-2017-0044

Rull V (2014) Time continuum and true long-term ecology: from theory to practice. Front Ecol Evol 2. https://doi.org/10.3389/fevo.2014.00075

Rundel PW, Graham EA, Allen MF et al (2009) Environmental sensor networks in ecological research. New Phytol 182:589–607. https://doi.org/10.1111/j.1469-8137.2009.02811.x

Sabatini FM, Burrascano S, Keeton WS et al (2018) Where are Europe's last primary forests? Divers Distrib 24:1426–1439. https://doi.org/10.1111/ddi.12778

Santopuoli G, di Cristofaro M, Kraus D et al (2019) Biodiversity conservation and wood production in a Natura 2000 Mediterranean forest. A trade-off evaluation focused on the occurrence of microhabitats. IForest – Biogeosciences For 12:76. https://doi.org/10.3832/ifor2617-011

Santopuoli G, Temperli C, Alberdi I et al (2020) Pan-European sustainable forest management indicators for assessing climate-smart forestry in Europe 1. Can J For Res. https://doi.org/10.1139/cjfr-2020-0166

Seddon N, Turner B, Berry P et al (2019) Grounding nature-based climate solutions in sound biodiversity science. Nat Clim Chang 9:84–87. https://doi.org/10.1038/s41558-019-0405-0

Seidl R, Rammer W, Spies TA (2014) Disturbance legacies increase the resilience of forest ecosystem structure, composition, and functioning. Ecol Appl 24:2063–2077. https://doi.org/10.1890/14-0255.1

Seidl R, Spies TA, Peterson DL et al (2016) REVIEW: searching for resilience: addressing the impacts of changing disturbance regimes on forest ecosystem services. J Appl Ecol 53:120–129. https://doi.org/10.1111/1365-2664.12511

Seymour RS, Hunter ML (1999) Principles of ecological forestry. In: Hunter ML (ed) Maintaining biodiversity in Forest ecosystems. Cambridge University Press, Cambridge, pp 22–62

Šimpraga M, Ghimire RP, Van Der Straeten D et al (2019) Unravelling the functions of biogenic volatiles in boreal and temperate forest ecosystems. Eur J For Res 138:763–787. https://doi.org/10.1007/s10342-019-01213-2

Steppe K, Sterck F, Deslauriers A (2015) Diel growth dynamics in tree stems: linking anatomy and ecophysiology. Trends Plant Sci 20:335–343

Tognetti R, Scarascia Mugnozza G, Hofer T (2017) Mountain watersheds and ecosystem services: balancing multiple demand of forest management in head-watersheds. European Forest Institute

Torresan C, del Río M, Hilmers T et al (2020) Importance of tree species size dominance and heterogeneity on the productivity of spruce-fir-beech mountain forest stands in Europe. For Ecol Manag 457:117716. https://doi.org/10.1016/j.foreco.2019.117716

Torresan C, Benito Garzon M, O'Grady M et al (2021) A new generation of sensors and monitoring tools to support climate-smart forestry practices. Can J For Res. https://doi.org/10.1139/cjfr-2020-0295

Tscharntke T, Clough Y, Wanger TC et al (2012) Global food security, biodiversity conservation and the future of agricultural intensification. Biol Conserv 151:53–59. https://doi.org/10.1016/j.biocon.2012.01.068

Valentini R, Marchesini LB, Gianelle D et al (2019) New tree monitoring systems: from industry 4.0 to nature 4.0. Ann Silvic Res 43:84–88. https://doi.org/10.12899/asr-1847

Verkerk PJ, Costanza R, Hetemäki L et al (2020) Climate-smart forestry: the missing link. For Policy Econ 115:102164. https://doi.org/10.1016/j.forpol.2020.102164

Vose JM, Peterson DL, Domke GM, et al (2018) Forests. U.S. Global Change Research Program

Willis KJ, Bailey RM, Bhagwat SA, Birks HJB (2010) Biodiversity baselines, thresholds and resilience: testing predictions and assumptions using palaeoecological data. Trends Ecol Evol 25:583–591. https://doi.org/10.1016/j.tree.2010.07.006

Wolfslehner B, Linser S, Pülzl H et al (2016) Forest bioeconomy – a new scope for sustainability indicators. European Forest Institute

Zellweger F, De Frenne P, Lenoir J et al (2019) Advances in microclimate ecology arising from remote sensing. Trends Ecol Evol 34:327–341. https://doi.org/10.1016/j.tree.2018.12.012

Chapter 2
Defining Climate-Smart Forestry

Andrew Weatherall, Gert-Jan Nabuurs, Violeta Velikova,
Giovanni Santopuoli, Bożydar Neroj, Euan Bowditch, Christian Temperli,
Franz Binder, L'ubica Ditmarová, Gabriela Jamnická, Jerzy Lesinski,
Nicola La Porta, Maciej Pach, Pietro Panzacchi, Murat Sarginci,
Yusuf Serengil, and Roberto Tognetti

Abstract Climate-Smart Forestry (CSF) is a developing concept to help policy-makers and practitioners develop focused forestry governance and management to adapt to and mitigate climate change. Within the EU COST Action CA15226, CLIMO (Climate-Smart Forestry in Mountain Regions), a CSF definition was developed considering three main pillars: (1) adaptation to climate change, (2) mitigation of climate change, and (3) the social dimension. Climate mitigation occurs through carbon (C) sequestration by trees, C storage in vegetation and soils, and C substitution by wood. However, present and future climate mitigation depends on

A. Weatherall (✉)
National School of Forestry, University of Cumbria, Cumbria, UK
e-mail: andrew.weatherall@cumbria.ac.uk

G.-J. Nabuurs
Wageningen Univ & Res, Wageningen, Netherlands
e-mail: gert-jan.nabuurs@wur.nl

V. Velikova
Institute of Plant Physiology and Genetics – Bulgarian Academy of Sciences, Sofia, Bulgaria

G. Santopuoli · R. Tognetti
Centro di Ricerca per le Aree Interne e gli Appennini (ArIA),
Università degli Studi del Molise, Campobasso, Italy
e-mail: giovanni.santopuoli@unimol.it; tognetti@unimol.it

B. Neroj
Bureau for Forest Management and Geodesy, Sekocin Stary, Poland
e-mail: bozydar.neroj@zarzad.buligl.pl

E. Bowditch
Inverness College UHI, University of the Highlands and Islands, Inverness, UK
e-mail: euan.bowditch.ic@uhi.ac.uk

C. Temperli
Swiss Federal Institute for Forest, Snow and Landscape Research WSL – Forest Resources
and Management, Birmensdorf, Switzerland
e-mail: christian.temperli@wsl.ch

© The Author(s) 2022
R. Tognetti et al. (eds.), *Climate-Smart Forestry in Mountain Regions*, Managing
Forest Ecosystems 40, https://doi.org/10.1007/978-3-030-80767-2_2

the adaptation of trees, woods, and forests to adapt to climate change, which is also driven by societal change.

Criteria and Indicators (C & I) can be used to assess the climate smartness of forestry in different conditions, and over time. A suite of C & I that quantify the climate smartness of forestry practices has been developed by experts as guidelines for CSF. This chapter charts the development of this definition, presents initial feedback from forest managers across Europe, and discusses other gaps and uncertainties, as well as potential future perspectives for the further evolution of this concept.

2.1   Introduction

Anthropogenic climate change has been described as the "*defining issue of our time*" (United Nations 2020). This chapter and the whole book will focus on one potential solution of how to manage our trees, woods, and forests to enable them to adapt to and mitigate climate change for the benefit of human society and wider biodiversity. This is Climate-Smart Forestry (CSF).

F. Binder
Sachgebiet Schutzwald und Naturgefahren, Bayerische Landes-anstalt für Wald und Forstwirtschaft, Freising, Germany
e-mail: Franz.Binder@lwf.bayern.de

L. Ditmarová · G. Jamnická
Institute of Forest Ecology, Slovak Academy of Sciences, Zvolen, Slovakia
e-mail: ditmarova@ife.sk; jamnicka@ife.sk

J. Lesinski
Department of Forest Biodiversity, University of Agriculture, Krakow, Poland
e-mail: jerzy.lesinski@urk.edu.pl

N. L. Porta
Department of Sustainable Agroecosystems and Bioresources, IASMA Research and Innovation Centre, Fondazione Edmund, Mach, Trento, Italy
e-mail: nicola.laporta@fmach.it

M. Pach
Department of Forest Ecology and Silviculture, Faculty of Forestry, University of Agriculture in Krakow, Kraków, Poland
e-mail: rlpach@cyf-kr.edu.pl

P. Panzacchi
Facoltà di Scienze e Tecnologie, Libera Università di Bolzano, Bolzano/Bozen, Italy

M. Sarginci
Duzce Univ, Forestry Fac, Duzce, Turkey
e-mail: muratsarginci@duzce.edu.tr

Y. Serengil
Dept Watershed Management, Istanbul Univ, Fac Forestry, Istanbul, Turkey
e-mail: serengil@istanbul.edu.tr

2.1.1   Why Do we Need Climate Smart Forestry?

Like all specialist disciplines, forest management is replete with jargon, a language that is helpful to the subject expert, but alienating to policymakers, the public, and even practitioners (who may not always keep up with the latest scientific developments in their field). Recent examples of jargon to describe land management approaches, including forestry, are "ecosystem services" and "natural capital." Some jargon such as "nature-based solutions" seem accessible and obvious to the user, as to a certain extent the phrase describes the purpose, but others such as "rewilding" clearly mean many different things to many people. Even more established apparently descriptive phrases can be deceptive in their complexity. For example, although "sustainable forest management" is a term that forestry academics and researchers understand and are able to expand as a definition that:

> aims to maintain and enhance the economic, social and environmental values of all types of forests, for the benefit of present and future generations (FAO 2020).

Most ordinary forest workers and users (stakeholders) are more likely to describe it as a way of managing trees so that when some of them are harvested, the forest survives. This is in fact closer to the first published definition of sustainability itself, which derives from "Sylvicultura Oeconomica," (von Carlowitz 1713) which described "the sustainable management of forest resources."

It could be argued that sustainable forest management (SFM) already addresses climate change by maintaining and enhancing environmental values for the benefit of present and future generations. However, a challenge of SFM, and also of an ecosystem services approach, is that managers aim to fulfill many objectives simultaneously. Inevitably there are trade-offs, which means that some attempt to value, or rank, objectives in terms of priorities is necessary. For those who believe that climate change is the greatest challenge of our times, CSF is an approach that identifies the adaptation of trees, woods, and forests to climate change and the use of forestry to mitigate climate change as the priority for SFM, so that other ecosystem services can be provided now and in the future.

2.1.2   Definition and Approaches to Climate Smart Forestry

This chapter derives from the work of Working Group 1 in an EU Co-operation in Science and Technology (COST) Action, CA15226, Climate Smart Forestry in Mountain Regions (CLIMO). It comprises a brief review of the literature concerning the novel concept of CSF, a definition developed in a participatory approach led by the working group, an introduction to using criteria and indicators familiar from SFM approaches for CSF, an analysis of gaps and uncertainties in the definition and approaches, a consideration of the perspective of forest management and finally, an indication of how the CSF process should develop to become more than just another piece of jargon, but a tool to enable policymakers and practitioners to protect, improve, and enhance the management of our trees woods and forests in our climate changed world.

2.2   A Brief History of Climate Smart Forestry

To be able to put CSF in context, it is important to know the evolution of our understanding of the role of forests in the global climate. Keeling (1960) suggested that the observed seasonal trend in CO_2 in the atmosphere (i.e., the zigzag of the Keeling Curve) was the result of net photosynthesis in the northern hemisphere. Tans et al. (1990) improved our understanding of the role that global and especially northern hemisphere forests were playing in the global carbon cycle.

Forests gained a lot of attention because of their large C pools in biomass and soil (Dixon et al. 1994), especially as C turnover time in forest ecosystems is much longer than agricultural and grassland areas (Harmon 1992). Thus, it became an urgent issue to determine the amount of C sequestration and fluxes in biomass and soil pools after large areas of deforestation in tropical forests in the 1980s (Kimmins 1997).

The early and mid-1990s became the time of negotiations working toward the Kyoto Protocol, the first worldwide legally binding agreement aimed at reducing global greenhouse gas emissions. With information about the role of forests being incomplete and scarce at the time, debates in Kyoto swung between encompassing global forests in a binding agreement to completely omitting their role due to the lack of insight and genuine concerns that forests might be used for greenwashing (i.e., the role of forests was confined to strictly human-induced activities). It was believed that these activities would be clearly discernible (well monitored) and their role limited. Article 3.3 (and 3.4) stated:

> .. *direct human induced land use change and forestry activities limited to afforestation, reforestation, and deforestation since 1990...* (UNFCC 1997).

Article 3.4. specified additional measures in forest management. However, since the overall reduction target was very small, there were fears that forests would be used for obscuring this small target, rather than actually reducing emissions from other sectors. Therefore, a long period of uncertainty about rules and the role of forests began. An IPCC special report on Land Use, Land Use Change and Forestry (LULUCF; Watson et al. 2000) only increased the controversy, partly because of its complexity and partly because of the very large potentials identified. For example, under Article 3.4 alone, a potential reduction of 9 Gt CO_2/y in 2040 was identified. For comparison: in the mid-1990s, the total global greenhouse gas (GHG) emissions from fossil fuels were around 26 Gt CO_2/y. However, because of the complexity of rules and guidelines and the low overall reduction targets, not much happened in the land use sector as a result of the Kyoto Protocol until forests were specifically included in Article 5 of the Paris Agreement (UNFCC 2015) and subsequently, the accounting rules for the inclusion of LULUCF in climate targets (Korosuo et al. 2020) were published.

The scope for activities in the land use sector widened, in part because of a narrow focus on mitigation and also because of controversy over large-scale monoculture plantations for afforestation. Nature-based solution was the term that attempted to capture biodiversity and social issues at the same time, defined by the International Union for Conservation of Nature as follows:

> *Actions to protect, sustainably manage and restore natural or modified ecosystems that address societal challenges effectively and adaptively, simultaneously providing human well-being and biodiversity benefits* (Cohen-Shacham et al. 2016).

According to the definition of the Institute of Development Studies:

> *being "climate smart'" describes an organization's ability to manage existing and future climate change risks while taking advantage of opportunities associated with climate change* (IDS 2007).

In agricultural sciences, this term was firstly adopted as Climate Smart Agriculture (CSA) by the FAO at the Hague Conference on Agriculture, Food Security and Climate Change in 2010 (FAO 2013) and refined by Lipper et al. (2014), but can be traced back to the 1990s when the growing awareness of farmers needing to adapt to new constraints due to climate change was first recognized (Easterling et al. 1992). The CSA definition integrates three main elements: (1) productivity, which refers to the sustainable increase of agricultural productivity and incomes from crops, livestock, and fish, without negative impact on the environment; (2) adaptation to climate change refers to make production systems more resilient and better able to withstand extreme weather events; and (3) mitigation – referring the reduction and/or removal of the greenhouse gases released by agriculture (The World Bank 2016).

The term "climate smart forestry" was first launched in 2008 (Nitschke and Innes 2008) and the CSF concept was first used in 2015 (Nabuurs et al. 2015) and since then has been modified through interactions with multiple stakeholders providing input to develop the concept (Bowditch et al. 2020; Kauppi et al. 2018; Nabuurs et al. 2017; Nabuurs et al. 2018; Verkerk et al. 2020; Yousefpour et al. 2018). CSF can arguably be seen as a category of Nature-Based Solutions with a focus on forests and forestry, which increasingly provides evidence on the effects of climate change (Schelhaas et al. 2003).

In summary, CSF initially developed as a similar concept to the CSA concept of FAO (FAO 2013) with a focus on using forestry to mitigate climate change (rather than adapting forest to climate change), while considering regional differences (Nabuurs et al. 2018).

2.3   A Definition from the EU COST Action Climate Smart Forestry in Mountain Regions

In 2016, an EU Co-Operation in Science and Technology (COST) Action was established to develop a concept of CSF with a particular focus on European mountain forests (Tognetti 2017). The aim of Working Group 1 within this COST Action Climate Smart Forestry in Mountain Regions (CLIMO) was to translate the CSA concept to forestry developing a definition and selecting Criteria and Indicators C&I) for CSF (COST Action CA15226 2016).

The new CSF definition was developed on three main thematic areas: 1) mitigation, 2) adaptation, and 3) social dimension and integrates the three-dimensions of sustainable development (economic, social, and environmental) (COST Action CA15226 2016; Bowditch et al. 2020) (Fig. 2.1).

Within the framework of the COST Action CLIMO, a wide range of experts with different expertise contributed to the development of a new CSF definition through interactive discussions during and between three separate meetings of Working Group 1 and cross-Working Group engagement (Bowditch et al. 2020), involving representatives from 28 countries (http://climo.unimol.it/). It was specifically intended that this definition should be no longer than one page for ease of sharing (Fig. 2.2).

Some of the aspects underlying the definition have been studied experimentally. Jandl et al. (2018) focusing on climate smart management strategies for Austrian forests, examined and evaluated carbon dynamics in the stem biomass and soils. The authors concluded that the production of long-living wood products is the preferred implementation of CSF, and the production of bioenergy is suitable as a by-product of high-value forest products (Jandl et al. 2018). However, CSF measures can vary from country to country and region to region depending on different circumstances (e.g. socioecological and technological framework, climate change impacts, and cultural aspects), and the success of CSF requires the balance between them (Verkerk et al. 2020). Indeed, case studies in three European regions (Spain, Czech Republic, and Republic of Ireland) differ in the composition and history of their forests and forest sectors, clearly demonstrating that CSF mitigation measures need to consider local- or country-specific conditions (Nabuurs et al. 2018).

Fig. 2.1 The main pillars of Climate-Smart Forestry

Climate-smart forestry is sustainable adaptive forest management and governance to protect and enhance the potential of forests to adapt to, and mitigate climate change. The aim is to sustain ecosystem integrity and functions and to ensure the continuous delivery of ecosystem goods and services, while minimising the impact of climate-induced changes on forests on well-being and nature's contribution to people.

Adaptation measures forests that maintain or improve their ability to grow under current and projected climatic conditions and increase their resistance and resilience. The adaptive capacity to changes in climate and to the timing and size of climate-induced disturbances (e.g., fire, extreme storm events, pests and diseases) can be enhanced by promoting genetic, compositional, structural, and functional diversity at both stand and landscape scales. This includes facilitating natural regeneration and planting of native as well as non-native tree species, genetic variants and individuals that are considered to be adapted to future conditions. Increased connectivity assists the migration of forest species.

Mitigation of climate change by forests is a combination of carbon sequestration by trees, carbon storage by forest ecosystems, especially soils, and forest derived products, such as structural timber, and by carbon substitution -directly by replacing fossil fuels with bioenergy and indirectly through use of wood to substitute for higher carbon footprint materials.

The *social dimension* of forestry holds many aspects, from the involvement of stakeholders from local communities, and their conflicts over land use or for the access to skills and technology, to global forest governance challenges. Climate change may jeopardize forest ecosystem functioning and brings social and economic consequences for people, which may modify priorities of ecosystem services at various scales. Assessment for ecosystem services could be a tool making this process more efficient with respect to indicators relevant for governance regime and actors involved.

In summary, **climate-smart forestry** should enable both forests and society to transform, adapt to and mitigate climate-induced changes.

Fig. 2.2 Climate smart forestry (CSF) definition from the EU COST CA15226, Climate Smart Forestry in Mountain Regions (Bowditch et al. 2020)

2.4 Criteria and Indicators for the Assessment of Climate-Smart Forestry

2.4.1 Assessing Climate Smart Forestry

Recent advances on the concept of CSF in Europe have encouraged the development of tools and approaches to measure its effects on forest health, function, and productivity.

Concepts such as CSF are only meaningful if they are developed with suitable C&I to monitor whether the principles outlined in the definition are being adopted over time. Indicators need to balance ease of collection against being as detailed as possible, but general enough to be widely applicable. For CSF, an indicator is a variable, generally quantitative, that enables one to describe the status of forests and forestry as well as trends in forest development. It needs to be applicable in as many forest ecosystems and methods of forest management as possible allowing comparisons across temporal and spatial scales.

2.4.2 Criteria and Indicators for Sustainable Forest Management

Rather than reinventing the wheel, the COST Action participants first evaluated the existing pan-European C&I for SFM (Santopuoli et al. 2016; Wolfslehner and Baycheva-Merger 2016). In the past 30 years, as a result of several initiatives about sustainable development, numerous sets of C&I for SFM have been proposed worldwide (Castañeda 2000; Linser et al. 2018). In Europe, the main driving force involved in the implementation of C&I for SFM is FOREST EUROPE, a multi-stakeholder participatory process currently involving 46 European countries and the European Union (EU) as signatory bodies. Since the 1990s, seven Ministerial Conferences have taken place (Fig. 2.3), within which C&I for SFM were defined and adopted.

The first set of C&I for SFM was approved at the Lisbon pan-European conference of 1998 (MCPFE 2001), as were the "Pan-European Operational Level Guidelines for SFM" that became the basis for the development of the forest certification scheme Programme for the Endorsement of Forest Certification (PEFC) (Rametsteiner and Simula 2003).

The first set of C&I was improved at the Vienna Ministerial Conference (MCPFE 2003), and subsequently updated in Madrid 2015 (Forest Europe 2015). This robust process has currently led to 6 criteria, 34 quantitative indicators (Table 2.1), and 11 qualitative indicators covering all aspects of SFM.

Although their implementation is not legally binding, the Pan-European C&I for SFM generated a broad variety of responses among FOREST EUROPE signatory bodies and were formally adopted by the signatory bodies as a policy framework for

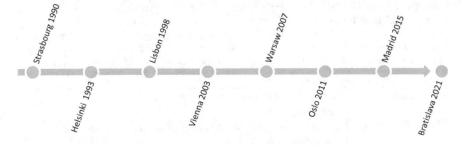

Fig. 2.3 Forest Europe Ministerial Conference on the Protection of forests in Europe (Former MCPFE) timeframe. The eighth Ministerial Conference will take place in April 2021 in Bratislava (Slovakia)

forest management concerns. The Pan-European C&I for SFM are collected in a harmonized way, are broadly accepted by policy makers, cover the most important forest ecosystem services, and are publicly available. This makes them a suitable basis for further development toward an indicator set for the assessment of CSF at a European scale.

2.4.3   From Sustainable Forest Management to Climate-Smart Forestry Indicators

The COST Action participants assessed these SFM C&I and judged twenty-five indicators to be highly relevant to CSF, four new indicators were also identified by the CLIMO participants. As a result, a total of 29 indicators were selected as suitable to assess climate adaptation and mitigation by CSF (Bowditch et al. 2020).

Some challenges for C&I implementation across signatory countries are still evident (Santopuoli et al. 2016; Wolfslehner and Baycheva-Merger 2016), even if they provide great support to assess many aspects of SFM, and most of them are useful to support the assessment of CSF (Bowditch et al. 2020; Santopuoli et al. 2020).

Selecting or developing new indicators to assess CSF requires a multidisciplinary approach that covers all the aspects of SFM, which are related to climate change. Beyond modelling approaches (Mäkelä et al. 2012; Pretzsch et al. 2014; Zeller and Pretzsch 2019) that provide useful information on long-term forest growth to promote adaptive forest management, CSF should support forest decision-makers and managers to help adapt to and mitigate climate change while maintaining long-term ecosystem service provision. For example, focusing on C storage by prioritizing soil sustainability and extending the life cycle of timber products through the circular bioeconomy.

Focusing on the climate smart vision, 10 out of 34 quantitative indicators are the most recurrent indicators used for monitoring the effects of climate change on forest resources (Santopuoli et al. 2020). Particularly important were the indicators 1.4

Table 2.1 Criteria and Indicators for Sustainable Forest Management. Sources: Updated Pan-European Indicators for Sustainable Forest Management, as adopted by the FOREST EUROPE Expert Level Meeting 30 June – 2 July 2015, Madrid, Spain. Accessed (https://foresteurope.org), June 2020. Accessed 25 Jan 2021. Five qualitative indicators for forest policy are followed by the 6 criteria each with a qualitative indicator and one or more quantitative indicators (34 in all)

	No.	Indicator	
Forest policy and governance	1	National Forest Programmes or equivalent	
	2	Institutional frameworks	
	3	Legal/regulatory framework: National (and/or subnational) and international commitments	
	4	Financial and economic instruments	
	5	Information and communication	
Criteria	No.	Indicator	Full text
Criterion 1: Maintenance and appropriate enhancement of forest resources and their contribution to global carbon cycles	C.1	Policies, institutions, and instruments to maintain and appropriately enhance forest resources and their contribution to global carbon cycles	
	1.1	Forest area	Area of forest and other wooded land, classified by forest type and by availability for wood supply, and share of forest and other wooded land in total land area
	1.2	Growing stock	Growing stock on forest and other wooded land, classified by forest type and by availability for wood supply
	1.3	Age structure and/or diameter distribution	Age structure and/or diameter distribution of forest and other wooded land, classified by availability for wood supply
	1.4	Forest carbon	Carbon stock and carbon stock changes in forest biomass, forest soils, and in harvested wood products
Criterion 2: Maintenance of forest ecosystem, health, and vitality	C.2	Policies, institutions, and instruments to maintain forest ecosystems health and vitality	
	2.1	Deposition and concentration of air pollutants	Deposition and concentration of air pollutants on forest and other wooded land
	2.2	Soil condition	Chemical soil properties (pH, CEC, C/N, organic C, base saturation) on forest and other wooded land related to soil acidity and eutrophication, classified by main soil types
	2.3	Defoliation	Defoliation of one or more main tree species on forest and other wooded land in each of the defoliation classes
	2.4	Forest damage	Forest and other wooded land with damage, classified by primary damaging agent (abiotic, biotic, and human induced)
	2.5	Forest land degradation	Trends in forest land degradation

(continued)

Table 2.1 (continued)

	No.	Indicator	
Criterion 3: Maintenance and encouragement of productive functions of forests (wood and nonwood)	C.3	Policies, institutions, and instruments to maintain and encourage the productive functions of forests	
	3.1	Increment and fellings	Balance between net annual increment and annual fellings of wood on forest available for wood supply
	3.2	Roundwood	Quantity and market value of roundwood
	3.3	Nonwood goods	Quantity and market value of nonwood goods from forest and other wooded land
	3.4	Services	Value of marketed services on forest and other wooded land
Criterion 4: Maintenance, conservation, and appropriate enhancement of biological diversity in forest ecosystems	C.4	Policies, institutions, and instruments to maintain, conserve, and appropriately enhance the biological diversity in forest ecosystems	
	4.1	Diversity of tree species	Area of forest and other wooded land, classified by number of tree species occurring
	4.2	Regeneration	Total forest area by stand origin and area of annual forest regeneration and expansion
	4.3	Naturalness	Area of forest and other wooded land by class of naturalness
	4.4	Introduced tree species	Area of forest and other wooded land dominated by introduced tree species
	4.5	Deadwood	Volume of standing deadwood and of lying deadwood on forest and other wooded land
	4.6	Genetic resources	Area managed for conservation and utilization of forest tree genetic resources (in situ and ex situ genetic conservation) and area managed for seed production
	4.7	Forest fragmentation	Area of continuous forest and of patches of forest separated by nonforest lands
	4.8	Threatened forest species	Number of threatened forest species, classified according to IUCN red list categories in relation to total number of forest species
	4.9	Protected forests	Area of forest and other wooded land protected to conserve biodiversity, landscapes, and specific natural elements, according to MCPFE categories
	4.10	Common forest bird species	Occurrence of common breeding bird species related to forest ecosystems
Criterion 5: Maintenance and appropriate enhancement of protective functions in forest management (notably soil and water)	C.5	Policies, institutions, and instruments to maintain and appropriately enhance the protective functions in forest management	
	5.1	Protective forests – Soil, water, and other ecosystem functions – infrastructure and managed natural resources	Area of forest and other wooded land designated to prevent soil erosion, preserve water resources, maintain other protective functions, protect infrastructure, and manage natural resources against natural hazards

(continued)

Table 2.1 (continued)

	No.	Indicator	
Criterion 6: Maintenance of other socioeconomic functions and conditions	C.6	Policies, institutions, and instruments to maintain other socioeconomic functions and conditions	
	6.1	Forest holdings	Number of forest holdings, classified by ownership categories and size classes
	6.2	Contribution of forest sector to GDP	Contribution of forestry and manufacturing of wood and paper products to gross domestic product
	6.3	Net revenue	Net revenue of forest enterprises
	6.4	Investments in forests and forestry	Total public and private investments in forests and forestry
	6.5	Forest sector workforce	Number of persons employed and labor input in the forest sector, classified by gender and age group, education, and job characteristics
	6.6	Occupational safety and health	Frequency of occupational accidents and occupational diseases in forestry
	6.7	Wood consumption	Consumption per head of wood and products derived from wood
	6.8	Trade in wood	Imports and exports of wood and products derived from wood
	6.9	Wood energy	Share of wood energy in total primary energy supply, classified by origin of wood
	6.10	Recreation in forests	The use of forests and other wooded land for recreation in terms of right of access, provision of facilities, and intensity of use

"Carbon stock," 4.1 "Tree species composition," 2.4 "Forest damages," and 6.9 "Energy from wood resources." The number of indicators that resulted useful to support CSF increased significantly when all aspects of SFM, particularly the socioeconomic aspects, are considered (Bowditch et al. 2020). Overall, indicators belonging to the criteria "Forest Biological Diversity" and "Forests Health and Vitality" are considered particularly important to manage forests according to a climate smart approach (Fig. 2.4).

Finally, four new indicators, concerning the forest structure, were suggested by CLIMO participants during the CLIMO meetings (Bowditch et al. 2020). Monitoring these indicators (management system, slenderness coefficient, and tree crown distribution both vertical and horizontal) allows to observe the impacts of forest management on the forest productivity and growth, as well as the delivery of ecosystem services, supporting CSF evaluation (Fig. 2.4). These indicators can be evaluated through remote sensing and thus are particularly important, because they can be monitored frequently providing timely forest inventory data (e.g., Giannetti et al. 2020).

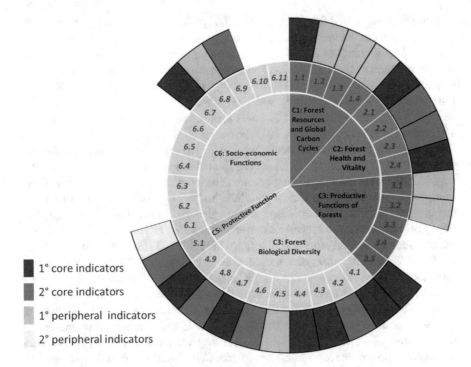

Fig. 2.4 Indicators relevant for assessing Climate-Smart Forestry. The set of indicators refers to Vienna 2003, since some of the indicators from the updated set (Madrid 2015) require further development and testing for consideration (Bowditch et al. 2020)

2.5   A Critical Analysis of the Definition, Gaps, and Uncertainties

The definition of the CSF concept derived from the COST Action (Fig. 2.1, Bowditch et al. 2020) was an important development in reasserting that the climate adaption of forests is a vital component, in part because this is necessary to secure future climate mitigation by forestry. This definition also recognized the importance of the social dimension. However, CSF is an evolving concept and this definition marks the current stage in its development, not an end point. In particular, it is important to recognize that the definition is derived by a group working from a European perspective on climate smart forestry in mountain regions.

2.5.1   Gaps and Uncertainties

When scaling-up to global level and beyond mountain environments, several issues need to be reconsidered. For example, a future definition of CSF should cover a more global climate change context, reducing emissions from forest degradation and deforestation is one of the most important ways of combatting climate change.

It has been suggested that in areas more likely to maintain optimal growth conditions for forests in the long term, forest management should make use of their mitigation capacity (Jandl et al. 2015). The definition does not emphasize that SFM already provides climate mitigation and that the role of CSF in enhancing this may benefit from some positive feedbacks. For example, the growth rate of some European forests has accelerated and total stand volume prior to final cutting is reached now much earlier than 100 years ago, thus increasing the C sequestration and substitution (Mäkinen et al. 2017; Pretzsch et al. 2014; Pretzsch et al. 2019; Socha et al. 2017; Socha and Staniaszek 2015). However, the potential for further mitigation of climate change is uncertain, since there is a lack of sufficient knowledge of how elevated CO_2 and temperature and changing weather pattern will affect tree growth, nor are there historical parallels (Yousefpour et al. 2012). Furthermore, the impacts of pests, diseases, and abiotic threats are uncertain. For example, European countries have recently experienced a series of noticeable forest disturbances, such as several storms in the fall/winter 2017–2018, extended drought in 2018 and 2019 with subsequent bark beetle outbreaks, and disastrous wild forest fires. Cumulative evidence proves that CC is contributing to the increased frequency and intensity of these forest disturbances (Forest Europe 2019). Thus, while individual trees may grow faster, forest resilience may decline, so that the overall level of C sequestration is reduced. This is why trade-offs and conflicts between adaptation and mitigation measures should be considered (Böttcher et al. 2009) and why adaptation appears before mitigation in the COST Action definition (Bowditch et al. 2020).

It is important to have a definition of climate smart forestry, just as there is one for CSA; however, for adaptation and mitigation measures to be truly successful, an approach that considers the entire land use system is required. It is therefore necessary to harmonize forest management under climate change with respective measures in agriculture, wildlife conservation, and any other objectives with implications for land management and the bioeconomy. This does not mean that the concept of CSF is wrong, but a gap exists in defining how it fits within a wider climate smart landscapes approach.

The focus in the definition on social dimension is a new look at the problem, but it deserves more attention (Scheffers et al. 2017). In particular, indicators and analysis methods to measure and depict potential trade-offs between fostering adaptation and mitigation and ecosystem service provision need to be further developed. For example, the economic costs for adaptation and mitigation treatments need to be quantified in order to device CSF scenarios that are economically feasible in the long term. For example, in many mountain areas, the protection efficiency against rockfall, avalanches, and landslides must be ensured when currently nonautochthonous tree species or provenances are introduced to adapt forests to the future climate. Another example is the potential trade-off between adaptation or mitigation and the provision of the forests recreational, cultural, and tourism services. The role of professionals (scientists and forest managers) involved in education and clarification of climate change processes can be important not only for forest owners but also for the society as the whole and therefore for public acceptance of climate smart forestry (Laakkonen et al. 2018).

2.6   Developing a Forest Manager Vision of CSF

An important step for implementing CSF on any type of forest policy is ensuring that concepts are accessible and translatable into practice for forest managers (Groot et al. 2010; Sousa-Silva et al. 2018; Bowditch et al. 2019). They play a key role in adopting the mechanisms of policy and turning them into common and best practice, representing a broad behavioral change that can have wide-ranging benefits (Nichiforel 2010; Carmon-Torresa et al. 2011; Raymond et al. 2016). However, the science-policy-practice interface has been difficult to navigate with many still emphasizing a large disconnect in communication (Nijnik et al. 2016). The CSF definition and indicators are a first step in introducing forest management specifically focused on climate change response. Although these were developed by a range of forest professionals, only a small number of managers were involved; therefore, engaging managers was viewed as a crucial stage of the process of refining and testing the accessibility and relevance of the work (Bowditch et al. 2020).

In an online survey, forest managers from 14 European and neighboring countries were asked to critique the CSF definition and indicators from a management perspective. Representatives from each country involved in the CLIMO project disseminated the survey to public, private, community, or other relevant forest management entities within the country to capture the range of perspectives and challenges.

2.6.1   Forest manager's Response

Forty-seven percent of all managers viewed climate change as a critical or high risk to management aims and objectives; however, 42% viewed it as a medium risk and 11% considered it to be a low or nonexistent risk. Around 41% of managers believed that they were equipped with the tools and knowledge to respond to climate change, 40% were unsure, and the remainder did not believe they were equipped. Examples of contrasting options included:

"we have knowledge but constraints outside our control prevents us from effective delivery"

"there are more threats than ever before but as professionals would rise to the challenge through constant pursuit of knowledge"

A main challenge identified by forest managers is the ability to turn knowledge into action and management approaches with constraints ranging from systemic national forestry policy and management, to capacity to deliver on aims at stand level due to available time, resources, and bureaucratic barriers.

The CSF definition presented to managers was generally well-received with 62% saying it was accessible, clear, and relevant, but 38% either saying it was too complex or that they did not understand the definition. The majority of the negative responses were in countries where the definition had to be translated into the native language with a possibility of some meanings and phrases being lost in translation.

Although 37% of managers found the definition either very useful or useful with 54% finding it moderately or marginally useful, 9% found the definition not useful. Examples of contrasting responses are:

"it is succinct and clear and brings together useful aims"

"A definition should also include the economic dimension, long-term profitability"

"would become lost in the busy job of a manager but would be good as reference during design and operational phases".

"distant from the realities of management in the field".

The CSF list of indicators based on the pan-European Criteria and Indicators for Sustainable Forest Management was well received by the forest managers, who all acknowledged it represented a comprehensive set of management concerns. Despite the positive attitude toward the indicators, most managers highlighted the limited scope of using indicators in management plans, as the current systems (national, regulatory, and company) were not compatible to integrate into plans. Managers further highlighted that there were too many indicators, which would be time consuming to measure, additionally managers pointed out that they did not have the knowledge or resources to measure most indicators. *"Tree species composition"* and *"natural regeneration"* were identified as the most important Sustainable Forest Management (SFM) indicators, whereas the *"slenderness coefficient"* and *"roundwood"* were ranked as the least important. *"Erosion prevention and maintenance of soil health"* were the top ranked ecosystem services indicators followed by *"water and air purification."* Ranked least important were *"pharmaceuticals and biochemicals"* followed by *"food."* Managers suggested that the indicators could be streamlined or modified for different forest types or objectives. The current list was unrealistic to implement but considered appropriate as a checklist and a broader list that could be classified into different areas of management.

2.6.2   Refinement of Definition and Indicators

The main suggestions to improve the definition and indicators focused around economic and social factors. Most notably profitability or revenue from management and transport, and the relationship with GHG emissions. Further clarification on the C cost of producing different forest products and bringing them to market was highlighted by a cross-section of managers, emphasizing the importance of integrating life cycle knowledge into management decisions (Karvonen et al. 2017). It was also suggested that measurement of the benefits of direct fossil fuel substitution from forestry could explicitly translate another element of the definition into an indicator (Münnich Vass 2017). The use of technology was also mentioned as a potential indicator to track integration and use, which either benefits or hinders CSF adaptation (Biggs et al. 2010; Ghaffariyan et al. 2017).

Support of communities and rural areas was mentioned widely by respondents as an indicator that could evaluate the importance of a forest to the local area and wider rural economy. Greater recognition of small landowners and their management needs, as well as recognizing contributions to climate change was viewed as important locally and landscape wide to encourage investment in CSF. The level of public awareness of forest management and services was identified as a potentially powerful social indicator, which demonstrates the current disconnect between forestry sector and society about the role of forests and forestry including their benefit to the wider environment (Upton et al. 2015; Seidl et al. 2016).

A key theme emerged that addressed wider issues of communication among policy, science, and practice, which highlighted the need to integrate explicit climate change adaptation and mitigation goals into grants and incentives (Opdam et al. 2013; Fischer et al. 2015; Blades et al. 2016). This was further supported by a range of forest managers expressing the need to challenge traditional silviculture and approaches to forest management, as well as considering other land uses such as agriculture in joined-up approaches:

"We cannot be afraid of having healthy discussion that challenges traditional management's compatibility with current goals"

Training and education also emerged as a common theme:

"there needs to be a commitment to training those future professionals and current professionals in climate and resilience thinking and practice".

Other managers identified that scenario planning within management plans and at higher levels would be crucial to climate change responses (Jandl et al. 2018):

"Local climate change scenarios that address fine scale change will be really important for managers and provide guidance for planning and redundancies".

Scenario-driven analyses would give managers response pathways to follow in case of unexpected or unprecedented events affecting the productivity and integrity of their forests.

In general, the definition was viewed as a positive start by the majority of forest managers who saw it as a vision statement to reference broad aims and only lacked wording on economic implications. The indicators were identified by managers as a set of tools that could potentially have practical relevance for their work. However, the indicators required clear instructions and tools for them to be implemented into management plans. The next step would be to trial a set of indicators with forest managers to assess the ease of use and interpretation to inform current data and/or create new baselines.

2.7 Future Perspectives for CSF

With the definition of CSF in this chapter and in Bowditch et al. (2020) and numerous previous applications (e.g., Nabuurs et al. 2017; Yousefpour et al. 2018; Jandl et al. 2018), the concept of CSF is established in forest science. The next step will be to implement CSF in practice. This encompasses balancing adaptation, mitigation, and ES provision from the stand to the European scale, and working with international partners to expand the definition to suit a global understanding of the concept. While in some cases, all three aspects of CSF may be considered in management decisions at the stand scale, other circumstances may require prioritizing for one or the other at the landscape scale. Decisions on such sparing versus sharing strategies may depend on topography, structure of the forest landscape, forest industry and administration, and other circumstances in different countries and regions. CSF needs to link global priorities with specific local conditions. A clear definition of CSF and its implementation in practical forest management can contribute to this link.

The implementation of climate smart management decisions should be embedded within the cyclical adaptive management process of planning, implementing, monitoring, evaluating, and revising CSF management (Walters 1986). A forward-looking rather than reactive approach should be adopted for planning (Yousefpour et al. 2017). This involves considering climate and other environmental and socioeconomic conditions expected for the future as well as their uncertainties in decision making. Results from species distribution models may provide a basis for the selection of candidate tree species to grow under future conditions (e.g., Hanewinkel et al. 2012), whereas dynamic forest development models may deliver understanding on successional dynamics and management and disturbance impacts under climate change scenarios (Temperli et al. 2020; Reyer et al. 2015; Seidl et al. 2017; Gutsch et al. 2018). Specifically, these models can be used to evaluate potential CSF scenarios, including schemes for natural regeneration and planting (assisted migration), and generally deliver management targets for forward looking adaptive managers at a broad range of spatial and temporal scales (Pretzsch et al. 2008; Yousefpour et al. 2018; Jandl et al. 2018). In addition, a database of "best practices" from individual forest management agencies, regions, and countries may serve as useful decision tools to promote CSF management.

Indicator system to measure mitigation, adaptation, and ES provision, such as the one suggested in this chapter based on C& I for sustainable forest management by Forest Europe (Forest Europe 2015), need to be constantly updated to tackle upcoming challenges. With C sinks in European forests being limited (Nabuurs et al. 2013), mitigation strategies need to also focus on storing C in wood products and buildings and thereby substituting fossil fuel–intensive energy sources. Hence, indicators to quantify mitigation need to go beyond the C sequestered in the tree biomass and the soil, but also include the wood value chain (Verkerk et al. 2020). Challenging questions on system boundaries need to be resolved in that regard (Sandin et al. 2016). The CSF aspect of adaptation is often captured indirectly as the so-called adaptive capacity of forests and the forestry industry (Lindner et al. 2010; Irauschek et al. 2017). Indicators on provenances, tree species and stand and forest type diversity, as well as on the density of forest road networks and the regulatory and economic boundaries of forest enterprises are inter alia used for this purpose. A step forward would be to assess adaptation directly by quantifying the difference between the current and a targeted state of the forest. This may include measuring the progress of assisted migration of climate change-adapted provenances and (native and nonnative) tree species (Bolte et al. 2009). Indicators could be the percentage of a drought-adapted provenance or tree species, or forest structural parameters that measure disturbance resistance and resilience (Bryant et al. 2019; Temperli et al. 2020). These difference-indicators could be advantageous for a more targeted adaptation process, but may also create challenges with regard to comparability across stands, landscapes, or countries, because management targets need to be defined specifically for each spatial entity. Efforts to further harmonize indicators internationally are pivotal for climate smart policy making at European levels (Alberdi et al. 2016).

Evaluating and revising CSF strategies completes the adaptive management cycle. Evaluation needs to assess whether targeted ES can be provided sustainably (also considering social and economic aspects) as forests adapt to climate change and novel tree species compositions emerge. Thereby climate change may also create opportunities. Expanding deciduous trees in subalpine conifer forests may offer a broader spectrum of site-adapted tree species that can be promoted following timber harvesting or natural disturbances. This may benefit management toward heterogeneous stand structures and thus the long-term maintenance of the forest's protective function against rockfall and landslides (Bebi et al. 2016), as well as positive effects on soil water availability and water cycling at the landscape scale. Moreover, forest stands with high levels of genetic diversity and species richness may improve ecosystem service provision including the production of raw materials, medical resources, tourism, recreation, and aesthetic, cultural, and spiritual experiences. The CSF concept offers the opportunity to connect agriculture and forestry in submountain regions to create an effective (integrated) climate smart management system of whole areas. CSF decisions must consider uncertainties (i.e., by promoting a range of candidate tree species) as CSF paradigms of today may shift in the next decades as we learn from the effects of past management. Further developments of the CSF concept need to ensure that it remains flexible and dynamic such that it can be applied to a broad range of environmental and socioeconomic conditions in an uncertain future.

In summary, CSF is a continuously evolving concept; the definition presented here from COST Action CA15226 Climate Smart Forestry in Mountain Regions and use throughout this book aims to help policymakers and practitioners develop focused governance and management through which forests can adapt and mitigate climate change, while continuing to deliver wide benefits to society (Bowditch et al. 2020).

References

Alberdi I, Michalak R, Fischer C et al (2016) Towards harmonized assessment of European forest availability for wood supply in Europe. For Policy Econ 70:20–29. https://doi.org/10.1016/j.forpol.2016.05.014

Bebi P, Bugman H, Lüscher P et al (2016) Auswirkungen des Klimawandels auf Schutzwald und Naturgefahren. In: Pluess AR, Augustin S, Brang P (eds) Wald im Klimwandel – Grundlagen für Adaptationsstrategien. Bundesamt für Umwelt, Bern; Eidg. Forschungsanstalt WSL, Birmensdorf; Haupt, Bern, Stuttgart, Wien

Biggs R, Westley FR, Carpenter SR (2010) Navigating the back loop: fostering social innovation and transformation in ecosystem management. Ecol Soc 15(2):9

Blades JJ, Klos PZ, Kemp KB (2016) Forest managers' response to climate change science: evaluating the constructs of boundary objects and organizations. Forest Ecol Manag 360:376–387. https://doi.org/10.1016/j.foreco.2015.07.020

Bolte A, Ammer C, Lof M et al (2009) Adaptive forest management in Central Europe: climate change impacts, strategies and integrative concept. Scand J For Res 24:473–482. https://doi.org/10.1080/02827580903418224

Böttcher H, Barbeito I, Reyer CH et al (2009) Role of forest management in fighting climate change. Forest management work group report. In: Karjalainen T, Lindner M, Niskanen A et al (eds) Joensuu Forestry Networking Week 2009. Fighting Climate Change: Adapting Forest. Policy and Forest Management in Europe. Group Work Reports and Conclusions. Working Papers of the Finnish Forest Research Institute 135:41–53

Bowditch EAD, McMorran R, Bryce R et al (2019) Perception and partnership: developing forest resilience on private estates. For Policy Econ 99:110–122

Bowditch E, Santopuoli G, Binder F et al (2020) What is climate-smart forestry? A definition from a multinational collaborative process focused on mountain regions of Europe. Ecosyst Serv. https://doi.org/10.1016/j.ecoser.2020.101113

Bryant T, Waring K, Sánchez Meador A et al (2019) A framework for quantifying resilience to forest disturbance. Front For Glob Change 2. https://doi.org/10.3389/ffgc.2019.00056

Carmon-Torresa C, Parra-López C, Groot JCJ et al (2011) Collective action for multi-scale environmental management: achieving landscape policy objectives through cooperation of local resource managers. Landsc Urban Plan 103:24–33

Castañeda F (2000) Criteria and indicators for sustainable forest management: international processes, current status and the way ahead. Unasylva 51:34–40

Cohen-Shacham E, Walters G, Janzen C et al (eds) (2016) Nature-based solutions to address global societal challenges. IUCN, Gland. xiii + 97pp

COST Action CA15226 (2016), CLIMO (Climate- Smart Forestry in Mountain Regions). Available via https://www.cost.eu/cost-action/climate-smart-forestry-in-mountain-regions/#tabs|Name:parties|https://www.cost.eu/cost-action/climate-smart-forestry-in-mountain-regions/#tabs|Name:parties. Accessed 25 June 2020

De Groot RS, Fisher B, Christie M et al (2010) Integrating the ecological and economic dimensions in biodiversity and ecosystem service valuation. In: The Economics of Ecosystems and Biodiversity (TEEB): Ecological and Economic Foundations. Earthscan. Available via http://library.wur.nl/WebQuery/wurpubs/401249

Dixon RK, Solomon AM, Brown S et al (1994) Carbon pools and flux of global forest ecosystems. Science 263(5144):185–190

Easterling WE, Rosenberg NJ, Lemon KM et al (1992) Simulations of crop responses to climate change: effects with present technology and currently available adjustments (the 'smart farmer' scenario). Agr Forest Meteorol 59:75–102. https://doi.org/10.1016/0168-1923(92)90087-K

FAO (2013) Climate-Smart Agriculture: Sourcebook. Food and Agriculture Organization of the United Nations, Rome

FAO (2020) Sustainable Forest Management. Available via http://www.fao.org/forestry/sfm/en/ Accessed 25 Jan 2020

Fischer J, Gardner TA, Bennett EM et al (2015) Advancing sustainability through mainstreaming a social-ecological systems perspective. Curr Opin Environ Sustain 14:144–149

Forest Europe (2015) Madrid ministerial declaration: 25 years together promoting sustainable Forest management in Europe. Madrid, p 10

Forest Europe (2019) Pro-active management of forests to combat climate change driven risks Policies and measures for increasing forest resilience & climate change adaptation

Ghaffariyan, MR, Brown, M, Acuna, M et al (2017) An international review of the most productive and cost effective forest biomass recovery technologies and supply chains. Renew Sust Energ Rev 74:145–158. Available via https://www.sciencedirect.com/science/article/pii/S1364032117302174 Accessed Mar 7 2019

Giannetti F, Puletti N, Puliti S et al (2020) Assessment of UAV photogrammetric DTM-independent variables for modelling and mapping forest structural indices in mixed temperate forests. Ecol Indic 117:106513

Gutsch M, Lasch-Born P, Kollas C et al (2018) Balancing trade-offs between ecosystem services in Germany's forests under climate change. Environ Res Lett 13:045012. https://doi.org/10.1088/1748-9326/aab4e5

Hanewinkel M, Cullmann DA, Schelhaas M-J et al (2012) Climate change may cause severe loss in the economic value of European forest land. Nat Clim Chang. https://doi.org/10.1038/nclimate1687

Harmon ME (1992) Long-term experiments on log decomposition at the HJ Andrews Experimental Forest (Vol. 280). US Department of Agriculture, Forest Service, Pacific Northwest Research Station

IDS (2007) Towards 'Climate Smart' Organizations. In: Focus Research and analysis from the Institute Of Development Studies. Issue 02 Climate Change Adaptation. Available via https://opendocs.ids.ac.uk/opendocs/bitstream/handle/20.500.12413/2538/Towards%20Climate%20Smart%20Organisations%20IDS%20in%20Focus%202.8.pdf?sequence=1 Accessed 15 June 2020

Irauschek F, Rammer W, Lexer MJ (2017) Evaluating multifunctionality and adaptive capacity of mountain forest management alternatives under climate change in the eastern Alps. Eur J For Res 136:1051–1069. https://doi.org/10.1007/s10342-017-1051-6

Jandl R, Bauhus J, Bolte A et al (2015) Effect of climate-adapted Forest management on carbon pools and greenhouse gas emissions. In: Whitehead D (ed) Climate change and carbon sequestration, Current Forestry reports. Springer, p 7

Jandl R, Ledermann T, Kindermann G et al (2018) Strategies for climate-smart forest management in Austria. Forests 9:1–15. https://doi.org/10.3390/f9100592

Karvonen J, Halder P, Kangas J et al (2017) Indicators and tools for assessing sustainability impacts of the forest bioeconomy. For Ecosyst 4(1):2

Kauppi P, Hanewinkel M, Lundmark L et al (2018) Climate smart forestry in Europe. European Forest Institute

Keeling CD (1960) The concentration and isotopic abundances of carbon dioxide in the atmosphere. Tellus 12(2):200–203

Kimmins H (1997) Balancing act. Environmental issues in forestry, 2nd edn. UBC Press, Vancouver

Korosuo A, Vizzarri M, Pilli R et al (2020) Forest reference levels under Regulation (EU) 2018/841 for the period 2021–2025, EUR 30403 EN, Publications Office of the European Union, Luxembourg. doi:https://doi.org/10.2760/27529

Laakkonen A, Zimmerer R, Kähkönen T et al (2018) Forest policy and economics Forest owners' attitudes toward pro-climate and climate-responsive forest management. Forest Policy Econ 87:1–10. https://doi.org/10.1016/j.forpol.2017.11.001

Lindner M, Maroschek M, Netherer S et al (2010) Climate change impacts, adaptive capacity, and vulnerability of European forest ecosystems. Forest Ecol Manag 259:698–709. https://doi.org/10.1016/j.foreco.2009.09.023

Linser S, Wolfslehner B, Bridge SR et al (2018) 25 years of criteria and indicators for sustainable forest management: how intergovernmental C&I processes have made a difference. Forests 9(9):578. https://doi.org/10.3390/f9090578

Lipper L, Thornton P, Campbell B et al (2014) Climate-smart agriculture for food security. Nat Clim Chang 4:1068–1072. https://doi.org/10.1038/nclimate2437

Mäkelä A, Río MD, Hynynen J et al (2012) Using stand-scale forest models for estimating indicators of sustainable forest management. For Ecol Manag 285:164–178. https://doi.org/10.1016/j.foreco.2012.07.041

Mäkinen H, Yue C, Kohnle U (2017) Site index changes of scots pine, Norway spruce and larch stands in southern and Central Finland. Agr Forest Meteorol 237–238:95–104. https://doi.org/10.1016/j.agrformet.2017.01.017

MCPFE (2001) Pan-European indicators for sustainable Forest management. Third ministerial conference for protection of forests in Europe, Lisbon

MCPFE (2003) Improved pan-european indicators for sustainable forest management. Vienna, Austria

Münnich Vass M (2017) Renewable energies cannot compete with forest carbon sequestration to cost-efficiently meet the EU carbon target for 2050. Renew Energy 107:164–180

Nabuurs G-J, Lindner M, Verkerk PJ et al (2013) First signs of carbon sink saturation in European forest biomass. Nat Clim Chang 3:792–796. https://doi.org/10.1038/nclimate1853

Nabuurs G-J, Delacote P, Ellison D et al (2015) A new role for forests and the forest sector in the EU post-2020 climate targets. European Forest Institute

Nabuurs GJ, Delacote P, Ellison D et al (2017) By 2050 the mitigation effects of EU forests could nearly double through climate smart forestry. Forests 8(12):484. https://doi.org/10.3390/f8120484

Nabuurs G-J, Verkerk PJ, Schelhaas M-J et al (2018) Climate-smart forestry: mitigation impacts in three European regions. From science to policy 6. European Forest Institute

Nichiforel L (2010) Forest owners' attitudes towards the implementation of multi-functional forest management principles in the district of Suceava, Romania. Ann For Res 53(1):71–80

Nijnik M, Nijnik A, Brown I (2016) Exploring the linkages between multifunctional forestry goals and the legacy of spruce plantations in Scotland. Can J For Res 46(10):1247–1254

Nitschke CR, Innes JL (2008) Integrating climate change into forest management in south-Central British Columbia: an assessment of landscape vulnerability and development of a climate-smart framework. For Ecol Manag 256(3):313–327

Opdam P, Iverson J, Zhifang N et al (2013) Science for action at the local landscape scale 1439–1445

Pretzsch H, Grote R, Reineking B et al (2008) Models for forest ecosystem management: a European perspective. Ann Bot 101:1065–1087. https://doi.org/10.1093/aob/mcm246

Pretzsch H, Biber P, Schütze G et al (2014) Forest stand growth dynamics in Central Europe have accelerated since 1870. Nat Commun 5:4967. https://doi.org/10.1038/ncomms5967

Pretzsch H, del Río M, Biber P et al (2019) Maintenance of long – term experiments for unique insights into forest growth dynamics and trends : review and perspectives. Eur J For Res 138(1):165–185. https://doi.org/10.1007/s10342-018-1151-y

Rametsteiner E, Simula M (2003) Forest certification—an instrument to promote sustainable forest management? J Environ Manag 67:87–98

Raymond CM, Bieling C, Fagerholm N (2016) The farmer as a landscape steward: comparing local understandings of landscape stewardship, landscape values, and land management actions. Ambio 45(2):173–184

Reyer CPO, Bugmann H, Nabuurs G-J et al (2015) Models for adaptive forest management. Reg Environ Chang:1–5. https://doi.org/10.1007/s10113-015-0861-7

Sandin G, Peters G, Svanström M (2016) Life cycle assessment of Forest products: challenges and solutions. Springer International Publishing

Santopuoli G, Ferranti F, Marchetti M (2016) Implementing criteria and indicators for sustainable Forest Management in a decentralized setting: Italy as a case study. J Environ Policy Plan:18. https://doi.org/10.1080/1523908X.2015.1065718

Santopuoli G, Temperli C, Alberdi I et al (2020) Pan-European sustainable Forest management indicators for assessing climate-smart forestry in Europe. Can J For Res. https://doi.org/10.1139/cjfr-2020-0166

Scheffers BR, Meester L, De BTCL et al (2017) The broad footprint of climate change from genes to biomes to people. Nature 354(6313)

Schelhaas MJ, Nabuurs G-J, Schuck A (2003) Natural disturbances in the European forests in the 19th and the 20th centuries. Glob Chang Biol 9:1620–1633

Seidl R, Spies TA, Peterson DL et al (2016) Searching for resilience: addressing the impacts of changing disturbance regimes on forest ecosystem services. J Appl Ecol 53(1):120–129

Seidl R, Thom D, Kautz M et al (2017) Forest disturbances under climate change. Nat Clim Chang 7:395–402. https://doi.org/10.1038/nclimate3303

Socha J, Staniaszek J (2015) Długookresowe trendy w dynamice wzrostu wysokości sosny zwyczajnej w Puszczy Niepołomickiej. Acta Agraria et Silvestria LIII:49–60. https://doi.org/10.2478/kultura-2013-0012

Socha J, Bruchwald A, Neroj B (2017) Aktualna i potencjalna produkcyjność siedlisk leśnych Polski dla głównych gatunków lasotwórczych. Kraków

Sousa-Silva R, Verbist B, Lomba Â et al (2018) Adapting forest management to climate change in Europe: linking perceptions to adaptive responses. Forest Policy Econ 90:22–30

Tans PP, Fung IY, Takahashi T et al (1990) Observational constraints on the global atmospheric CO2 budget. Science 247:1431–1438

Temperli C, Blattert C, Stadelmann G (2020) Trade-offs between ecosystem service provision and the predisposition to disturbances: a NFI-based scenario analysis. For Ecosyst 7:27. https://doi.org/10.1186/s40663-020-00236-1

The World Bank (2016). Climate-Smart Agriculture Indicators. World Bank Group Report number 105162-GLB

Tognetti R (2017) Climate-smart forestry in mountain regions COST action CA15226. Impact 3:29–31

UNFCC (2015) Adoption of the Paris agreement, 21st conference of the parties. United Nations, Paris

UNFCCC (1997) Kyoto Protocol to the United Nations framework convention on climate change adopted at COP3 in Kyoto, Japan, on 11 December 1997. Available via http://unfccc.int/resource/docs/cop3/07a01.pdf Accessed 24 Jan 2021

United Nations (2020) Available via https://www.un.org/en/sections/issues-depth/climate-change/ Accessed 22 Jan 2021

Upton V, Dhubháin ÁN, Bullock C (2015) Are forest attitudes shaped by the extent and characteristics of forests in the local landscape? Soc Nat Resour 28(6):641–656

Verkerk PJ, Costanza R, Hetemäki L et al (2020) Climate-smart forestry: the missing link. Forest Policy Econ 115:102164. https://doi.org/10.1016/j.forpol.2020.102164

von Carlowitz HC (1713) Sylvicultura Oeconomica, oder Haußwirthliche Nachricht und Naturmäßige Anweisung zur Wilden Baum Zucht, Leipzig

Walters CJ (1986) Adaptive management of renewable resources. McGraw Hill, New York

Watson RT, Noble IR, Bolin B et al (2000) Land use, land-use change and forestry: a special report of the intergovernmental panel on climate change. Cambridge University Press

Wolfslehner B, Baycheva-Merger T (2016) Evaluating the implementation of the pan-European criteria and indicators for sustainable forest management – a SWOT analysis. Ecol Indic 60:1192–1199. https://doi.org/10.1016/j.ecolind.2015.09.009

Yousefpour, R, Jacobsen, JB, Thorsen BJ et al (2012). A review of decision-making approaches to handle uncertainty and risk in adaptive forest management under climate change. 1–15. DOI: https://doi.org/10.1007/s13595-011-0153-4

Yousefpour R, Temperli C, Jacobsen JB et al (2017) A framework for modeling adaptive forest management and decision making under climate change. Ecol Soc 22(4):40. https://doi.org/10.5751/ES-09614-220440

Yousefpour R, Augustynczik ALD, Reyer CP et al (2018) Realizing mitigation efficiency of European commercial forests by climate smart forestry. Sci Rep 8(1):1–11

Zeller L, Pretzsch H (2019) Effect of forest structure on stand productivity in central European forests depends on developmental stage and tree species diversity. Forest Ecol Manag 434:193–204. https://doi.org/10.1016/j.foreco.2018.12.024

Chapter 3
Assessment of Indicators for Climate Smart Management in Mountain Forests

M. del Río, H. Pretzsch, A. Bončina, A. Avdagić, K. Bielak, F. Binder, L. Coll, T. Hilmers, M. Höhn, M. Kašanin-Grubin, M. Klopčič, B. Neroj, M. Pfatrisch, B. Stajić, K. Stimm, and E. Uhl

Abstract This chapter addresses the concepts and methods to assess quantitative indicators of Climate-Smart Forestry (CSF) at stand and management unit levels. First, the basic concepts for developing a framework for assessing CSF were reviewed. The suitable properties of indicators and methods for normalization, weighting, and aggregation were summarized. The proposed conceptual approach considers the CSF assessment as an adaptive learning process, which integrates scientific knowledge and participatory approaches. Then, climate smart indicators

M. del Río (✉)
INIA, Forest Research Centre, Madrid, Spain

iuFOR, Sustainable Forest Management Research Institute, University of Valladolid & INIA, Valladolid, Spain
e-mail: delrio@inia.es

H. Pretzsch · T. Hilmers · M. Pfatrisch
Chair of Forest Growth and Yield Science, School of Life Sciences Weihenstephan, Technical University of Munich, Freising, Germany
e-mail: hans.pretzsch@tum.de; torben.hilmers@tum.de

A. Bončina · M. Klopčič
Biotechnical Faculty, Department of Forestry and Renewable Forest Resources, University of Ljubljana, Ljubljana, Slovenia
e-mail: Andrej.Boncina@bf.uni-lj.si; Matija.Klopcic@bf.uni-lj.si

A. Avdagić
Faculty of Forestry, Department of Forest Management Planning and Urban Greenery, University of Sarajevo, Sarajevo, Bosnia and Herzegovina
e-mail: a.avdagic@sfsa.unsa.ba

K. Bielak
Department of Silviculture, Institute of Forest Sciences, Warsaw University of Life Sciences, Warsaw, Poland
e-mail: Kamil.Bielak@wl.sggw.pl

F. Binder
Bavarian State Institute of Forestry (LWF), Freising, Germany
e-mail: Franz.Binder@lwf.bayern.de

59
R. Tognetti et al. (eds.), *Climate-Smart Forestry in Mountain Regions*, Managing Forest Ecosystems 40, https://doi.org/10.1007/978-3-030-80767-2_3

were applied on long-term experimental plots to assess CSF of spruce-fir-beech mixed mountain forest. Redundancy and trade-offs between indicators, as well as their sensitivity to management regimes, were analyzed with the aim of improving the practicability of indicators. At the management unit level, the roles of indicators in the different phases of forest management planning were reviewed. A set of 56 indicators were used to assess their importance for management planning in four European countries. The results indicated that the most relevant indicators differed from the set of Pan-European indicators of sustainable forest management. Finally, we discussed results obtained and future challenges, including the following: (i) how to strengthen indicator selections and CSF assessment at stand level, (ii) the potential integration of CSF indicators into silvicultural guidelines, and (iii) the main challenges for integrating indicators into climate-smart forest planning.

3.1 Introduction

In many countries worldwide, a transition from the paradigm of sustainable management focused on wood production (von Carlowitz 1713) toward multi-criteria forest ecosystem management is observed (Lindner 2000; Bolte et al. 2009; Messier et al. 2013, 2015; Bončina et al. 2019). The main causes for this paradigmatic shift (Yaffee 1999) are related to the enhanced need for various ecosystem services

L. Coll
Department of Agriculture and Forest Engineering (EAGROF), University of Lleida, Lleida, Spain

Joint Research Unit CTFC-AGROTECNIO, Solsona, Spain
e-mail: lluis.coll@udl.cat

M. Höhn
Department of Botany, Faculty of Horticultural Science, SZIU, Budapest, Hungary
e-mail: Hohn.Maria@kertk.szie.hu

M. Kašanin-Grubin
Institute for Chemistry, Technology and Metallurgy, University of Belgrade, Belgrade, Serbia
e-mail: mkasaningrubin@chem.bg.ac.rs

B. Neroj
Bureau for Forest Management and Geodesy, Sekocin Stary, Poland
e-mail: bozydar.neroj@zarzad.buligl.pl

B. Stajić
Faculty of Forestry, University of Belgrade, Belgrade, Serbia
e-mail: branko.stajic@sfb.bg.ac.rs

K. Stimm · E. Uhl
Chair of Forest Growth and Yield Science, School of Life Sciences Weihenstephan, Technical University of Munich, Freising, Germany

Bavarian State Institute of Forestry (LWF), Freising, Germany
e-mail: kilian.stimm@tum.de; kilian.stimm@lwf.bayern.de; enno.uhl@tum.de

beyond forest products, such as recreation, protection of biodiversity (De Groot et al. 2002), but also the finding that diverse forests may have higher stability and recover capability in view of environmental threats (Knoke et al. 2008; Biber et al. 2015). The tools for monitoring, assessing, and managing forest ecosystems originally developed from sustainable wood production forestry (Hundeshagen 1826, 1828, Speidel 1972, pp. 162–164). In view of the paradigm shift, they need to be adapted to the extended scope and multiple criteria of forest ecosystem analyses and forest management (Pretzsch et al. 2008; Schwaiger et al. 2019; Hilmers et al. 2020). Examples for such an extension are criteria and related indicators for addressing biodiversity (Schulze et al. 2004; Geburek et al. 2010; Heym et al. 2021) or nutrients balance (Stupak et al. 2011), needed for sustainability.

Sustainability indicators quantify the state and the development of specific aspects of forest ecosystems and management in order to describe, assess, and manage forests regarding ecological, economical, and socioeconomical criteria (Azar et al. 1996; Pretzsch and Puumalainen 2002). Climate smartness has been introduced as a new concept for sustainable forest management (SFM) in view of climate change (Bowditch et al. 2020). According to Bowditch et al. (2020), Climate-Smart Forestry (CSF) is defined as "sustainable adaptive forest management and governance to protect and enhance the potential of forests to adapt to, and mitigate climate changes. The aim is to sustain ecosystem integrity and functions and to assure the continuous delivery of ecosystem goods and services (ESs), while minimizing the impact of climate-induced changes on mountain forests on well-being and nature's contribution to people".

This can be perceived as a new dimension of forest management, protection, health, and stability in terms of the current European perspective of sustainability (MCPFE 1993; Mayer 2000), which strengthens the delivery of ESs. In order to make it operational for monitoring and management purposes, climate smartness may be characterized by criteria and quantitative indicators (Pretzsch 2009, pp. 536–537).

The aim of this chapter is to evaluate criteria and indicators for CSF assessment at stand and management unit level. In detail, we (i) review existing approaches for CSF assessment, (ii) develop a list of indicators for climate smartness quantification at stand level, (iii) exemplarily apply a set of climate smartness indicators at stand level to mixed mountain forests, (iv) review concepts to integrate criteria and indicators of CSF in forest management planning; and (v) discuss the developed approaches and concepts in order to evaluate and demonstrate their potential impact on adaptive forest ecosystem management in terms of Lindner (2000) and Bolte et al. (2009). Notice that Chapter 4 of this book (Temperli et al. 2021) further derives the idea of smartness criteria and indicators at the spatial units beyond the stand and forest management unit level, i.e., at the regional or national scales.

3.2 Concepts for Assessing Climate-Smart Forestry at Stand and Forest Management Unit Level

The assessment of CSF can be done at different spatial scales, from stand or management unit levels, both directly linked to forest practice, to large scales such as regional, national, or global, which are more relevant for forest policy issues. Criteria and indicators (C&I) selected by Bowditch et al. (2020) in the framework of (CSF) definition were based on the Pan-European C&I for sustainable forest management (SFM), which are suitable to address adaptation to and mitigation of climate change (see also Chap. 2 of this book; Weatheral et al. 2021). Some few more indicators were added to the existing concept. The assessment of C&I of SFM and CSF have been widely developed at large scales, such as national scale (Wijewardana 2008; Pülzl et al. 2012; Santopuoli et al. 2020). However, the selection of indicators and their assessment, including their standardization and weighting for aggregation to a smartness composite indicator, should be adapted at the scale they are going to be used. Here, we focus on stand and forest management unit levels.

3.2.1 Indicator Selection

When selecting C&I, there are several recommended characteristics to be considered (e.g., Vacik and Wolfslehner 2004; Hagan and Whitman 2006; Reed et al. 2006), which might be more or less relevant depending on the goals and the scale of application. Among them, the following properties can be highlighted for CSF assessment at stand and forest management unit level:

– *Relevance* – the indicator is closely related with the criteria, with sound scientific information that support this relation (e.g., carbon stocks in aboveground biomass).
– *Sensitivity* – the indicator provides a measure so that changes in the indicator directly reflect observed changes in the climate smartness criteria. They can be linear (positive or negative) or nonlinear. As the aim is to characterize climate smartness of forest management, it is important that the indicator is sensitive to different management options.
– *Practicality* – the indicator is easily estimated from the available information or can simply integrate existing information, i.e., at stand level from forest inventories, remote sensing images, or visual assessments without need for additional analyses.
– *Understandability and utility* – the indicator is clearly understandable and interpretable by different users and can be easily applied in forest practice.

Other characteristics, like the indicator providing a direct measure instead of using a surrogate function ("validity" in Vacik and Wolfslehner 2004), may be less relevant at stand or landscape level. For some functions covered by the concept of CSF, it is not always possible to provide direct indicators at stand level as it would

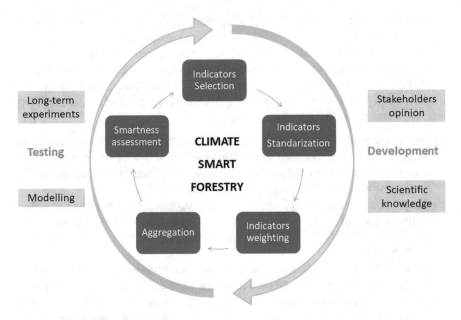

Fig. 3.1 Adaptive learning process for developing the framework to assess CSF at stand and management unit levels

require additional measurements, analysis, or even destructive sampling, which could reduce their practicality and utility. For instance, for assessing biodiversity, the number of large trees or microhabitats is often used as surrogate of flora and fauna diversity (Winter and Möller 2008; Alberdi et al. 2013).

In some cases, there is a trade-off between practicality and scientific rigorousness of indicators. Generally, indicators developed based on local context (bottom-up approach) prioritize the practicality, while indicators derived from expert and scientific knowledge (top-down approach) are generally more rigorous (Reed et al. 2006). However, the practicality, understandability, and utility are indeed key characteristics for the implementation of C&I for CSF assessment. Therefore, indicators based on top-down approach should be tested and evaluated in a local context. This means that adaptive learning processes for indicator development and assessment are recommended ways to improve the robustness and utility of methods (Reed et al. 2006) (Fig. 3.1).

3.2.2 Indicator Normalization

To compare the values of different indicators and to aggregate them into a composite indicator that summarizes the complement of several criteria, in our case for the CSF at stand and management unit (landscape) level, it is necessary to normalize or standardize the different indicators as they may be defined in different bit comparable units.

According to Pollesch and Dale (2016), three aspects can be considered when normalizing indicators: (i) the indicator bearing, i.e., whether an increase in the indicator means an approach to the "ideal" (i.e., optimal, theoretical) value or more distance; (ii) whether the normalization is internal or not, i.e., based on the data set; and (iii) the normalization scheme or method used. Different methods of normalization have been presented for standardizing indicators (e.g., Pollesch and Dale 2016); here, we summarized them in three groups:

(a) Ratio and z-score normalization methods. Ratio normalizations use the minimum, the maximum, or both values from the data set, whereas the z-score normalization is based on the mean and standard deviation of the data set. In this group, the normalization is therefore internal.
(b) Target normalization schemes or goal standardization, which use a baseline and/or target values for transformation (different functions can be used). The advantage of the target normalization schemes is that they can be used with various data sets.
(c) Benchmarking normalization function or value function approach, which assigns a normalized value to each indicator value based on existing knowledge (scientific knowledge, expert knowledge, questionnaires, etc.). It can be done by the direct rating, difference standard sequence technique, or mid-value method. As with the target normalization schemes, this method is not internal.

The method used for standardization is relevant as it can strongly influence the results and final climate smartness assessment (Talukder et al. 2017). When developing a framework for CSF assessment at the stand and forest management unit level, several normalization methods can be jointly applied for indicators, depending on data sources, knowledge, and indicator nature. However, aiming to develop a CSF assessment process that can be broadly applied internal methods should be avoided.

3.2.3   Weighting and Aggregating

Once indicators are assessed, a common option is to aggregate them into a composite indicator, which reflects the status of the object under evaluation. In some cases, different sub-indicators are also aggregated in a composite indicator linked to criteria. However, such aggregation is not always well accepted as the final value can involve loss of meaning and other disadvantages (OECD 2008, pp. 14). One common option is to avoid aggregation by the use of graphical summary of indicators, e.g., wheel or amoeba diagrams (Reed et al. 2006).

The way to weight the different indicators is probably the most challenging task when using composite indicators. Different weighting and aggregation methods to develop composite indicators were recently reviewed by Greco et al. (2019). Here, we briefly summarized the most relevant aspects and methods for developing C&I of CSF at stand and forest management unit level.

The simplest option is *not weighting*, i.e., giving the same value to each indicator/sub-indicator and then average or sum them. In this case, it can be particularly important to aggregate first the sub-indicators of a given indicator (same dimension) in a unique value or even all the indicators linked to a given criterion. This means that the final weight of some sub-indicators will vary. This method is often applied due to its objectivity and simplicity in spite of neglecting different relevance of indicators and correlations among them.

One option to weight indicators is to focus on data sources and nature of indicators, assigning higher weight to indicators based on more trustworthy and sound data (Freudenberg 2003). In the case of CSF, it is reasonable to consider to what extent an indicator is linked to adaptation and mitigation issues, giving more weight to indicators which are directly and accurately related to them, e.g., carbon stocks related to mitigation. However, the best approaches to avoid biases related to indicators' nature and data availability are those based on participatory processes. There are different participatory approaches such as the *budget allocation process*, in which participants have to distribute "n" points among indicators; the *analytic hierarchy process* based on pairwise comparisons of importance expressed on an ordinal scale; and *conjoint analysis* based on participant's preferences. The participatory approaches are difficult to implement when the aim of the C&I assessment is not clearly communicated or when there are too many indicators (Greco et al. 2019). Regarding C&I of CSF, it may be challenging for the participants to balance the different components of the CSF definition, i.e., sustainability, adaptation, mitigation, and ecosystem service provision (Bowditch et al. 2020).

Other options consider the relationship among indicators/sub-indicators in the weighting process or *data-driven weights*. These methods are based on different statistical methods, such as correlation analysis, multiple linear regression, principal component analysis (PCA), or data envelopment analysis (DEA). For example, the factor loadings of the first component of the PCA can be used as weights of the single indicators.

Regarding the aggregation, which is the final step in developing a composite indicator, different classification approaches are introduced in literature. Following the review by Greco et al. (2019), they can be divided in compensatory and non-compensatory aggregation, besides other mixed strategies. In *compensatory* approaches, for instance, using averages (arithmetic, geometric, etc.), a low value of one indicator can be compensated by a high value of another indicator. This approach bears the risk of hiding existing trade-offs between indicators resulting in undesirable incoherencies with the applied weighting. Using geometric averages instead of arithmetic averages can reduce the compensability among indicators (OECD 2008). *Non-compensatory* methods based on multi-criteria decision analysis avoid compensations among indicators and inconsistencies with the weighting process and thus involve a more complex analysis. Consequently, the method has not received a wide application to natural resource management outside of theoretical studies. While the compensatory technique provides a sound measure of overall performance of a given system (e.g., forest system), the non-compensatory technique alerts decision makers to presence of particularly poor performance with respect to individual criteria (cf. Jeffreys 2004).

3.2.4 Framework for CSF Assessment at Stand and Management Unit Level

To build up a framework for assessing CSF involves all steps, described above, from selection to aggregation of indicators into a composite CSF indicator (Fig. 3.1). In each step, different options with varying degrees of complexity can be selected, which can result in different weaknesses and strengths of the process and finally in different smartness assessments. Thus, any developed framework should be tested several times and iteratively refined until reaching a consolidated version, i.e., the development should be an adaptive learning process.

Science-based indicators and normalization and aggregation methods frequently derive in complex approaches, which later can be hardly applied in forest practice (top-down approaches). Contrary, other approaches focus on end-users' perceptions and local context to guarantee further application (bottom-up) but which can fail in assessment accuracy. Following Reed et al. (2006), an iterative learning process, which integrates top-down and bottom-up approaches, may result in a scientifically rigorous and feasible final framework.

Focusing on CSF assessment at stand and management unit level, any approach may unquestionably consider the integration of forest managers through participatory methods to warranty applicability. The extensive scientific knowledge on forest dynamics and management can assure the reliability of the process. On the other hand, information provided by long-term experiments in mountain forests (Pretzsch et al. 2019, 2021) as well as the more sophisticated and accurate forest models and decision support systems (Mäkelä et al. 2012) can help to test and improve the developed framework (Fig. 3.1). In the following paragraphs, we draft an approach for developing a framework to assess CSF at stand and management unit levels.

3.3 Assessment of CSF in Mountain Forest Stands: Exemplified by Norway Spruce-Silver Fir-European Beech Mixed Stands

3.3.1 Development of C&I Framework for Assessing Indicators of CSF at Stand Level

A forest stand is the smallest unit where forest management activities are decided on and implemented. Type and intensity of the management activities (e.g., thinning type, regeneration) depend on the management objectives and the current status of forest stands. Objectives may be manifold like timber production and/or forest for recreation or protection. Here, we describe an approach for assessing CSF at stand level when climate smartness (e.g., adaptation, mitigation) is intended to act as a general management strategy. The method presented can be generally used for assessing CSM at stand level. Through subsequent evaluations, the effect of management on the development of climate smartness can be monitored.

The approach was developed by using data from 12 long-term plots in the Bavarian Alps for assessing CSF in mixed stands of Norway spruce (*Picea abies* L- Karst), silver fir (*Abies alba* Mill.), and European beech (*Fagus sylvatica* L.) in mountain areas. Later, it was adapted to mixed mountain forests in other regions using six long-term plots in Bosnia and Herzegovina and two plots in Slovenia, as well tested in long-term experimental plots. However, the developed framework can be readjusted to other forest types, management systems, and regions by adapting the normalization of indicators/sub-indicators to specific characteristics of the respective region.

3.3.1.1 Selection of Indicators

We selected a subset of climate smartness indicators (Bowditch et al. 2020) that relate to stand-level characteristics (Table 3.1). A standardized protocol for data recording and assessment was set up (Pfatrisch 2019). This includes the definition of up to five quantitatively measurable or ratable characteristics of the indicator (sub-indicators) (Table 3.1). In our study, detailed yield data from long-term experimental plots were used, but the protocol is also applicable using yield data from common forest inventories and some additional information, which can be easily compiled in the field.

The values of the stand-specific indicator/sub-indicators were derived from existing measurements and from estimations in situ following standardized procedures (e.g., Level I protocol for 2.3 defoliation (Forest Europe 2015)). Some indicator values were assessed on species level (e.g., 4.3 naturalness) and then aggregated at stand level. Others are only evaluated on stand level (e.g., 1.2 growing stock).

3.3.1.2 Normalization

The indicator values need to be normalized to compare different sub-indicators and to aggregate them. The basic principle of the assessment was to reference the plot-specific values of the sub-indicators' characteristics in relation to reference values derived from existing information and knowledge. For most of the sub-indicators, target normalization schemes (goal standardization) were employed, using the target values either as a maximum or minimum threshold or as a mean reference value. For the other indicators/sub-indicators, the direct-rating approach (benchmarking normalization function approach) was used.

The transforming functions used in the target normalization schemes were linear, following three main patterns depending on the indicator bearing and reference values. When the benchmarking value represents the maximum value desired an increasing function was used, having the optimum at the maximum value of 1 (Fig. 3.2a), e.g., the maximum aboveground carbon stock expected for N.

Table 3.1 Selected climate-smart indicators and corresponding characteristics of assessment (sub-indicators), required plot data

Nr	Indicator	Sub-indicators	Abbrev.	Required plot data
1.2	Growing stock	Growing stock	G_1.2	Growing stock in m³/ha
1.3	Diameter distribution	Diameter/age distribution	Dd_1.3	Diameter distribution in defined classes
1.4	Carbon stock	Carbon Stock	C_1.4.1	Carbon stock in C t/ha
		Development of Carbon Stock	C_1.4.2	10-year change of carbon stock C t/ha
		Substitution	C_1.4.3	Total quantity of carbon substitution in the last 10 years by products from fellings
2.3	Defoliation	Defoliation	Def_2.3	Estimated needle/leaf loss of five dominant trees per species
2.4	Forest damage	Risk probability	Dam_2.4.1	Estimated risk probability of different forest damages
		Impact of damage	Dam_2.4.2	Estimated impact of forest damages
2.5	Stability	Slenderness coefficient	Stb_2.5.1	Slenderness coefficient
		Tree height	Stb_2.5.2	Tree height in m
		Stock density	Stb_2.5.3	Stock density (yield table related)
3.1	Increment and fellings	Increment	IF_3.1.1	Annual increment in m³/ha
		Fellings	IF_3.1.2	Average annual fellings in m³/ha
		Effect on growing stock	IF_3.1.3	Annual relative rate toward target growing stock
4.1	Tree sp. suitability	Tree species suitability	Sp_4.1	Site suitability of occurring tree species weighted by species-specific basal area proportion
4.2	Regeneration	Regenerated area	Reg_4.2.1	Area proportion of regeneration in %
		Height of regeneration	Reg_4.2.2	Area related height of the regeneration in cm
		Density of regeneration	Reg_4.2.3	Plant density of regeneration in plants/ha
		Regeneration potential	Reg_4.2.4	Number of tree species in regeneration and main stand
		Browsing	Reg_4.2.5	Estimated damage by browsing
4.3	Naturalness	Naturalness (stand establishment)	Nat_4.3.1	Type of stand historic regeneration and species choice
		Naturalness (sp. composition)	Nat_4.3.2	Tree species basal area in % and dominance % rate in the regeneration
		Soil scarification	Nat_4.3.3	Impact factor for and scarification of soil
4.4	Introduced tree sp.	Introduced tree species	Int_4.4	Tree species stem number in %

(continued)

Table 3.1 (continued)

Nr	Indicator	Sub-indicators	Abbrev.	Required plot data
4.5	Deadwood	Quantity of deadwood (total)	Dead_4.5.1	Estimated deadwood quantity
		Standing deadwood volume	Dead_4.5.2	Estimated volume of standing deadwood
		Decomposition rate	Dead_4.5.3	Percentage of quantity in different decomposition classes
		Light exposure	Dead_4.5.4	Estimated percentage in three exposure steps
4.6	Genetic resources	Phenotypic similarity	Gen_4.6.1	Similarity level by species and species proportion in stem number
		Gen conservation	Gen_4.6.2	Method of stand regeneration
4.8	Threatened forest sp.	Threatened forest species	Thr_4.8	Number of stems in %
4.91/2	Distribution of tree crowns	Crown layers (vertical)	Ver_4.9.1	Crown layers
		Canopy level (horizontal)	Hor_4.9.2	Canopy level/crown closure
6.10	Accessibility	Distance to road	Acc_6.10.1	Shortest distance to next forest road
		Road density	Acc_6.10.2	Road density within the surrounding 100 ha

spruce-silver fir-E. beech mixture in Bavaria is 360 C t·ha^{-1}. This value was derived from unmanaged long-term yield trials located in the Bavarian Alps. When the optimum represents a minimum value, a decreasing function was applied (Fig. 3.2b), e.g., difference between the "ideal" size distribution and observed distribution, for which no difference is the best value (1). In other cases, the reference value represents a maximum within a range, with an increasing function below this reference and a decreasing function above this (Fig. 3.2c), e.g., optimum growing stock for rich sites is 350 m^{3}ha^{-1} (Bayerische Staatsforsten 2018). Independently of the pattern, when the reference value benchmarks a regional mean value, it is correlated to the smartness value of 0.5, e.g., for volume increment, the average value in Bavaria is used as reference for mean smartness 0.5. When necessary, the functions were truncated in order to assign a 0 or a 1 beyond established limits (Fig 3.2d). For instance, for the coefficient of slenderness as stability indicator, below 40 always means the highest smartness (1) and above 120 always the lowest (0), assigning a mean smartness (0.5) to a coefficient of slenderness of 80 (Pretzsch 2009). In some cases, only a one-sided truncation was applied.

Due to practicality, some indicators were estimated by direct rating. This method was applied when required measurements for indicator estimation would involve long time-consuming and expensive work or when the indicator expresses a qualitative aspect that can be assessed by discrete classes. For example, the sub-indicator browsing damage was assessed in the field classifying the damage in four classes

Fig. 3.2 Transforming function
types for indicators normalization

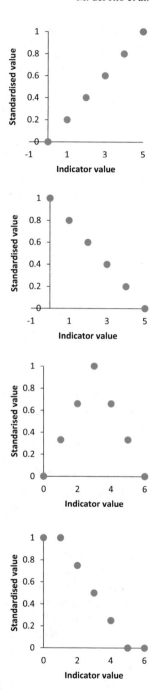

from high (0) when most of the trees were affected by wild game to low (1) in case of absence or only single, scattered damages in the stand.

The data base for the determination of the necessary reference values were obtained from various sources (e.g., forest inventories, soil/hazard maps, silvicultural guidelines, literature). These reference values can be index values, specific limits, or region-specific values. For the indicators/sub-indicators that were standardized using a region-specific value, this value was adapted when the approach was extended to mixed mountain forests in Bosnia and Herzegovina and in Slovenia. It is important to consider the regional character of references to be able to classify the plot-specific climate smartness at regional level. This enables a comparison of assessments of climate smartness values of different stands at different study sites and also over time.

3.3.1.3 Description of Indicators

The indicator "growing stock" (G 1.2) was evaluated by the measured merchantable wood of the respective plot or forest stand. For the evaluation, the current growing stock was set in relation to the stock targeted for the area. In the case study, for the Bavarian Alps, this was 350 m^3 and $300 \text{ m}^3 \text{ ha}^{-1}$, respectively, on productive and less productive sites according to the management goal of the Bavarian State Forest Enterprise (Bayerische Staatsforsten 2018) for continuous cover forest management. The transforming process followed the function in Figure 3.2c.

The current diameter distribution (Dd_1.3) was compared to the ideal diameter distribution for mixed mountain forests indicating a stable structural diversity (Bayerische Staatsforsten 2018) (50% in class 7–20 cm; 25% in 21–40 cm; 12,5% in 41–60 cm; 6,25% in 61–80 cm; 3,13% in >80 cm). Transforming was done using a declining function (Fig. 3.2b).

The indicator "1.4 carbon stock" was composed of three sub-indicators. Firstly, carbon stock itself (C_1.4.1) was calculated by applying species-specific biomass expansion factors to the growing stock of merchantable wood (Forrester et al. 2017). The reference value was 360 t ha^{-1}, reflecting a mean maximum value within fully stocked mountain mixed forest in Bavaria. Transforming used an increasing function. Secondly, the development of the carbon stock within the last 10 years period was referenced against the initial carbon stock. The application of an increasing transformation function led to higher smartness values with higher rates of recent carbon sequestration. In case of substitution (C_1.4.3), savings in terms of carbon release through substituting materials and fossil fuel were considered. The amount of harvested timber within the last 10 years period was converted into substituted carbon amounts by applying specific factors for roundwood and fuelwood reported by Hofer et al. (2007). As reference for a mean, a 10-year substitution effect of 16.09 t ha^{-1} C was used. This value was derived from an analysis of Klein and Schulz (2012), who investigated the substitution effect based on timber harvest information from 2003 to 2008 in Bavaria. The transformation process followed a right-side truncated increasing function.

Direct rating was applied to defoliation (Def_2.3), which was assessed by classifying the percentage of needle or leaf loss of five dominant tree per species. The classification referred to the graduation according to Forest Europe (2015). Estimations were first species-specific. In the second phase, the species-specific values were weighted by the percentage of basal area of the species and aggregated to a mean plot value.

"Forest damage" (2.4) combined the risk probability (Dam_2.4.1) of each possible risk (e.g., windthrow, bark beetle, snow breakage) and its impact (Dam_2.4.2) on plot level. Possible risks were derived from hazard maps or the previous occurrence of damages. The appraisal was based on expert knowledge and used classes from very high (smartness value = 0) to very low (smartness value 1). The impact was evaluated considering the impact on vitality, stability, and quality, which could have different weighting if necessary. Finally, a mean value for smartness was attained by averaging the damage-specific values. The third sub-indicator evaluated the number of possible damages (Dam_2.4.3).

The slenderness coefficient (Stb_2.5.1), tree height (Stb_2.5.2), and stocking density (Stb_2.5.3) were assessed within the indicator stability (2.5). Concerning the slenderness coefficient, species-specific values were weighted by their basal area proportion and then transformed by a two-sided truncated function. In literature, the value 80 for slenderness coefficient is reported as benchmark (Pretzsch 2009) with lower values indicating higher stability and higher values indicating less stability. Tree height was assumed to indicate higher stability with values below 20 m (mean value of the indicator scale) and less stability with higher values, respectively (Rottmann 1986). Transforming thus followed a decreasing function (Fig. 3.2b). Lastly, stocking density was classified into three classes (smartness values 0, 0.5, 1) by indexing the stocking density against yield table values. Classes considered higher stability at very low and very high stocking densities (Rottmann 1986).

"Increment and felling" (3.1) consisted of the three sub-indicators increment (IF_3.1.1), fellings (IF_3.1.2) and the mutual effect of both toward the target growing stock (IF_3.1.3). In case of increment and felling, the respective current values were benchmarked to 9.3 m^3 ha^{-1} $year^{-1}$, representing a mean value in mountain mixed forests (Hilmers et al. 2019a). The transforming process used an increasing function (Fig. 3.2a). The effect toward the target growing stock was assessed by calculating the annual relative trend rate of stock change. Positive values indicated an approaching trend and negative values, a diverging trend. The rates were classified into five levels of smartness.

Occurring tree species were appointed to one of three classes of site suitability (unsuitable, suitable, and optimal) in sub-indicator Sp_4.1. The suitability was assessed using information about growing conditions and literature (e.g., Otto 2000; Schütt et al. 2002). The species-specific value was weighted by its basal area proportion.

"Regeneration (4.2)" was divided into five sub-indicators. As regeneration, all plants below 7 cm diameter at breast height were considered. Firstly, the regenerated area (Reg_4.2.1) concerned the proportion of regenerated area of the entire plot. Transformation followed an increasing function using 100% as maximum. Secondly, the mean height of the regeneration (Reg_4.2.2) was related to the maximum browsing height, indicating a trusted regeneration. Values were converted by an increasing function; values above the threshold were capped. Thirdly, the observed density of regeneration (Reg_4.2.3) was related to general species-specific plant densities of artificially regenerated stands. Values above twice the number of the reference were truncated during a linear increasing transformation. Regeneration potential (Reg_4.2.4) evaluated the number of tree species found in regeneration against the number of species in the main stand. Again, the linear transformation function was cut at numbers of species in the generation, doubling the number of species in the main stand. Lastly, the damage by browsing (Reg_4.2.5) was categorized into four classes adapted from StMELF (2017) with higher smartness at less browsing damage.

The naturalness of stand establishment (Nat_4.3.1) grouped the evaluated stand into classes, which were defined by the proportion of natural and artificial regeneration and the closeness of involved species to the potentially natural vegetation (adopted from MacDicken 2015). Groups ranged from natural regeneration with naturally occurring tree species to artificial planting of non-autochthonous species. The naturalness of species composition (Nat_4.3.2) (Riedel et al. 2017) considered the current composition within two layers of a stand, i.e., the understory/regeneration (height < 4 m) and main stand (height > 4 m). The layer which was in future silvicultural focus received a double counting. The composition within the layers was grouped into classes defined by the proportion of species belonging to natural vegetation. Within sub-indicator Nat_4.3.3 (soil scarification), the affectation of the stand by different agents (cattle trampling, tracks, waste deposition, fertilization, forest roads) (Beer 2003) was reducing the maximum achievable smartness value. To each factor, a specific negative value was assigned and multiplied by a three-level intensity factor (three levels).

The indicator "Introduced tree species" (Int_4.4) classified occurring tree species into five categories of invasiveness according to Spellmann et al. (2015), ranging from species of natural vegetation to invasive species causing harm to natural vegetation and humans. Each tree species was weighted by its stem number proportion giving the same weight independently from tree size.

Smartness related to deadwood (4.5) considered the amount and structural characteristics of deadwood for biodiversity reasons. Four sub-indicators were addressed. The first total amount of deadwood (Dead_4.5.1) considered standing and lying deadwood. The amount was classified into five groups, whereas group borders were drawn using reported functional group-specific minimum amounts (Bauer et al. 2005; Moning et al. 2009). Solely standing deadwood was evaluated by the second sub-indicator (Dead_4.5.2). Here, a threshold of 15 m^3 ha^{-1} was used indicating a prerequisite for the occurrence of the three-toed woodpecker species (*Picoides* sp.) (Bütler et al. 2004). An increasing function was applied for smartness-value transformation. The proportion of decomposition degrees was addressed with sub-indicator Dead_4.5.3. Higher smartness values were achieved when all decomposition degree classes according to Lachat et al. (2014) were evenly distributed. Thus, transformation followed a decreasing function (Fig. 3.2b). As different light exposure situations of deadwood were relevant in terms of habitat provision, the distribution of deadwood amounts was classified into three light exposure classes (Dead_4.5.4) by assessing the crown closure degree above deadwood. The measured values were transformed as in the previous sub-indicator, whereas the optimal distribution was not equal between classes.

"Genetic resources" (4.6) were indirectly assessed through five classes of phenotypic similarity (Gen_4.6.1) of each tree species (Priehäusser 1958). Species-specific values were weighted by the species proportions of the total stem number. Genetic conservation (Gen_4.6.2) as second sub-indicator was evaluated by assigning the plot to one of five classes. Classes considered both, the genetic resources of the main stand and the management approach of regeneration (Kätzel and Becker 2014; Konnert et al. 2015).

The indicator "Threatened forest species" (4.8) recognized the occurrence of locally endangered red list species within the plot using the IUCN database. Classification followed the definition by Forest Europe (2015) of increasing imminence. The occurrence of a species belonging to the class of most endangered species determined the smartness value.

The "Distribution of tree crowns" was evaluated by determining visually or quantitatively the vertical layering (Ver_4.9.1) and the proportion of horizontal crown coverage (Hor_4.9.2) (Pretzsch 2009). Vertical layering was assessed using three scales (mono-layered, double-layered, multilayered). In case of crown coverage, a full coverage of the plot area was assumed as possible maximum value.

Accessibility (6.10) was of interest for forest economical and recreational purposes. Here, assessment was guided by economic criteria. In the first step, the minimum distance of the plot to a forest road (distance to road, Acc_6.10.1) was quantified and classified considering the distance dependent applicable most efficient transportation system. Secondly, the general road density (Acc_6.10.2) in terms of running meters per ha was estimated using a circular sample centered within the plot. A reference of 25 running meters per hectare was used as reference. The transforming process used an optimum within a range algorithm (Fig. 3.2c).

3.3.2 Indicator Assessment in Spruce-Fir-Beech Mixed Forest Stands

The selected indicators were assessed in 20 long-term experimental plots of spruce-fir-beech mixed mountain forests. We selected this forest type as a model example as it represents the most frequent and relevant mountain forest in Central and Eastern Europe (Hilmers et al. 2019a). The long-term experimental plots represent managed and unmanaged stands of these mixed mountain forests. In Table 3.2, the main characteristics of the studied long-term plots are presented. However, in most of the plots, there were no felling during the last 10 years (period used for estimation of time-dependent indicators).

Figure 3.3 shows the mean and standard deviation of the 36 sub-indicators and indicators from the values estimated on the 20 plots. On average, the greatest values (smartest) were found for sub-indicators related to the criteria "Biological

Table 3.2 Long-term experimental plots in mixed mountain forests used to assess CSF indicators. Main stand variables in the last survey are included. N, tree number per ha; BA, stand basal area; V, volume; PAIV, periodical mean annual stem volume increment

Plot	Country	Altitude m.a.s.l.	N Trees·ha⁻¹	BA m²·ha⁻¹	V m³·ha⁻¹	PAIV m³·ha⁻¹·year⁻¹
1	Germany	1271	257	37.7	518.9	6.1
2	Germany	1463	362	43.7	570.8	4.7
3	Germany	1235	319	56.4	896.1	9.5
4	Germany	1091	241	23.8	334.7	4.6
5	Germany	1091	493	36.4	455.7	3.9
6	Germany	1281	378	42.8	598.5	7.7
7	Germany	1281	433	80.7	1284.9	14.5
8	Germany	1294	590	41.0	475.9	13.3
9	Germany	860	854	45.0	546.1	7.8
10	Germany	934	1259	20.3	211.1	7.3
11	Germany	934	696	22.7	326.3	7.7
12	Germany	884	659	53.8	833.4	11.4
13	Bosnia & Herzegovina	1110	701	38.1	390.1	10.2
14	Bosnia & Herzegovina	1280	538	40.3	425.9	10.5
15	Bosnia & Herzegovina	1320	468	39.6	521.3	11.6
16	Bosnia & Herzegovina	1400	297	33.9	477.7	7.0
17	Bosnia & Herzegovina	1220	377	44.2	538.1	9.7
18	Bosnia & Herzegovina	1320	431	38.5	454.9	8.0
19	Slovenia	1421	500	60.8	925.2	13.3
20	Slovenia	1375	650	52.5	738.2	13.7

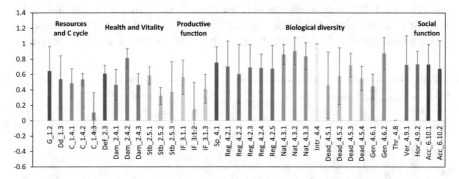

Fig. 3.3 Mean and standard deviation of the 36 sub-indicators and indicators representing five different criteria estimated in the 20 experimental plots in mixed mountain forests

diversity." The lowest values were obtained for sub-indicators related to "productive functions" (C_1.4.3 and IF_3.1.2), due to the absence of felling during the studied period in most of the plots. For most of the indicators/sub-indicators, the variability among studied long-term experimental plots was rather high. Two exceptions were the indicators for introduced (Intr_4.4) and threatened species (Thr_4.8), which showed no variation at all. All plots reveal the best rating regarding introduced species and the lowest rating regarding threatened species. This indicates that for the considered spruce-fir-beech mixed forests, these indicators were not very relevant. However, we kept them in the list of indicators, as in other stands or other types of forests they may have higher relevance. In this way, they may provide useful information for comparison with other less natural forests. The accessibility sub-indicators (Acc_6.1.1 and Acc_6.1.2) were estimated only in 13 experimental plots.

For a more understandable assessment of CSF at stand level, the different sub-indicators of a given indicator were aggregated. As the first option, equal weighting was evaluated. But taking the nature and difficulty of accurate estimation of some sub-indicators into account (Sect. 3.3.1.3), it was decided to apply a different weighting of indicators (C_1.4, Stb_2.5, IF_3.1, Reg_4.2, Nat_4.3, Dead_4.5). This weighting was based on the information content and accuracy of sub-indicators and on positive and negative correlations among sub-indicators of a given indicator (Sect. 3.4.3). Such correlations revealed some redundancy and trade-offs between different aspects of climate smartness. Nevertheless, the two weighting options resulted in similar indicator values (results not shown).

Figure 3.4 depicts that for most of the 16 indicators, the mean value of the 20 experimental plots reached or exceeded the value of 0.5 (average or greater smartness). The highest values were again observed for indicators related to biological diversity, especially those referring to species composition (Sp_4.1, Nat_4.3, Intr_44), except for threatened species (Thr_4.8). The mean value of the indicator related to carbon stocks (C_1.4) was below 0.5. This indicated that in most of the plots, the mitigation capacity was not as high as possible in this type of forest. Furthermore, these low values can be explained by the high reference value used for carbon stocks and by the low amount of carbon in products (substitution) due to the lack of felling, which also resulted in a low value of indicator IF_3.1. Another indicator with a mean below 0.5 was stability (Stb_2.5), due to the high stand density and mean height (Fig. 3.3), which creates high risk of windthrow and snow breakage.

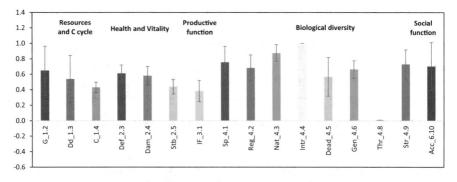

Fig. 3.4 Mean and standard deviation of the 16 weighted and aggregated indicators estimated for the 20 experimental plots in mixed mountain forests

3.3.3   Redundancy and Trade-offs Among Indicators

The values obtained for most of the indicators on the experimental plots were used to analyze whether there is some redundancy among indicators as well for detecting the presence of trade-offs between different aspects of climate smartness. For this analysis, the sub-indicators Intr_4.4 and Thr_4.8 were removed from the analysis as they showed a constant value in all the plots. The same was applied to Acc_6.10.1 and Acc_6.10.2 sub-indicators because they were not available for seven plots.

First, a correlation analysis was done among sub-indicators belonging to indicators with several sub-indicators (Fig. 3.5). The Spearman's rank order correlation was applied as some sub-indicators did not follow a normal distribution. As the abovementioned, the sub-indicators of some indicators showed significant positive correlations, which suggest that some of them could be left out, reducing the efforts of field work. For example, this occurred for the first three sub-indicators of the deadwood indicator. As the sub-indicator decomposition rate (Dead_4.5.3) was

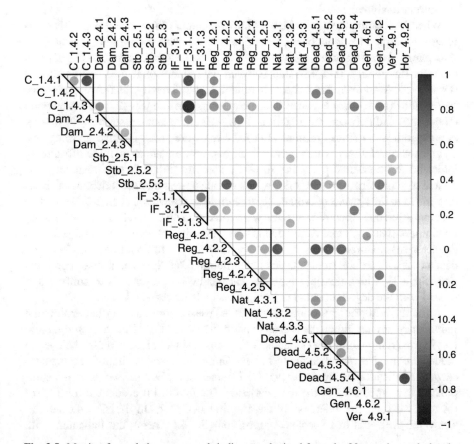

Fig. 3.5 Matrix of correlation among sub-indicators obtained from the 20 experimental plots in mixed mountain forests. Only significant Spearman correlations are shown ($p < 0.05$). Black triangles comprehend the correlations among sub-indicators of a given indicator

highly correlated to deadwood amount (Dead_4.5.1), the former, which is more difficult to be precisely assessed in the field, could be omitted. If an ever-greater simplification is needed, only the sub-indicator standing deadwood volume (Dead_4.5.2)
could be maintained, which is easily derivable from a standard forest stand inventory. Similarly, for regeneration either the sub-indicator height of regeneration
(Reg_4.2.2) or browsing (Reg_4.2.5) could be omitted. In other cases, the correlations between sub-indicators of a given indicator were negative. This indicated the
presence of some trade-offs and the importance of considering all of them, as it
happened for carbon sub-indicators (C_1.4.1 and C_1.4.2). It is important to note
that there are also some significant positive correlations between sub-indicators of
different indicators, as it occurred for C_1.4.3 and IF_3.1.2. Although it might suggest some redundancy, they should be maintained as they are expressing different
aspects of their respective indicators, which can be compensated by other sub-
indicators resulting in lack of correlation between indicators (as occurred between
C_1.4 and IF_3.1, Fig. 3.6). Notice that any conclusions regarding information content or redundancy of the indicators cannot be transferred to other forest types without further analyses.

When integrating the sub-indicators into indicators (Table 3.1), the positive cor-
relations among indicators of a given criteria (1–4) were not significant (Fig. 3.6).
Exceptions from this were the correlations between growing stock (G_1.2) and
diameter distribution (Dd_1.3) and between naturalness (Nat_4.3) and deadwood
(Dead_4.5). Moreover, for indicators related to biodiversity, there were negative
correlations (trade-offs) between tree species composition (Sp_4.1) and deadwood
(Dead_4.5) and between regeneration (Reg_4.2) and genetic resources (Gen_4.6).
Among indicators from different criteria, there were some positive and negative
significant correlations, which may indicate some redundancy and trade-offs among
indicators for measured plots. For instance, stability (Stb_2.5) was positively correlated to stand structure (Str_4.9), which could suggest that the indicator of structure added in the context of climate smart definition (Bowditch et al. 2020) could be
eventually left out. Accordingly, there were some evident trade-offs as those between
naturalness (Nat_4.3) and deadwood (Dead_4.5) with growing stocks (G_1.2) and
diameter distribution indicator (Dd_1.3). There were further trade-offs between
deadwood with carbon stocks (C_1.4), defoliation (Def_2.3), and species composition (Sp_4.1), which possibly indicate that deadwood presence is to some extent
related with the degree of stand decay in the stands investigated here.

An analysis of principal components (PCA) was performed to further explore the
redundancy among indicators and to explain the variability of the assessed indicators in mixed mountain forest stands. This statistical technique can also be used to
reduce the number of indicators to be used in the assessment, simplifying the subsequent application of the developed C&I framework. The first two principal components explained 54% of the total variance. The first factor accounted for 30% of
the total variance, the indicators of the criterion 1 (G_1.2, Dd_1.3, C_1.4), defoliation (Def_2.3), and tree species composition (Sp_4.1), being the indicators with

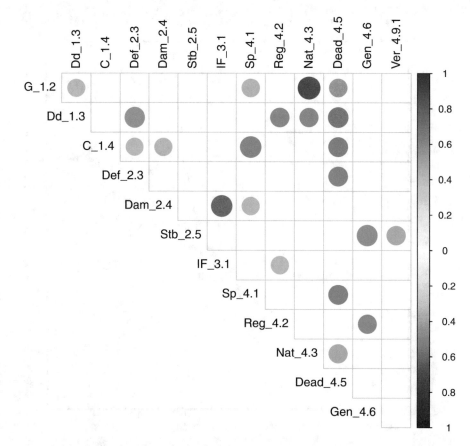

Fig. 3.6 Matrix of correlation among indicators obtained from the 20 experimental plots in mixed mountain forests. Only significant Spearman correlations are shown (p < 0.05); the larger the dot, the greater the correlation

higher positive loadings in these axes (Fig. 3.7), while deadwood (Dead_4.5) and naturalness (Nat_4.3) showed high negative loadings, which agrees with previous identified trade-offs. The second component explained 24% of the variability, with high positive loadings for stability (Stb_2.5) and genetic resources (Gen_4.6) and negative for increment and felling (IF_3.1) and regeneration (Reg_4.2).

In the biplot (Fig. 3.7), three groups of plots can be identified: the first group with high positive values in the first component (plots 13,14,15,16, 17, 18, 19, 20); the second group linked to the high values of indicators increment and felling and regeneration (plots 4, 10, 11), which are those plots with felling during the last 10 years; and the more dispersed third group with negative scores in the first component and positive in the second (plots 1, 2, 3, 5, 6, 7, 9, 12).

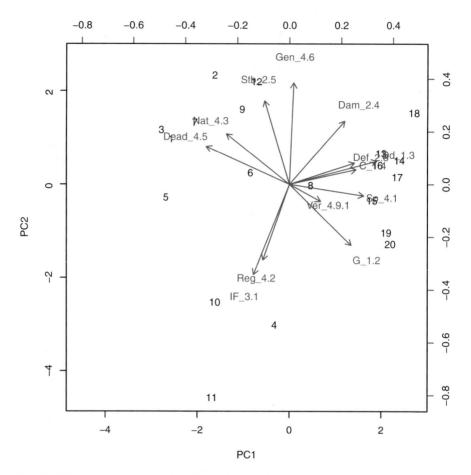

Fig. 3.7 Principal component analysis biplot showing the variation in plots (black numbers) and their relationships to indicators (blue arrows)

3.3.4 Assessing CSF in Spruce-Fir-Beech Mixed Stands

The aggregation of indicator values to a final score of climate smartness can simply be achieved by directly averaging the values. This method, although being objective, might not be the most appropriate, considering the number and information content of the indicators (see Sect. 3.2.3). Here, three methods of weighting were applied to obtain a composite indicator by averaging weighted indicators (compensatory aggregation method) in the 20 studied plots (Fig. 3.8).

(i) *Equal weighting or non-weighting.* All the indicators receive the same importance in the composite climate smartness indicator.

(ii) *Weighting by suitability for adaptation and mitigation monitoring.* In this option, if a given indicator is suitable for monitoring both aspects, adaptation and mitigation simultaneously, its weight is double than if it is suitable for

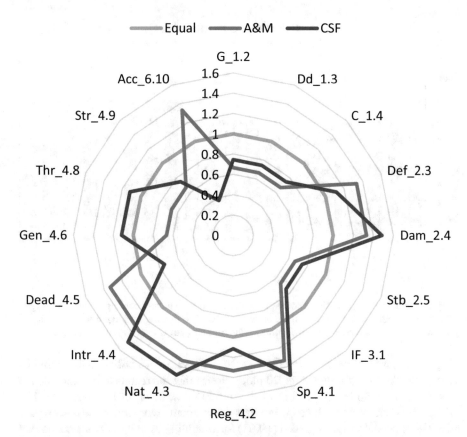

Fig. 3.8 Different weightings of the CSF indicators. Equal, same weight in all the indicators; A&M, weighting by capability to monitor suitability for adaptation and mitigation; CSF, weighting by the centrality for CSF (Bowditch et al. 2020)

monitoring only one of them. The suitability of the different indicators for assessing adaptation and mitigation forest management was based on the classification developed by Bowditch et al. (2020), who used an iterative participatory process involving various experts in forest-related fields from the Cost Action CLIMO.

(iii) *Weighting by the centrality for Climate-Smart Forestry*. In Bowditch et al. (2020), the most relevant indicators for assessing CSF were identified by a network analysis, which considered both the suitability of indicators to monitor adaptation and mitigation and the forest ecosystem services they address. They established four groups of indicators considering their degree of centrality, which were used for weighting purposes. The highest weight was assigned to the indicators belonging to the first core group (e.g., forest damage Dam_2.4) and the lowest weight to the second peripheral group (e.g., accessibility Acc_6.1) (Fig. 3.8).

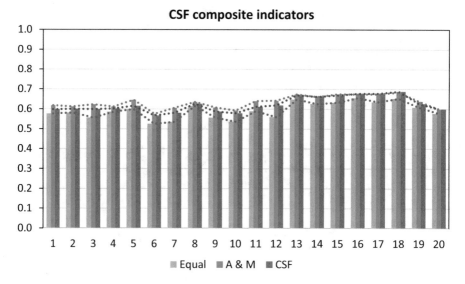

Fig. 3.9 Final climate smartness values of the 20 experimental plots according to the three weighting types. Equal, same weight in all the indicators; A&M, weighting by suitability for adaptation and mitigation monitoring; CSF, weighting by the centrality for CSF (Bowditch et al. 2020)

Figure 3.9 presents the resulting plot-specific CSF values according to the three different types of weighting for the 20 plots. Notice that the results do not include the indicator accessibility (Acc_6.1) as this indicator was not always available. The differences among the three weightings were small, with mean values of 0.59 (±0.04) for equal weighting, 0.63 (±0.03) for weighting by suitability for adaptation and mitigation monitoring, and 0.62 (±0.04) for weighting by centrality for CSF. The largest differences within weighting types were found for plots 3, 7, and 12, whereas in each case the highest values occur when using the second weighting.

In all cases, the CSF composite value is greater than 0.5 (Fig. 3.9), which represents the mean climate smartness following the used indicator normalization and weighting procedure. Concerning the CSF weighting type, the plot 18 showed the highest value (0.69) and plot 6 the lowest value (0.57). It can be observed that the highest values were reported for the Bosnian plots (plots 13–18), which are those with greater values in the indicators related to the first principal component (Dd_1.3, C_1.4, Def_2.3, Sp_4.1) (Fig. 3.7).

3.3.5 Sensitivity of CSF Indicators

To test the sensitivity of the indicators concerning different species composition, environmental changes, and management, data from additional long-term experimental plots in mountain forests in Bavaria were used (Table 3.3). Four plots representing different species composition were selected from the experimental site

Table 3.3 Geographical information and site characteristics of the 10 experimental plots. E, elevation (m a.s.l.); T, mean annual temperature (°C); P, annual precipitation (mm)

Experiment	N°. plots	Composition	Treatment	Period	No. of surveys	Longitude	Latitude	E	T	P
ZWI 111	4	E. beech; N. spruce; N. spruce-E. beech	Light-heavy thin. f. above, mixture portion	1954–2015	10	13°18′22″	49°3′57″	745	5	1270
FRY 129	6	N. spruce-E. beech-S. fir	Selection forestry; level of standing stock and threshold diameter	1980–2018	7	13°35′184″	48°51′19″	720	6.5	1200

ZWI -111 (Hilmers et al. 2019b), including one monospecific spruce plot, two monospecific beech plots (two thinning options), and one mixed spruce-beech plot. The experimental site FRY-129 (Pretzsch 2019) (6 plots) was chosen to compare the effect of different levels of growing stock (management) in uneven-aged spruce-fir-beech mixed stands. A more detailed information about the main stand characteristics of the long-term experimental plots can be found in Appendix 3.1.

For the chosen long-term experimental plots, the sub-indicators corresponding to indicators growing stock, diameter distribution, carbon stock, stability, increment and felling, and structure were estimated from inventory data during the monitoring period (Table 3.3). The sub-indicators were aggregated into the six indicators using the same weighting as in Sect. 3.3.2 in order to be comparable with the previous CSF assessment.

The effect of the species composition on selected indicators was in general larger than the effect of different growing stocks, reflected by higher variance between types (Fig. 3.10 left and right plots). By trend, in uneven-aged spruce-fir-beech mixed forests, the indicators showed higher values. The indicator growing stock (G_1.2) was very variable among and within plots, showing a decreasing trend with time in experimental plots with high standing volume (less removed volume) (Appendix 3.1). However, the spruce-fir-beech plots with lower growing stock and one beech plot, which maintained a lower growing stock, presented higher smartness values (Fig. 3.10b). This indicates that the selected reference value and normalization function penalize stands with high growing stocks.

The diameter distribution (Dd_1.3) and structure (Str_4.9) indicators were mainly influenced by species composition and age structure (Fig. 3.10c, d, k, and l), being greater for uneven-aged spruce-fir-beech mixed stands; medium for beech, probably to its strong shade tolerance; and lower for spruce-beech and spruce plots. It is noteworthy that in spruce-fir-beech mixed plots, there was a decreasing trend in Dd_1.3, but it was not observed for Str_4.9.

The indicators carbon stocks (C_1.4) and stability (Stb_2.5) did not vary largely among the different plots, being rather stable over time (Fig. 3.10e–h). The smartness value of the stability indicator was greater in spruce-fir-beech plots than in the other plots but in all cases lower than the medium smartness (0.5). For carbon, it ranged between 0.4 and 0.6. This agrees with the values shown in Figure 3.4 and suggests a low sensitivity of these two indicators for this type of mountain forest. The respective values might be readjusted in future applications by revising the reference values or/and changing the transforming functions.

The indicators increment and felling (IF_3.1) were sensitive to felling but not to species composition (Fig. 3.10i, j). However, the volatile changes observed suggest that the period of 10 years used for its evaluation influences the sensitivity. Using longer reference periods could result in more stable lines, which would reflect better long-term trends, which is more relevant for CSF. Accordingly, upscaling to the management unit would allow a better assessment of this indicator.

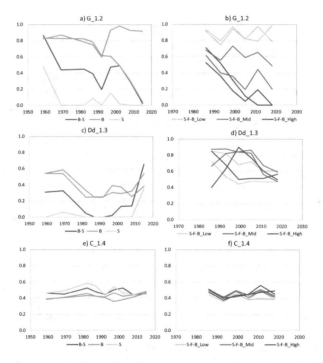

Fig. 3.10 Development of indicators for different stand compositions (B, beech; F, fir; S, spruce; mixed: BS and SFB) and growing stocks (low, middle, and high). (**a–b**), Growing stock G_1.2; (**c–d**), diameter distribution Dd_1.3; (**e–g**), carbon stock C_1.4; (**g–h**), stability Stb_2.5; (**i–j**), increment and fellings IF_3.1; (**k–l**), distribution of tree crowns Str_4.9

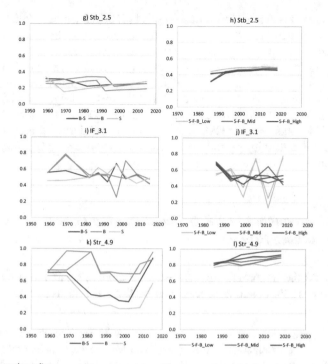

Fig. 3.10 (continued)

3.4   Importance of C&I of CSF in Forest Management Planning

3.4.1   Forest Planning and Climate-Smart Forestry

This section is focused on the importance of C&I of CSF in forest planning. The target scale is forest management unit level (FMU), since in many countries FMU is the most common spatial scale (combining stand and landscape) of forest planning (Cullotta et al. 2015). While in the past, the primary goal of traditional forest management planning was to ensure timber sustainability (Pommerening and Murphy 2004), nowadays forest planning can be understood as a tool to provide the desired ecosystem services for society and forest owners in a sustainable manner under socio-environmental changes. Due to climate change and increasing frequency of disturbances such as windthrows, ice storms, and insect attacks (Hanewinkel et al. 2012; Seidl et al. 2014; Nagel et al. 2017), forest planning needs to be adequately adapted to a changeable environment. This implies the continuous, repeated, and extensive monitoring to better understand the influence of climate change on forest dynamics, along with adapting forest management to the expected changes through managing uncertainties and risks. Beside adaptation, mitigation strategies are gaining more relevance in forest planning, since they may contribute greatly to enhancing forest carbon stores (Hof et al. 2017).

Indicators of sustainable forest management (SFM) (Bachmann 2002) have been traditionally applied in forest planning (Bončina 2001). Mostly, they are related to the status of forest stands (e.g., growing stock, stand volume increment, tree species composition), forest management activities (e.g., annual cut, proportion of natural regeneration in total regeneration), and impact of environmental change (e.g., sanitary felling). Climate change adaptation and mitigation strategies can be viewed as a risk component of SFM (Spittlehouse and Stewart 2003). Therefore, this calls to additional C&I. Indicators of CSF may have a substantial role in forest planning to better monitor and address the needs for adaptation and mitigation in forest management.

3.4.2 Involvement of CSF Indicators in the Forest Planning Process

To understand the importance of indicators for climate smart forest management planning, the whole planning process can be divided into five phases which are interconnected:

1. *Inventory, analyses, and evaluation*: to begin with the process of management planning, the current state of the respective FMU needs to be sampled and analyzed. The essential aspects of CSF can be recorded using the classical or further improved forest inventories (e.g., broadening their scope to include variables related to forest carbon pools and carbon sequestration, forest health, or biodiversity (Corona et al. 2011)). Thus, many above indicators at stand level (e.g., growing stock (1.2), regeneration (4.2), carbon stock (1.4), and stability (2.5)) (Table 3.1) are relevant for the assessment of CSF at FMU. For instance, some of them may indicate forests' response to climate change (e.g., damage level of forest stands, growth of stands and trees, regeneration pattern) or show response of forests to recent management activities carried out for adaptation of forests to climate change (e.g., tree species composition, diameter structure of forest stands). When dealing with those indicators, two aspects should be considered. Firstly, indicators are more powerful for the assessment of CSF if their current value is compared to their values from previous inventories. This enables insight into changes of structure and processes in forest stands. Secondly, the same indicators are useful for assessing various CSF aspects, e.g., impact of climate change on forest stands and effectiveness of forest management activities for adaptation.

2. *Defining (management) objectives*: management objectives should reflect demands of forest owners and society. Management objectives are operationalized through operational objectives. Thus, the desired structure and composition of forest stands are defined by silvicultural objectives. For CSF, it is important to search for forest stand composition and structure which will be adapted to

changeable climatic conditions and thus contribute to reducing the risks in forest management (see also Chap. 8 of this book; Pach et al. 2021). Silvicultural objectives are usually determined separately for the different forest site types; they are defined with selected stand parameters, which can be treated as indicators of CSF. Among them, tree species composition, stand structure, and target diameter of tree species are the most relevant.

3. *Modeling and elaborating scenarios:* based on the analysis and defined management goals, projections of potential forest development paths are undertaken which can be model aided. Usually, a number of different forest management are explored in a scenario analysis, and the best option under the given constraints and management objectives is identified (cf. Pommerening and Grabarnik 2019). Models of forest stand development are important for many purposes: i) adaptation of forests and forest management to climate change, ii) selection of optimal management strategies, and iii) reduction of risks and uncertainties in forest management. When modeling stand development, the same indicators might be applied as in planning phase 1. Scenarios are often focused on demographic changes of forest stands (important CSF indicators: diameter structure, age structure, stand density, etc.) under different management strategies (CSF indicators: cut intensity, silvicultural system) (e.g., Rosset et al. 2014).

4. *Defining management activities:* then, results of sampling and scenario analyses feed into the management plan as a set of silvicultural prescriptions to the given stand. The main part of management activities is focused on silvicultural activities (important CSF indicators: structure of planned harvesting, felling intensity, thinning concept, regeneration system etc.) and protection measures. All measures influence the structure and processes in forest stands, and therefore, their impacts reflect in changed values of CSF indicators related to the status of forest stands, which can be observed in the next forest inventory. Management decisions were made for other fields of forest management beside silviculture (e.g., forest road construction, forest protection, recreation). Better accessibility to forest areas and suitable harvesting technologies contribute to effective forest management when huge forest damages occur; therefore, they can serve as important CSF indicators as well.

5. *Monitoring of forests and forest management:* implemented management activities are usually registered, this being important for understanding how forest stands react to specific management activities under changeable environment. New experiences can be applied into future decision-making about forest management activities. Besides forest management, some other impacts on forest stands can occur. Among them, severe disturbances noticeably change the structure and composition of forest stands. Therefore, registration of sanitary felling is a part of forest management monitoring; the amount and structure of sanitary felling can serve as an important CSF indicator. Monitoring of sanitary felling (e.g., Klopcic et al. 2009) in a longer time period is substantial for understanding the susceptibility of forest stands to various agents of natural disturbances and for adapting forest management to reduce risks.

3.4.3 Estimation of Importance of CSF Indicators in Forest Planning at the Forest Management Unit Level

A list of possible CSF indicators available in forest planning was created, and the importance of indicators for assessing CSF at the level of forest management unit in four countries was estimated by Likert scale (1, not important for CSF at all; 2, not important; 3, neutral; 4, important; 5, extremely important). Assessment of indicators was based on the definitions of CSF (Bowditch et al. 2020) and on the possibilities to operationalize them in the forest management unit plan.

The scheme of European criteria of SFM was followed, but a much larger set of possible indicators was included into the analyses. It included most of the above-mentioned indicators for CSF assessment at stand level (Table 3.1) but without disaggregating them into sub-indicators. In total, a set of 56 parameters was estimated (Appendix 3.2). The importance of indicators was estimated in regard to their role in the planning process for:

– Understanding the influence of climate change on forests structure and stand dynamics in a FMU
– Analyzing the status of forests in a FMU in regard to the impact of climate change
– Modeling the development of forest stands in regard to the changed climatic conditions
– Evaluating the effectiveness of implemented silvicultural activities
– Determining the objectives and measures which will contribute to the adaptation and mitigation of forests and forest management in a FMU
– Monitoring the development of forest stands in regard to the implemented management activities and climate change

Ranking of indicators' importance for climate smart forest planning by representatives from four countries (Bosnia and Herzegovina, Poland, Slovenia, and Spain) shows that quite a number of indicators, which are not part of the European system of C&I of sustainable forest management, are very important in forest planning at the FMU level (Table 3.4). Tree species composition of natural regeneration was uniformly estimated as the most important indicator for CSF planning. It indicates capacity for adaptation of forest stands to climate change as well as the effectiveness of past forest management. Some indicators in the list are crucial for assessing the impact of extreme events on trees and forest stands as well as the susceptibility of stands to natural disturbances (e.g., forest damage, vitality status, amount, and structure of sanitary felling). Climate change may strongly influence the tree growth pattern; therefore, quite expectedly some indicators may be connected to tree and stand growth. Forest plans define the management activities for the next period. Consequently, it was to be expected that some management indicators were ranked as very important, e.g., planned and implemented silvicultural works, management system applied, and felling. Silviculture and cutting are the main tools for creating structure and composition which is adapted to climate change. This is why indicators describing forest stand structure and composition were assessed as highly

Table 3.4 Mean value of the importance of parameters for the CSF assessment of CSF in forest planning at the FMU level (only indicators with average value > = 4 is presented; indicators were assessed with ranks from 1 to 5) (CV – coefficient of variation in percentage)

Indicators	Mean value of importance	CV (%)
Tree species composition of natural regeneration	5.00	0.0
Forest damage	4.83	2.3
Regeneration (type of regeneration)	4.75	5.3
Vitality status of tree species/forest stands	4.67	4.8
Silvicultural works (planned and implemented)	4.67	4.8
Management system applied	4.58	5.5
Growth of trees and stands (e.g., diameter growth…)	4.58	5.5
Register of harvested trees in past planning period (tree species, dimension)	4.33	15.4
Tree species composition of single forest stands	4.33	5.1
Growth intensity of forest stands (volume increment/stand volume)	4.33	5.1
Increment and felling	4.25	5.9
Density of forest stands (basal area, tree number, SDI)	4.17	2.7
Protective forests – soil, water, and other ecosystem functions	4.17	18.7
Diversity of tree species	4.17	18.7
Damages of trees (stands) per agent (wind, snow…)	4.08	0.7
Growing stock	4.00	16.7
Amount and structure of sanitary felling according to the main agents	4.00	16.7

important (e.g., growing stock, tree species composition, stand density). In most European countries, forest planning supports multi-objective forest management oriented to providing various services. One indicator directly related to ecosystem services was included into the set of important indicators at the FMU level. Some of the indicators from the list (e.g., register of harvested trees) indicate that monitoring is an important part of CSF and planning.

3.5 Challenges and Perspectives

3.5.1 Refining the Selection of Indicators/Sub-indicators at Stand Level

The selection of indicators is an important step in the development of any assessment framework. Indicators can provide a reliable overview of the forest situation, allowing a comprehensible and transparent assessment of forest management (Blattert et al. 2017). Although Pan-European indicators for SFM were designed for application at the national scale, in this study they were adapted for their application at stand level. Suitable, quantifiable, or ratable sub-indicators were defined and forest-type and region-specific reference values and transforming functions assigned. However, the presented approach may not give the full picture of CSF as not all aspects have been addressed.

For example, protective functions, like protection against avalanches and rock-falls, as well as protection of soil not included, yet play an important role in mountain areas. Although these agents are known to highly depend on physiographic and site factors (e.g., slope, soil type, roughness of the forest floor), stand-level indicators related to the structure and composition of forest stands may also provide important information about the protective role of a stand (Blattert et al. 2017). These variables include the mean stand density, the basal area (or the average diameter at breast height), and the percentage of evergreen/deciduous species for rockfall protection (see *Rockfall Protection Index* in Cordonnier et al. 2013); the mean tree height, the canopy cover during the winter, and the stand density or basal area for protection against snow avalanches (see *Avalanche Protection Index* in Cordonnier et al. 2013); and the forest canopy cover (%) for landslide and erosion protection. Some of these parameters were here used in other indicators but not explicitly to assess the protective function.

Soils in native forest seldom experience significant disturbances which are more common for soils in other land-use systems; thus, the importance of soil characteristics is often underestimated in forest management practices and planning. However, climate change, atmospheric deposition, and/or deforestation can cause dramatic changes in the quality of forest soils, by altering the soil organic matter (Raison and Khanna 2011; Prietzel et al. 2020), and changes in hydrological processes which can enhance surface runoff and soil erosion, increase the recharge of groundwater, and cause the reduction of organic carbon, nitrogen, phosphorous, and exchangeable potassium, calcium, and magnesium (Pennock and van Kessel 1997). Furthermore, bedrock has a significant role in vegetation growth by regulating physical and chemical properties in soils (Hahm et al. 2014); it can also change the response of vegetation to climate factors (Jiang et al. 2020). Thus, some indicators related to soil properties could help to estimate the future forest growth and vitality and the need for adaptation under conditions of climate and/or land-use change. The most important soil characteristics for predictions of changes that can occur in forests due to land-use and/or climate change are texture, content of organic carbon, and available ions.

Mountain forests are also known to hold important biodiversity values, since they provide habitats for many animals and plant species of high community interest. Stand-level indicators related with the capacity of forests to sustain biodiversity are varied (Gao et al. 2015) and include the following: (*i*) the diversity of species of both the tree and the understory strata, which can be calculated using Shannon's index with basal area or plant cover as a measure of species relative abundances, respectively (Neumann and Starlinger 2001); (*ii*) the tree size diversity (i.e., structural diversity) (Staudhammer and LeMay 2001); (*iii*) the presence of large standing and lying deadwood (m^3/ha) and its decay class (fresh vs decay) (Lassauce et al. 2011); (*iv*) the abundance of large living trees (trees·ha^{-1}) (Vuidot et al. 2011); and (*v*) the presence and number of microhabitats in the trees such as cavities, bark pockets, cracks, sap runs, or trunk rots (Bütler et al. 2013). The first three types of indicators are included in the presented approach to assess CSF, but indicators for the last two groups might be added. In the last decade, some efforts have been made to compile biodiversity indicators into a single index (Geburek et al. 2010; Gonin et al. 2017) with the aim of providing forest managers with a simple tool of both: to evaluate the potential of a given forest stand to support diverse species and to identify the factors that can be improved through the implementation of forest management and planning strategies. Since sustaining biodiversity helps to maintain robust ecosystems, CSF calls for a detailed inclusion of biodiversity indicators.

As an integral part of the biological diversity, genetic variation safeguards adaptability of forest species and their populations to environmental changes and impacts by pests, diseases, and by climate change (El-Lakany et al. 2001). Accordingly, high adaptability based on biological variation definitely starts at the genetic level. Assessing and monitoring genetic resources in forests should be one of the main prerequisites for CSF. The impacts of silvicultural methods and the management practices on the genetic resources have only recently received increasing interest. DNA markers allow the initiation of different genetic surveys with the aim to estimate the quality of forest genetic resources. However, there is still low practical experience of these activities and have rarely been applied on a larger scale. Multiple genetic parameters like diversity indices of population (heterozygosity, allele frequencies, inbreeding coefficients) will enable to early detect potentially harmful changes of forest adaptability, before these appear at higher biodiversity levels, e.g., species or ecosystem (Fussi et al. 2016). To explore the evolutionary adaptability of populations in a specific environment and to get insights into the selection drivers, breeding programs or directed selections for climate-smart forests are needed. In the LIFEGENMON (http://www.lifegenmon.si/) project ending in 2020, a research group proposed to define respective optimal indicators and verifiers and to edit guidelines for a forest genetic monitoring system for selected tree species in different European countries and regions. This can serve as an early warning system to aid the assessment of a species response to environmental change at a long-term temporal scale and also be used for CSF assessment.

Providing space for recreation and human well-being is nowadays an important forest ecosystem service in mountain and other forests but especially in the urban and near urban forest areas (Pröbstl et al. 2009). Due to climate change and increased people's awareness about the importance of outdoor activity, the increased demands for especially warm-weather recreation activities (i.e., hiking, backpacking, picnicking, camping) may appear (Hand and Lawson 2018), triggering higher pressure on (mountain) forests in the future. Thus, regulating recreation is an important issue of forest management planning (Wilkes-Allemann et al. 2015), to address the trade-offs between recreation demand, timber production (Ahtikoski et al. 2011), and the provision of habitats for endangered plant and animal species (Rösner et al. 2014). Accessibility to (mountain) forests was recognized as a relevant indicator of recreational forest ecosystem service when evaluating CSF (see Table 3.1). Köchli and Brang (2005) used accessibility together with patch diversity, stand structure, and developmental stage of a stand to develop a recreation index. In addition, Edwards et al. (2012) evaluated recreation through visual attractiveness of forest stands by assessing 12 indicators of forest stand structure, such as tree sizes, spacing, visual penetration through the stand, deadwood, etc. Several indicators exposed in Köchli and Brang (2005) and Edwards et al. (2012) or their proxies are already on the current list of indicators to assess CSF, while others could possibly be added. Even if many indicators can be included into CSF assessment, one has to take the proportionality of data collection effort and the added informational value into account.

3.5.2 Strengthening CSF Assessment at Stand Level

Beyond the selection of proper and relevant indicators, they as well as the composite indicators need to be validated and readjusted to improve CSF assessment. This validation should be done at the different steps, from selection, normalization, and weighting to the aggregation (Singh et al. 2012). The developed framework for assessing CSF considers from the beginning the need for continuous updating by defining the framework as an adaptive learning process (Fig. 3.1). This chapter shows the first attempt to fulfill the different phases of the developed framework, but further efforts are needed until a satisfactory CSF assessment is reached. Linking the development of indicators framework to data collection efforts allowed us to have the first evaluation and propose improvements for future attempts.

Defining the right thresholds and transforming functions is a complex task, which needs further testing and readjustments. In Sect. 3.3.1.3, the regional thresholds for the different indicators and sub-indicators were set up from expert knowledge and literature. The use of target normalization based on such reference values has been recommended against other normalization methods when the indicator assessment is context dependent (Pollesch and Dale 2016), as occurs with CSF assessment. Hence, the specific thresholds used for single indicators need a regional reference. The first test of CSF assessment presented in this chapter made this obvious. For instance, the high reference value used for carbon stocks (C_1.4.1) derived from Bavarian sites resulted in low values of smartness for this indicator despite the rather high growing stocks in many plots. Complex uneven-aged mountain forests managed by a selection or irregular shelterwood system are characterized by very stable but medium values of aboveground productivity over time, and thus of carbon stock. However, they might have very positive long-term effect on soil organic carbon storage (Seidl et al. 2008), that was not investigated in this study. Similarly, some of the simple transforming functions could be revised. For growing stock (G_1.2), one possible improvement could be to change the slope of the transforming function in the right branch (Fig. 3.2c), which then results in a lower decrease in smartness when the difference to the reference value is caused by higher values compared to lower values.

In case of trade-offs between indicators, weighting is increasingly important. The varying but specific social and manager's demands concerning expected ecosystem services can thus better be considered. For example, if we consider the observed oppositional trade-off between the indicators' "naturalness" (Nat_4.3), which here is strongly related to regeneration and "growing stock" (G_1.2), the increase of one entails a reduction of the other. Depending on their focus, the managers need to decide on weighting. With a defined weighting of related indicators, a target-oriented forest management can then be planned and implemented more precisely.

Our evaluation regarding the weighting methods, which tested three different weighting options, did not provide a clear basis for decision to select one (Fig. 3.9). However, weighting by the centrality for CSF may be recommended. Forest management in mountain areas has to consider the large body of ecosystem services (Blattert et al. 2017). Weighting by centrality allows addressing the importance of different ecosystem services and reducing the possible inherent bias of selected indicators. Nevertheless, it is important to remark that the kind of normalization used introduces implicit weighting of indicators (Booysen 2002), by including thresholds and transforming functions which consider smartness. This can be observed in Figure 3.10, where the stability indicator (Stb_2.5) showed low values, although they represent different species composition and management, which might result in lower values of the composite indicator.

In future steps, other methods for weighting and aggregating should be tested to guarantee the robustness of the composite indicators. Thus, non-compensatory aggregations could be compared to the used compensatory ones. Multi-criteria decision analysis (MCDA) can be used to deal with possible trade-offs among indicators or overrepresenting indicators (e.g., several indicators for a given criteria) (Wolfslehner and Vacik 2011). Finally, sensitivity analyses can be used to determine the indicators influence on the composite indicator value, giving a better understanding of the whole process (Greco et al. 2019).

3.5.3 Use of Indicators of Climate Smartness for Development of Silvicultural Prescriptions

In the past, the development of silvicultural prescriptions and guidelines focused mainly on wood production; in the last few decades, additional aspects such as carbon sequestration, biodiversity, or recreation were integrated (Hilmers et al. 2020). Indicators and criteria of climate smartness may become essential additional aspects of silvicultural prescription in regions with increasing risk of drought, snow breakage, or storm (Churchill et al. 2013). No matter whether silvicultural prescriptions are derived and formulated normatively and qualitatively or based on scenario analyses, both approaches should consider the mitigation and adaptation aspects of the derived and prescribed silvicultural guidelines for a given region and forest type (D'Amato et al. 2011).

Quantitatively based indicators and criteria of climate smartness have the advantage that they may be implemented in forest stand simulators in addition to other criteria of an extended concept of sustainability (Kneeshaw et al. 2000). Consideration of climate smartness aspects becomes of increasing importance, as in the last decades forest science and forestry were faced with environmental impacts on forest ecosystems such as acid rain, increasing atmospheric ozone concentration, and eutrophic deposition as well as climate change. There was hardly any previous experience from experiments or monitoring how forestry may mitigate or adapt to such environmental changes (see Chap. 10 of this book; Tognetti et al. 2021). Field experiments are costly and very long-lasting; they are important but not sufficient to quickly provide forest management with recommendations for decision-making under environmental stress. Under such conditions, simulation models and model scenarios are often the only alternative for getting decision support. And stand or tree simulators, equipped with indicators and criteria of climate smartness, may be just the appropriate tool for developing new well through-thought silvicultural guidelines by scenario analyses (see also Chap. 8 of this book; Pach et al. 2021). The resulting quantitative silvicultural prescriptions may subsequently promote the transition from the analysis to the design of complex mixed-species stands and their increased implementation and successful regulation.

3.5.4 Prospects for Adapting the Set of Indicators for Climate Smart Forest Planning

There are many challenges for forest planning to address climate smartness. Issues related to how to manage and limit uncertainties and risks in forest management are probably the main ones. Traditional forest planning based on stable conditions is certainly not appropriate any more. The concept of adaptive, climate smart forest management also involving new silviculture strategies seems to be a more promising alternative.

Forest planning is an important tool for CSF operationalization as a merged part of SFM (Nabuurs et al. 2017). As previously mentioned, the European set of C&I of SFM is predominantly aimed at forest policy at national spatial level. But similar to stand level, they can be used in planning processes as soon they are operationalized. Nonetheless, additional indicators at the FMU need to be considered for CSF. The important indicators for climate smart forest planning as defined in our study are related to describing (1) forest management, (2) forest stand reaction to implemented forest management activities, (3) impact of extreme events on forest stands, and (4) capacity of forest stands (and management) for adaptation and mitigation.

The selected indicators (Table 3.4) are important in the whole planning process. By introducing the system of forest inventory based on permanent sample plots, the quality of information was strongly improved (Tomppo et al. 2010), as it enables insight into changes of forest stands. However, the role of indicators is not limited to understanding forest stand development only, since they are important for management decisions, too. CSF, similarly as SFM, should be understood as an active approach. There are many general suggestions about the adaptation of forest management to climate change and its mitigation potential, e.g., those related to the rotation length, silvicultural systems, and thinning regime (e.g., Ruiz-Peinado et al. 2013; Brang et al. 2014; Bravo et al. 2016; Sohn et al. 2016; Socha et al. 2017). However, the general suggestions should be adapted to the natural, economic, and social settings in single FMUs. As a consequence, indicators describing active forest management at the FMU level and its impact on forest stands are crucial for operational CSF.

A set of indicators can be applied in multi-criteria decision analysis (MCDA) (e.g., Duncker et al. 2012; Blattert et al. 2017) to support decision-making as well in the estimation of management effectiveness for providing CSF. This seems to be a promising approach for CSF planning. A forest management unit can be an appropriate spatial framework for applying MCDA.

In the concept of adaptive forest management, improved management activities can be understood as a "new experiment." This is why monitoring of forest stand response on various activities is crucial. For both – management activities and response of forest stands to them – indicators are needed. By integrating CSF assessment at stand and management unit, some indicators at stand level (Table 3.1) may increase their significance when being upscaled for providing information of spatial variability at forest management unit (e.g., growing stock, size distribution in even-aged structures, increment, and felling).

Long-term experimental plots can strongly support the development of adaptive forest management. This chapter shows an example of how experimental plots (Sect. 3.3.5) can be used for extracting information of the impact of different silvicultural options for climate smartness, as well as for evaluating indicator assessment. New adaptive forest management strategies to achieve CSF need to be tested scientifically, so collaborative experimental networks which cover different conditions (site, owners, management objectives, etc.) are required (Holmes et al. 2014). The application of the developed framework to broader networks of experimental plots, such as those presented in Chapter 5 of this book (Pretzsch et al. 2021), would enable to improve the framework and reach a robust system for climate smartness assessment.

Acknowledgments The authors would like to acknowledge networking support by the COST (European Cooperation in Science and Technology) Action CLIMO (Climate-Smart Forestry in Mountain Regions – CA15226) financially supported by the EU Framework Programme for Research and Innovation HORIZON 2020. Further, we would like to thank the Bayerische Staatsforsten (BaySF) for providing the observational plots and to the Bavarian State Ministry of Food, Agriculture, and Forestry for permanent support of the project W 07 "Long-term experimental plots for forest growth and yield research" (#7831-26625-2017). In Slovenia, the study was additionally supported by the research program P4-0059 Forest, forestry, and renewable forest resources. In Bosnia and Hercegovina, collecting data from plots is partly supported by the Ministry of Civil Affairs of Bosnia and Herzegovina.

Appendices

Appendix 3.1. Overview of Growth and Yield Characteristics of the 10 Long-Term Experimental Plots Used in the Evaluation of CSF Indicators Development (Sect. 3.3.5). B, E. beech; S, N. spruce; F, s. fir

Experiment	Plot	Composition	Survey period First-last	No. of surveys	Value	Remaining stand							Removal stand			PAIV
						N	hdom	ddom	hq	dq	BA	V	N	BA	V	
						ha⁻¹	m	cm	m	cm	m²ha⁻¹	m³ha⁻¹	ha⁻¹ year⁻¹	m²ha⁻¹y⁻¹	m³ha⁻¹y⁻¹	m³ha⁻¹y⁻¹
ZWI 111	2	B	1954–2015	10	mean	507	31.7	42.3	27.8	31.3	25.1	380.7	13	0.6	9.3	9.9
					min–max	139–1076	24.6–36.5	31–49	20.1–34.3	17.3–42.9	18.8–31.6	274.6–497.9	0–38	0–2.7	0–42.8	8.5–11.3
ZWI 111	3	S-B	1954–2015	10	mean	459	34.3	47.1	32.1	39.1	35.7	560.2	10	0.6	9.3	13.8
					min–max	214–945	27–39.6	34.5–56.6	23.3–37.6	24–49.5	31.3–40.9	390.5–715.2	0–28	0–1.9	0–33.2	11.7–15.7
ZWI 111	4	B	1954–2015	10	mean	628	32.1	41.1	27.9	28.6	29.8	464.8	14	0.3	4.3	11.4
					min–max	292–1436	24.5–38.7	29.3–50.8	18.8–35.2	14.8–39.5	26.5–35.8	275–694.3	1–50	0–0.9	0.2–10.5	9.5–13.7
ZWI 111	5	S	1954–2015	10	mean	397	37.1	49	35	41.8	44.2	715.6	8	0.7	10.7	16.9
					min–max	206–806	26.9–42.1	33.5–61.4	23.6–40.9	25–54.8	39.4–52.6	462.6–872.7	0–24	0–1.8	0–30.3	11.2–24.3

			First-last			ha^{-1}	m	cm	m	cm	m^2ha^{-1}	m^3ha^{-1}	$ha^{-1}\,year^{-1}$	$m^2ha^{-1}y^{-1}$	$m^3ha^{-1}y^{-1}$	$m^3ha^{-1}y^{-1}$
FRY 129	11	B-S-F	1980–2018	7	mean	513	33.3	52.3	23.7	31.4	29.8	404	7	1	14.6	10.3
					min - max	364–704	30.5–34.7	45.6–56.9	18.7–29	22.7–41.6	23.3–37.5	293.1–507.1	0–14	0–2.2	0–33.8	7.4–13.4
FRY 129	12	B-S-F	1980–2018	7	mean	511	35.3	57.2	25.2	33.6	36.7	520.2	5	0.8	12	11.2
					min - max	344–932	34.4–36.1	53.8–59.9	20.3–29.1	25.3–40.5	32.6–41.4	468.4–574.9	1–13	0.1–1.8	1.3–28.5	8.3–13.7
FRY 129	21	B-S-F	1980–2018	7	mean	755	33.6	53.3	20.6	25.8	31.9	427.6	12	0.8	11.7	10
					min - max	444–948	32.3–34.5	50.3–57.6	18–23.9	21.4–31.7	28.2–35.1	378.7–489.6	0–23	0–1.7	0–27.6	8.6–12.1
FRY 129	22	B-S-F	1980–2018	7	mean	787	37.2	64.2	24.3	31.6	45.4	651.4	7	0.4	5.8	13.2
					min - max	356–1100	35.3–39.3	57.9–71	21.8–28.3	26.6–39.2	32.5–59.9	476.5–873.1	1–13	0–2.1	0–31.3	9.9–15.6
FRY 129	31	B-S-F	1980–2018	7	mean	415	37.1	65	27.4	38.9	38.8	601.1	5	0.8	11.8	12.3
					min - max	284–688	35.9–38.1	61–68.2	24.2–30.3	32.1–45.5	33.3–42.6	507.3–661.6	0–13	0–2.1	0–33.6	10.1–14.9
FRY 129	32	B-S-F	1980–2018	7	mean	457	37.8	66.2	27.9	39.1	42.9	660.8	5	0.6	8.8	10.5
					min - max	240–740	36.6–39.3	61.8–69.9	25.2–31.8	34.8–46	35.4–48.4	543.9–749.8	0–11	0–2.1	0–33.5	8.4–12.6

Appendix 3.2. List of Indicators Assessed for Their Importance for Climate-Smart Forestry Planning

Criteria	Indicators
Forest resources	Forest area
	Growing stock
	Age structure
	Diameter distribution
	Forest carbon
Forest health and vitality	Deposition and concentration of air pollutants
	Soil condition
	Defoliation
	Forest damage
	Forest land degradation
Productive functions	Increment and felling
	Roundwood
	Non-woods goods
	Services
Forest biological diversity	Diversity of tree species
	Regeneration
	Naturalness
	Introduced tree species
	Deadwood
	Genetic resources
	Forest fragmentation
	Threatened forest species
	Protected forests
	Common forest bird species
Protective function	Protective forests – soil, water, and other ecosystem functions
Socioeconomic functions	Forest holdings
	Contribution of forest sector to GDP
	Net revenue
	Investments in forests and forestry
	Expenditure for services
	Forest sector force
	Occupational safety and health
	Wood consumption
	Trade in wood
	Wood energy
	Accessibility for recreation
	Cultural in spiritual values

(continued)

Criteria	Indicators
Other	Management system applied
	Slenderness coefficient
	Vertical structure of forest stands
	Horizontal distributions of tree crowns
	Tree species composition of natural regeneration
	Recruitment of trees above threshold (usually dbh = 8 or dbh = 10 cm)
	Amount and structure of sanitary felling according to the main agents)
	Register of harvested trees in past planning period (tree species, dimension)
	Growth of trees and stands (e.g., diameter growth…)
	Vitality status of tree species /forest stands
	Horizontal structure of forest stands (patchiness)
	Density of forest stands (basal area, number, SDI)
	Tree species composition of single forest stands
	Silvicultural works (planed and implemented)
	Damages of trees (stands) per agents (wind, snow…)
	Mortality rate of trees
	Growth intensity of forest stands (volume increment/stand volume)
	Timber quality of trees
	Register of natural disturbances in a FMU (windthrow, draughts…)

References

Ahtikoski A, Tuulentie S, Hallikainen V, Nivala V, Vatanen E, Tyrväinen L, Salminen H (2011) Potential trade-offs between nature-based tourism and forestry, a case study in Northern Finland. Forests 2:894–912

Alberdi I, Cañellas I, Hernández L et al (2013) A new method for the identification of old-growth trees in National Forest Inventories: application to *Pinus halepensis* Mill. stands in Spain. Ann For Sci 70:277–285

Azar C, Holmberg J, Lindgren K (1996) Socio-ecological indicators for sustainability. Ecol Econ 18(2):89–112

Bachmann P (2002) Forestliche Planung I/III. Professsur Forsteinrichtung und Waldwachstum ETH Zürich

Bauer HG, Bezzel E, Fiedler W (2005) Kompendium der Vögel Mitteleuropas – alles über Biologie, Gefährdung und Schutz. Aula-Verlag Wiebelsheim

Bayerische Staatsforsten (2018) Waldbauhandbuch Bayerische Staatsforsten. Richtlinie für die Waldbewirtschaftung im Hochgebirge. Version 01.00 (03/2018), Regensburg, p 141

Beer P (2003) Vegetationskundliche Untersuchungen und Entwicklung eines Verfahrens zur Bewertung ausgewählter Laubwaldflächen im Schweizer Mittelland, Geographischen Institut der Universität Zürich, Diplomarbeit, Juni 2003

Biber P, Borges JG, Moshammer R, Barreiro S, Botequim B, Brodrechtova Y, Eriksson LO (2015) How sensitive are ecosystem services in European forest landscapes to silvicultural treatment? Forests 6(5):1666–1695

Blattert C, Lemm R, Thees O, Lexer MJ, Hanewinkel M (2017) Management of ecosystem services in mountain forests: review of indicators and value functions for model based multi-criteria decision analysis. Ecol Indic 79:391–409

Bolte A, Ammer C, Löf M, Madsen P, Nabuurs GJ, Schall P, Rock J (2009) Adaptive forest management in central Europe: climate change impacts, strategies and integrative concept. Scand J For Res 24(6):473–482

Bončina A (2001) Concept of sustainable forest management evaluation in forestry planning at the forest management unit level: some experiences, problems and suggestions from Slovenian Forestry. Criteria and indicators for sustainable forest management at forest management unit level (Franc A ed.). EFI Proc 38:247–260

Bončina A, Simončič T, Rosset C (2019) Assessment of the concept of forest functions in Central European forestry. Environ Sci Pol 99:123–135

Booysen F (2002) An overview and evaluation of composite indices of development. Soc Indic Res 59(2):115–151

Bowditch E, Santopuoli G, Binder F, del Río M, La Porta N, Kluvankova T, Lesinski L, Motta R, Pach M, Panzacchi P, Pretzsch H, Temperli C, Tonon G, Smith M, Velikova V, Weatherall A, Tognetti T (2020) What is climate-smart forestry? A definition from a multinational collaborative process focused on mountain regions of Europe. Ecosyst Serv 43:101113

Brang P, Spathelf P, Larsen B, Bauhus J, Bončina A, Chauvin C, Drössler L, García-Güemes C, Heiri C, Kerr G, Lexer MJ, Mason B, Mohren F, Mühlethaler U, Nocentini S, Svoboda (2014) Suitability of close-to-nature silviculture for adapting temperate European forests to climate change. Forestry 87(4):492–503

Bravo F, del Río M, Bravo-Oviedo A, Ruiz-Peinado R, del Peso C, Montero G (2016) Forest management strategies and carbon sequestration. In: Bravo F, Lemay V, Jandl R (eds) Managing forest ecosystems: the challenge of climate change, 2nd edn. Springer, pp 251–275

Bütler R, Angelstam P, Ekelund P, Schlaepfer R (2004) Dead wood threshold values for the three-toed woodpecker presence in boreal and sub-alpine forest. Biol Conserv 119:305–318

Bütler R, Lachat T, Larrieu L, Paillet Y (2013) Habitat trees: key elements for forest biodiversity. In: Kraus D, Krumm F (eds) Integrative approaches as an opportunity for the conservation of forest biodiversity. European Forest Institute, pp 84–91

Churchill DJ, Larson AJ, Dahlgreen MC, Franklin JF, Hessburg PF, Lutz JA (2013) Restoring forest resilience: from reference spatial patterns to silvicultural prescriptions and monitoring. For Ecol Manag 291:442–457

Cordonnier T, Berger F, Elkin C, Lämas T, Martinez M (2013) ARANGE deliverable D2.2 – models and linker functions (indicators) for ecosystem services

Corona P, Chirici G, McRoberts RE (2011) Contribution of large-scale forest inventories to biodiversity assessment and monitoring. For Ecol Manag 262:2061–2069

Cullotta S, Bončina A, Carvalho-Ribeiro SM, Chauvin C, Farcy C, Kurttila M, Maetzke FG (2015) Forest planning across Europe: the spatial scale, tools, and inter-sectoral integration in land-use planning. J Environ Plan Manag 58:1384–1411

D'Amato AW, Bradford JB, Fraver S, Palik BJ (2011) Forest management for mitigation and adaptation to climate change: insights from long-term silviculture experiments. For Ecol Manag 262(5):803–816

De Groot RS, Wilson MA, Boumans RM (2002) A typology for the classification, description and valuation of ecosystem functions, goods and services. Ecol Econ 41(3):393–408

Duncker PS, Raulund-Rasmussen K, Gundersen P, Katzensteiner K, De Jong J, Ravn HP, Smith M, Eckmüllner O, Spiecker H (2012) How forest management affects ecosystem services, including timber production and economic return: synergies and trade-offs. Ecol Soc 17(4):50

Edwards DM, Jay M, Jensen FS, Lucas B, Marzano M, Montagné C, Peace A, Weiss G (2012) Public preferences for structural attributes of forests: towards a pan-European perspective. For Policy Econ 19:12–19

El-Lakany H, Winkel K, Hawtin G (2001) Forest genetic resources, conservation and management. FAO, DFSC, IPGRI. 2: In managed natural forests and protected areas (in situ) Int Plant Genetic Resources Institute. Rome, Italy

Forest Europe (2015) Background information for the updated pan-european indicators for sustainable forest management. https://foresteurope.org/wp-content/uploads/2016/10/3AG_UPI_Updated_Backgr_Info.pdf, latest access 2020-12-15

Forrester DI, Tachauer IHH, Annighoefer P, Barbeito I, Pretzsch H, Ruiz-Peinado R, Stark H, Vacchiano G, Zlatanov T, Chakraborty T, Saha S, Sileshi GW (2017) Generalized biomass and leaf area allometric equations for European tree species incorporating stand structure, tree age and climate. For Ecol Manag 396:160–175

Freudenberg M (2003) Composite indicators of country performance: a critical assessment. OECD science, Technology and industry working papers no. 2003/16. OECD Publishing, Paris

Fussi B, Westergren M, Aravanopoulos F, Baier R, Kavaliauskas D, Finzgar D, Alizoti P, Bozic G, Avramidou E, Konnert M, Kraigher H (2016) Forest genetic monitoring: an overview of concepts and definitions. Environ Monit Assess 188:493

Gao T, Nielsen AB, Hedblom M (2015) Reviewing the strength of evidence of biodiversity indicators for forest ecosystems in Europe. Ecol Indic 57:420–434

Geburek T, Milasowszky N, Frank G, Konrad H, Schadauer K (2010) The Austrian forest biodiversity index: all in one. Ecol Indic 10(3):753–761

Gonin P, Larrieu L, Deconchat M (2017) Index of Biodiversity Potential (IBP): how to extend it to mediterranean forests? Forêt Méditerranéenne 38(3):343–350

Greco S, Ishizaka A, Tasiou M et al (2019) On the methodological framework of composite indices: a review of the issues of weighting, aggregation, and robustness. Soc Indic Res 141:61–94

Hagan JM, Whitman AA (2006) Biodiversity indicators for sustainable forestry: simplifying complexity. J For 104(4):203–210

Hahm WJ, Riebe CS, Lukens CE, Araki S (2014) Bedrock composition regulates mountain ecosystems and landscape evolution. Proc Natl Acad Sci U S A 111:3338–3343

Hand MS, Lawson M (2018) Effects of climate change on recreation in the Northern Rockies Region. In: Halofsky JE, Peterson DL, Dante-Wood SK, Hoang L, Ho JJ, Joyce LA (eds) Climate change vulnerability and adaptation in the Northern Rocky Mountains [Part 2]. Gen. Tech. Rep. RMRS-GTR-374. U.S. Department of Agriculture, Forest Service, Rocky Mountain Research Station, Fort Collins, pp 398–433

Hanewinkel M, Cullmann DA, Schelhaas MJ, Nabuurs GJ, Zimmermann NE (2012) Climate change may cause severe loss in the economic value of European forest land. Nat Clim Chang 3:203–207

Heym M, Uhl E, Moshammer R, Dieler J, Stimm K, Pretzsch H (2021) Utilising forest inventory data for biodiversity assessment. Ecol Indic 121:107196

Hilmers T, Avdagic A, Bartkowicz L, Bielak K, Binder F, Boncina A, Dobor L, Forrester DI, Habi ML, Ibrahimspahic A, Jaworski A, Klopcic M, Matocic B, Nagel TA, Petras R, del Río M, Stajic B, Uhl E, Zlatanov T, Tognetti R, Pretzsch H (2019a) The productivity of mixed mountain forests comprised of *Fagus sylvatica*, *Picea abies*, and *Abies alba* across Europe. Forestry 92(5):512–522

Hilmers T, Steinacker L, Pretzsch H (2019b) Zur Abhängigkeit der Zuwachsverteilung im Plenterwald von der Art und Größe der Bäume. DVFFA, Sektion Ertragskunde Zwiesel, Tagungsband: 1–11. ISSN 1432-2609

Hilmers T, Biber P, Knoke T, Pretzsch H (2020) Assessing transformation scenarios from pure Norway spruce to mixed uneven-aged forests in mountain areas. Eur J For Res 139:567–584

Hof AR, Dymond CC, Mladenoff DJ (2017) Climate change mitigation through adaptation: the effectiveness of forest diversification by novel tree planting regimes. Ecosphere 8(11):e01981

Hofer P, Taverna R, Werner F, Kaufmann E, Thürig E (2007) CO2-Effekte der Schweizer Forst-und Holzwirtschaft. Szenarien künftiger Beiträge zum Klimaschutz. Bundesamt für Umwelt (BAFU), Bern, Forschungsbericht

Holmes TP, McNulty S, Vose JM, Prestemon JP, Harbin L (2014) A conceptual framework for adaptive forest management under climate change. In: Vose JM, Klepzig KD (eds) Climate change adaption and mitigation management options. A guide for natural resource managers in southern forest ecosystems. CRC Press/Taylor and Francis, pp 45–60

Hundeshagen JC (1826) Die Forstabschätzung auf neuen wissenschaftlichen Grundlagen. Verlag Heinrich Laupp, Tübingen

Hundeshagen JC (1828) Encyclopädie der Forstwissenschaft, 2. Aufl. Verlag Heinrich Laupp, Tübingen

Jeffreys I (2004) The use of compensatory and non-compensatory multi-criteria analysis for small-scale forestry. Small-Scale For 3:99–117

Jiang Z, Liu H, Wang H, Peng J, Meersmans J, Greem SM, Quine A, Wu X, Song Z (2020) Bedrock geochemistry influences vegetation growth by regulating the regolith water holding capacity. Nat Commun 11:2392

Kätzel R, Becker F (2014) Erhaltung und nachhaltige Nutzung forstlicher genressourcen. Eberswalder Forstliche Schriftenreihe 58

Klein D, Schulz C (2012) Die Kohlenstoffbilanz der Bayerische Forst und Holzwirtschaft / Bayerische Landesanstalt für Wald und Forstwirtschaft. – Abschlussbericht

Klopcic KM, Poljanec A, Gartner A, Boncina A (2009) Factors related to natural disturbances in mountain Norway spruce (*Picea abies*) forests in the Julian Alps. Ecoscience 16(1):48–57

Kneeshaw DD, Leduc A, Messier C, Drapeau P, Gauthier S, Par D, Carignan R, Doucet R, Bouthillier L (2000) Development of integrated ecological standards of sustainable forest management at an operational scale. For Chron 76(3):481–493

Knoke T, Ammer C, Stimm B, Mosandl R (2008) Admixing broadleaved to coniferous tree species: a review on yield, ecological stability and economics. Eur J For Res 127(2):89–101

Köchli DA, Brang P (2005) Simulating effects of forest management on selected public forest goods and services: a case study. For Ecol Manag 209:57–68

Konnert M, Müller D, Baier R, Huber G (2015) Konzept zum Erhalt und zur nachhaltigen Nutzung forstlicher Genressourcen in Bayern. Bayerische Amt für forstliche Saat und Pflanzenzucht

Lachat T, Brang P, Bolliger M, Bollmann K, Brändli U, Bütler R, Herrmann S, Schneider O, Wermelinger B (2014) Totholz im Wald - Entstehung, Bedeutung und Förderung. WSL-Merkblatt für die Praxis, Mai, Nr 52

Lassauce A, Paillet Y, Jactel H, Bouget C (2011) Deadwood as a surrogate for forest biodiversity: meta-analysis of correlations between deadwood volume and species richness of saproxylic organisms. Ecol Indic 11:1027–1039

Lindner M (2000) Developing adaptive forest management strategies to cope with climate change. Tree Physiol 20(5–6):299–307

MacDicken K (2015) FRA 2015 - terms and definitions. Forest Resources Assessment Working Paper 180

Mäkelä A, del Rio M, Hynynen J, Hawkins MJ, Reyer C, Soares P, van Oijen M, Tome M (2012) Using stand-scale forest models for estimating indicators of sustainable forest management. For Ecol Manag 285:164–178

Mayer P (2000) Hot spot: forest policy in Europe: achievements of the MCPFE and challenges ahead. Forest Policy Econ 1(2):177–185

MCPFE (1993) Resolution H1: general guidelines for the sustainable management of forests in Europe. Proceedings of 2nd Ministerial Conference on the Protection of Forests in Europe, Helsinki, Finland, p 5

Messier C, Puettmann KJ, Coates KD (eds) (2013) Managing forests as complex adaptive systems: building resilience to the challenge of global change. Routledge

Messier C, Puettmann K, Chazdon R, Andersson KP, Angers VA, Brotons L, Levin SA (2015) From management to stewardship: viewing forests as complex adaptive systems in an uncertain world. Conserv Lett 8(5):368–377

Moning C, Bussler H, Müller J (2009) Ökologische Schlüsselwerte in Bergmischwäldern als Grundlage für eine nachhaltige Forstwirtschaft. In: Wissenschaftliche Reihe, Nationalparkverwaltung Bayerischer Wald, Dezember 2009

Nabuurs G, Delacote P, Ellison D, Hetemäki L, Id ML (2017) By 2050 the mitigation effects of EU forests could nearly double through climate smart. Forests 8:484

Nagel TA, Mikac S, Dolinar M, Klopčič M, Keren S, Svoboda M, Diaci J, Bončina A, Paulic V (2017) The natural disturbance regime in forests of the Dinaric Mountains: a synthesis of evidence. For Ecol Manag 388:29–42

Neumann M, Starlinger F (2001) The significance of different indices for stand structure and diversity in forests. For Ecol Manag 145:91–106

OECD, European Commission, Joint Research Centre (2008) Handbook on constructing composite indicators: methodology and user guide. OECD Publishing

Otto HJ (2000) Standortsansprüche der wichtigsten Waldbaumarten, 8. Aufl. Auswertungs- und Informationsdienst für Ernährung, Landwirtschaft und Forsten e.V, Aschaffenburg

Pach M, Bielak K, Bončina A, Coll L, Höhn M, Kasanin-Grubin M, Lesiński J, Pretzsch H, Skrzyszewski J, Spathelf P, Tonon G, Weatherall A, Zlatanov T (2021) Climate-smart silviculture in mountain regions. In: Managing Forest Ecosystems, Vol. 40, Tognetti R, Smith M, Panzacchi P (Eds): Climate-Smart Forestry in Mountain Regions. Springer Nature, Switzerland, AG

Pennock DJ, van Kessel C (1997) Clear-cut forest harvest impacts on soil quality indicators in the mixedwood forest of Saskatchewan, Canada. Geoderma 75:13–32

Pfatrisch M (2019) Indicators to assess the climate smartness of mixed mountain forests containing Picea, Abies and Fagus. Masterthesis, Technische Universität München, DVFFA MWW-MA 2246, p 139

Pollesch NL, Dale VH (2016) Normalization in sustainability assessment: methods and implications. Ecol Econ 130:195–208

Pommerening A, Grabarnik P (2019) Individual-based methods in forest ecology and management. Springer, p 411

Pommerening A, Murphy ST (2004) A review of the history, definitions and methods of continuous cover forestry with special attention to afforestation and restocking. Forestry 77:27–44

Pretzsch H (2009) Forest dynamics, growth and yield. Springer, Berlin, p 664

Pretzsch H (2019) Transitioning monocultures to complex forest stands in Central Europe: principles and practice. Burleigh Dodds Science Publishing Limited. https://doi.org/10.19103/as.2019.0057.14

Pretzsch H, Puumalainen J (2002) Up- and downscaling: EU/ICP Level I and II. In: Puumalainen J, Kennedy P, Folving S (eds) Forest biodiversity - assessment approaches for Europe S.127: 50–61

Pretzsch H, Grote R, Reineking B, Rötzer T, Seifert S (2008) Models for forest ecosystem management: a European perspective. Ann Bot 101:1065–1087

Pretzsch H, Hilmers T, Uhl E, et al (2021) Efficacy of trans-geographic observational network design for revelation of growth pattern in mountain forests across Europe. In: Managing Forest Ecosystems, Vol. 40, Tognetti R, Smith M, Panzacchi P (Eds): Climate-Smart Forestry in Mountain Regions. Springer Nature, Switzerland, AG

Pretzsch H, del Río M, Biber P, Arcangeli C, Bielak K, Brang P, Dudzinska M, Forrester DI, Klädtke J, Kohnle U, Ledermann Th, Matthews R, Nagel J, Nagel R, Nilsson U, Ningre F, Nord-Larsen Th, Wernsdörfer H, Sycheva E (2019) Maintenance of long-term experiments for unique insights into forest growth dynamics and trends: review and perspectives. European Journal of Forest Research 138(1):165–185 https://doi.org/10.1007/s10342-018-1151-y

Priehäusser G (1958) Die Fichten-Variationen und -kombinationen des Bayerischen Waldes nach phänotypischen Merkmalen mit Bestimmungsschlüssel. In: Forstwiss Cbl, Nr. 77: 151–171

Prietzel J, Falk W, Reger B, Uhl E, Pretzsch H, Zimmermann L (2020) Half a century of Scots pine forest ecosystem monitoring reveals long-term effects of atmospheric deposition and climate change. Glob Chang Biol 26(10):5796–5815

Pröbstl U, Elands B, Wirth V (2009) Forest recreation and nature tourism in Europe: context, history and current situation. In: Bell S, Simpson M, Tyrväinen L, Sievänen T, Pröbstl U (eds) European forest recreation and tourism: a handbook. Taylor and Francis, pp 12–32

Pülzl H, Prokofieva I, Berg S, Rametsteiner E, Aggestam F, Wolfslehneret B (2012) Indicator development in sustainability impact assessment: balancing theory and practice. Eur J For Res 131:35–46

Raison RJ, Khanna PK (2011) Possible impacts of climate change on forest soil health. In: Singh B, Cowie A, Chan K (eds) Soil health and climate change. Soil biology, vol 29. Springer, Berlin/Heidelberg

Reed MS, Fraser EDG, Dougill EJ (2006) An adaptive learning process for developing and applying sustainability indicators with local communities. Ecol Econ 59(4):406–418

Riedel T, Henning P, Kroiher F, Polley H, Schmitz F, Schwitzgebel F (2017) Die Dritte Bundeswaldinventur – BWI 2012 – Inventur und Auswertungsergebnisse. Johann Heinrich von Thünen-Institut

Rösner S, Mussard-Forster E, Lorenc T, Müller J (2014) Recreation shapes a "landscape of fear" for a threatened forest bird species in Central Europe. Landsc Ecol 29(1):55–66

Rosset C, Schütz JP, Günter M, Gollut C (2014) WIS.2 – a sustainable forest management decision support system. Math Comput For Natl Res Sci 6(2):89–100

Rottmann M (1986) Wind- und Sturmschäden im Wald. J.D. Sauerländer's Verlag

Ruiz-Peinado R, Bravo-Oviedo A, López-Senespleda E, Montero G, del Río M (2013) Do thinnings influence biomass and soil carbon stocks in Mediterranean maritime pinewoods? Eur J Forest Res 132:253–262

Santopuoli G, Temperli C, Alberdi I, Barbeito I, Bosela M, Bottero A, Klopčič M, Lesinski L, Panzacchi P, Tognetti R (2020) Pan-European sustainable forest management indicators for assessing Climate-Smart Forestry in Europe. Can J For Res 0(0):1–10. https://doi.org/10.1139/cjfr-2020-0166

Schulze CH, Waltert M, Kessler PJA, Pitopang R, Veddeler D, Mühlenberg M, Gradstein SR, Leuschner C, Steffan-Dewenter I, Tscharntke T (2004) Biodiversity indicator groups of tropical land-use systems: comparing plants, birds, and insects. Ecol Appl 14(5):1321–1333

Schütt P, Schuck HJ, Stimm B, Aas G, Baasch R, Blaschke H, Dobner M, Krug E, Maier J, Schill H, Schütt P, Schuck HJ, Stimm B (eds) (2002) Lexikon der Baum- und Straucharten. Nikol Verlagsgesellschaft mbh & Co.KG, Hamburg

Schwaiger F, Poschenrieder W, Biber P, Pretzsch H (2019) Ecosystem service trade-offs for adaptive forest management. Ecosyst Serv 39:100993

Seidl R, Rammer W, Lasch P, Badeck FW, Lexer MJ (2008) Does conversion of even-aged, secondary coniferous forests affect carbon sequestration? A simulation study under changing environmental conditions. Silva Fenn 42:369–386

Seidl R, Schelhaas M-J, Rammer W, Verkerk PJ (2014) Increasing forest disturbances in Europe and their impact on carbon storageç. Nat Clim Chang 4(9):806–810

Singh RK, Murty HR, Gupta SK, Dikshit AK (2012) An overview of sustainability assessment methodologies. Ecol Indic 15(1):281–299

Socha J, Bruchwald A, Neroj B (2017) Aktualna i potencjalna produkcyjność siedlisk leśnych Polski dla głównych gatunków lasotwórczych. Kraków

Sohn JA, Saha S, Bauhus J (2016) Potential of forest thinning to mitigate drought stress: a meta-analysis. For Ecol Manag 380:261–273

Speidel G (1972) Planung im Forstbetrieb. Verlag Paul Parey, Hamburg, Berlin, 267 pp

Spellmann H, Vor T, Bolte A, Ammer C (2015) Potenziale und Risiken eingeführter Baumarten - Baumartenportraits mit naturschutzfachlicher Bewertung. Universitätsverlag Göttingen

Spittlehouse DL, Stewart RB (2003) Adaptation to climate change in forest management. BC J Ecosyst Manage: 4/1: art1

Staudhammer CL, LeMay VM (2001) Introduction and evaluation of possible indices of stand structural diversity. Can J For Res 31:1105–1115

StMELF (2017) Anweisung - für die Erstellung der Forstlichen Gutachten zur Situation der Waldverjüngung 2018 / Bayerisches Staatsministerium für Ernährung, Landwirtschaft und Forsten (StMELF). Ludwigstraße 2, 80539 München – Anleitung

Stupak I, Lattimore B, Titus BD, Smith CT (2011) Criteria and indicators for sustainable forest fuel production and harvesting: a review of current standards for sustainable forest management. Biomass Bioenergy 35(8):3287–3308

Talukder B, Hipel KW, vanLoon GW (2017) Developing composite indicators for agricultural sustainability assessment: effect of normalization and aggregation techniques. Resources 6:66

Temperli C, Santopuoli G, Bottero A, et al (2021) National Forest Inventory data to evaluate Climate-Smart Forestry. In: Managing Forest Ecosystems, Vol. 40, Tognetti R, Smith M, Panzacchi P (Eds): Climate-Smart Forestry in Mountain Regions. Springer Nature, Switzerland, AG

Tognetti R, Valentini R, Belelli Marchesini L, Gianelle D, Panzacchi P, Marshall JD (2021) Continuous monitoring of tree responses to climate change for smart forestry – a cybernetic web of trees. In: Managing Forest Ecosystems, Vol. 40, Tognetti R, Smith M, Panzacchi P (Eds): Climate-Smart Forestry in Mountain Regions. Springer Nature, Switzerland, AG

Tomppo E, Gschwantner T, Lawrence M, McRoberts RE (2010) National forest inventories. Pathways for common reporting. Springer, New York

Vacik H, Wolfslehner B (2004) Entwicklung eines Indikatorenkatalogs zur Evaluierung einer nachhaltigen Waldbewirtschaftung auf betrieblicher Ebene. Schweizerische Zeitschrift fur Forstwesen 155(11):476–486

von Carlowitz HC (1713) Sylvicultura Oekonomica oder Haußwirthliche Nachricht und Naturmäßige Anweisung zur wilden Baum-Zucht. JF Braun, Leipzig

Vuidot A, Paillet Y, Archaux F, Gosselin F (2011) Influence of tree characteristics and forest management on tree microhabitats. Biol Conserv 144(1):441–450

Weatherall A, Nabuurs G-J, Velikova V, et al (2021) Defining Climate-Smart Forestry. In: Managing Forest Ecosystems, Vol. 40, Tognetti R, Smith M, Panzacchi P (Eds): Climate-Smart Forestry in Mountain Regions. Springer Nature, Switzerland, AG

Wijewardana D (2008) Criteria and indicators for sustainable forest management: the road travelled and the way ahead. Ecol Indic 8(2):115–122

Wilkes-Allemann J, Pütz M, Hirschi C (2015) Governance of forest recreation in urban areas: analysing the role of stakeholders and institutions using the institutional analysis and development framework. Environ Policy Gov 25(2):139–156

Winter S, Möller GC (2008) Microhabitats in lowland beech forests as monitoring tool for nature conservation. For Ecol Manag 255:1251–1261

Wolfslehner B, Vacik H (2011) Mapping indicator models: from intuitive problem structuring to quantified decision-making in sustainable forest management. Ecol Indic 11(2):274–283

Yaffee SL (1999) Three faces of ecosystem management. Conserv Biol 13:713–725

Chapter 4
National Forest Inventory Data to Evaluate Climate-Smart Forestry

Christian Temperli, Giovanni Santopuoli, Alessandra Bottero, Ignacio Barbeito, Iciar Alberdi, Sonia Condés, Thomas Gschwantner, Michal Bosela, Bozydar Neroj, Christoph Fischer, Matija Klopčič, Jerzy Lesiński, Radoslaw Sroga, and Roberto Tognetti

Abstract National Forest Inventory (NFI) data are the main source of information on forest resources at country and subcountry levels. This chapter explores the strengths and limitations of NFI-derived indicators to assess forest development with respect to adaptation to and mitigation of climate change, that is, the criteria of Climate-Smart Forestry (CSF). We reflect on harmonizing NFI-based indicators across Europe, use literature to scrutinize available indicators to evaluate CSF, and apply them in 1) Switzerland, where CSF is evaluated for NFI records and simulation model projections with four management scenarios; 2) 43 selected European

C. Temperli (✉) · A. Bottero · C. Fischer
Swiss Federal Institute for Forest, Snow and Landscape Research WSL,
Birmensdorf, Switzerland
e-mail: christian.temperli@wsl.ch; alessandra.bottero@wsl.ch; christoph.fischer@wsl.ch

G. Santopuoli · R. Tognetti
Dipartimento Agricoltura, Ambiente e Alimenti, Universita` degli Studi del Molise,
Campobasso, Italy
e-mail: giovanni.santopuoli@unimol.it; tognetti@posta.unimol.it

I. Barbeito
Southern Swedish Forest Research Center, Swedish University of Agricultural Sciences,
Alnarp, Sweden

Université de Lorraine, AgroParisTech, INRA, UMR Silva, Nancy, France
e-mail: ignacio.barbeito@slu.se

I. Alberdi
Instituto Nacional de Investigación y Tecnología Agraria y Alimentaria, Madrid, Spain
e-mail: alberdi.iciar@inia.es

S. Condés
Department of Natural Systems and Resources, School of Forest Engineering and Natural
Resources, Universidad Politécnica de Madrid, Madrid, Spain
e-mail: sonia.condes@upm.es

R. Tognetti et al. (eds.), *Climate-Smart Forestry in Mountain Regions*, Managing
Forest Ecosystems 40, https://doi.org/10.1007/978-3-030-80767-2_4

countries, for which the indicators for Sustainable Forest Management (SFM) are used. The indicators were aggregated to composite indices for adaptation and mitigation and to an overall CSF rating. The Swiss NFI records showed increased CSF ratings in mountainous regions, where growing stocks increased. Simulations under business-as-usual management led to a positive CSF rating, whereas scenarios of increased harvesting decreased either only adaptation or both mitigation and adaptation. European-level results showed increases in CSF ratings for most countries. Negative adaptation ratings were mostly due to forest damages. We discuss the limitations of the indicator approach, consider the broader context of international greenhouse gas reporting, and conclude with policy recommendations.

4.1 Introduction

Climate-Smart Forestry (CSF) has been suggested as forest management concept with the goal to combine 1) the adaptation of forests to climate change, 2) the mitigation of climate change through the sequestration of atmospheric carbon by trees, and 3) the maintenance of forest ecosystem service provision (Bowditch et al. 2020). Previous applications of the CSF concept mainly focused on mitigation potentials at the national to European scale using available literature (Nabuurs et al. 2017) or simulations of forest development under various management scenarios (Nabuurs et al. 2018; Jandl et al. 2018; Yousefpour et al. 2018). A stand-scale application of the CSF concept has been developed by participants of the COST Action CLIMO (see Chap. 3 of this book: del Río et al. 2021) using a comprehensive set of indicators to evaluate both the adaptation and mitigation potential of mixed spruce, fir, and beech forests in Europe (Pfatrisch 2019). However, a comprehensive

T. Gschwantner
Federal Research and Training Centre for Forests, Natural Hazards and Landscape BFW, Vienna, Austria
e-mail: thomas.gschwantner@bfw.gv.at

M. Bosela
Faculty of Forestry, Technical University in Zvolen, Zvolen, Slovakia
e-mail: ybosela@tuzvo.sk

B. Neroj · R. Sroga
Bureau for Forest Management and Geodesy, Sekocin Stary, Poland
e-mail: bozydar.neroj@zarzad.buligl.pl; radoslaw.sroga@zarzad.buligl.pl

M. Klopčič
Biotechnical Faculty, Department of Forestry and Renewable Forest Resources, University of Ljubljana, Ljubljana, Slovenia
e-mail: matija.klopcic@bf.uni-lj.si

J. Lesiński
Department of Forest Biodiversity, University of Agriculture, Krakow, Poland
e-mail: jerzy.lesinski@urk.edu.pl

assessment method that simultaneously accounts for the three CSF aspects of adaptation, mitigation, and sustainable provision of ecosystem goods and services (ES) at a national to European scale is still lacking.

Assessments of CSF at this national to continental scale is important for forest policy making (Verkerk et al. 2020). Forests play an important role in fulfilling the reduction goals of greenhouse gas emissions as per the Paris agreement (Seddon et al. 2019). Thus, national governments are encouraged to device forest policy recommendations to mitigate climate change (FAO 2018). This can be achieved with management that favors tree and forest growth such that more CO_2 is sequestered from the atmosphere than released from the forest through respiration, decay of deadwood, and the decay and burning of harvested wood products (Köhl et al. 2020). This may encompass the prolongation of cutting cycles to sequester carbon in the living biomass. This, in turn, may be in conflict with policies that aim at raising the capacity of forests to adapt to climate change by reducing rotation lengths and cutting cycles. Shorter management intervals may be applied as part of conversion strategies to promote more drought- and disturbance-resistant tree species and a higher species diversity in currently species-poor forests. However, forest management to increase forest growth through stand density reduction (thinning) may also be in concert with the goal of carbon sequestration (Lindner et al. 2010; Diaconu et al. 2017; Jandl et al. 2019). Reducing timber harvesting may conflict with policies to sequester carbon in wood products and to substitute fossil fuel–intensive energy and building material (Sathre and O'Connor 2010). Management for adaptation may conflict with nature conservation goals, such as the retention of old-growth forest structures. Mitigation through carbon assimilation in standing tree biomass may collide with the need for ensuring advanced regeneration and stability in forests that protect against rockfall and avalanches in mountain areas (Brang et al. 2006). Hence, adaptation, mitigation, and the provision of ES need to be balanced in climate-smart management recommendations, also at a regional to national scale.

NFIs provide reliable and robust data and indicators on a regular basis that are representative at the national scale and cover time periods of several decades in many countries (Alberdi et al. 2016b). Thus, they allow the identification of forest development trajectories that resulted from climate-smart (active or passive) forest policy making. Most NFIs record information at the plot level, such as silvicultural treatments, and others at the tree level, such as tree species, stem diameter, tree height, or health status (Tomppo et al. 2010). By identifying the drivers of past forest development and CSF indicators, we may be able to derive climate-smart policy recommendations that simultaneously promote adaptation, mitigation, and ES provision.

The Pan-European criteria and indicators for Sustainable Forest Management (SFM) were established as a basic tool for defining, promoting, and monitoring SFM across Europe, with the last updated set being endorsed by the seventh Ministerial Conference in Madrid in 2015 (Forest Europe 2015a). These 34 quantitative and 11 qualitative indicators are organized in 6 criteria, are broadly accepted, and are publicly available at the national level through the reports on the State of

Europe's Forest (SoEF, Forest Europe 2015b, https://foresteurope.org/state-europes-forests-2015-report/, Accessed 8 December 2020). They cover a broad range of aspects on the state of forests (e.g., growing stock, age class distribution, and forest type) and their functions in terms of timber and the production of nonwood goods, biodiversity conservation, protection of ecosystem functions, and socioeconomy. The broad thematic coverage of the SFM indicators may allow their application as national-scale proxies for the contribution of forests to climate change mitigation and their capacity to adapt to climate change.

The goal of this chapter is to explore the possibilities and potential limitations of NFI-based indicators to quantify the mitigation and the adaptive capacity of European forests. To this end, we use the available literature to scrutinize NFI-based indicators, including the ones of Forest Europe for SFM, for their suitability as proxies for mitigation and adaptive capacity (Sect. 4.2 of this chapter). We highlight the advances in the harmonization of the indicators based on the NFIs from different countries, with different recording methods, and the necessary considerations with respect to comparing indicator development across countries (Sect. 4.3). To test this approach, we evaluate forest development with respect to mitigation and adaptive capacity in two cases: 1) for the five biogeographic NFI production regions of Switzerland and 2) for selected European countries. In Sect. 4.4, we describe the calculation of indicators and how estimates for adaptive capacity and mitigation were derived. Sections 4.5 and 4.6 contain the results for Switzerland and the selected European countries, respectively. We critically evaluate the approach and identify areas for improvement in Sect. 4.7 and, thus, interpret the results with respect to the broader context of international greenhouse gas reporting and global climate dynamics (Sect. 4.8). We conclude with management and policy recommendations in Sect. 4.9.

4.2   Indicators to Quantify Adaptation and Mitigation, a Review

Adaptation and mitigation are considered the most important management measures to counteract climate change and its negative impacts on forests (Spittlehouse 2005; Nabuurs et al. 2018) and society. Despite the increased awareness among forest decision makers and managers to promote adaptation and mitigation strategies, there are still large uncertainties on how to evaluate the effects of their implementation. Differences in socialeconomic and environmental conditions, challenges in data collection, and the analysis of climate change impacts in general are mentioned by many authors as the main causes of these uncertainties (Seidl and Lexer 2013; Forsius et al. 2016; Viccaro et al. 2019). Participatory approaches involving diverse expert and stakeholder groups were often suggested as a viable tool to overcome the uncertainties in measuring the effects of management to adapt forest ecosystems to climate change (Nelson et al. 2016). It is a very common approach to select and rank SFM alternatives (Santopuoli et al. 2012; Paletto et al. 2014;

Pastorella et al. 2016b) through the collection and evaluation of stakeholder opinions. Nevertheless, evaluations based on participatory approaches could be subjective, depending on the stakeholder experiences and priorities.

For this reason, the development of an objective method to use SFM indicators for assessing adaptation, mitigation, and CSF is strongly required. Criteria and indicators (C&I) for SFM represent the most important tools to assess the sustainability of forest management at the pan-European scale (Santopuoli et al. 2016; Baycheva-Merger and Wolfslehner 2016). However, it is crucial to select a subset of indicators from the whole C&I set that is a suitable measure for adaptation to and mitigation of climate change. Here, we conducted a literature review to highlight those indicators from the pan-European C&I set that have been frequently considered suitable to assess adaptation and mitigation management measures in forest ecosystems.

The search was conducted in February 2020 using the Scopus® database (https://scopus.com) through two queries, one for adaptation and one for mitigation using the following Boolean search terms:

- (TITLE-ABS-KEY (climate AND adaptation) AND TITLE-ABS-KEY (sustainable AND forest AND management) AND TITLE-ABS-KEY (indicator))
- (TITLE-ABS-KEY (climate AND mitigation) AND TITLE-ABS-KEY (sustainable AND forest AND management) AND TITLE-ABS-KEY (indicator))

We did not use constraints about the year of publication, but we excluded non-English as well as not relevant articles, that is, publications not strictly focused on the use of SFM indicators. For this reason, all papers were accurately screened, duplicates removed, and SFM indicators used to assess adaptation and mitigation, respectively, were identified and recorded.

A total of 50 papers were extracted from the Scopus® database, 20 for adaptation, and 30 for mitigation. During the screening phase, we discarded 32 papers that were considered not relevant. We counted the occurrence of SFM indicators in the remaining 18 articles, 6 for adaptation, and 12 for mitigation, respectively. All papers, but one in 1997, were published in the period 2011–2017.

Scrutinizing the 18 articles revealed that 22 out of 34 indicators were suitable to assess adaptation and mitigation (Fig. 4.1). In particular, 13 out of 22 were useful to assess both adaptation and mitigation, while 4 indicators were mentioned only for adaptation (i.e., 4.2, 6.3, 6.4, 6.11) and 5 only for mitigation (i.e., 1.1, 3.3, 4.3, 6.5, 6.9).

Overall, authors highlighted 17 indicators as suitable to assess adaptation, among which *tree species composition, roundwood, forest damage,* and *deadwood* were the most frequent. In particular, authors stressed that biodiversity conservation strongly supports adaptive management strategies (Klenk et al. 2015). According to many authors, all the 17 indicators could provide support for forest decision and policy makers (Hlásny et al. 2014). These indicators may raise awareness among forest managers and practitioners about climate change impacts and promote the implementation of adaptive management at local level (Seidl and Lexer 2013). Moreover, they provide support to researchers for developing new and more appropriate scenarios for the sustained provision of ecosystem services (Klenk et al. 2015; Hlásny et al. 2017).

Energy from wood resources and *carbon stock* were the most frequent among the 18 indicators mentioned to assess the mitigation of forest management measures. Particular attention is given to the forest carbon stock, highlighting that forest management can increase the carbon storage in forests (Colombo et al. 2012). In addition, forest management plays a crucial role in producing high-quality timber products, allowing to store carbon for years to decades, while promoting ecological sustainability through the production of wood energy (Paletto et al. 2017; Buonocore et al. 2019).

In conclusion, the literature review highlighted that *carbon stock* is in absolute terms the most mentioned SFM indicator, followed by *tree species composition* and *roundwood quantity*. Slightly less mentioned were *forest damage*, *growing stock*, and *deadwood*, while *energy from wood* was only mentioned for mitigation issues.

4.3   National Forest Inventories: Harmonization of Mitigation and Adaptation Indicators

NFIs are one of the main data sources for national-, continental-, and global-scale assessments of forest resources and their sustainability (McRoberts et al. 2012). Due to increasing information needs, the scope of NFIs has been broadening to include new variables (Tomppo et al. 2010; Vidal et al. 2016). However, the estimates produced by different countries frequently lack international comparability due to differences in applied definitions, sampling designs, and measurement protocols (Lawrence et al. 2010; Alberdi et al. 2016a). The European National Forest Inventory Network (ENFIN) carried out numerous research projects to develop tools for comparable results at the international level (Vidal et al. 2016). Comparability of European NFIs can be achieved by defining "reference definitions" and by establishing "bridging functions" as tools to calculate harmonized estimations from national inventories (Ståhl et al. 2012; Alberdi et al. 2016a). Thus, many indicators were subject to harmonization at the European scale, such as forest area (Vidal et al. 2008; Gabler et al. 2012), growing stock (Vidal et al. 2008; Tomter et al. 2012; Gschwantner et al. 2019), and others, like stem quality and increment, that were investigated to identify harmonization opportunities (Bosela et al. 2016).

From the ten most frequent indicators for adaptation and mitigation in Fig. 4.1, five are typically provided by the NFIs: carbon stock, tree species composition, forest damage, growing stock, and deadwood volume. A Europe-wide, harmonized database of tree species distribution was elaborated using NFI data (~375,000 sample plots) and additional information to create "The European Atlas of Tree Species: modelling, data and information on tree species" (de Rigo et al. 2016). However, it is important to mention that the effect of monitored area and plot design on the probability of discovery needs further research.

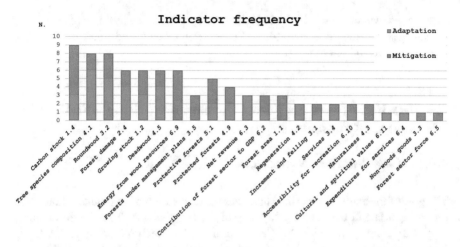

Fig. 4.1 Indicator frequency for adaptation (17 blue bars) and mitigation (18 orange bars) according to the overall frequency

Growing stock is one of the most analyzed variables for harmonization. Vidal et al. (2008) published a reference definition, establishing its components and thresholds. Differences in field measurements at the sample tree-level (e.g., minimum diameter at breast height) and tree compartments included in volume models (stump, stem top, or branches) cause the main differences. Tomter et al. (2012) presented case studies for six European countries and more recently, Gschwantner et al. (2019) published harmonized growing stock estimates at the European scale.

Deadwood volume is another key variable in terms of harmonization. Woodall et al. (2009), Chirici et al. (2011), and Rondeux et al. (2012) established a reference definition and performed case studies. However, further analysis on the performance of bridging functions and estimation methods is desirable.

The total carbon stock is composed of carbon in a) all living above-ground biomass, including stem, stump, branches, bark, seeds, and foliage; b) all biomass of living roots; c) all nonliving woody biomass not contained in the litter, either standing, lying on the ground, or in the soil; d) all nonliving biomass with a diameter less than the minimum diameter for deadwood; and e) organic carbon in mineral and organic soils. From these components, only the carbon of living above-ground biomass can be considered as harmonized (Avitabile and Camia 2018). To complement the picture on the potential of forests to sequester carbon, the assessment of wood quality and assortment structure is essential (Bosela et al. 2016). This allows to prioritize between climate policies for in situ carbon storage, for biomass harvests to generate energy and for harvesting quality construction timber (Obersteiner et al. 2010; Böttcher et al. 2012). Harmonization of the assessment of wood quality and timber assortments among NFIs of European countries is necessary for NFI data to be used for such prioritization at the European scale.

Finally, the indicator of forest areas with damages has not yet been harmonized across NFIs. Kovac et al. (2020) recognize it as a key indicator for the assessment of the conservation status. However, there are open questions, such as the severity of damages considered or the required proportion of damaged trees, to classify a plot as undamaged or damaged.

4.4 Methods To Assess Forest Development Using NFI-Data and CSF Indicators

4.4.1 Case Study 1: Switzerland

The goal of the analysis described here was the calculation of aggregated indices for the mitigation and the adaptive capacity of the five Swiss NFI production regions (Fig. 4.2). Forests in Switzerland are dominated by beech (*Fagus sylvatica* L.) at low elevations (<600 m a.s.l.), with the proportion of fir (*Abies alba* Mill.) and spruce (*Picea abies* L.) increasing toward the tree line, where stone pine (*Pinus cembra* L.), larch (*Larix decidua* Mill.), Scots pine (*Pinus sylvestris* L.), and mountain pine (*Pinus mugo* Turra.) dominate (Cioldi et al. 2010). Past management favored conifers predominantly in the Jura, the Plateau, and in the Prealps, but more recent changes in management have reduced the area covered by pure conifer forests by 8% since 1985 (Brändli and Röösli 2015).

Aggregated indices for adaptive capacity and mitigation were used to evaluate forest management or policies in regions, at the country level and under specific

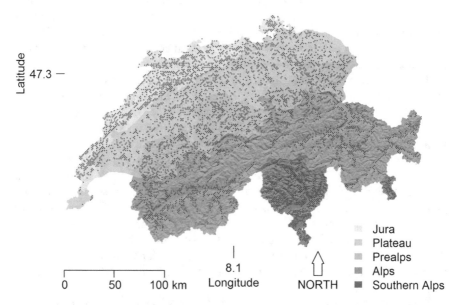

Fig. 4.2 Location of NFI sample plots (red dots) in Switzerland. The color shading shows the borders of the five NFI production regions of Switzerland

Fig. 4.3 Conceptual
diagram showing how
forest development or
forestry that leads to
positive changes (Δ) in
adaptive capacity
(adaptation) and mitigation
is considered climate smart
(i.e., scenario S1)

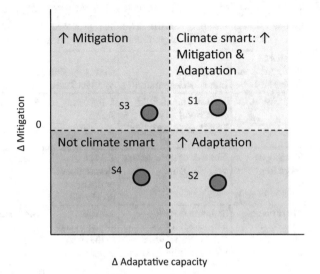

scenarios (S1–S4; Fig. 4.3). Positive changes in adaptive capacity and mitigation were considered as climate smart (S1). Adaptive capacity at the cost of mitigation (S2) vice versa (S3) and negative changes in both adaptive capacity and mitigation (S4) were not considered as climate smart. The following steps were necessary to arrive at the aggregate indicators for adaptive capacity (adaptation) and mitigation:

1. Selection of indicators: We used a set of indicators that was available from the Swiss NFI to quantify the provision of a range of ecosystem services (Blattert et al. 2017; Temperli et al. 2020) and that corresponded well to a subset of the Forest Europe SFM indicators (Table 4.1).

2. Indicator calculation and aggregation: Growing stock, density of large trees, diversity of tree diameters at breast height (DBH), tree species diversity, and deadwood volume were calculated at the level of sample plots and then averaged to the 5 NFI production regions. Sustainability was assessed as the difference between harvested wood and increment at the level of production regions. Avalanche and rockfall protection (API and RPI, respectively) were assessed with indicators combining stand- and site-related factors to estimate a ratio of the current stand parameters and those needed for optimal protection (Berger and Dorren 2007; Cordonnier et al. 2013). The proportions of sample plots with API and RPI >0.95 were used as estimates for avalanche and rockfall protection at the regional and national levels. The indicators were categorized to represent the provision of ecosystem services: the maintenance of resources, the production of timber, the provision of biodiversity, and the provision of protection against rockfall and avalanches. See Table 4.1 for the categorization and note that the indicators only partially correspond to the Forest Europe C&I for SFM.

Calculation of change in mitigation and adaptive capacity: The regional-/national-level indicator values were scaled between the minimum and maximum values

Table 4.1 Correspondence between indicators available from the Swiss NFI and the Forest Europe SFM indicators, together with the proposed categorization for ecosystem service representation and the weights representing their suitability as proxies for adaptation and mitigation

Available indicator from Swiss NFI	Indicator description	Corresponding SFM Indicator	Ecosystem service	Adaptation (weight)	Mitigation (weight)
Growing stock	Volume (m³ ha⁻¹) of stem wood >12 cm DBH within bark (incl. stump and top)	1.2 growing stock	Resources	0	1
Density of large trees	Number of trees per ha >80 cm DBH	1.3 age structure and/ or diameter distribution	Biodiversity	1	0
DBH diversity	Shannon diversity of basal area in 4 cm DBH classes (note: highly correlated with growing stock at regional scale)	1.3 age structure and/ or diameter distribution	Biodiversity	1	0
Sustainability	Difference between harvested wood and increment (m³ ha⁻¹ yr.⁻¹)	3.1 increment and felling	Production	1	0
Tree species diversity	Shannon diversity based on basal area share of species	4.1 diversity of tree species	Biodiversity	1	0
Deadwood	Volume (m³ ha⁻¹) of standing and lying deadwood (fine and coarse)	4.5 deadwood	Biodiversity	0.5	0.5
Avalanche protection	Proportion of sample plots within the SilvaProtect avalanche protection perimeter with avalanche protection index >0.95	5.1 protective forests – soil, water, and other ecosystem functions – infrastructure and managed natural resources	Protection	0.5	0.5
Rockfall protection	Proportion of sample plots within the SilvaProtect rockfall protection perimeter with rock fall protection index >0.95	5.1 protective forests – soil, water, and other ecosystem functions – infrastructure and managed natural resources	Protection	0.5	0.5

observed in different regions and time steps. Each indicator change was then calculated as the difference between the scaled regional indicator values at the beginning and at the end of the observed time period. The changes were then averaged across the indicators with weights that describe the indicators' contribution to mitigation and/or adaptation (Table 4.1). The different values of weights impact the contribution of each indicator to mitigation and adaptation as follows: with a value of 0, the indicator has no contribution to mitigation or adaptation; with a value of 1, the indicator contributes entirely to mitigation or adaptation; with a weight of 0.5, the indicator contributes equally to mitigation and adaptation. These weights were derived during several CLIMO workshops by a varied group of forestry and forest ecology experts.

We calculated changes in mitigation or adaptation *retrospectively* for the 5318 NFI sample plots that were visited during NFI2 (1993–95), NFI3 (2004–06), and NFI4 (2009–17) (Abegg et al. 2014; Traub et al. 2017) and *prospectively* using simulation results of the NFI-based scenario model MASSIMO for years 2016–2106 under 4 contrasting management scenarios (Stadelmann et al. 2019). These scenarios were developed during a previous project to assess potential timber yields in Swiss forests. They included 1) a business-as-usual (BAU) scenario that assumed a continued increase in growing stock in poorly accessible mountain forests and a decrease in growing stock in the Plateau region. 2) Under a so-called constant stock scenario, harvest is increased or decreased to maintain a constant growing stock as observed in NFI4 in all production regions. 3) The conifer scenario promoted conifers (mainly Norway spruce) in the regeneration to meet future increases in the demand for construction wood and assumed a reduction of growing stock to 300 m3 ha − 1 until 2046 and then an increase to 300–330 m3 ha − 1, depending on the region. 4) Under the energy scenario, timber production was maximized to meet the increasing demands for energy wood and wood-based chemicals. Target diameters were assumed to be of little importance under this scenario, such that growing stock was reduced until 2046 to 200–300 m^3 ha^{-1}, depending on the region, and thereafter growing stock was held constant (for details, see Stadelmann et al. 2016).

4.4.2   Case Study 2: Selected EU Countries

The proposed methodological approach to evaluate forest development with regard to Climate-Smart Forestry indicators at the European level included two main steps. 1) Assign a weight for each pan-European SFM indicator based on a literature review. 2) Display the trend over time of aggregated indicators describing how forestry may mitigate climate change and how the capacity of forests to adapt to climate change develops across European countries.

Step 1): Following the literature review, a subset of indicators was selected (see Sect. 4.2) from the current pan-European set of C&I (Table 4.1). The subset included the indicators that were mentioned at least 3 times (6 was the maximum value obtained from the literature review). The weights were assigned through a pairwise comparison as per the Analytic Hierarchy Process (AHP, Saaty 1980). The AHP is frequently used in environmental and forest sectors as a decision support tool (Kuusipalo and Kangas 1994; Ananda and Herath 2003; Wolfslehner et al. 2005; Santopuoli et al. 2016).

To implement the pairwise comparison, first, the relative priority of the indicators (RP) was assessed as follows:

$$RP = \left(\frac{Cit_{np}}{Cit_{pmax}} \right) \left(\frac{1}{Cit_1} + \frac{1}{Cit_2} + \cdots + \frac{1}{Cit_n} \right)$$

where

RP is the relative priority

Cit_{np} is the number of publications that mention the focal indicator

Cit_{pmax} is the maximum number of times that one of the 12 indicators was mentioned (i.e., 6 for *tree species composition* and 6 for *carbon stock* for adaptation and mitigation, respectively),

Cit_1 is the total number of indicators mentioned by the same author for the first time

Cit_2 is the total number of indicators mentioned by the same author for the second time

Cit_n is the total number of indicators mentioned by the same author for the n time

For example, the indicator 1.4 *carbon stock* was mentioned in three papers ($Cit_{np} = 3$) among those used for the adaptation review. The total number of indicators mentioned by the first paper was seven ($Cit_1 = 7$), while the second and third papers mentioned a total of six ($Cit_2 = 6$) and four ($Cit_3 = 4$) indicators, respectively. Considering the Cit_{pmax} of six, the RP for carbon stock was 0.280.

Subsequently, the RPs were used to create the reciprocal matrix (Saaty 1980; Kangas et al. 1993; Mendoza and Prabhu 2000) for adaptation and mitigation separately to obtain the Eigenvector for each indicator. The overall priority was calculated for each indicator, considering the ratio of the number of articles, 6 and 12 for adaptation and mitigation respectively, and the total number of the articles (18) multiplied by the Eigenvectors (i.e., overall priority = $0.33*Eigen_{Adaptation} + 0.67* Eigen_{Mitigation}$). The indicators *energy from wood resources, carbon stock,* and *tree species composition* yielded the highest overall priority (ranks 1–3), while *forest area, net revenue,* and *deadwood* the lowest (ranks 10–12), reflecting their frequency of association with adaptation and mitigation in the literature (Table 4.2).

Step 2): The calculation of the aggregate indices for adaptation and mitigation was based on the data reported in the State of Europe's Forests (SoEF) database. First of all, all available data were downloaded for each indicator for the years 1990, 2000, 2005, 2010, and 2015. Subsequently, four pairwise comparisons (i.e., 2000 vs. 1990; 2005 vs. 2000; 2010 vs. 2005; 2015 vs. 2010) were made for each indicator, assessing the relative trend (i.e., percentage of changes) at country level.

Table 4.2 Indicator weights for the subset of indicators used in the evaluation of adaptation and mitigation across Europe. The reported values represent the Eigenvectors obtained through the pairwise comparison (Saaty 1980) for adaptation (Eigen$_{Adaptation}$) and mitigation (Eigen$_{Mitigation}$) and the overall priority. The rank reflects the overall priority values

Indicator	Indicator name	Eigen$_{Adaptation}$	Eigen$_{Mitigation}$	Overall priority	Rank
1.1	Forest area	–	0.049	0.033	10
1.2	Growing stock	0.040	0.034	0.036	7
1.4	Carbon stock	0.059	0.240	0.180	2
2.4	Forest damage	0.097	–	0.032	11
3.2	Roundwood	0.202	0.025	0.084	4
3.5	Forests under management plans		0.069	0.046	6
4.1	Tree species composition	0.463	–	0.154	3
4.2	Regeneration		0.050	0.033	8
4.5	Deadwood	0.040	–	0.013	12
5.1	Protective forests		0.120	0.080	5
6.3	Net revenue	0.099	–	0.033	9
6.9	Energy from wood resources	–	0.414	0.276	1

Three out of 46 pan-European countries (Holy See, Monaco and Russian Federation) have been excluded, because no data were available for some of the indicators and years.

To arrive at a measure for adaptation capacity and mitigation, the relative changes observed for each indicator were multiplied with the overall priority value (Table 4.2) for adaptation and mitigation separately. The obtained values were then summed for each country and for each observed period and displayed in a scatter plot, within which adaptation was on the x-axes and mitigation on the y-axes. A European-level estimate was calculated as the average value for adaptation and mitigation.

4.5 Results of the Swiss Case Study

The time period between NFI2 and NFI4 was characterized by increasing growing stock throughout Switzerland, except for the Plateau region, where it decreased (Fig. 4.4). This is also reflected by the sustainability indicator that decreased the most in the Plateau, with the dip in 2006 (NFI3) being due to the storm Lothar in 1999. Deadwood and number of stems >80 cm DBH increased, whereas species diversity remained mostly unchanged. DBH diversity decreased in the Plateau and the Prealps, while it increased in the other regions. Protection against avalanches and rockfall generally increased in the Alps and the Southern Alps.

The aggregated indicators for change in adaptation and mitigation (Fig. 4.5) reflected the increasing growing stock in the Alps and the Southern Alps, where adaptation and mitigation increased in total and for most criteria. In contrast, the reduction in the production-related indicator (sustainability) in the Jura, the Plateau, and the Prealps resulted in an overall negative CSF evaluation, which was accentuated in the Plateau by the decrease in growing stock.

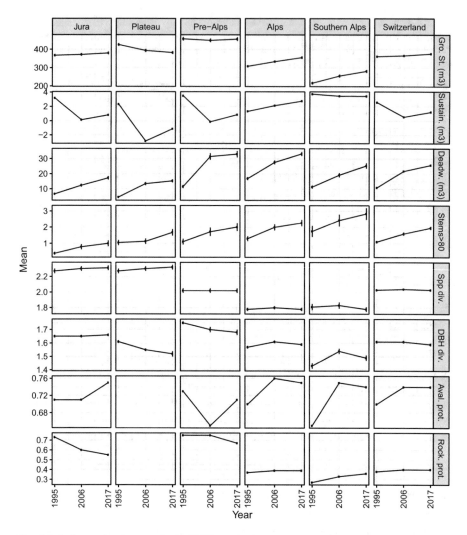

Fig. 4.4 Indicator development in the 5 NFI production regions and for the whole of Switzerland (columns) from NFI2 (1993–1995) to NFI3 (2004–2006) to NFI4 (2009–2017). Error bars show standard errors of the mean across sample plots. Note that avalanche and rockfall protection was only calculated for sample plots within the protection forest perimeter, which does not overlap with the Plateau region

The development of indicators under the four management scenarios in Switzerland reflected the scenario specifications (Fig. 4.6). Growing stock along with deadwood and the number of stems with DBH >80 cm continued to increase under the BAU scenario in most parts of Switzerland, except for the Plateau (not shown). The increased harvesting under the conifer and the energy scenarios until 2046 resulted in the sustainability indicator to drop sharply at the beginning of the simulations and to recover after 2046, when harvesting was reduced again.

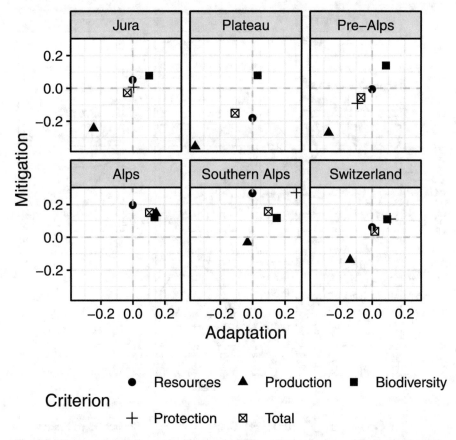

Fig. 4.5 Change in adaptation and mitigation by NFI production regions of Switzerland and the whole country, SFM criteria, and average of all indicators (Total). Note that the scale is relative to the range of regional indicator values. Positive changes in adaptation and mitigation (top-right quadrant) are considered as "climate-smart"

DBH diversity decreased under all scenarios, but predominantly under the conifer and energy scenarios, in which increased timber harvesting resulted in more homogeneous forest structures. Species diversity dropped similarly under all scenarios, probably due to a modelling artifact: the number of species in the regeneration data (<12 cm DBH) was lower than in the record of trees (>12 cm DBH) used to initialize simulations. Avalanche protection decreased and rockfall protection increased due to a simulated increase in average DBH, to which avalanche protection is negatively and rockfall protection positively related (Cordonnier et al. 2013).

BAU management was the most climate-smart option, considering the whole of Switzerland after 90 years (2106) of simulated forest development (Fig. 4.7). BAU benefited all SFM criteria and thus led to a positive change in both adaptation and mitigation. Increased harvesting under the three other scenarios reduced the

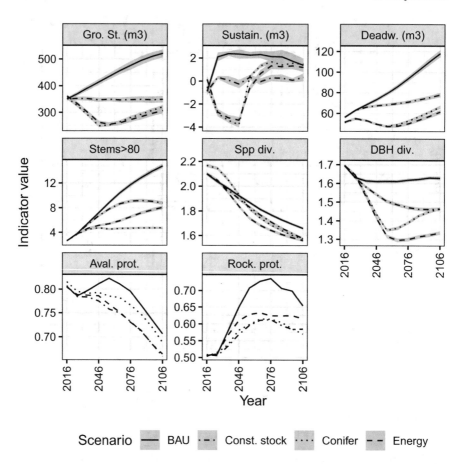

Fig. 4.6 Indicator development for Switzerland under four forest management scenarios. Shaded bands show standard errors of the mean across sample plots and 20 replications of model simulations

resources, biodiversity, and protection criteria and thus decreased only adaptation (constant stock) or both adaptation and mitigation (Conifer and Energy). The sustainability indicator recovered by 2106, such that the production criterion benefitted both adaptation and mitigation under all scenarios.

4.6   Results of the European Case Study

Overall, a positive trend was observed in the period 1990 and 2000 for mitigation (Fig. 4.8). Changes in the adaptive capacity of forests were mixed, showing a slightly positive trend with more countries on the right side of the graph than on the left side. Only few countries show a clear positive trend for both adaptation and mitigation. For example, Belarus (BY), for which an increase in the proportion of

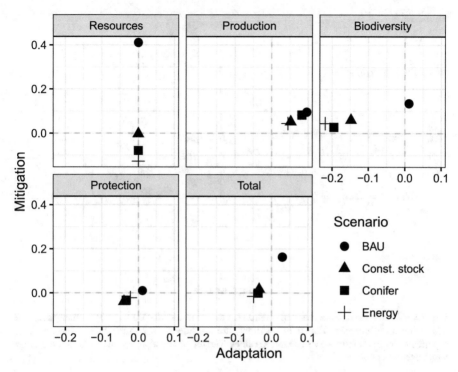

Fig. 4.7 Change in adaptation capacity and mitigation for the whole of Switzerland and for years 2016–2106 by simulated management scenario, SFM criteria, and averaged overall indicators (Total)

forests covered by a management plan was observed in 2000, and Iceland (IS), where an increased afforestation rate was observed. Moreover, IS experienced increases in both growing stock and area of mixed forest, which are the most important indicators for adaptation.

The negative trend observed for adaptation depends mostly on the increase in forest damages between 1990 and 2000, especially for countries such as Serbia (RS) and Liechtenstein (LI), which showed the lowest negative values. The main causes for the high damages in Serbia were insects and diseases and in LI, it was wildlife browsing. Negative results were observed also for the Netherlands (NL) due to the increased forest damage caused by fire. The average trend over all countries (EU, which includes also countries outside the European union) was negative with respect to changes in adaptation (−0.11) due to the strongly negative values in RS, LI, and NL. Without the negative outliers, the average values of EU are positive with 0.03 for adaptation and 0.08 for mitigation.

In the period 2000–2005, results showed a positive trend in mitigation, with many countries moving to the right side of the scatter plot, and few countries that showed a negative mitigation trend (Fig. 4.9). Concerning adaptation, the negative trend mainly depended on the increase in forest damages. Nevertheless, this increase was strongly affected by methodological reasons rather than by an actual increase in

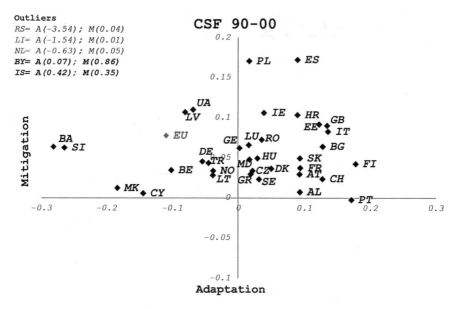

Fig. 4.8 Change in adaptation and mitigation for the period 1990–2000. Positive changes in both adaptation and mitigation (top-right quadrant) are considered as climate smart. The countries reported as outliers are those with extreme values. The red diamond, reported as EU, is the average value over all countries

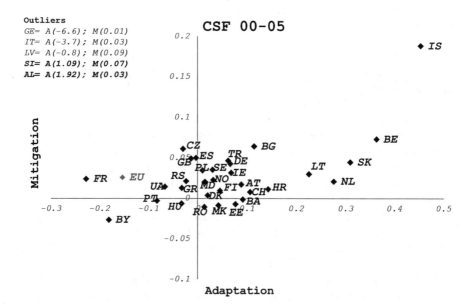

Fig. 4.9 Change in adaptation and mitigation for the period 2000–2005. See caption of Fig. 4.8 for details

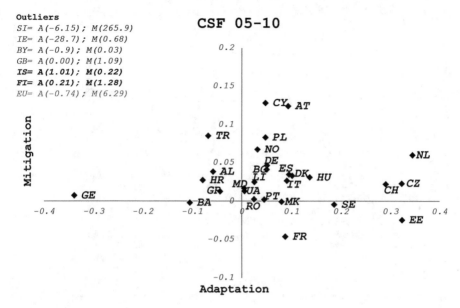

Fig. 4.10 Change in adaptation and mitigation for the period 2005–2010. See caption of Fig. 4.8 for details

forest damages. For example, in Georgia (GE), since 2005, the national report has also included abiotic and biotic causes for assessing forest damages in addition to forest fire. Similarly, in Italy (IT), other causes are reported only for the year 2005, which strongly affected the trend. Differences in data collection and reporting affected the evaluation at both, country and European levels.

Even though most countries showed very small changes, positive trends were observed. For example, the increased area of mixed forests and the increased net revenue are the main indicators that highlight the positive adaptation trend for Slovenia (SI). In Albania (AL), such an indicator is the deadwood amount, which increased strongly in the 2000–2005 period, while increases of timber production and growing stock were observed in IS.

In the period 2005–2010, most countries showed a positive trend for both mitigation and adaptation (Fig. 4.10), with few cases for which a negative trend was rather strong, particularly for adaptation. The main causes can be found in the reduced cover of mixed forests with respect to pure forests in SI and BY. In 2005, there were 1,182,000 ha of mixed forests (2 or more tree species) in SI and 61,000 ha of pure forests. In 2010, mixed forests covered 1,065,500 ha, while pure forests suddenly covered 181,500 ha. We are unable to explain this large increase in pure forest cover, which resulted in the ratio of mixed to pure forest to drop from 19.37 to 5.87 between 2005 and 2010, thus contributing to the low adaptation value. The cause for the low adaptation values in Ireland (IE) was an increase in forest damages between 2005 and 2010. Positive trends mainly reflected increases in net revenue and timber production as for IS. Concerning mitigation, wood for energy was the most important indicator and increased in Finland (FI) and SI, where also an increase of protection role by forest was registered.

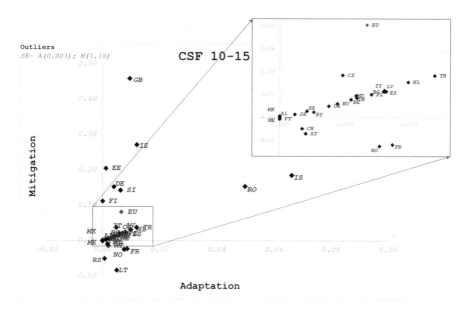

Fig. 4.11 Change in adaptation and mitigation for the period 2010–2015. See caption of Fig. 4.8 for details

For the period 2010 and 2015, results show a positive trend, with few countries showing a negative trend for mitigation (Fig. 4.11). Nevertheless, there is one critical point that hinders a correct evaluation, which is the lack of data for 2015 in most countries and for many indicators. Most of the indicators are provided by the NFIs, and the reporting frequency of the NFIs is often not the same as for the SoEF report (Marchetti et al. 2018). For this reason, not all indicators were available at the time of the SoEF publication.

The available indicators include *forest area, growing stock, carbon stock,* and *energy from wood.* Changes in these indicators show positive trends, especially for SK, which reported a considerable increase in wood for energy supply (+297%) between 2010 and 2015.

4.7   Critical Evaluation of Indicators and Potential for Improvement

This chapter described a subset of SFM indicators that can be used for assessing the potential for adaptation and mitigation. However, the catalog of variables monitored in NFIs, their field protocols, and definitions vary between countries (Tomppo et al. 2010). Therefore, some counties experienced difficulties to report data according to terms and definitions established by Forest Europe, for example, regarding forest available for wood supply (Alberdi et al. 2016b). These issues have at least partly led to gaps in the available Forest Europe SFM indicator data for some countries and years. To avoid the dependence on missing data, the method applied here returns the results based on the cumulative effect of available indicators multiplied by their relative value. This requires at least one indicator to be available for at least two subsequent inventory periods. The missing data were considered as null. Missing data can be compared with no change values between two subsequent inventory periods, because in both cases, the value of change is zero and thus does not add to the cumulative value. Even though a "missing data" entry is different from a "no-change" value, in practice, it results in no development in both cases. The lack of data can be considered one of the most hindering factors for developing and implementing CSF strategies (Santopuoli et al. 2020). Thus, monitoring, assessing, and accurate reporting are crucial for supporting CSF policy making, and researchers should facilitate methods for data collection and analysis.

To improve data availability, the timber assortment structure of the European forests as well as the use of harvested timber in the managed forests is crucial to gain a holistic picture on the potential of forests in carbon management (i.e., mitigation). Such information is particularly important to assess the carbon sequestration in wood products and the effect of wood products and wood-based energy to substitute fossil energy sources (Obersteiner et al. 2010; Braun et al. 2016; Köhl et al. 2020). However, only a few European countries assess wood quality and even less countries quantify timber assortments in their NFIs (Bosela et al. 2016). For example, in Switzerland (first case study in this chapter), the consideration of carbon sequestration in wood products and substitution of fossil fuel–intensive energy may have changed the CSF assessment for the Plateau region, where overharvesting (negative change in sustainability indicator) resulted in the negative overall CSF assessment due to decreasing growing stock (Blattert et al. 2020). Currently, many European NFIs are not yet capable of reporting on timber assortments in a harmonized way and further developments toward more integrated and harmonized NFI-based indicators on carbon sequestration and substitution are necessary at the European level.

Forests provide multiple ES to the society. However, the NFI-based indicators mainly cover provisioning and regulating services and indicators on sociocultural ES categories are lacking in many NFIs, but see, for example, NFIs of Denmark (DK) and United Kingdom (UK, Edwards et al. 2011). One of the most important ES from the last category is tourism and outdoor recreation in forests (Sievänen et al. 2013), with recreational tourism evolving from a niche market to a mainstream element of global tourism with annual growth rates of 10–30 percent (Bell et al. 2009; FAO 2010). In the last decades, several studies have provided insights on stand attributes that define forest recreational attractiveness (Giergiczny et al. 2014). Most of these studies found that visitors in Europe prefer mixed over monospecific forests (Gundersen and Frivold 2008; Paletto et al. 2013; Giergiczny et al. 2015; Grilli et al. 2016; Hegetschweiler et al. 2017; Pelyukh et al. 2019) and uneven-aged forests with at least a few large trees (Ribe 2009; Filyushkina et al. 2017). A high depth of view, that is, high visual penetration, was also preferred (Heyman 2012). The perception of deadwood varies and likely needs to be differentiated. Visitors prefer root plates and highly degraded deadwood pieces over fresh harvesting residues (Rathmann et al. 2020). The presence of deadwood is more appreciated when visitors are informed about the ecological function of deadwood (Pastorella et al. 2016a; Gundersen et al. 2017).

The marginal rate of substitution (MRS) reflects one's willingness to pay for visiting the forest concerned (Bańkowski 2019). The MRS index could be expressed by the distance that somebody is willing to travel in order to reach the desired forest. Higher values of the MRS index correspond to a higher recreational attractiveness of the forest. This index has been developed for all forests in Poland by Bańkowski (2019), who described forest recreational attractiveness using attributes, such as forest age, age diversification (even- or uneven-aged), forest type (monospecific and mixed: two-specific, multispecific), diversification of the tree-species composition, abundance of advance growth and/or undergrowth, height of the ground vegetation cover, silvicultural system applied, and touristic infrastructure and facilities. The mentioned indicators for these stand attributes determining visual attractiveness can be derived from data collected in most NFIs. They can thus be used to complement the indicators for adaptation and mitigation, allowing for balanced assessments that also consider sociocultural ecosystem services.

4.8 Inventory-Based Assessments of CSF in a Broader Context

The case studies presented in this chapter provide a basis for future assessment of climate smartness of European forests and its change over time. They present a possible way to use forestry statistics, either those reported in SoEF or other aggregated data provided by NFIs, to assess the adaptation and mitigation potential of European forests. Working with the data of a single country, such as Switzerland in our first case study, has the advantage of the data being consistent and harmonized. This allowed comparisons of indicator developments among regions and, together with the MASSIMO forest development model, potential forest management scenarios could be evaluated in terms of adaptation and mitigation. This evaluation of management scenarios showed that BAU management that continues harvests below the mean annual increment in Swiss mountain regions performed best in terms of adaptation and mitigation. Hence, this illustrates a potential trade-off between timber harvesting and CSF.

Based on the statistics reported in SoEF, the results of our second case study suggest that the forests in most European countries positively contribute to both adaptation and mitigation, with a few exceptions. However, there are changes in the potential over time, which might be partly attributed to a change in the accuracy of provided data over time, as well as due to a change in the methodological approach. For example, a country may have started to use NFI data to report to SoEF instead of less statistically sound data from other sources during the time period of 1990–2015. Another factor that strongly affects CSF indicator development across Europe is the occurrence of large-scale disturbances, including windstorms followed by bark beetle outbreaks (Seidl et al. 2017), which certainly was decisive for the CSF assessment of some countries. Finally, the selection of indicators, which was limited to those that are reported by the European countries to SoEF, may have affected the overall assessment of the adaptation and mitigation potential.

In the following, we discuss the broader context, in which large-scale assessments of CSF need to be interpreted. As a recent study by Naudts et al. (2016) suggested, wood harvesting and change in species composition in European forests had led to a carbon debt of 3.1 petagrams. Although there has been considerable afforestation since 1750, many originally broadleaved forests were replaced with coniferous species. From a more recent perspective, there has been considerable land cover changes in Europe since the 1990s (Huang et al. 2020). Abandonment of agricultural land and transition to forest was suggested to have a widespread cooling effect in western and central Europe, whereas in eastern Europe, this same land use transition had a warming effect. The opposite climate system response between the western and eastern parts of Europe is explained by the stronger contribution of a reduced surface albedo after reforestation in eastern Europe and a lower evaporative cooling of soils on eastern Europe (Huang et al. 2020). In general, afforestation and reforestation are considered to be the main processes driving forest-related carbon sequestration. Increasing the productivity of forests and thus increasing growing stock may also contribute strongly to carbon sequestration and thus climate change mitigation (Pretzsch et al. 2014). However, this entails a presumably higher disturbance risk (McDowell et al. 2020) and recently, there have been signs that the carbon sink in European forests saturated (Nabuurs et al. 2013). Hence, focusing on adaptation rather than mitigation may be of higher priority, where growing stocks are already high.

There is a strong debate whether a no-management approach is better for carbon sequestration than its managed counterparts (Van Deusen 2010; Griscom et al. 2017; Luyssaert et al. 2018; Baldocchi and Penuelas 2019; Grassi et al. 2019). Luyssaeret et al. (2018) found that if the current forest cover in Europe is sustained, the additional climate benefits from forest management are only modest and relevant at local level, but not at the global level. The authors of the study further suggest that Europe should not rely on forest management to mitigate climate change and should support the recommendation of focusing on forest adaptation. However, Grassi et al. (2019) put this into perspective by highlighting that our knowledge on management effects on CO_2 sequestration is still poor and that premature recommendations should be avoided. Further, Baldocchi and Peñuelas (2019) suggested that the rates and amount of net carbon uptake are low compared to the CO_2 released by fossil fuels combustion. They also point out that management of forests focused on carbon sequestration can cause unintended consequences and should be considered with caution. In response to that, Griscom et al. (2019) argued that ecosystem-based options have so far been underinvestigated and that there are some positive examples of policies that successfully integrated both fossil fuel emission reduction and natural climate solutions. Yousefpour et al.

(2018) suggested that an economically efficient CSF policy would prioritize carbon sequestration (with no-management option being dominant) in northern, eastern, and central European countries, where lower wood prices, high labor, and harvesting costs or a mixture thereof prevail. In contrast, forests in the west should be harvested to provide wood for the substitution of fossil fuels and carbon-intensive materials.

The fifth report by IPCC (IPCC 2014) suggested that forest management strategies should 1) lead to a sustained yield of timber, fiber, and energy, 2) maintain or even increase carbon stock in forests (including in the soil), and 3) preserve forest biodiversity. As pointed out by Hisano et al. (2018), biodiversity may aid in mitigating climate change impacts on biodiversity itself, because more diverse forests can be more resilient and thus can potentially better adapt to a changing climate. More diverse ecosystems can likely better maintain ecosystem functioning through enhanced facilitation effects (Hisano et al. 2018; Jactel et al. 2018; Ammer 2019). However, a recent study by Sabatini et al. (2019) showed a highly variable relationship between species richness and carbon stock at the stand scale in European temperate forests. They further suggested that maximizing cobenefits between carbon and biodiversity may require stand scale approaches to reach positive effects at the landscape scale.

These studies show that there are still important knowledge gaps on the effectiveness of adaptation and mitigations management and their consequences for ecosystem service provision. Forest inventory–based monitoring and scenario assessments, such as those that have been exemplified in this chapter, have a strong potential to fill some of these knowledge gaps. In particular, biodiversity-related indicators on regeneration and naturalness may gain importance. However, this requires further harmonization across countries and continents. Furthermore, ground-based assessments should be complemented with other data sources (e.g., remote sensing and eddy covariance flux towers) and projections from dynamic ecosystem models to capture factors, such as albedo, biodiversity-productivity relationships, disturbance vulnerability, and economic realities.

4.9 Conclusions and Outlook

To assess the vulnerability of forests to both gradual and sudden switches in environmental conditions, and to evaluate the adaptive capacity of forests and forestry to climate change, integration of multiscale forest ecological studies with repeated NFIs is required. NFIs are relevant as they are the primary data source for reporting on forest resilience and vulnerability to European institutions. To sustain forest-adaptive strategies and provide climate-smart indicators (Bowditch et al. 2020), integration of NFIs, experimental forest management plots, and other networks, such as ICP forest (e.g., Trotsiuk et al. 2020), are especially important due to the time elapsed between management actions and stand responses. Combination with airborne forest observation may provide additional information on stand structure, such that reliable biomass estimation and biodiversity monitoring at the stand-to-landscape scales are possible (see Chap. 11 of this book: Torresan et al. 2021).

Climate-smart approaches in forestry are dynamically connected with the sustainable delivery of forest products and ecosystem services (Nabuurs et al. 2015, 2017; Yousefpour et al. 2018). Mitigation measures undertaken in forests (reducing emissions from deforestation and forest degradation and enhancing forest carbon sinks and product substitution) will be effective in the long term (Kauppi et al. 2018); therefore, a European forest–related policy should be dynamically extended as well (Verkerk et al. 2020), to address climate targets in the forestry sector (Nabuurs et al. 2018). Adaptation measures in forests will be required to secure the continued delivery of forest ecosystem services. NFIs are, therefore, key tools for implementing, monitoring, evaluating, and revising CSF management practices at the landscape level, as well as policies at the country and European level.

In this context, spatial forestry planning may opt for a concentration of high-yield wood extraction in productive forest areas, and an expansion of forest land spared for other services and conservation (sensu Ceddia et al. 2014). A simultaneous implementation of intensification and conservation, however, needs to be accompanied by strategies that carefully regulate land use. This can be supported by flexible incentives and support policies at the regional scale and by strengthening co-management regimes in local communities. Straightforward identification of adaptation and mitigation options, land suitability ratings, and cost-benefit analysis that all can be supported with NFI data are necessary for risk assessments.

A mosaic of forest landscapes with mixed management strategies, combining productive forest units (e.g., short-rotation coppices) with high-nature-value forest areas (e.g., old-growth forests), can be envisaged for some areas in Europe (e.g., Schall et al. 2020), while in others, production and conservation goals can be integrated at relatively small forest areas (Kraus and Krumm 2013). Forest inventory data can assist in delineating these areas. The spatial and temporal distribution of wood harvesting should consider the patterns of environmental disturbances and extreme events that historically occurred within a specific region. Smartness indicators can be used to determine land-use intensity, underlying utilization benefits versus conservation options (e.g., separation of nature and silviculture vs. integration of ecosystem services and natural capital; abrupt changes in management zones vs. spatial continuity of mixed systems; intensification for equilibrium vs. maintenance of resilience). Harmonized statistics and agreed indicators are also necessary for applying decision-making processes and reporting climate-related measures linked with EU directives and regulations.

The scope of NFIs has broadened to satisfy increasing information requirements (Tomppo et al. 2010; Alberdi et al. 2017), collecting among other things (e.g., biodiversity, disturbance impacts, or nonwood forest products), information on carbon stocks, emissions and removals, and on anthropogenic forest-related emissions and land use activities (Tomppo et al. 2010). While there is a potential for forest mitigation in Europe by increasing carbon sequestration and through substitution effects (Grassi et al. 2019), their effect on the climate system is limited due to beginning of carbon sink saturation (Nabuurs et al. 2013) and shifts towards more frequent disturbances and younger forests (McDowell et al. 2020). Silvicultural treatments to adapt forest composition and structure to expected new climatic conditions and disturbance regimes should be prioritized over mitigation measures, especially for highly stocked forests with low carbon sink capacity and where vulnerabilities to disturbances are high (Luyssaert et al. 2018). Forest inventory data may assist in identifying and reconciling potential trade-offs among adaptation, mitigation, and management objectives to provide ecosystem services (Gutsch et al. 2018; Temperli et al. 2020).

Although production, adaptation, and mitigation practices tend to be approached separately due to a variety of technical, political, and socioeconomic constraints, the forest infrastructure may allow more holistic management, if designed, inventoried, and managed appropriately. The necessity to monitor both forest productivity, forest health and other ecosystem services, such as increasingly demanded recreational and tourism related services, suggests that future forest inventories should focus on testing and implementing indicators related to agreed criteria for Climate-Smart Forestry, balancing the different components. Information provided by NFIs will be critical for developing indicators to identify trade-offs between policies that may increase the vulnerability to disturbances (e.g., carbon sequestration in old-growth forests), strategies to increase forest resilience, timber production, recreation, biodiversity conservation, and other ecosystem services. In this context, NFI data are critical for monitoring the impacts of climate change and the effectiveness of Climate-Smart Forestry measures. Even though NFIs have originally been designed at the national or subnational levels, the emerging synergies of Climate-Smart Forestry at the European scale call for the synchronization of forest inventories and reporting schemes. Coordinated multipurpose forest inventories are needed for monitoring valuable multifunctional forest ecosystems.

References

Abegg M, Brändli U-B, Cioldi F (2014) Fourth national forest inventory – result tables and maps on the internet for the NFI 2009-2013 (NFI4b). www.lfi.ch (28.10.2015)

Alberdi I, Gschwantner T, Bosela M et al (2016a) Harmonisation of data and information on the potential supply of wood resources. In: Vidal C, Alberdi IA, Hernández Mateo L, Redmond JJ (eds) National Forest Inventories: assessment of wood availability and use. Springer International Publishing, Cham, pp 55–79

Alberdi I, Michalak R, Fischer C et al (2016b) Towards harmonized assessment of European forest availability for wood supply in Europe. For Policy Econ 70:20–29. https://doi.org/10.1016/j.forpol.2016.05.014

Alberdi I, Vallejo R, Álvarez-González JG et al (2017) The multi-objective Spanish National Forest Inventory. For Syst 26:e04S. https://doi.org/10.5424/fs/2017262-10577

Ammer C (2019) Diversity and forest productivity in a changing climate. New Phytol 221:50–66. https://doi.org/10.1111/nph.15263

Ananda J, Herath G (2003) The use of analytic hierarchy process to incorporate stakeholder preferences into regional forest planning. For Policy Econ 5:13–26. https://doi.org/10.1016/S1389-9341(02)00043-6

Avitabile V, Camia A (2018) An assessment of forest biomass maps in Europe using harmonized national statistics and inventory plots. For Ecol Manage 409:489–498. https://doi.org/10.1016/j.foreco.2017.11.047

Baldocchi D, Penuelas J (2019) The physics and ecology of mining carbon dioxide from the atmosphere by ecosystems. Glob Chang Biol 25:1191–1197. https://doi.org/10.1111/gcb.14559

Bańkowski J, Sroga R, Basa K et al (2019) Koncepcja zagospodarowania turystycznego dla leśnego kompleksu promocyjnego "lasy doliny baryczy" -przykładowy operat turystyczny. In: Czerniak A (ed) Turystyka i rekreacja w lasach Państwowego Gospodarstwa Leśnego Lasy Państwowe na przykładzie Dolnego Śląska. Bogucki Wydawnictwo Naukowe, Poznań, Poland, pp 85–164

Baycheva-Merger T, Wolfslehner B (2016) Evaluating the implementation of the Pan-European criteria and indicators for sustainable forest management – a SWOT analysis. Ecol Indic 60:1192–1199. https://doi.org/10.1016/j.ecolind.2015.09.009

Bell S, Simpson M, Tyrväinen L et al (2009) European forest recreation and tourism : a handbook. Taylor & Francis

Berger F, Dorren LKA (2007) Principles of the tool Rockfor.net for quantifying the rockfall hazard below a protection forest. Schweiz Z Für Forstwes 158:157–165. https://doi.org/10.3188/szf.2007.0157

Blattert C, Lemm R, Thees O et al (2017) Management of ecosystem services in mountain forests: review of indicators and value functions for model based multi-criteria decision analysis. Ecol Indic 79:391–409. https://doi.org/10.1016/j.ecolind.2017.04.025

Blattert C, Lemm R, Thürig E et al (2020) Long-term impacts of increased timber harvests on ecosystem services and biodiversity: a scenario study based on national forest inventory data. Ecosyst Serv 45:101150. https://doi.org/10.1016/j.ecoser.2020.101150

Bosela M, Redmond J, Kučera M et al (2016) Stem quality assessment in European National Forest Inventories: an opportunity for harmonised reporting? Ann For Sci 73:635–648. https://doi.org/10.1007/s13595-015-0503-8

Böttcher H, Verkerk PJ, Gusti M et al (2012) Projection of the future EU forest CO2 sink as affected by recent bioenergy policies using two advanced forest management models. GCB Bioenergy 4:773–783. https://doi.org/10.1111/j.1757-1707.2011.01152.x

Bowditch E, Santopuoli G, Binder F et al (2020) What is climate-smart forestry? A definition from a multinational collaborative process focused on mountain regions of Europe. Ecosyst Serv 43:101113. https://doi.org/10.1016/j.ecoser.2020.101113

Brändli U-B, Röösli B (2015) Resources. In: Forest Report 2015: Conditions and use of Swiss forests. Swiss Federal Office for the Environment FOEN, Bern, and Swiss Federal Institute for Forest. Snow and Landscape Research WSL, Birmensdorf, pp 29–42

Brang P, Schönenberger W, Frehner M et al (2006) Management of protection forests in the European Alps: an overview. For Snow Landsc Res 80:23–44

Braun M, Fritz D, Weiss P et al (2016) A holistic assessment of greenhouse gas dynamics from forests to the effects of wood products use in Austria. Carbon Manag 7:271–283. https://doi.org/10.1080/17583004.2016.1230990

Buonocore E, Paletto A, Russo GF, Franzese PP (2019) Indicators of environmental performance to assess wood-based bioenergy production: a case study in northern Italy. J Clean Prod 221:242–248. https://doi.org/10.1016/j.jclepro.2019.02.272

Ceddia MG, Bardsley NO, Gomez-y-Paloma S, Sedlacek S (2014) Governance, agricultural intensification, and land sparing in tropical South America. Proc Natl Acad Sci 111:7242–7247. https://doi.org/10.1073/pnas.1317967111

Chirici G, McRoberts RE, Winter S et al (2011) Harmonization tests. In: Chirici G, Winter S, McRoberts RE (eds) National Forest Inventories: contributions to Forest biodiversity assessments. Springer Netherlands, Dordrecht, pp 121–190

Cioldi F, Baltensweiler A, Brändli U-B et al (2010) Waldressourcen. In: Brändli U-B (ed) Schweizerisches Landesforstinventar. Ergebnisse der dritten Erhebung 2004–2006.

Eidgenössische Forschungsanstalt für Wald, Schnee und Landschaft WSL. Bundesamt für Umwelt, BAFU, Birmensdorf, Bern, pp 31–114

Colombo SJ, Chen J, Ter-Mikaelian MT et al (2012) Forest protection and forest harvest as strategies for ecological sustainability and climate change mitigation. For Ecol Manage 281:140–151. https://doi.org/10.1016/j.foreco.2012.06.016

Cordonnier T, Berger F, Elkin CM, et al (2013) ARANGE deliverable D2.2: models and linker functions (indicators) for ecosystem services. ARANGE - Grant no. 289437- advanced multifunctional forest management in European mountain ranges

de Rigo D, Caudullo G, Houston Durrant T, San-Miguel-Ayanz J (2016) The European Atlas of Forest Tree Species: modelling, data and information on forest tree species. In: San-Miguel-Ayanz J, de Rigo D, Houston Durrant T, Mauri A (eds) European Atlas of Forest Tree Species. Publ. Off. EU, Luxembourg, p e01aa69+

del Río M, Pretzsch H, Bončina A, et al (2021) Assessment of indicators for climate smart management in mountain forests. In: Managing Forest Ecosystems, Vol. 40, Tognetti R, Smith M, Panzacchi P (Eds): Climate-Smart Forestry in Mountain Regions. Springer Nature, Switzerland, AG

Diaconu D, Kahle H-P, Spiecker H (2017) Thinning increases drought tolerance of European beech: a case study on two forested slopes on opposite sides of a valley. Eur J For Res 136:319–328. https://doi.org/10.1007/s10342-017-1033-8

Edwards D, Jensen FS, Marzano M et al (2011) A theoretical framework to assess the impacts of forest management on the recreational value of European forests. Ecol Indic 11:81–89. https://doi.org/10.1016/j.ecolind.2009.06.006

FAO (2010) Global Forest Resources Assessment 2010 – Main Report – FAO Forestry Paper 163. Food and Agriculture Organization of the United Nations, Rome

FAO (2018) Climate change for forest policy-makers – an approach for integrating climate change into national forest policy in support of sustainable forest management – Version 2.0. FAO Forestry Paper no. 181. Licence: CC BY-NC-SA 3.0 IGO, Rome

Filyushkina A, Agimass F, Lundhede T et al (2017) Preferences for variation in forest characteristics: Does diversity between stands matter? Ecol Econ 140:22–29. https://doi.org/10.1016/j.ecolecon.2017.04.010

Forest Europe (2015a) Madrid Ministerial Declaration: 25 years together promoting Sustainable Forest Management in Europe, Madrid, p 10

Forest Europe (2015b) State of Europe's Forests 2015 Report. Available from https://foresteurope.org/stateeuropes-forests-2015-report/. Forest Europe, Madrid, Spain

Forsius M, Akujärvi A, Mattsson T et al (2016) Modelling impacts of forest bioenergy use on ecosystem sustainability: Lammi LTER region, southern Finland. Ecol Indic 65:66–75. https://doi.org/10.1016/j.ecolind.2015.11.032

Gabler K, Schadauer K, Tomppo E et al (2012) An enquiry on forest areas reported to the global forest resources assessment—is harmonization needed? For Sci 58:201–213. https://doi.org/10.5849/forsci.10-060

Giergiczny M, Valasiuk SV, De Salvo M, Signorello G (2014) Value of Forest Recreation. Meta-analyses of the European Valuation Studies. Ekon Śr:77–83

Giergiczny M, Czajkowski M, Żylicz T, Angelstam P (2015) Choice experiment assessment of public preferences for forest structural attributes. Ecol Econ 119:8–23. https://doi.org/10.1016/j.ecolecon.2015.07.032

Grassi G, Cescatti A, Matthews R et al (2019) On the realistic contribution of European forests to reach climate objectives. Carbon Balance Manag 14:8. https://doi.org/10.1186/s13021-019-0123-y

Grilli G, Jonkisz J, Ciolli M, Lesinski J (2016) Mixed forests and ecosystem services: Investigating stakeholders' perceptions in a case study in the Polish Carpathians. For Policy Econ 66:11–17. https://doi.org/10.1016/j.forpol.2016.02.003

Griscom BW, Adams J, Ellis PW et al (2017) Natural climate solutions. Proc Natl Acad Sci 114:11645–11,650. https://doi.org/10.1073/pnas.1710465114

Griscom BW, Lomax G, Kroeger T et al (2019) We need both natural and energy solutions to stabilize our climate. Glob Chang Biol 25:1889–1890. https://doi.org/10.1111/gcb.14612

Gschwantner T, Alberdi I, Balázs A et al (2019) Harmonisation of stem volume estimates in European National Forest Inventories. Ann For Sci 76:24. https://doi.org/10.1007/s13595-019-0800-8

Gundersen VS, Frivold LH (2008) Public preferences for forest structures: A review of quantitative surveys from Finland, Norway and Sweden. Urban For Urban Green 7:241–258. https://doi.org/10.1016/j.ufug.2008.05.001

Gundersen V, Stange EE, Kaltenborn BP, Vistad OI (2017) Public visual preferences for dead wood in natural boreal forests: The effects of added information. Landsc Urban Plan 158:12–24. https://doi.org/10.1016/j.landurbplan.2016.09.020

Gutsch M, Lasch-Born P, Kollas C et al (2018) Balancing trade-offs between ecosystem services in Germany's forests under climate change. Environ Res Lett 13:045012. https://doi.org/10.1088/1748-9326/aab4e5

Hegetschweiler KT, Plum C, Fischer C et al (2017) Towards a comprehensive social and natural scientific forest-recreation monitoring instrument—a prototypical approach. Landsc Urban Plan 167:84–97. https://doi.org/10.1016/j.landurbplan.2017.06.002

Heyman E (2012) Analysing recreational values and management effects in an urban forest with the visitor-employed photography method. Urban For Urban Green 11:267–277. https://doi.org/10.1016/j.ufug.2012.02.003

Hisano M, Searle EB, Chen HYH (2018) Biodiversity as a solution to mitigate climate change impacts on the functioning of forest ecosystems. Biol Rev 93:439–456. https://doi.org/10.1111/brv.12351

Hlásny T, Mátyás C, Seidl R et al (2014) Climate change increases the drought risk in Central European forests: What are the options for adaptation? For J 60:5–18. https://doi.org/10.2478/forj-2014-0001

Hlásny T, Barka I, Kulla L et al (2017) Sustainable forest management in a mountain region in the Central Western Carpathians, northeastern Slovakia: the role of climate change. Reg Environ Change 17:65–77. https://doi.org/10.1007/s10113-015-0894-y

Huang B, Hu X, Fuglstad G-A et al (2020) Predominant regional biophysical cooling from recent land cover changes in Europe. Nat Commun 11:1–13. https://doi.org/10.1038/s41467-020-14,890-0

IPCC (2014) Climate Change 2014: Synthesis Report. Contribution of Working Groups I, II and III to the Fifth Assessment Report of the Intergovernmental Panel on Climate Change [Core Writing Team, R.K. Pachauri and L.A. Meyer (eds.)]. IPCC, Geneva.

Jactel H, Gritti ES, Drössler L et al (2018) Positive biodiversity-productivity relationships in forests: climate matters. Biol Lett 14. https://doi.org/10.1098/rsbl.2017.0747

Jandl R, Ledermann T, Kindermann G et al (2018) Strategies for Climate-Smart Forest Management in Austria. Forests 9:592. https://doi.org/10.3390/f9100592

Jandl R, Spathelf P, Bolte A, Prescott CE (2019) Forest adaptation to climate change—is non-management an option? Ann For Sci 76:48. https://doi.org/10.1007/s13595-019-0827-x

Kangas J, Laasonen L, Pukkala T (1993) A method for estimating forest landowner's landscape preferences. Scand J For Res 8:408–417. https://doi.org/10.1080/02827589309382787

Kauppi PE, Sandström V, Lipponen A (2018) Forest resources of nations in relation to human well-being. PLoS One 13:e0196248. https://doi.org/10.1371/journal.pone.0196248

Klenk NL, Larson BMH, Mcdermott CL (2015) Adapting forest certification to climate change. Wiley Interdiscip Rev. Clim Change 6:189–201. https://doi.org/10.1002/wcc.329

Köhl M, Ehrhart H-P, Knauf M, Neupane PR (2020) A viable indicator approach for assessing sustainable forest management in terms of carbon emissions and removals. Ecol Indic 111:106057. https://doi.org/10.1016/j.ecolind.2019.106057

Kovac M, Gasparini P, Notarangelo M et al (2020) Towards a set of national forest inventory indicators to be used for assessing the conservation status of the habitats directive forest habitat types. J Nat Conserv 53:125747. https://doi.org/10.1016/j.jnc.2019.125747

Kraus D, Krumm F (eds) (2013) Integrative approaches as an opportunity for the conservation of forest biodiversity. European Forest Institute

Kuusipalo J, Kangas J (1994) Managing biodiversity in a forestry environment I Manejando la bio-diversidad en un ambiente de bosque. Conserv Biol 8:450–460. https://doi.org/10.1046/j.1523-1739.1994.08020450.x

Lawrence M, McRoberts RE, Tomppo E et al (2010) Comparisons of National Forest Inventories. In: Tomppo E, Gschwantner T, Lawrence M, McRoberts RE (eds) National Forest Inventories – Pathways for Common Reporting. Springer, New York, pp 19–32

Lindner M, Maroschek M, Netherer S et al (2010) Climate change impacts, adaptive capacity, and vulnerability of European forest ecosystems. For Ecol Manage 259:698–709. https://doi.org/10.1016/j.foreco.2009.09.023

Luyssaert S, Marie G, Valade A et al (2018) Trade-offs in using European forests to meet climate objectives. Nature 562:259–262. https://doi.org/10.1038/s41586-018-0577-1

Marchetti M, Vizzarri M, Sallustio L et al (2018) Behind forest cover changes: is natural regrowth supporting landscape restoration? Findings from Central Italy. Plant Biosystems - An International Journal Dealing with all Aspects of Plant Biology 152:524–535. https://doi.org/10.1080/11263504.2018.1435585

McDowell NG, Allen CD, Anderson-Teixeira K et al (2020) Pervasive shifts in forest dynamics in a changing world. Science 368. https://doi.org/10.1126/science.aaz9463

McRoberts RE, Tomppo EO, Schadauer K, Ståhl G (2012) Harmonizing National Forest Inventories. For Sci 58:189–190. https://doi.org/10.5849/forsci.12-042

Mendoza GA, Prabhu R (2000) Development of a methodology for selecting criteria and indicators of sustainable forest management: a case study on participatory assessment. Environ Manag 26:659–673. https://doi.org/10.1007/s002670010123

Nabuurs G-J, Lindner M, Verkerk PJ, et al. (2013) First signs of carbon sink saturation in European forest biomass. Nat Clim Change advance online publication. https://doi.org/10.1038/nclimate1853

Nabuurs G-J, Delacote P, Ellison D, et al. (2015) A new role for forests and the forest sector in the EU post-2020 climate targets. From Science to Policy 2. European Forest Institute.

Nabuurs G-J, Delacote P, Ellison D et al (2017) By 2050 the Mitigation Effects of EU Forests Could Nearly Double through Climate Smart Forestry. Forests 8:484. https://doi.org/10.3390/f8120484

Nabuurs G-J, Verkerk PJ, Schelhaas M-J, et al. (2018) Climate-Smart Forestry: mitigation impacts in three European regions. European Forest Institute.

Naudts K, Chen Y, McGrath MJ et al (2016) Europe's forest management did not mitigate climate warming. Science 351:597–600. https://doi.org/10.1126/science.aad7270

Nelson HW, Williamson TB, Macaulay C, Mahony C (2016) Assessing the potential for forest management practitioner participation in climate change adaptation. For Ecol Manage 360:388–399. https://doi.org/10.1016/J.FORECO.2015.09.038

Obersteiner M, Böttcher H, Yamagata Y (2010) Terrestrial ecosystem management for climate change mitigation. Curr Opin Environ Sustain 2:271–276. https://doi.org/10.1016/j.cosust.2010.05.006

Paletto A, Meo ID, Cantiani MG, Maino F (2013) Social Perceptions and Forest Management Strategies in an Italian Alpine Community. Mt Res Dev 33:152–160. https://doi.org/10.1659/MRD-JOURNAL-D-12-00115.1

Paletto A, De Meo I, Di Salvatore U, Ferretti F (2014) Perceptive analysis of the Sustainable Forest Management (SFM) through the cognitive maps. For - Riv Selvic Ed Ecol For 11:125–137. https://doi.org/10.3832/efor1245-011

Paletto A, De Meo I, Grilli G, Nikodinoska N (2017) Effects of different thinning systems on the economic value of ecosystem services: A case-study in a black pine peri-urban forest in Central Italy. Ann For Res 60:313–326. https://doi.org/10.15287/afr.2017.799

Pastorella F, Avdagić A, Čabaravdić A et al (2016a) Tourists' perception of deadwood in mountain forests. Ann For Res 59:311–326. https://doi.org/10.15287/afr.2016.482

Pastorella F, Giacovelli G, Maesano M et al (2016b) Social perception of forest multifunctionality in southern Italy: The case of Calabria Region. J For Sci 62:366–379. https://doi.org/10.17221/45/2016-JFS

Pelyukh O, Paletto A, Zahvoyska L (2019) Comparison between people's perceptions and preferences towards forest stand characteristics in Italy and Ukraine. Ann Silvic Res 43:4–14. https://doi.org/10.12899/asr-1786

Pfatrisch M (2019) Indicators to assess the climate smartness of mixed mountain forests containing Picea, Abies and Fagus. Master Thesis, TU München.

Pretzsch H, Biber P, Schütze G et al (2014) Forest stand growth dynamics in Central Europe have accelerated since 1870. Nat Commun 5:4967. https://doi.org/10.1038/ncomms5967

Pretzsch H, Forrester DI, Bauhus J (eds) (2017) Mixed-species forests: ecology and management. Springer-Verlag, Berlin Heidelberg

Rathmann J, Sacher P, Volkmann N, Mayer M (2020) Using the visitor-employed photography method to analyse deadwood perceptions of forest visitors: a case study from Bavarian Forest National Park, Germany. Eur J For Res 139:431–442. https://doi.org/10.1007/s10342-020-01260-0

Ribe RG (2009) In-stand scenic beauty of variable retention harvests and mature forests in the U.S. Pacific Northwest: The effects of basal area, density, retention pattern and down wood. J Environ Manage 91:245–260. https://doi.org/10.1016/j.jenvman.2009.08.014

Rondeux J, Bertini R, Bastrup-Birk A et al (2012) Assessing deadwood using harmonized national forest inventory data. For Sci 58:269–283. https://doi.org/10.5849/forsci.10-057

Saaty TL (1980) The analytic hierarchy process. McGraw-Hill, New York

Sabatini FM, de Andrade RB, Paillet Y et al (2019) Trade-offs between carbon stocks and biodiversity in European temperate forests. Glob Chang Biol 25:536–548. https://doi.org/10.1111/gcb.14503

Santopuoli G, Requardt A, Marchetti M (2012) Application of indicators network analysis to support local forest management plan development: a case study in Molise, Italy. IForest - Biogeosciences For 5:31–37. https://doi.org/10.3832/ifor0603-009

Santopuoli G, Marchetti M, Giongo M (2016) Supporting policy decision makers in the establishment of forest plantations, using SWOT analysis and AHPs analysis. A case study in Tocantins (Brazil). Land Use Policy 54:549–558

Santopuoli G, Temperli C, Alberdi I et al (2020) Pan-European Sustainable Forest Management indicators for assessing Climate-Smart Forestry in Europe. Can J For Res. https://doi.org/10.1139/cjfr-2020-0166

Sathre R, O'Connor J (2010) Meta-analysis of greenhouse gas displacement factors of wood product substitution. Environ Sci Policy 13:104–114. https://doi.org/10.1016/j.envsci.2009.12.005

Schall P, Heinrichs S, Ammer C et al (2020) Can multi-taxa diversity in European beech forest landscapes be increased by combining different management systems? J Appl Ecol 57:1363–1375. https://doi.org/10.1111/1365-2664.13635

Seddon N, Sengupta S, Garcia Espinosa M et al (2019) Nature-based solutions in nationally determined contributions. IUCN and University of Oxford. Gland, Switzerland and Oxford

Seidl R, Lexer MJ (2013) Forest management under climatic and social uncertainty: Trade-offs between reducing climate change impacts and fostering adaptive capacity. J Environ Manage 114:461–469. https://doi.org/10.1016/j.jenvman.2012.09.028

Seidl R, Thom D, Kautz M et al (2017) Forest disturbances under climate change. Nature Climate Change 7:395–402. https://doi.org/10.1038/nclimate3303

Sievänen T, Edwards D, Fredman P et al (eds) (2013) Social indicators in the forest sector in Northern Europe: a review focusing on nature-based recreation and tourism. TemaNord, Nordic Council of Ministers, Copenhagen

Spittlehouse DL (2005) Integrating climate change adaptation into forest management. For Chron 81:691–695

Stadelmann G, Herold A, Didion M et al (2016) Holzerntepotenzial im Schweizer Wald: Simulation von Bewirtschaftungsszenarien. Schweiz Z Forstwes 167:152–161. https://doi.org/10.3188/szf.2016.0152

Stadelmann G, Temperli C, Rohner B et al (2019) Presenting MASSIMO: a management scenario simulation model to project growth, harvests and carbon dynamics of Swiss forests. Forests 10:94. https://doi.org/10.3390/f10020094

Ståhl G, Cienciala E, Chirici G et al (2012) Bridging national and reference definitions for harmonizing forest statistics. For Sci 58:214–223. https://doi.org/10.5849/forsci.10-067

Temperli C, Blattert C, Stadelmann G et al (2020) Trade-offs between ecosystem service provision and the predisposition to disturbances: a NFI-based scenario analysis. For Ecosyst 7:27. https://doi.org/10.1186/s40663-020-00236-1

Tomppo E, Gschwantner T, Lawrence M, McRoberts RE (eds) (2010) National Forest Inventories: Pathways for Common Reporting. Springer, Dordrecht

Tomter SM, Gasparini P, Gschwantner T et al (2012) Establishing Bridging Functions for Harmonizing Growing Stock Estimates: Examples from European National Forest Inventories. For Sci 58:224–235. https://doi.org/10.5849/forsci.10-068

Torresan C, Luyssaert S, Filippa G, Imangholiloo M, Gaulton R (2021) Remote sensing technologies for assessing climate-smart criteria in mountain forests. In: Managing Forest Ecosystems, Vol. 40, Tognetti R, Smith M, Panzacchi P (Eds): Climate-Smart Forestry in Mountain Regions. Springer Nature, Switzerland, AG

Traub B, Meile R, Speich S, Rösler E (2017) The data storage and analysis system of the Swiss National Forest Inventory. Comput Electron Agric 132:97–107. https://doi.org/10.1016/j.compag.2016.11.016

Trotsiuk V, Hartig F, Cailleret M et al (2020) Assessing the response of forest productivity to climate extremes in Switzerland using model–data fusion. Glob Chang Biol 26:2463–2476. https://doi.org/10.1111/gcb.15011

Van Deusen P (2010) Carbon sequestration potential of forest land: management for products and bioenergy versus preservation. Biomass Bioenergy 34:1687–1694. https://doi.org/10.1016/j.biombioe.2010.03.007

Verkerk PJ, Costanza R, Hetemäki L et al (2020) Climate-smart forestry: the missing link. For Policy Econ 115:102164. https://doi.org/10.1016/j.forpol.2020.102164

Viccaro M, Cozzi M, Fanelli L, Romano S (2019) Spatial modelling approach to evaluate the economic impacts of climate change on forests at a local scale. Ecol Indic 106. https://doi.org/10.1016/j.ecolind.2019.105523

Vidal C, Lanz A, Tomppo E et al (2008) Establishing forest inventory reference definitions for forest and growing stock: a study towards common reporting. Silva Fenn 42. https://doi.org/10.14214/sf.255

Vidal C, Alberdi I, Hernández L, Redmond JJ (eds) (2016) National forest inventories: assessment of wood availability and use. Springer International Publishing

Wolfslehner B, Vacik H, Lexer MJ (2005) Application of the analytic network process in multi-criteria analysis of sustainable forest management. For Ecol Manage 207:157–170. https://doi.org/10.1016/j.foreco.2004.10.025

Woodall CW, Rondeux J, Verkerk PJ, Ståhl G (2009) Estimating dead wood during national forest inventories: a review of inventory methodologies and suggestions for harmonization. Environ Manag 44:624–631. https://doi.org/10.1007/s00267-009-9358-9

Yousefpour R, Augustynczik ALD, Reyer CPO et al (2018) Realizing mitigation efficiency of European commercial forests by climate smart forestry. Sci Rep 8:345. https://doi.org/10.1038/s41598-017-18,778-w

Chapter 5
Efficacy of Trans-geographic Observational Network Design for Revelation of Growth Pattern in Mountain Forests Across Europe

H. Pretzsch, T. Hilmers, E. Uhl, M. del Río, A. Avdagić, K. Bielak, A. Bončina, L. Coll, F. Giammarchi, K. Stimm, G. Tonon, M. Höhn, M. Kašanin-Grubin, and R. Tognetti

Abstract Understanding tree and stand growth dynamics in the frame of climate change calls for large-scale analyses. For analysing growth patterns in mountain forests across Europe, the CLIMO consortium compiled a network of observational plots across European mountain regions. Here, we describe the design and efficacy of this network of plots in monospecific European beech and mixed-species stands of Norway spruce, European beech, and silver fir.

H. Pretzsch (✉) · T. Hilmers
Chair of Forest Growth and Yield Science, Department of Life Science Systems, TUM School of Life Sciences, Technical University of Munich, Freising, Germany
e-mail: hans.pretzsch@tum.de; torben.hilmers@tum.de

E. Uhl · K. Stimm
Chair of Forest Growth and Yield Science, Department of Life Science Systems, TUM School of Life Sciences, Technical University of Munich, Freising, Germany

Bavarian State Institute of Forestry (LWF), Freising, Germany
e-mail: enno.uhl@tum.de; kilian.stimm@tum.de; kilian.stimm@lwf.bayern.de

M. del Río
INIA, Forest Research Centre, Madrid, Spain

iuFOR, Sustainable Forest Management Research Institute, University of Valladolid & INIA, Valladolid, Spain
e-mail: delrio@inia.es

A. Avdagić
Faculty of Forestry, Department of Forest Management Planning and Urban Greenery, University of Sarajevo, Sarajevo, Bosnia and Herzegovina
e-mail: a.avdagic@sfsa.unsa.ba

K. Bielak
Department of Silviculture, Institute of Forest Sciences, Warsaw University of Life Sciences, Warsaw, Poland
e-mail: kamil.bielak@wl.sggw.pl

© The Author(s) 2022
R. Tognetti et al. (eds.), *Climate-Smart Forestry in Mountain Regions*, Managing Forest Ecosystems 40, https://doi.org/10.1007/978-3-030-80767-2_5

First, we sketch the state of the art of existing monitoring and observational approaches for assessing the growth of mountain forests. Second, we introduce the design, measurement protocols, as well as site and stand characteristics, and we stress the innovation of the newly compiled network. Third, we give an overview of the growth and yield data at stand and tree level, sketch the growth characteristics along elevation gradients, and introduce the methods of statistical evaluation. Fourth, we report additional measurements of soil, genetic resources, and climate smartness indicators and criteria, which were available for statistical evaluation and testing hypotheses. Fifth, we present the ESFONET (European Smart Forest Network) approach of data and knowledge dissemination. The discussion is focussed on the novelty and relevance of the database, its potential for monitoring, understanding and management of mountain forests toward climate smartness, and the requirements for future assessments and inventories.

In this chapter, we describe the design and efficacy of this network of plots in monospecific European beech and mixed-species stands of Norway spruce, European beech, and silver fir. We present how to acquire and evaluate data from individual trees and the whole stand to quantify and understand the growth of mountain forests in Europe under climate change. It will provide concepts, models, and practical hints for analogous trans-geographic projects that may be based on the existing and newly recorded data on forests.

A. Bončina
Biotechnical Faculty, Department of Forestry and Renewable Forest Resources,
University of Ljubljana, Ljubljana, Slovenia
e-mail: andrej.boncina@bf.uni-lj.si

L. Coll
Department of Agriculture and Forest Engineering (EAGROF), University of Lleida,
Lleida, Spain

Joint Research Unit CTFC-AGROTECNIO, Solsona, Spain
e-mail: lluis.coll@udl.cat

F. Giammarchi · G. Tonon
Faculty of Science and Technology, Free University of Bolzano-Bozen, Bolzano, Italy
e-mail: francesco.giammarchi@unibz.it; giustino.tonon@unibz.it

M. Höhn
Faculty of Horticultural Science, Department of Botany, SZIU, Budapest, Hungary
e-mail: hohn.maria@kertk.szie.hu

M. Kašanin-Grubin
Institute for Chemistry, Technology and Metallurgy, University of Belgrade, Belgrade,
Serbia
e-mail: mkasaningrubin@chem.bg.ac.rs

R. Tognetti
Dipartimento di Agricoltura, Ambiente e Alimenti, Università degli Studi del Molise,
Campobasso, Italy
e-mail: tognetti@unimol.it

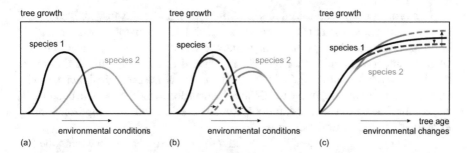

Fig. 5.1 The effect of changing environmental conditions on species growth. (**a**) Environmental changes may be detrimental for species 1 but advantageous for species 2 due to better match of fundamental niche. (**b**) The fundamental niches are modified by interspecific synecological effects. (**c**) The course of growth of trees and species-specific ranking may be modified within the lifetime of trees due to environmental changes

5.1 Assessing the Climate Sensitivity of the Growth of European Mountain Forests

Environmental changes (indicated by an arrow in Fig. 5.1a) may promote a species with a more suitable fundamental niche (species 2) and reduce the growth of species 1 (Fig. 5.1a). Tree growth indicates the effects of climate and other environmental conditions on the production, adaptation, and mitigation of forest stands. In Figure 5.1, we indicate the fundamental niche by a monocausal gradient along the abscissa and the growth as indicator for fitness on the ordinate.

In mixed stands, the species also encounter interspecific competition with additional synecological disadvantages, as indicated by a narrowing of the fundamental to the real niche in Figure 5.1b. Due to their longevity, trees may be exposed to environmental changes over centuries and modify the course of their growth, as well as species-specific competition and facilitation (Fig. 5.1c).

For trees in the northern latitudes or in the higher elevations of mountain areas, this means that they change their growth due to the modified potential growing conditions, and in addition, they may face the new competition effects by other species of the ecosystem. Repeated observations in permanent plots are, therefore, necessary to confirm or confute the status of monitored trees and their growth trends over time (Franklin 1989). These plots may be also useful for re-examining ecological theories (e.g. disturbance ecology, forest succession) and temporal series (e.g. biomass accumulation, tree mortality) in the framework of environmental changes (van Mantgem and Stephenson 2007; Harmon and Pabst 2015).

In mountain forests, at the edge of their ecological amplitude, little changes of environmental conditions may trigger strong non-linear effects on tree growth superimpose by additional competition effects due to strengthening of neighbours, which grow in the proximity and are better adapted to the new conditions (Pretzsch et al. 2020a, b).

This project deals with the effects of climate changes on the growth dynamics of mountain forest ecosystems that are so far much less understood than the ones in northern latitudes, although they may undergo even worse changes regarding the stability and ecosystem service provision (Tognetti et al. 2017). In addition to the non-linear reaction pattern at the left or right branch of the ecological niche, mountain forests are very susceptible to climate changes due to their harsh site conditions, slopes, mechanical unstable conditions, remoteness, and thus limiting accessibility to mitigating silvicultural measures.

The concept and data acquisition should provide the basis for answering the following questions and scrutinizing the following hypotheses:

(i) How is the state of the productivity, vitality, and climate smartness of mountain forests in Europe?
(ii) How did the stand productivity of mountain forests change in the recent centuries according to records from long-term experiments and information extracted from increment core analyses?
(iii) How did the growth of the main tree species change in the recent centuries and were any changes of tree growth depending on the elevation above sea level?

This chapter presents the interdisciplinary database and trans-geographic plot network, underlying recent research articles (Hilmers et al. 2019; Pretzsch et al. 2020a, b; Torresan et al. 2020; del Río et al. 2021).

5.2   State of the Art of Monitoring and Observational Approaches

The analysis of forest growth is one of the fundamentals of forestry and forest science, so there are different well-established approaches for obtaining the required data from forest stands (e.g. Kangas and Maltamo 2006; Pretzsch 2009; Ferretti and Fischer 2013). Here, we briefly introduce the main concepts and approaches, underlining the most relevant aspects of data sources, to properly analyse the growth trends and responses to disturbances and extreme weather events in mountain forests under global climate change. We address different *organization levels* across both *temporal and spatial scales*, and we present selected *methodological approaches*.

Organizational Level The most common approach used for analysing long-term growth trends is the tree level, since tree coring allows easily obtaining long tree-ring series. However, it is not possible to evaluate the climate change impacts on forest dynamics without addressing the growth at stand level, which implicitly considers ingrowth and mortality. When stand-level data are not available, tree-level data covering stand size distribution can be an acceptable compromise, since growth response of trees to climate, disturbance, and extreme events varies significantly among social classes of trees (Pretzsch et al. 2018). Given that biomass accumulation in forests depends on the balance between growth (carbon sequestration) and mortality (carbon loss) of trees, monitoring changes at an individual tree level may enable a better understanding of forest-climate feedback and post-disturbance dynamics (see Chap. 10 of this book: Tognetti et al. 2021). Studies at lower levels, such as organ or cell growths, may provide additional information to better understand forest growth variation with climate change, but in general, they are not feasible for large samples. On the other hand, applications for forest management and biodiversity conservation may require detailed data representing large spatial extents, which can be obtained through remote sensing (see Chap. 11 of this book: Torresan et al. 2021).

Spatial Scale Forest growth data can be gathered from local to global scale. Local or regional growth trends analysis can be very relevant for their use at these scales and especially when the study area represents the rear edge of the species distribution (e.g. Dorado-Liñán et al. 2020; Hernández et al. 2019). However, trans-geographic networks across large areas are needed to identify general patterns and main drivers of growth trends in regard to gradual and episodic environmental stress (Gazol and Camarero 2016).

Temporal Scale Two aspects related to temporal scale must be considered, the *temporal resolution* and the *continuity*, i.e. temporal vs. permanent plots. Regarding the former, daily and intra-annual tree growth may reveal useful information about climate drivers and growth, but for analysing growth trends and forest dynamics, annual resolution is generally the most robust option. Lower resolution such as 5- or 10-year periods may not always well describe the growth patterns; however, long-term series can also provide unique information on forest growth trends (Pretzsch et al. 2014; Hilmers et al. 2019). Nevertheless, temporal resolution is often linked to continuity, since long-term series at stand level inevitably involves permanent plots. Advantages and disadvantages of temporal and permanent plots have been frequently discussed (e.g. Gadow 1999), but unquestionably, for exploring forest growth trends and understanding climate change consequences, long-term experimental plots, where stand history has been recorded, offer invaluable information (Pretzsch et al. 2019). Therefore, in this respect, long-term experiments so far outperform the information potential of inventory plots that have been increasingly established during the last two decades under the umbrella of National Forest Inventories in Europe.

Methodological Approach According to the used concept, experimental approaches can be classified as observational or manipulated ones. Traditionally, manipulated experiments with a statistical design and control of factors have been accepted as the correct way to identify the causal effects. However, the increasing capacity of obtaining large amounts of data strengthens the ability of observational approaches for testing hypotheses, so they currently are an essential source to analyse global environmental problems at a large spatial-temporal scale (Sagarin and Pauchard 2010). In forests, the most critical part of observational data is that often the stand management history is unknown, although this issue can be overcome by long-term monitoring. Observational approaches have been classified as inventory-based or exploratory methods (Bauhus et al. 2017, pp. 53–64). Inventoried-based approaches follow different sampling designs, generally systematic sampling, with the aim to gain in representativeness of the whole studied population. On the contrary, exploratory approaches distribute samples along gradients of specific factors to study their causal relationship. For our aim, spatial distribution of samples/plots may be designed to cover different site conditions, as tree and stand growth, as well as the impact of climate change, are strongly dependent on them. For this, transects along environmental gradients are particularly useful (Pretzsch et al. 2014). In mountain areas, altitudinal transects are generally the most efficient option (Ettinger et al. 2011).

Ideally, to study growth trends and responses to extreme events, data should cover all kinds of site conditions, across a large geographical area, during a long period of time, and focussing at least on tree and stand level, and at annual resolution, but of course this is not realistic. Often there are available data, which cover a large spatial scale but short temporal scale or vice versa, so integrating approaches are needed (Sagarin and Pauchard 2010), like the network presented in Section 5.3. In Section 5.7, further discussion of advantages and disadvantages of different approaches is included, with special emphasis on long-term experimental plots.

There are several examples of large forest growth databases based on different approaches. Ruiz-Benito et al. (2020) reviewed the available data sources in Europe for modelling climate change impacts on forests, including growth databases, such as the following: National Forest Inventories (Tomppo et al. 2010); the ICP forests European network (Ferretti and Fischer 2013); the DEIMS-SDR, including the Long-Term Research sites (LTER) (Wohner et al. 2019); the International Tree-Ring Data Bank (ITRDB) (Grissino-Mayer and Fritts 1997); networks of long-term experiments, like the Northern European Database of Long-Term Experiments (NOLTFOX) and the worldwide ForestGEO network (Anderson-Teixeira et al. 2015); and different remote sensing data sources.

Some of these datasets have their origin in institutional collaboration among countries, but the increasing number of initiatives for sharing research data, as the recent Global Forest Biodiversity Initiative (GFB), is remarkable (Liang et al. 2016). Although nowadays the publication of research data is often demanded by many funding institutions and publishing houses, these initiatives suppose a good opportunity for large-scale analysis, as they compile the information in a common platform.

5.3 The CLIMO Design of Transnational Observational Network

5.3.1 Study Design and Data Used

Tree species distribution and competitiveness in mountain forest ecosystems are strongly determined by geographic and topographic factors (Fig. 5.2). It is expected that climate change may affect the growth performance of tree species in mountain regions differently but possibly leading to a modification of the fitness and

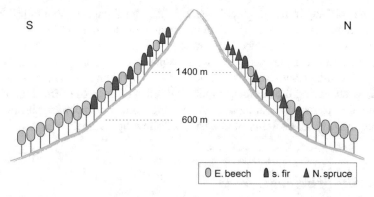

Fig. 5.2 Elevation and aspect are the main factors shaping species distribution and stand composition in mountain forests

subsequently to a change in tree species composition and distribution (Becker and Bugmann 2001). A comprehensive view about the general performance and a potentially recent change in performance of European mountain forests is widely missing. Two empirical studies were designed in the frame of CLIMO to improve the knowledge of historic and recent growth dynamics in mountain forests. We selected two most common types of mountain forests. In study 1, we investigated mountain mixed forests, comprising Norway spruce, silver fir, and European beech, while in case of study 2, we focused on monospecific beech and beech-dominated mixed stands. Both studies were intended to analyse the growth and growth trends on stand and tree level. Short- and long-term growth trends were analysed against various factors concerning stand structure (e.g. tree species composition, density, diameter distribution) and site factors (e.g. climate, elevation).

Two different data sources were utilized to create a transnational observational network and to compile respective datasets for the analyses. Concerning mixed mountain forests, existing stand- and tree-level data from repeated inventoried long-term observational plots across European mountain regions were collected (Sect. 5.3.4). In contrast, the network of temporary plots representing geographic and elevation gradients in European beech-dominated stands were established and inventoried. Additionally, tree cores were obtained in both cases (study 1 and 2). Tree cores were used to analyse species-specific growth trends and growth reaction to drought events as well as to reconstruct recent stand-level performance on temporary plots applying the method described by Heym et al. (2018).

The analysis using temporary plots in beech-dominated stands followed two main research lines (RL). The first (RL1) focused on the effect of stand structural parameters on beech performance at tree and stand level. The second (RL2) intended to reveal the effect of elevation on growth rate and growth temporal trends. In the former case, two plots per site were established having similar elevation and growing conditions and differing in stand structural characteristics. In the latter case, two similarly structured stands at different elevations, but growing in similar conditions, were sampled. In some sites, both RLs were combined, i.e. two structures at each elevation.

When designing trans-geographic studies, used local datasets should follow the common standards. In particular, when involving new (temporary) sample data, the common standards for site selection, data sampling, and data analysis are a prerequisite to facilitate analyses and to reduce post-processing effort. Following the keystones of common standards guarantees (i) the strengthening of statistical analysis options by enhancing the parameter specific number of degrees of freedom, (ii) the comparability of results with those from existing studies, (iii) confirmability of the analyses, and (iv) the usability of the data for follow-up studies.

5.3.2   Site Selection Criteria

Before plot establishment, in situ criteria for site and stand selection need to be defined. They have to be deduced from the study-specific research questions and hypotheses. These criteria have to delineate the subject of research and identify which factors to be included in the analyses are necessarily kept constant and which are allowed to vary. In study 2, which utilized the newly established plots, site selection was limited to mountainous regions. However, stand selection per site was more in-depth, requiring a specific current stand age range for all plots. Concerning the RL1, the two plots of a single site were allowed only to differ in stand structural characteristics (e.g. density, species composition) whereas keeping site conditions and elevation constant. Concerning the RL2, two similar structured stands, having same topographic features, but only discriminated in elevation (min. 200 m), had to be selected per site (Table 5.1, cf. Fig. 5.3).

5.3.3   Plot Metadata

After plot establishment, a precise and detailed description of the plot and topographic characteristics of plots, as well as environmental conditions, is necessary (Table 5.2). Coordinates and plot shape information guarantee the permanent identification of the single plot location. Topographic characteristics should be as detailed as possible and provide at least information about the factors needed for data analysis. The degree of detail concerning information on soil conditions and historic and current climatic characteristics is again dependent on the aim of the analyses.

5.3.4   Tree Inventory and Dendrochronology

The set of tree-specific variables to be collected per plot depends on the detail needed for the intended analyses. In CLIMO, empirical studies concerned productivity and structure on both tree and stand level. Thus, beside the standard stand data, also single-tree information is required to address and interlink both levels

Table 5.1 Exemplary site selection criteria for temporal plots used in CLIMO study 2

Category	Criteria	RL1	RL2
Geographic	Specification of location	European mountain regions[a]	European mountain regions[a]
	Elevation	Equal for two plots per site	Min. 200 m in difference between two plots per site
	Site factors (slope, aspect, soil type)	Constant for min. Two plots per site	Constant for min. Two plots per site
Forest stand	Species composition	Monospecific beech stands – basal area of beech ≥90% Beech-dominated mixed stands – basal area of beech >30% and <70%	Monospecific beech stands – basal area of beech ≥90%
	Stand age	Similar for the dominant trees and between 70 and 100 years	Similar for the dominant trees and between 70 and 100 years
	Management history	Unmanaged for at least 15 years	Unmanaged for at least 15 years
	Plot size	Min. 0.1 ha, min. 50 trees per plot	Min. 0.1 ha, min. 50 trees per plot
	Stand structure	Different between two plots per site	Constant between two plots per site

[a]Mountain region followed the respective national definition – mountain definition can be constrained by a combination of elevation and ruggedness (Kapos et al. 2000) or by ruggedness of terrain only, irrespective of elevation (Körner et al. 2011)

(Table 5.3). Additional information can still be planned, once monitoring plots have been established, such as repeated observations of reproductive structures, phenological phases, physiological conditions, and mortality rates, to select novel indicators for assessing climate smartness of forests over time.

Temporary plots, measured for the first time, like in the case of study 2, generally lack information about recent stand growth. However, stand-level increment can be reconstructed with the help of tree-ring chronologies derived from tree cores (Heym et al. 2018). As annual increment varies among trees of different social classes (del Rio et al. 2014; Torresan et al. 2020), it is important that the sampled trees cover the whole spectrum of the stand diameter distribution (Cherubini et al. 1998). To consider mortality when estimating stand productivity, dead trees have to be included in tree inventory, also estimating the probable year of death. In managed stands, an inventory of stumps and an estimate of the year of thinning improve the accuracy of reconstruction. However, selecting stands that have not been managed during the last 15 years may improve the stand growth reconstruction.

5.4   Network, Locations, Site Characteristics

In total, the trans-geographic network made it possible to collect and homogenize data from 159 observational plots in 14 countries across Europe (Fig. 5.3, Table 5.4). Plots are located mainly in fully stocked, un-thinned, or slightly thinned forest

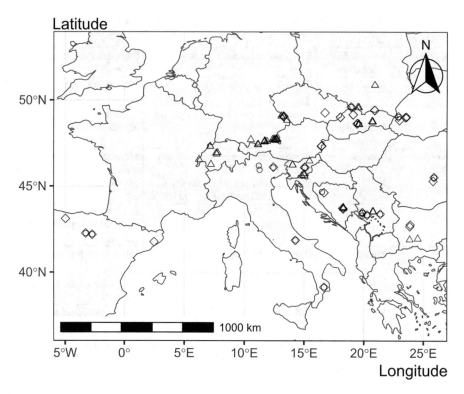

Fig. 5.3 Location of the 89 long-term observational plots in mixed mountain forests (triangles) and 72 temporary observational plots in monospecific stands of beech in mountain areas (rhombuses; $n = 48$) and mixed mountain forests (circles; $n = 22$) of 14 countries. The dataset covers mountain forests in Bosnia and Herzegovina, Bulgaria, Czech Republic, Germany, Hungary, Italy, Poland, Romania, Serbia, Slovakia, Slovenia, Spain, Switzerland, and Ukraine

stands that reflect natural dynamics and climatic variability. The dataset covered the mountain forests in Bosnia and Herzegovina, Bulgaria, Czech Republic, Germany, Hungary, Italy, Poland, Romania, Serbia, Slovakia, Slovenia, Spain, Switzerland, and Ukraine. The observational network comprises 89 long-term plots in mixed mountain forests mainly consisting of European beech (*Fagus sylvatica* L.), Norway spruce (*Picea abies* (L.) Karst), and silver fir (*Abies alba* Mill.), which have been under observation for at least 30 years. In addition, 70 temporary observational plots were established (see Sect. 5.3.3), representing 48 monospecific stands of European beech and 22 mixed mountain forests. In the latter case, European beech was mainly mixed with Norway spruce and silver fir, but studied plots included other admixed species as well, such as Scots pine (*Pinus sylvestris* L.), sycamore maple (*Acer pseudoplatanus* L.), European larch (*Larix decidua* Mill.), and European hornbeam (*Carpinus betulus* L.). Except for Scots pine and sycamore maple, these minor species, however, represent less than 10% of the stand basal area. All the study sites are located in mountain regions, from Picos de Europa (Spain) in the west to the Southern Carpathians (Romania) in the east, and from the Tatras (Poland) in the north to the Apennines (Italy) in the south. Elevations vary

Table 5.2 Exemplary list of single plot-specific information (metadata)

Information	Criteria		Unit
Location	Site identification	Unique site identifier	
	Plot identification	Unique plot identifier	
	Coordinates	Latitude/longitude	Specific to coordinate system, deviation north
	Plot shape		
	Contact		
Topographic information	Aspect		Degree, letters
	Slope (°)		Degree
	Position in the slope		Category
	Slope length		
	Elevation (m a.s.l.)		m a.s.l.
Information on soil and climate	Parental material	International nomenclature	
	Soil depth		
	Climate data		
Other specifics	Stand age	Estimates, inventory	Year
	Further remarks		

Table 5.3 Variables of single-tree measurements, exemplary for study 2

Variable	Unit/number	Sample size (CLIMO study)
Date of survey		
Tree number	1–899 for inner plot trees, ≥900 for border tree	Full inventory
Tree coordinates	Local X, Y coordinates in m	Full inventory
Diameter at breast height	cm (one decimal)	Full inventory
Tree height	m (one decimal)	Full inventory
Height of crown base	m (one decimal)	Full inventory
Crown radii	m (one decimal)	Full inventory, min. 4 cardinal directions
Specifics remarks (e.g. damages, sample tree for coring)	Descriptive (character type)	Specific trees
Tree cores	2 per tree, perpendicular at 1.3 m	15 dominant beeches, 15 trees covering the rest of the diameter distribution, in case of mixed stands 15 trees of mixed species covering the diameter distribution

Table 5.4 Geographical information and site characteristics of the 159 observational plots

Country	Period	Composition	Coordinates		Site characteristics		
			Latitude	Longitude	E	T	P
Bosnia and Herzegovina	Long-term	Mixed	43°47′54.6″N	18°16′49.6″E	1006	7.7	1179
Bosnia and Herzegovina	Long-term	Mixed	43°45′18″N	18°18′11.2″E	1257	6.7	1333
Bosnia and Herzegovina	Long-term	Mixed	43°44′55.8″N	18°15′03.2″E	1291	6.6	1354
Bosnia and Herzegovina	Long-term	Mixed	43°44′49.1″N	18°15′54″E	1192	7.0	1293
Bosnia and Herzegovina	Long-term	Mixed	43°46′27″N	18°17′54.4″E	1166	7.0	1277
Bulgaria	Long-term	Mixed	41°55′06.8″N	23°50′29.7″E	1569	2.6	1066
Bulgaria	Long-term	Mixed	41°57′55.6″N	24°31′14″E	1391	3.3	956
Germany	Long-term	Mixed	47°43′47.6″N	10°32′24.9″E	826	7.2	1426
Germany	Long-term	Mixed	47°37′19.6″N	11°53′59.8″E	1136	6.6	1615
Germany	Long-term	Mixed	47°35′38.2″N	11°41′41.1″E	1271	4.7	2281
Germany	Long-term	Mixed	47°45′51.3″N	12°29′44.8″E	939	5.8	1936
Germany	Long-term	Mixed	47°42′14.4″N	12°26′47.9″E	927	5.1	2000
Germany	Long-term	Mixed	47°42′12.9″N	12°28′26.3″E	860	6.8	1646
Germany	Long-term	Mixed	47°26′15.7″N	11°09′57.3″E	1463	4.5	1745
Germany	Long-term	Mixed	47°25′59.8″N	11°09′48.5″E	1768	2.0	2182
Germany	Long-term	Mixed	47°44′10.6″N	12°21′51.4″E	902	5.1	2236
Germany	Long-term	Mixed	47°43′02.2″N	12°42′15″E	934	6.1	1810
Germany	Long-term	Mixed	47°43′02.2″N	12°42′15″E	934	6.1	1810
Germany	Long-term	Mixed	47°42′50.1″N	12°42′27.3″E	973	6.1	1810
Germany	Long-term	Mixed	47°42′50.1″N	12°42′27.3″E	973	6.1	1810
Germany	Long-term	Mixed	47°26′52.2″N	11°07′24.6″E	1235	4.8	1470
Germany	Long-term	Mixed	47°26′52.2″N	11°07′24.6″E	1235	4.8	1470
Germany	Long-term	Mixed	47°26′52.2″N	11°07′24.6″E	1235	4.8	1470
Germany	Long-term	Mixed	47°26′52.2″N	11°07′24.6″E	1235	4.8	1470
Germany	Long-term	Mixed	47°42′56.6″N	12°40′09.7″E	884	6.8	1707
Germany	Long-term	Mixed	47°36′04″N	11°39′43.9″E	1091	6.1	1998
Germany	Long-term	Mixed	47°36′04″N	11°39′43.9″E	1091	6.1	1998
Germany	Long-term	Mixed	47°36′04″N	11°39′43.9″E	1091	6.1	1998
Germany	Long-term	Mixed	47°36′04″N	11°39′43.9″E	1091	6.1	1998
Germany	Long-term	Mixed	47°39′18.1″N	11°43′13.3″E	1281	6.1	2059
Germany	Long-term	Mixed	47°39′18.1″N	11°43′13.3″E	1281	6.1	2059
Germany	Long-term	Mixed	47°39′18.1″N	11°43′13.3″E	1281	6.1	2059
Germany	Long-term	Mixed	47°39′18.1″N	11°43′13.3″E	1281	6.1	2059
Germany	Long-term	Mixed	48°51′19.2″N	13°35′18.4″E	743	6.8	1072

(continued)

Table 5.4 (continued)

Country	Period	Composition	Coordinates		Site characteristics		
			Latitude	Longitude	E	T	P
Germany	Long-term	Mixed	48°51′19.2″N	13°35′18.4″E	743	6.8	1072
Germany	Long-term	Mixed	48°51′19.2″N	13°35′18.4″E	743	6.8	1072
Germany	Long-term	Mixed	48°51′19.2″N	13°35′18.4″E	743	6.8	1072
Germany	Long-term	Mixed	48°51′19.2″N	13°35′18.4″E	743	6.8	1072
Germany	Long-term	Mixed	48°51′19.2″N	13°35′18.4″E	743	6.8	1072
Germany	Long-term	Mixed	49°05′55.1″N	13°05′30.1″E	951	5.2	1339
Germany	Long-term	Mixed	49°05′55.1″N	13°05′30.1″E	951	5.2	1339
Germany	Long-term	Mixed	49°05′55.1″N	13°05′30.1″E	951	5.2	1339
Germany	Long-term	Mixed	49°05′55.1″N	13°05′30.1″E	951	5.2	1339
Germany	Long-term	Mixed	49°05′18.9″N	13°17′41.7″E	1037	4.3	1402
Germany	Long-term	Mixed	49°05′18.9″N	13°17′41.7″E	1037	4.3	1402
Germany	Long-term	Mixed	49°05′53.7″N	13°15′07.1″E	787	5.8	1344
Germany	Long-term	Mixed	49°05′59.4″N	13°14′59″E	779	6.5	1294
Germany	Long-term	Mixed	47°37′57.5″N	11°41′23.2″E	1294	5.4	2163
Poland	Long-term	Mixed	50°53′39.5″N	20°54′09.5″E	501	6.5	731
Poland	Long-term	Mixed	50°53′25.6″N	20°53′56″E	600	6.1	791
Poland	Long-term	Mixed	50°53′52.1″N	20°54′22.3″E	425	6.8	684
Poland	Long-term	Mixed	49°35′38.9″N	19°28′42.6″E	1015	5.0	1403
Poland	Long-term	Mixed	49°35′36.7″N	19°33′24.2″E	972	5.2	1377
Poland	Long-term	Mixed	49°35′36″N	19°33′12.1″E	966	5.2	1373
Poland	Long-term	Mixed	49°35′50.7″N	19°28′36.6″E	902	5.5	1334
Poland	Long-term	Mixed	49°35′35.5″N	19°33′39.2″E	982	5.2	1383
Poland	Long-term	Mixed	49°35′37.6″N	19°33′42.7″E	958	5.3	1368
Poland	Long-term	Mixed	49°35′24.6″N	19°34′07.2″E	1087	4.8	1447
Slovakia	Long-term	Mixed	48°38′34.1″N	19°32′21.8″E	803	5.5	780
Slovakia	Long-term	Mixed	48°46′22.1″N	20°44′36.3″E	773	5.6	862
Slovakia	Long-term	Mixed	48°46′18.8″N	20°43′32.3″E	738	5.7	840
Slovakia	Long-term	Mixed	48°47′23.8″N	20°40′07.3″E	621	6.2	769
Slovakia	Long-term	Mixed	48°45′35.1″N	20°42′56.9″E	845	5.3	906
Slovakia	Long-term	Mixed	48°36′57″N	19°33′57.6″E	693	6.6	796
Slovakia	Long-term	Mixed	48°37′26.1″N	19°35′59.9″E	786	6.2	854
Slovakia	Long-term	Mixed	48°37′55.6″N	19°34′17.4″E	733	6.4	821
Slovenia	Long-term	Mixed	45°45′13.7″N	14°59′42.2″E	909	6.9	1751
Slovenia	Long-term	Mixed	45°45′13.7″N	14°59′42.2″E	909	6.9	1751
Slovenia	Long-term	Mixed	45°45′13.7″N	14°59′42.2″E	909	6.9	1751
Slovenia	Long-term	Mixed	45°39′51.8″N	15°00′25.3″E	910	6.9	1756
Slovenia	Long-term	Mixed	45°39′51.8″N	15°00′25.3″E	910	6.9	1756

(continued)

Table 5.4 (continued)

Country	Period	Composition	Coordinates		Site characteristics		
			Latitude	Longitude	E	T	P
Slovenia	Long-term	Mixed	45°39′51.8″N	15°00′25.3″E	910	6.9	1756
Slovenia	Long-term	Mixed	45°37′21.4″N	14°48′52.9″E	917	6.9	1757
Slovenia	Long-term	Mixed	46°29′14.3″N	15°27′18.5″E	970	6.0	1464
Slovenia	Long-term	Mixed	46°14′49.6″N	14°03′40.3″E	1426	4.7	2770
Slovenia	Long-term	Mixed	46°14′56″N	14°03′40.2″E	1375	4.9	2738
Slovenia	Long-term	Mixed	46°14′55.6″N	14°02′44.1″E	1443	4.7	2780
Slovenia	Long-term	Mixed	46°15′02.5″N	14°02′43.9″E	1421	4.7	2767
Slovenia	Long-term	Mixed	46°15′08.5″N	14°02′34.8″E	1375	4.9	2738
Switzerland	Long-term	Mixed	46°52′33.5″N	7°41′14.9″E	899	7.2	1390
Switzerland	Long-term	Mixed	46°52′33.5″N	7°41′14.9″E	899	7.2	1390
Switzerland	Long-term	Mixed	46°52′33.5″N	7°41′14.9″E	899	7.2	1390
Switzerland	Long-term	Mixed	46°57′34.5″N	7°46′25.2″E	890	7.1	1448
Switzerland	Long-term	Mixed	46°57′34.5″N	7°46′25.2″E	890	7.1	1448
Switzerland	Long-term	Mixed	47°20′05.4″N	7°09′53.1″E	790	7.4	1302
Switzerland	Long-term	Mixed	47°20′15″N	7°09′05.4″E	558	8.8	1140
Switzerland	Long-term	Mixed	47°20′15″N	7°09′05.4″E	558	8.8	1140
Switzerland	Long-term	Mixed	46°56′45.6″N	7°39′42.5″E	981	6.7	1477
Switzerland	Long-term	Mixed	46°33′31.8″N	6°13′18.8″E	1364	4.8	1796
Bosnia and Herzegovina	Temporary	Mixed	44°38′30″N	16°39′36.1″E	725	11.6	937
Bosnia and Herzegovina	Temporary	Mixed	43°43′28″N	18°17′09″E	1300	8.3	992
Bosnia and Herzegovina	Temporary	Mixed	44°41′09.1″N	16°29′40.5″E	663	11.4	1028
Germany	Temporary	Mixed	49°05′08.4″N	13°18′23.5″E	1120	6.5	1078
Italy	Temporary	Mixed	41°52′14.4″N	14°16′51.3″E	1332	11.3	692
Italy	Temporary	Mixed	39°09′13.5″N	16°39′53″E	1182	11.3	969
Italy	Temporary	Mixed	46°12′06.4″N	11°12′36.9″E	1271	9.1	932
Italy	Temporary	Mixed	46°05′56″N	12°25′49″E	1100	8.0	1057
Italy	Temporary	Mixed	41°52′11″N	14°17′26″E	1289	11.0	692
Poland	Temporary	Mixed	49°37′10.3″N	18°55′07.5″E	528	9.0	1128
Poland	Temporary	Mixed	49°37′33.2″N	18°55′11.8″E	665	8.2	1128
Romania	Temporary	Mixed	45°32′15.3″N	25°52′51.2″E	1251	6.4	624
Serbia	Temporary	Mixed	43°19′06.1″N	19°51′50.6″E	1227	8.1	823
Serbia	Temporary	Mixed	43°24′22.3″N	21°22′41.1″E	691	8.7	668
Serbia	Temporary	Mixed	43°29′12″N	19°51′38″E	1221	8.2	839
Serbia	Temporary	Mixed	43°21′01.8″N	20°15′17.6″E	1470	6.7	821
Slovakia	Temporary	Mixed	48°40′40.7″N	19°28′12.6″E	1180	6.1	889

(continued)

Table 5.4 (continued)

Country	Period	Composition	Coordinates		Site characteristics		
			Latitude	Longitude	E	T	P
Slovenia	Temporary	Mixed	46°05′33.7″N	15°03′44.3″E	1030	8.4	1223
Spain	Temporary	Mixed	42°16′11.4″N	3°16′05.4″W	1525	9.5	631
Spain	Temporary	Mixed	42°12′03″N	2°43′07″W	1390	10.7	575
Ukraine	Temporary	Mixed	49°01′09″N	23°28′10″E	763	7.9	965
Ukraine	Temporary	Mixed	48°51′12″N	22°58′60″E	1084	6.4	1063
Bosnia and Herzegovina	Temporary	Monospecific	44°38′38.7″N	16°40′06.4″E	524	12.7	937
Bosnia and Herzegovina	Temporary	Monospecific	44°41′07″N	16°29′43.2″E	669	11.4	1028
Bosnia and Herzegovina	Temporary	Monospecific	43°42′25″N	18°15′44″E	1292	8.4	992
Bosnia and Herzegovina	Temporary	Monospecific	43°44′41″N	18°13′21″E	1680	6.1	1022
Bulgaria	Temporary	Monospecific	42°40′21″N	23°51′03″E	1050	9.8	521
Bulgaria	Temporary	Monospecific	42°46′45″N	23°52′52″E	1350	8.1	539
Bulgaria	Temporary	Monospecific	42°40′23″N	23°51′07″E	1000	10.1	521
Czech Republic	Temporary	Monospecific	49°17′06.6″N	16°44′21.4″E	490	9.6	517
Czech Republic	Temporary	Monospecific	49°17′05.1″N	16°44′24.3″E	485	9.6	517
Czech Republic	Temporary	Monospecific	49°10′17.3″N	19°04′54.5″E	767	7.7	1037
Czech Republic	Temporary	Monospecific	49°10′53.1″N	19°05′40″E	1131	5.7	1037
Czech Republic	Temporary	Monospecific	49°10′39″N	19°05′33.7″E	1146	5.8	1037
Czech Republic	Temporary	Monospecific	49°02′08.3″N	18°01′07.5″E	415	10.0	753
Czech Republic	Temporary	Monospecific	49°01′24.4″N	18°01′30.7″E	620	8.9	753
Germany	Temporary	Monospecific	49°03′45.9″N	13°16′17.2″E	720	8.6	1078
Germany	Temporary	Monospecific	49°03′49.3″N	13°16′06.1″E	695	8.7	1078
Hungary	Temporary	Monospecific	47°21′46.3″N	16°29′12.2″E	640	10.0	602
Hungary	Temporary	Monospecific	47°21′10.4″N	16°26′16.6″E	840	9.0	638
Italy	Temporary	Monospecific	39°09′08″N	16°40′12.3″E	1182	11.1	969
Italy	Temporary	Monospecific	45°57′43.7″N	11°16′26.6″E	1274	8.8	1064
Italy	Temporary	Monospecific	46°07′08″N	12°25′47″E	1090	8.0	1057
Poland	Temporary	Monospecific	49°37′20.8″N	18°54′52.6″E	520	9.0	1128
Poland	Temporary	Monospecific	49°37′25.18″N	18°55′28.65″E	691	8.2	1128
Poland	Temporary	Monospecific	49°25′58.7″N	20°54′11.2″E	830	7.7	814

(continued)

Table 5.4 (continued)

Country	Period	Composition	Coordinates		Site characteristics		
			Latitude	Longitude	E	T	P
Poland	Temporary	Monospecific	49°25′35.5″N	20°53′56.8″E	860	7.6	814
Poland	Temporary	Monospecific	49°26′01.8″N	20°52′46.4″E	1020	6.8	814
Poland	Temporary	Monospecific	49°25′54.7″N	20°52′40.8″E	1032	6.7	814
Romania	Temporary	Monospecific	45°19′15.3″N	25°48′19.3″E	970	8.1	588
Romania	Temporary	Monospecific	45°32′14″N	25°53′01.7″E	1277	6.3	624
Serbia	Temporary	Monospecific	43°29′24.2″N	19°52′06.9″E	949	9.6	839
Serbia	Temporary	Monospecific	43°21′03.2″N	20°15′18.7″E	1470	6.7	821
Serbia	Temporary	Monospecific	43°20′47.4″N	20°15′37.4″E	1650	5.7	819
Serbia	Temporary	Monospecific	43°24′22.5″N	21°22′41.7″E	695	8.7	668
Slovakia	Temporary	Monospecific	48°38′56.1″N	19°24′43.1″E	750	8.5	763
Slovakia	Temporary	Monospecific	48°38′58.4″N	19°24′58.1″E	750	8.5	763
Slovenia	Temporary	Monospecific	46°06′56.1″N	15°03′42.8″E	600	10.6	1223
Slovenia	Temporary	Monospecific	46°05′38.3″N	15°03′58.7″E	1070	8.2	1223
Spain	Temporary	Monospecific	42°16′23.6″N	3°16′12.6″W	1526	9.5	631
Spain	Temporary	Monospecific	43°07′40″N	4°58′30″W	1279	10.0	900
Spain	Temporary	Monospecific	43°07′38″N	4°58′48″W	1474	9.0	900
Spain	Temporary	Monospecific	42°12′05″N	2°43′19″W	1430	10.5	575
Spain	Temporary	Monospecific	42°12′06″N	2°43′34″W	1430	10.5	575
Spain	Temporary	Monospecific	41°46′32″N	2°27′24″E	1186	11.6	711
Spain	Temporary	Monospecific	41°46′30″N	2°27′26″E	1183	11.7	711
Ukraine	Temporary	Monospecific	49°02′01″N	23°33′11″E	586	8.8	946
Ukraine	Temporary	Monospecific	49°02′01″N	23°33′16″E	578	8.9	946
Ukraine	Temporary	Monospecific	49°03′48″N	22°53′44″E	892	7.2	1000
Ukraine	Temporary	Monospecific	49°03′41″N	22°53′15″E	908	7.2	1000

E elevation (m a.s.l.), T mean annual temperature (°C), P total annual precipitation (mm). Climate data is shown as mean of the period 1971–2000 (CRU database; Harris et al. 2020)

from around 600 m a.s.l. to more than 1600 m a.s.l., both in monospecific and mixed stands. The sites cover a large range of climate conditions with mean temperatures between 2.6 and 10.2 °C and annual precipitation between 517 and 2780 mm (Fig. 5.4).

5.5   Stand Growth

The 159 observational plots included in the network are located in such a way that they represent a broad environmental gradient (Figs. 5.5 and 5.6, Table 5.4). Table 5.5 gives an overview of the characteristics of the forest stands, which are included in the network. Even though the observation period of the temporary

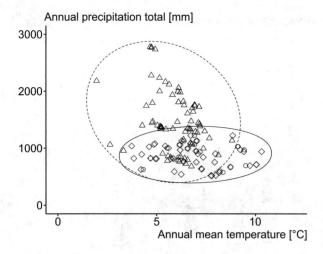

Fig. 5.4 Scatter plot of the total annual precipitation in mm and mean annual temperature in °C of the 89 long-term observational plots in mixed mountain forests (triangles) and 72 temporary observational plots in monospecific stands of beech (rhombuses) and mixed mountain forest plots (circles). Ellipses represent a convex hull of the temporary plots (solid line) and long-term observational plots in mixed mountain forests (dashed)

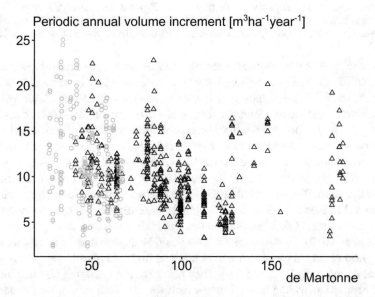

Fig. 5.5 Stand periodic annual volume increment, m^3 ha^{-1} $year^{-1}$ over de Martonne aridity index, mm °C^{-1} (de Martonne 1926). Long-term observational plots are shown with black triangles; temporary observational plots are shown with grey circles. Note that the values for periodic annual volume increments on the temporary plots were derived retrospectively (Heym et al. 2017, 2018)

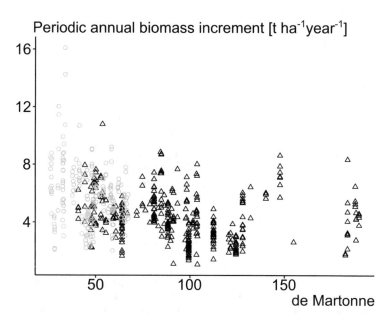

Fig. 5.6 Stand periodic annual biomass increment, t ha⁻¹ year⁻¹ over de Martonne aridity index, mm °C⁻¹ (de Martonne 1926). Long-term observational plots are shown with black triangles; temporary observational plots are shown with grey circles. Biomass was calculated by generalized biomass allometric equations (Forrester et al. 2017). Note that the values for periodic annual biomass increments on the temporary plots were derived retrospectively (Heym et al. 2017, 2018)

observational plots by retrospective calculation (see Sect. 5.3.1) only dates back to the year 2007, the data of the long-term experimental plots in particular offer a broad data basis for investigating productivity in terms of merchantable volume over bark and biomass of mountain forests at elevations of 600–1600 m a.s.l. in Europe (Fig. 5.2). The observation horizon of the long-term experimental plots reaches back several decades in most cases (Fig. 5.7) and, therefore, the dataset could be also used to answer the question of whether stand growth has changed in recent decades and how these changes are related to changes in climate conditions or forest structure (Torresan et al. 2020). The dataset is also suitable for testing if there was a shift in species-specific productivity of European beech, Norway spruce, or silver fir over recent decades and whether this shift was influenced by the mixture (Hilmers et al. 2019). Furthermore, the data could be used to answer questions, such as how productivity depends on site conditions and how these vary across geographical regions (Figs. 5.5 and 5.6). They also permit quantifying tree neighbourhood effects and analysing how climate affects the mechanisms of forest dynamics as well as inferring cause-effect relationships between disturbance events (e.g. droughts, outbreaks) and demographic changes (i.e. mortality and recruitment) (Lutz 2015).

Table 5.5 Overview of the stand characteristics of the 159 observational plots in mountain areas by study area

Country	Composition	Observation period	No. plots	No. Trees			Basal area (m² ha⁻¹)			Volume (m³ ha⁻¹)			Biomass (t ha⁻¹)			Volume increment (m³ ha⁻¹ year⁻¹)			Biomass increment (t ha⁻¹ year⁻¹)		
				min	median	max	min	median	max	min	median	max	min	median	max	min	median	max	min	median	max
Bosnia and Herzegovina	Mixed	Long-term	5	291	459	676	33.7	38.9	43.8	413.8	499.7	622.8	184.8	232.3	292.5	7.2	11.6	11.8	4.2	4.8	5.6
Bulgaria	Mixed	Long-term	2	459	536	614	38.9	55.8	72.8	499.6	827.8	1155.9	278.0	388.0	498.1	9.7	13.6	17.4	4.4	5.9	7.4
Germany	Mixed	Long-term	41	73	471	1259	12.7	40.2	80.7	194.1	546.1	1284.9	89.4	235.9	488.2	3.9	8.0	16.0	1.6	3.9	7.9
Poland	Mixed	Long-term	10	404	534	1080	37.4	42.4	46.9	498.0	615.8	744.1	237.1	351.8	458.8	8.0	9.8	14.4	4.0	5.0	7.7
Slovakia	Mixed	Long-term	8	154	256	513	22.3	41.8	60.6	405.5	726.7	1107.7	151.7	288.3	454.4	9.3	16.7	20.8	2.2	6.7	10.8
Slovenia	Mixed	Long-term	13	296	496	925	37.6	47.8	61.2	626.9	791.8	925.2	295.4	362.4	452.1	5.4	10.5	14.5	1.1	4.4	8.0
Switzerland	Mixed	Long-term	10	367	488	1192	24.9	33.7	56.6	252.0	489.3	866.3	124.4	191.9	468.2	6.6	14.8	19.4	2.9	5.5	8.7
Bosnia and Herzegovina	Mixed	Temporary	3	469	644	892	20.6	33.6	46.6	188.5	338.6	563.0	128.5	262.7	386.3	8.0	11.1	14.1	5.9	7.4	8.0
Germany	Mixed	Temporary	1	849	849	849	58.7	58.7	58.7	936.1	936.1	936.1	400.7	400.7	400.7	16.6	16.6	16.6	5.1	5.1	5.1
Italy	Mixed	Temporary	5	394	647	938	26.2	46.4	82.5	255.1	542.7	970.1	153.5	342.1	613.3	6.9	9.2	17.2	3.8	4.1	8.3
Poland	Mixed	Temporary	2	622	656	690	33.5	39.6	45.7	484.2	544.9	605.6	232.3	282.0	331.7	14.6	15.0	15.4	5.0	5.7	6.4
Romania	Mixed	Temporary	1	498	498	498	59.1	59.1	59.1	928.8	928.8	928.8	387.8	387.8	387.8	17.4	17.4	17.4	6.2	6.2	6.2
Serbia	Mixed	Temporary	4	305	450	950	43.4	44.1	45.7	479.8	733.7	747.0	320.5	368.3	403.0	7.9	9.3	10.3	3.1	4.1	5.2
Slovakia	Mixed	Temporary	1	549	549	549	61.0	61.0	61.0	979.6	979.6	979.6	401.4	401.4	401.4	21.9	21.9	21.9	6.5	6.5	6.5
Slovenia	Mixed	Temporary	1	545	545	545	59.3	59.3	59.3	846.5	846.5	846.5	479.3	479.3	479.3	10.9	10.9	10.9	5.3	5.3	5.3
Spain	Mixed	Temporary	2	984	1102	1221	51.4	53.8	56.2	454.4	479.6	504.8	293.6	299.0	304.4	8.8	9.0	9.2	3.8	4.0	4.1
Ukraine	Mixed	Temporary	2	427	455	483	56.4	57.9	59.3	780.1	879.0	978.0	457.5	488.7	520.0	14.2	16.4	18.6	5.8	7.4	9.1
Bosnia and Herzegovina	Monospecific	Temporary	4	299	572	781	26.3	31.0	45.6	252.9	380.9	497.7	199.7	273.2	423.3	5.7	9.4	13.6	4.3	6.5	8.0
Bulgaria	Monospecific	Temporary	3	510	551	1064	30.2	39.4	42.6	390.5	427.9	544.7	256.7	305.1	380.4	10.6	12.6	12.9	4.8	6.1	6.2

(continued)

Table 5.5 (continued)

Country	Composition	Observation period	No. plots	No. Trees			Basal area (m² ha⁻¹)			Volume (m³ ha⁻¹)			Biomass (t ha⁻¹)			Volume increment (m³ ha⁻¹ year⁻¹)			Biomass increment (t ha⁻¹ year⁻¹)		
				min	median	max	min	median	max	min	median	max	min	median	max	min	median	max	min	median	max
Czech Republic	Monospecific	Temporary	7	292	501	1061	22.2	47.5	53.6	322.1	633.5	1029.8	198.3	426.3	505.3	5.4	10.9	18.3	3.1	6.0	7.8
Germany	Monospecific	Temporary	2	766	919	1072	44.1	45.2	46.2	489.9	586.0	682.0	368.6	370.5	372.5	9.4	12.3	15.3	4.1	4.9	5.7
Hungary	Monospecific	Temporary	2	241	277	313	23.7	26.3	28.9	316.8	358.5	400.2	222.3	245.0	267.7	14.6	15.5	16.3	9.1	9.9	10.7
Italy	Monospecific	Temporary	3	428	637	920	30.6	53.6	77.6	227.3	579.2	816.9	232.8	308.2	517.5	4.2	7.3	20.9	2.0	3.1	7.6
Poland	Monospecific	Temporary	6	350	532	787	37.1	38.1	41.3	404.5	524.8	719.5	303.8	338.3	400.2	6.3	10.7	15.2	3.0	5.1	6.1
Romania	Monospecific	Temporary	2	641	684	728	42.6	43.4	44.1	558.2	588.4	618.7	360.4	372.8	385.2	13.1	14.7	16.3	6.6	6.6	6.7
Serbia	Monospecific	Temporary	4	466	474	520	44.1	44.3	58.8	545.9	742.6	742.8	403.4	408.7	577.4	6.0	7.8	10.0	3.2	4.2	5.9
Slovakia	Monospecific	Temporary	2	636	711	786	48.2	51.6	54.9	749.2	907.2	1065.1	412.5	445.2	478.0	17.8	18.4	19.0	6.9	7.1	7.2
Slovenia	Monospecific	Temporary	2	370	440	510	38.6	46.5	54.5	479.4	640.9	802.4	366.0	442.2	518.3	4.1	6.9	9.7	2.9	3.7	4.5
Spain	Monospecific	Temporary	7	466	960	1774	29.0	34.8	55.8	198.0	311.5	413.2	233.1	277.2	432.2	2.3	5.1	9.7	2.1	3.0	6.5
Ukraine	Monospecific	Temporary	4	458	465	533	39.0	51.1	54.0	479.6	740.9	789.7	355.4	482.3	520.8	10.4	13.1	14.8	5.3	6.1	6.5

Biomass was calculated by generalized biomass allometric equations (Forrester et al. 2017)

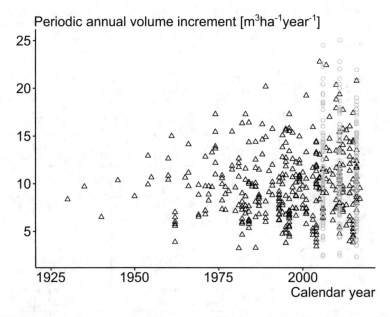

Fig. 5.7 Stand periodic annual volume increment, m³ ha⁻¹ year⁻¹ over the calendar years. Long-term observational plots are shown with black triangles; temporary observational plots are shown with grey circles. Note that the values for periodic annual volume increments on the temporary plots were derived retrospectively (Heym et al. 2017, 2018)

5.6 Tree Growth

From trees on the temporary plots and from trees in the buffer zone around the long-term plots, cores of 2437 European beech, 479 Norway spruce, and 907 silver fir trees were collected. Table 5.6 provides an overview of the tree-ring data. Some of the trees were up to 600 years old at the time of data collection, so that growth over several centuries could be analysed (Fig. 5.8, Table 5.6). With this dataset, questions can be answered as to whether the growth of the trees has changed over the long term and if there were species-specific changes in the growth trends during the last centuries (Figs. 5.8 and 5.9; Pretzsch et al. 2020a, b). Questions can also be answered at the single-tree level as to how tree growth changed along the site gradient and if there were different tree-growth patterns in mixed compared to monospecific stands (see Chap. 6 of this book: Pretzsch et al. 2021). Insights on radial growth responses to climate dynamics and extreme events in mixed versus pure stands of these species can also be gathered at specific sites of the network (e.g. Versace et al. 2020), as well as address questions on competitive interactions (e.g. Versace et al. 2019). Tree-growth series from mixed mountain forests can be also used to explore spatial-temporal patterns of the intra- and interspecific growth synchrony related to climate variation during the past century, which can help to identify between species temporal complementarity and the role of mixed mountain forests in the frame of climate change (del Río et al. 2021).

Table 5.6 Overview of the tree-ring data; dbh and age are related to the year 2017

Country	Species	Composition	No. plots	No. Trees	dbh (cm)			Height (m)			Age (years)			Diameter increment (mm year⁻¹)		
					min	median	max	min	median	max	min	median	max	min	median	max
Bosnia and Herzegovina	Beech	Mixed	5	87	15.6	52.5	167.0	9.1	28.6	35.3	65	199	569	0.09	2.60	15.57
Bulgaria	Beech	Mixed	2	47	23.0	35.8	56.6	19.6	26.0	41.5	90	115	151	0.18	2.82	13.87
Germany	Beech	Mixed	10	188	12.5	47.2	95.2	11.7	27.8	36.4	57	220	514	0.12	2.12	15.10
Italy	Beech	Mixed	5	127	8.4	28.3	73.0	7.6	19.7	37.4	50	106	212	0.09	2.53	16.58
Poland	Beech	Mixed	2	63	9.3	30.1	51.5	11.1	27.6	33.5	52	79	117	0.10	3.34	12.05
Romania	Beech	Mixed	1	25	10.2	28.3	40.2	12.5	28.6	34.3	72	104	154	0.08	2.09	11.06
Serbia	Beech	Mixed	6	132	12.1	54.8	76.6	13.3	32.7	39.8	61	151	353	0.01	3.14	15.00
Slovakia	Beech	Mixed	8	284	8.0	40.7	94.6	13.0	32.8	44.2	33	136	316	0.01	2.42	29.22
Slovenia	Beech	Mixed	5	67	18.3	58.0	118.2	17.1	28.2	42.3	95	179	539	0.12	2.26	13.64
Spain	Beech	Mixed	2	56	7.8	23.6	40.9	4.9	17.2	23.9	50	79	145	0.00	2.49	16.50
Switzerland	Beech	Mixed	1	9	40.9	49.9	68.8	17.1	26.8	32.6	74	113	160	0.36	3.68	16.78
Ukraine	Beech	Mixed	2	65	12.4	42.0	79.9	9.1	30.2	39.7	65	120	214	0.04	2.91	12.86
Bosnia and Herzegovina	Beech	Monospecific	4	58	19.4	34.7	53.2	11.7	21.1	35.0	53	100	351	0.12	2.64	16.09
Bulgaria	Beech	Monospecific	3	91	11.0	29.0	49.5	12.4	24.5	29.5	51	85	136	0.08	3.08	12.46
Czech Republic	Beech	Monospecific	7	220	7.0	35.1	83.3	4.3	24.0	44.0	29	115	389	0.01	2.14	17.55
Germany	Beech	Monospecific	2	56	8.9	32.3	58.7	7.8	23.1	35.3	67	96	117	0.10	3.14	11.52
Hungary	Beech	Monospecific	2	42	22.2	38.8	46.6	23.3	26.6	29.8	70	91	107	0.04	3.82	13.32
Italy	Beech	Monospecific	4	109	8.0	29.5	81.8	6.5	18.8	32.4	59	125	199	0.06	2.30	18.08
Poland	Beech	Monospecific	6	208	7.0	34.4	66.8	8.2	27.0	39.1	41	118	298	0.07	2.71	12.95
Romania	Beech	Monospecific	2	59	16.6	34.8	56.9	15.8	27.4	31.8	69	102	134	0.10	3.04	12.19
Serbia	Beech	Monospecific	3	55	14.7	34.3	81.3	11.1	28.2	35.0	65	203	299	0.06	1.60	10.97
Slovakia	Beech	Monospecific	2	47	11.4	38.6	64.3	10.8	34.2	46.1	37	100	122	0.20	3.62	12.91
Slovenia	Beech	Monospecific	2	53	10.5	45.1	63.2	7.7	26.9	32.4	38	173	308	0.11	2.31	10.06

Spain	Beech	Monospecific	7	181	7.0	28.9	50.9	5.7	17.4	25.8	36	141	318	0.01	1.78	16.85
Ukraine	Beech	Monospecific	4	108	10.2	40.7	84.4	9.8	27.9	39.3	60	100	308	0.06	3.30	15.24
Bosnia and Herzegovina	Fir	Mixed	4	72	12.4	65.6	198.0	5.9	28.3	35.9	17	230	591	0.10	3.92	21.10
Bulgaria	Fir	Mixed	2	52	27.2	46.1	69.6	21.1	29.8	42.3	53	100	142	0.36	3.99	15.18
Germany	Fir	Mixed	9	169	34.2	57.9	124.5	17.2	34.2	46.7	110	209	521	0.06	2.53	19.41
Italy	Fir	Mixed	6	145	7.0	43.2	82.5	4.5	23.9	35.6	23	155	239	0.01	2.87	17.38
Poland	Fir	Mixed	2	20	16.4	35.8	54.1	16.5	27.3	35.7	60	75	136	0.14	4.47	13.85
Serbia	Fir	Mixed	5	114	17.2	61.0	76.6	11.3	33.1	38.5	60	131	349	0.01	4.40	23.40
Slovakia	Fir	Mixed	7	254	19.4	55.2	105.7	15.6	37.5	49.5	90	151	256	0.06	3.00	34.24
Slovenia	Fir	Mixed	4	37	41.4	59.9	118.2	22.3	28.5	46.0	118	198	539	0.14	2.94	16.14
Switzerland	Fir	Mixed	1	11	44.1	64.0	103.1	23.1	30.0	44.0	46	115	169	0.30	6.32	24.72
Ukraine	Fir	Mixed	2	33	12.7	43.3	74.8	10.1	30.8	39.9	77	107	221	0.05	3.22	13.52
Bosnia and Herzegovina	Spruce	Mixed	3	43	9.9	136.0	200.0	4.3	29.6	39.6	22	245	582	0.08	4.65	20.16
Bulgaria	Spruce	Mixed	1	28	41.9	57.9	71.4	32.9	39.2	43.2	78	104	129	0.74	4.34	16.75
Germany	Spruce	Mixed	10	225	26.5	56.5	102.2	17.2	34.5	46.1	75	214	501	0.04	2.64	18.19
Italy	Spruce	Mixed	1	1	16.5	16.5	16.5	15.7	15.7	15.7	82	82	82	0.29	1.76	4.38
Poland	Spruce	Mixed	2	25	14.7	34.2	51.5	15.3	29.0	39.1	46	70	89	0.07	4.05	19.40
Romania	Spruce	Mixed	1	11	21.5	49.7	72.9	19.4	36.0	42.9	68	88	97	0.18	5.59	17.50
Serbia	Spruce	Mixed	3	46	16.3	40.3	81.2	10.7	30.2	41.0	39	117	321	0.00	3.11	18.26
Slovakia	Spruce	Mixed	2	44	21.2	54.4	78.8	21.2	41.9	49.9	51	116	141	0.19	3.91	23.21
Slovenia	Spruce	Mixed	5	42	24.2	52.0	92.4	21.7	31.2	45.5	113	150	429	0.11	3.23	19.69
Switzerland	Spruce	Mixed	1	14	38.4	58.7	76.9	25.3	31.6	41.8	66	138	188	0.06	4.02	28.40

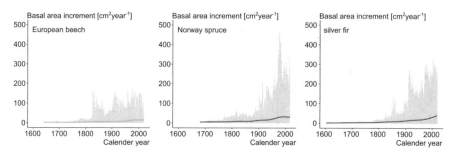

Fig. 5.8 Single-tree basal increment in cm² year⁻¹ over the calendar years of European beech, Norway spruce, and silver fir in the long-term and temporary observational forest plots. Coloured lines were generated by fitting a loess curve

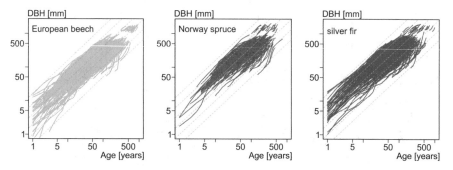

Fig. 5.9 Stem diameter-tree age trajectories of the 2437 European beech, 479 Norway spruce, and 907 silver fir trees during the last few centuries, in double logarithmic representation. Most trees show a linear increase in stem diameter with progressing age (reference lines $\ln\ln(d) = a_0 + a_1\ln\ln(\text{age})$ with $a_0 = 1$ and varying a_0) and no asymptotic growth curve pattern

5.7 Growth Characteristics Analysed Along Elevation Gradients

The data collected from the observational plots cover a wide range along an elevation gradient of ~600–1600 m a.s.l. (Figs. 5.10, 5.11 and 5.12). Both at stand level (Fig. 5.10; Hilmers et al. 2019) and at single-tree level (Fig. 5.11; Pretzsch et al. 2020a, b), it can be investigated if there are any growth trends along the elevation gradient and whether these trends have changed during the last century. Furthermore, the dataset offers the possibility to analyse if the elevation-dependent growth trends were different in the mixed compared to the monospecific stands and how the mixture, i.e. the interspecific competition, influences such changes (Figs. 5.11 and 5.12). Accordingly, intra- and interspecific growth synchrony in response to inter-annual fluctuations, as an indicator of tree species dependence on climate variability, can vary along the elevation gradient (del Río et al. 2021).

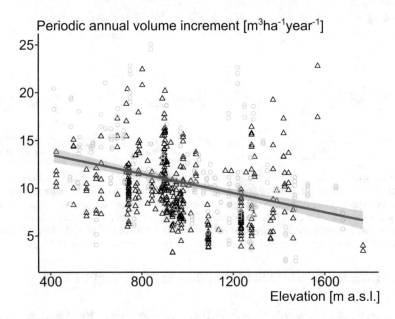

Fig. 5.10 Stand periodic annual volume increment (PAIV) in m³ ha⁻¹ year⁻¹ over elevation in m a.s.l. Black triangles represent long-term observational mixed mountain forests; grey circles show the PAIV of the temporary observational plots in monospecific beech stands and mixed mountain forests. The blue line is based on a linear model and the grey area indicates the 95% confidence interval. Note that the values for PAIV on the temporary plots were derived retrospectively (Heym et al. 2017, 2018)

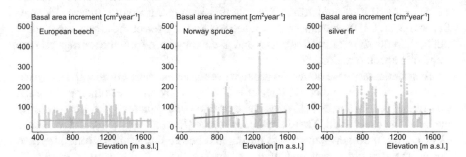

Fig. 5.11 Single-tree basal area increment in cm² year⁻¹ over elevation in m a.s.l. of European beech, Norway spruce, and silver fir in the long-term and temporary observational forest plots. Coloured lines are based on linear models

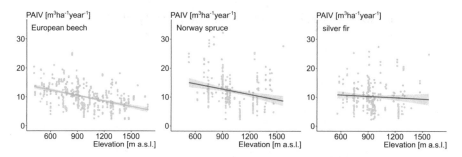

Fig. 5.12 Stand periodic annual volume increment (PAIV) in m^3 ha^{-1} $year^{-1}$ over elevation in m a.s.l of European beech, Norway spruce, and silver fir in the long-term and temporary observational forest plots. The periodic annual volume increment of the three tree species was scaled using the species share derived from SDI proportions (Pretzsch and Biber 2016). The coloured lines are based on linear models and the grey areas indicate the 95% confidence interval. Note that the values for PAIV on the temporary plots were derived retrospectively (Heym et al. 2017, 2018)

5.8 Concept of Statistical Evaluation of Drought Events

The analysis of tree growth-related data (e.g. tree-ring width, basal area increment, or stable isotope signatures) with the aim of detecting significant event years related to critical conditions for growth, such as droughts, firstly requires the use of reliable climate indices allowing their identification. Simpler indices can be applied, such as the de Martonne index (de Martonne 1926), which only requires temperature and precipitation data. In addition, more developed indices including also potential evapotranspiration, such as the PDSI (Palmer Drought Severity Index) (Palmer 1965) or the SPEI (Standardized Precipitation-Evapotranspiration Index) (Vicente-Serrano et al. 2010), have been developed to carefully select those years that stand out both in the short term (few years) and in the long term from the average climate trend of a specific region.

To infer the response of trees to drought events, an increasingly adopted approach is that of the Lloret indices (Lloret et al. 2011). They allow estimating, for each considered year in a tree-ring series, three components explaining the growth during and after the event itself. These are the resistance (*Rt*), recovery (*Rc*), and resilience (*Rs*), i.e. the ratio between growth during and before the drought event, after and during the drought event, and after and before the drought event, respectively. Conceptually, resistance represents the ability of a tree to cope with the stressful event during the same growing season, by possibly maintaining the previous growth rates, and hence is the most significant parameter to evaluate. Resilience and recovery are, on the other hand, related to the ability of the tree to sustain precedent growth rates in the short term after the event. It is, therefore, necessary to carefully select the time frame, in terms of years, of the average pre- and post-drought growth rate and to adjust it to species-specific requirements related to ecophysiology. The selection of the samples to be included in the resilience analysis is also of paramount importance. For instance, depending on the position of the tree crown within the vertical layering of the stand, the growth response to a drought event might differ consistently. Thus, separating dominant from dominated individuals and proceeding to differential modelling of their growth patterns are crucial to obtain consistent results. Recently, the Lloret indices have been integrated in an R package, called "pointRes" (Van der Maaten-Theunissen et al. 2015), which allows detecting both negative and positive pointer years, information that can be combined with that of drought indices to improve the accuracy of the investigation and to allow evaluating the growth response also during particularly favourable growing conditions. Normally, the resistance, resilience, and recovery analyses can be carried out on specific, pre-selected event years. It might be of interest, however, also to apply the above-mentioned Lloret indices to continuous, long-term series in order to investigate the response of trees to increasingly frequent drought events, which are affecting not only the growth patterns in the short term but also in the mid- and likely in the long term. In this context, it becomes relevant the removal of autocorrelation from the tree-growth series, using for instance ARMA (autoregressive moving average) or ARIMA (autoregressive integrated moving average) models and calculating Rt, Rc, and Rs on the residuals (Fig. 5.13).

When studying forests' response to drought events, it is possible to work at different spatial scales, i.e. tree level and stand level. Therefore, it is fundamental to consider the hierarchical and non-independent structure of the data and to assess possible nesting effects. To this extent, mixed models with random and fixed effects proved to be a valid solution, as they allow considering the above-mentioned issues, e.g. separating site- and plot-related variability from that linked to the factors of interest, such as species mixing and forest vertical and horizontal structure.

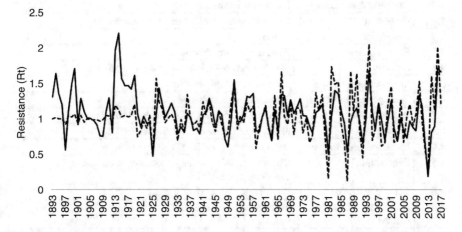

Fig. 5.13 Resistance index (Rt) calculated on a sample beech tree. The bold line represents the Rt calculated on basal area increment (BAI) directly, whereas the dashed line is the same index calculated on the BAI residuals after ARMA modelling

5.9 Climate Smartness

5.9.1 Assessing Climate-Smart Indicators

In order to assess climate smartness of forestry on stand level, experimental plots were evaluated on the basis of selected indicators. These were based on the Forest Europe C&I (Forest Europe 2015) set for sustainable forest management (https://foresteurope.org/sfm-criteria-indicators/). The assessment included a subset of 17 indicators, which were recorded on monospecific beech as well as mixed beech, spruce, and fir stands (Table 5.7). The selected indicators are characterized in particular by an area reference and a simple assessment method, which can be standardized for any region. Each indicator is characterized by measurable characteristics of assessment, which allow a more detailed evaluation (Table 5.7; see Chap. 6 of this book: Pretzsch et al. 2021). The assessment follows the procedure developed by Pfatrisch (2019).

In a recent exercise on literature review and multicriteria analysis, Santopuoli et al. (2020) highlighted that a subset of 10 indicators, from the current Pan-European C&I set (Forest Europe 2015), were more frequently used to assess adaptation and mitigation management strategies of forest ecosystems. In particular, carbon stock (mitigation), tree species composition (adaptation and mitigation), and forest damage (adaptation and mitigation) showed the highest priority value.

Table 5.7 Selected climate-smart indicators for field assessment according to the method of Pfatrisch (2019)

Nr	Indicator	Characteristic of assessment (n)
1.2	Growing stock	Growing stock (1)
1.3	Diameter distribution	Diameter distribution (1)
1.4	Carbon stock	Carbon Stock, development, substitution (3)
2.3	Defoliation	Defoliation (1)
2.4	Forest damage	Risk, impact, number of damages (3)
2.5	Stability	Slenderness coefficient, tree height, stand density (3)
3.1	Increment and felling	Increment, felling, relation (3)
4.1	Tree species composition	Tree species composition (1)
4.2	Regeneration	Area, height, density, potential, browsing (5)
4.3	Naturalness	Naturalness (sp. comp.), soil scarification (2)
4.4	Introduced tree species	Introduced tree species (1)
4.5	Deadwood	Quantity, standing volume, decomposition rate, light exposure, stand age (5)
4.6	Genetic resources	Phenotypic similarity, gen conservation (2)
4.8	Threatened forest species	Threatened forest species (1)
4.91/2	Distribution of tree crowns	Canopy level (horizontal), crown layers (vertical) (2)
6.1	Accessibility (for recreation)	Distance to road, road density (2)

Each indicator had 1–5 characteristics of assessment

5.9.2 European Dataset of Climate-Smart Indicators

The assessment of the climate-smart indicators was carried out on 46 trial plots in six different countries. The experimental plots cover a large elevation gradient from 40 to 1463 m above sea level as well as broad spatial gradients and different management types (Table 5.8). The resulting European-wide dataset allows an analysis of climate smartness at European and regional level with country- and stand-specific properties. In Chap. 6 of this book (Pretzsch et al. 2021), a more detailed information about climate smartness assessment is presented using spruce-fir-beech mixed mountain forests as case study.

5.9.3 Linking Yield and Climate-Smart Indicators: Research Objectives

The presented approach of climate smartness assessment is a promising way to analyse and estimate substantive yield data concerning climate smartness in forest management. With the evaluated effects, an optimized list of indicators can contribute to a more effective and sustainable climate-smart forestry. An iterative participatory process with input from experts (CLIMO) has recently provided the basis for developing a comprehensive and shared definition of climate-smart forestry (Bowditch et al. 2020). On the European level, climate-smart forestry indicators supported by network analysis approach can help to deepen the understanding about the development of mountain mixed forests along a broad geographical gradient. Linking yield data from long-term experimental plots with climate-smart indicators can provide valuable insights in forest dynamics with special regard to climate smartness. The obtained results can provide the basis for improvement of a future-oriented European forest management as well as regional silvicultural guidelines.

5.10 Soils

Soils in native forests rarely experience significant disturbances that are more common for soils in other land-use systems. However, climate change and/or deforestation can cause dramatic changes in the quality of forest soils. Alteration of quality or quantity of soil organic matter in forest soils is one of the most important consequences of climate change (Raison and Khanna 2011). Deforestation causes change in hydrological processes, which can enhance surface runoff and soil erosion, increase the recharge of groundwater and cause the reduction of organic carbon, nitrogen, phosphorus, and exchangeable potassium, calcium, and magnesium (Pennock and van Kessel 1997). The rate of soil degradation in such changed conditions depends largely on the type of bedrock, which was until recently considered of subordinate significance compared to climate and pedological characteristics (Jiang et al. 2020). Not only bedrock has a significant role in vegetation growth by

Table 5.8 Plot and yield data of analysed experimental plots, including species composition, type of management, and climate-smart indicator values

Country	No. of plots N	Species composition	Type of management		Elevation m a.s.l.	No. of trees N ha⁻¹	Basal area m² ha⁻¹	Volume m³ ha⁻¹	Periodic annual volume increment m³ ha⁻¹ y⁻¹	Climate-smart indicator Value
Bosnia and Herzegovina	10	Beech; spruce-fir-beech	Selective cutting	*mean*	1275	469	39.1	468	9.5	2.55
				min-max	1110–1400	297–701	33.9–44.2	390.1–538.1	7.0–11.6	2.5–2.6
Germany	12	Spruce-fir-beech	Combined shelterwood and femel-coupe system; set aside	*mean*	1135	545	42	587.7	8.2	2.3
				min-max	860–1463	241–1259	20.3–80.7	211.1–1284.9	3.8–14.5	1.9–2.7
Poland	4	Beech	Selective cutting; set aside	*mean*	936	561	38.9	547.1	7.4	2.3
				min-max	830–1032	350–804	37.7–41.7	436.3–677.1	4.7–15.4	2.0–2.4
Scotland	15	Sitka spruce; pine; oak; pine-oak	Shelterwood; clearcutting	*mean*	210	953	37.3	424.6	----	2.1
				min-max	40–450	88–2867	12.0–72.9	97.2–1163.8		1.5–2.9
Slovenia	2	Spruce-fir-beech	Selective cutting; set aside	*mean*	1432	575	56.7	831.7	13.5	2.6
				min-max	1421–1443	500–650	56.0–60.8	738.2–925.2	13.3–13.7	2.5–2.7
Spain	3	Pine; oak; pine-oak	Selective cutting	*mean*	787	1731	50.9	368.5	13.8	1.9
				min-max	765–815	998–2397	40.2–59.3	263.6–454.8	9.3–16.8	1.7–2.1
All	46			*mean*	963	806	44.2	537.9	10.5	2.0
				min-max	40–1463	88–2867	12.0–80.7	211.1–1284.9	3.8–14.5	1.5–2.9

regulating physical and chemical properties in soils, but it can also change the response of vegetation to climate factors (Jiang et al. 2020).

The multilayer and comprehensive network of beech in European mountains (study 2; see Sect. 5.4) contains soil data that can be used for spatial and temporal analyses of soil quality, testing existing and establishing new erosion indices, and predicting landform processes in the case of climate and land-use changes. This dataset contains a total of 76 soil samples from 20 pure beech forest stands from Bosnia and Herzegovina, Bulgaria, Czech Republic, Germany, Italy, Poland, Romania, Serbia, Slovakia, Slovenia, and Spain. Five types of bedrock were identified: limestone (8 plots, 28 samples), sandstone (5 plots, 20 samples), granite (5 plots, 20 samples), quartzite (1 plot, 4 samples), and andesite (1 plot, 4 samples). The Manual for Sampling and Analysis of Soils (Carter and Gregorich 2007) was followed for soil sampling and analyses. Soil samples were collected at 4 depths along the soil profiles (0–10, 10–20, 20–40, and 40–80 cm). The following soil characteristics were determined on all samples: pH; electrical conductivity; redox potential; content of $CaCO_3$; content of aggregates >2 mm; grain size composition; content of organic carbon and nitrogen; concentration of available ions Al, Ca, Mg, K, Na, Mn, and Fe; content of major and minor elements; and a concentration of carbonate, nitrite, sulphate, nitrate, bromide, oxalate, phosphate, fluoride, and chloride ions.

One of the most apparent results is that bedrock properties play a significant role in beech forest soil characteristics (Fig. 5.14). The PCA separated stands on limestone with the highest content of organic carbon (average 4.51%), pH of 6.2, and high content of exchangeable Ca, Mg, and K (24.04 ppm, 9.24 ppm, and 13.35 ppm, respectively) from sandstone and granite substrates. Soils on sandstone have 4 times and granite soils have 3.6 times lower content of organic carbon. Soil pH for both sandstone and granite bedrock is 4.9, and the exchangeable K is 0.6 times lower than limestone soils. Quartzite and andesite soils have similar characteristics to granite soils but cannot be further discussed due to the small number of samples. A group of soils with the bedrock that is classified as limestone are not typical carbonate rocks but are limestone moraine and calcareous sandstone, and although they

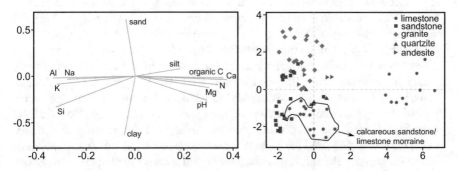

Fig. 5.14 Principal component analyses of pure beech forest soils developed on a range of bedrock types

have some characteristics of limestone soils, they are also by their textural and mineralogical properties close to sandstones (Fig. 5.14).

Obtained results are a good indication that soils on different bedrocks respond differently to environmental changes. Further investigations are aimed towards exploring existing soil erosion indices and developing new indicators that would be more appropriate for forest soils. Such information would be of great assistance for forest management and possible mitigation measures against climate change.

5.11 Genetic Resources

Evaluating genetic structure of forest stands provides insights into in situ surviving potential of tree species and their population. High genetic variation and elevated diversity of alleles assume long-term persistence, providing a good chance for populations to adapt to changing environments (Falk et al. 2006). In the past three decades, molecular techniques have become more and more available to determine the quality and quantity of the genetic diversity preserved within and among populations. Studies have shown the overall high genetic potential in forest tree populations as most of the species exhibited high within-population genetic diversity and relatively low divergence among adjacent populations (Porth and El-Kassaby 2014). Moreover, genetic variation is associated with the geographic distribution of the alleles and depends on the past historical events and demographic processes of populations. Accordingly, the levels of population genetic variation exhibit spatial and temporal differences that should be taken into consideration in any evaluation, like nature conservation activity or when forest management is designed (van Dam 2002).

Mountain forests are among the most exposed ecosystems to climate change (Becker and Bugmann 2001; Nogués-Bravo et al. 2007). Due to the fast environmental shifts, selective pressure acts on the genetic constitution of mountain tree species. European beech, as one of the most widespread trees of the European mountain forests, is presumed to be heavily affected (Kramer et al. 2010). However, beech has been thoroughly investigated because of its wide distribution and due to the great economic importance, so that a wealth of ecological, palaeoecological, and genetic data is available from the literature (Buiteveld et al. 2007; Magri 2008). All these data provide a good start to evaluate the genetic diversity of beech plots established within CLIMO along the distribution range of the species.

Twenty to twenty-five European beech trees were sampled from 12 plots of study 2, including nine countries (Bosnia, Bulgaria, Romania, Hungary, Slovakia Poland, Germany, Italy, and Spain). Six highly variable nuclear microsatellite markers revealed an equilibrated distribution of the genetic variation along the CLIMO plots. As expected, the genetic distance increased with the geographic distance among the study plots, and the highest variation was detected within several beech plots of the western Balkan Peninsula. This region is considered to be an important glacial refuge and a genetic hotspot for many broadleaf species including beech

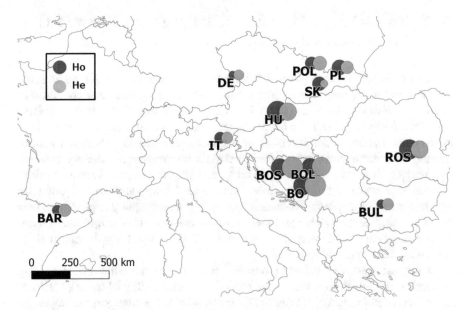

Fig. 5.15 Observed (Ho) and expected heterozygosity (He) of the 12 CLIMO European beech plots. The size of the circle is proportional with the extent of the genetic variation detected in each sample plot. BUL: Bulgaria; BOS, BOL, BO: Bosnia; ROS: Romania; HU: Hungary; SK: Slovakia; POL, PL: Poland; DE: Germany; IT: Italy, BAR: Spain

(Magri et al. 2006). Along with the evaluation of the genetic data in several sampling plots like that from Romania, Slovakia, and Hungary, observed heterozygosity values were higher than the expected heterozygosity values (Fig. 5.15). This means the abundance of heterozygote trees, which might be explained with the sampling strategy, as dominant trees were preferably collected and, most probably, these individuals were in a higher rate heterozygous. Moreover, Petit et al. (2001) reported a pronounced trend of increase in observed heterozygosity with the ageing of the stands. Another explanation should refer to management activity that might have altered the natural genetic composition of the stand. Genetic indices correlated with the climate variables provide further aspects on the frequency distribution of alleles and the environmental variables.

Molecular analysis, detecting neutral genetic variation, is a helpful tool in characterizing the variation of the experimental plots of forest trees and to determine the relationships between groups of individuals or among different plots. However, it should be emphasized that loci with adaptive significance likely were not sampled by these techniques and, accordingly, important adaptive variants of the samples may show different patterns. The selective pressures that affect the populations' adaptiveness are more difficult to be precisely identified, because of the huge genome of forest trees. Mapping of species' entire genome will identify more coding genes and candidates explaining adaptive variation (van Dam 2002).

5.12 Trans-Geographic Database of Long-Term Forest Plots in Mountainous Areas

On the one hand, the establishment of long-term experiments is very laborious, but on the other hand, they provide the unique insight into growth and condition of individual trees and the whole stand. The longer the observation period associated with interdisciplinary data, the better scientific and practical potentials are to study given natural phenomena. Therefore, it is important to make researchers aware of existing long-term plots, report their metadata for judgement whether they are appropriate for a special research question, and get eco-political support for their maintenance over longer time periods. Beyond the CLIMO networks of long-term and temporary plots (Sect. 5.3.4), here we present an exemplary approach on how to make public and available and memorize the existence of a unique set of long-term experiments that answered our research questions, but it may be essential for answering future questions as well.

In the last years, the study of mountain forests has been the focus of increasing research efforts in Europe due to the recognized vulnerability of these systems to the effects of global change (EEA 2012). This has led to the emergence of important research networks, such as the MOUNTFOR initiative (a Project Centre of the European Forest Institute, http://mountfor.fmach.it/) or more recently the SENSFOR and CLIMO Cost Actions (ES1203, CA15226). However, the interest of the European forest science community in mountain forests is not new, as shown by the large number of long-term research plots that have been established across the major mountain ranges of Europe during the last decades. Despite some of the long-term data derived from these plots that have been used in relevant scientific or technical publications (e.g. Martín-Benito et al. 2008; Pretzsch et al. 2015), the existence of the vast majority of these experiments is only known at the local level. Then, its potential use for trans-geographic assessments of forest growth trends (among other issues) is insufficiently exploited. In order to expand the visibility of these long-term experiments and promote future collaborative research initiatives among European scientists, the CLIMO consortium started an action aimed at compiling metadata about existing long-term forest plots established in European mountains. With this purpose, a number of CLIMO representatives conducted different surveys in their respective countries of origin to collect descriptive information on these forest plots, in particular with respect to the following items: the objective, the location (including geographical coordinates and elevation), the year of establishment, the surface, the main tree species and stand age (if even-aged), a short description of the data collected, and the organization and contact details of the person responsible of the plots and related publications (only if available in English). To be included in the surveys, experiments might be established at the latest by the end of the twentieth century and located in mountainous areas. However, these criteria were not applied restrictively (in part because the concept of "mountainous area" varies between a country and another). Accordingly, permanent plots established more recently, but with the aim of evaluating mountain forest responses to climatic changes, were also considered as well as plots located in lowlands under the frame of studies conducted

Fig. 5.16 Location of the long-term forest plots included in the ESFONET database

along elevation gradients. Plots established under the frame of large national/ regional forest inventories were not included.

All the collected information was compiled in a common database. In total, the database comprised metadata of 554 permanent plots distributed along 15 countries and covering the major European mountain ranges (Fig. 5.16). A total of 29 species were represented in the plots, with *Picea abies*, *Fagus sylvatica*, and *Pinus sylvestris* being the most dominant (Fig. 5.17). Most of the plots were established during the second half of the twentieth century and about two-third of them were located above 500 meters above sea level (Fig. 5.17). The dataset can be consulted online via a user-friendly interactive application designed with the *Shiny* R package (https://climoproject.shinyapps.io/CLIMO_dataset/). The application code is stored in an open repository (GitHub) and allows easy implementation of new data.

Under the frame of the CLIMO Cost Action, this database is considered the basis on which to build a reliable European Smart Forest Network (ESFONET) of long-term experiments strategically distributed across environmental gradients that may provide reliable data to assess the capacity of mountain forests to adapt to climatic changes. The way data is collected and organized determines the level of quality and the accessibility of information, in turn being critical to the success of the network. Implementing systematic procedures for data processing (including data quality evaluation and validation) and testing is, therefore, essential for eliminating errors and reducing uncertainty before data are transferred to a final database (quality assurance, QA) and for controlling the quality of submitted data stored in the final database (quality control, QC).

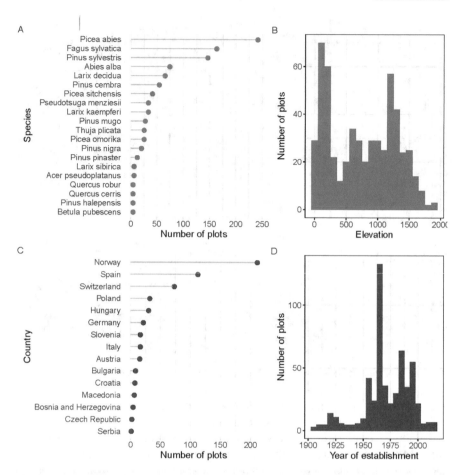

Fig. 5.17 Overview of some of the information collected in the database: (**a**) the 20 species most present in the database (ordered according to the number of plots in which they are present), (**b**) distribution of the plots according to the elevation, (**c**) number of plots located in the participating countries, and (**d**) distribution of the plots according to the year in which they were established

5.13 Discussion and Conclusion

5.13.1 *Exploiting Scattered Long-Term Experiments for Assessing Stand Growth, Resistance, and Climate Smartness by Pooling and Overarching Evaluation of Data*

Despite the credo that substantial information and knowledge require big datasets, overarching pooling and evaluation (Liang et al. 2016; Hilmers et al. 2019; Pretzsch et al. 2019) are still an exception in forest ecology or silviculture. For instance, long-term experiments on thinning were established and treated in a way standardized by, e.g. IUFRO (Becking 1953; Johann 1993); however, their evaluation is often restricted to the country or even to the regional level. The tremendous efforts and costs of maintaining long-term experiments may restrain the respective institutions from offering and exchanging their valuable data before having it exploited for their own purposes and publications. However, the evaluations at the state or national level are often of limited value. Samples and evaluations restricted to the state or country level may cover just a small portion of the natural distribution and niche of a tree species. Thus, general and reliable knowledge and relationships between growth and site conditions require trans-geographical evaluations. The broader the range of covered site conditions, the sample size, and the genetic diversity, the more robust and overarching the results for understanding, modelling, and scenario analyses and predictions (Liang et al. 2016; Zeller et al. 2018; Pretzsch et al. 2020a, b).

On the one hand, overarching networks are relevant for monitoring anthropogenic impacts on forest ecosystems and tree species. Local observations may indicate changes, but not their dependency on large-scale environmental conditions, and any shifts in the distribution ranges. On the other hand, overarching networks may contribute to adaptation and mitigation of forest ecosystems to climate change; overarching evaluations and robust growth-site relationships have gained higher importance under climate change. Many studies contribute to better understand the suitability of a given tree species under the expected future climate in, e.g. Central Europe by studying its performance in the north or south, where the species faces already at present the climate predicted for Central Europe in the future (Hanewinkel et al. 2014). This kind of trans-geographical studies is important for detection and remedy of climate changes along latitudinal gradients in the lowlands. And they are even more important along elevation gradients in the mountain forests, where environmental changes may happen in much more narrow space and in shorter time spans with severe consequences for a plenitude of socio-economic forest functions and services (Seidl et al. 2019; Hilmers et al. 2020).

5.13.2   The Information Potential of Long-Term Versus Inventory Plots

The pros and cons of long-term experiments compared with temporary plots have been questioned repeatedly (Gadow 1999; Nagel et al. 2012). Often, long-term experiments have been abandoned in order to cut costs. The common reasons for giving up long-term experiments are that forest areas with long-term experiments have to be left out from regular forest operations and that their establishment and survey are costly. In addition, it hardly fits to the contemporary funding organizations and the *zeitgeist* that it requires a couple of years for getting the first results.

Long-term experiments are also available for other ecosystems. In ecological research (LTER), agriculture and grassland (Rothamsted Research), soil science (LTE), or agroecosystems (LTAE), long-term observations have similar importance

(Blake 1999; Redman et al. 2004; Körschens 2006) as in forest ecology. Compared with other ecological fields, long-term experiments in forests are even under higher pressure due to their particular longevity and high space consumption. As forest land is often disregarded by politicians and other decision makers and extensively transformed into agricultural areas, urban building, and infrastructures, long-term experiments are frequently sacrificed and valuable long time series of survey discontinued.

It is an often heard, but misleading argument, that forest inventories that have been established during the last few decades at national or enterprise levels can substitute for the information potential of long-term experiments (Gadow 1999). Certainly, temporary plots of forest inventories may be harnessed by innovative Big Data methods, such as geospatial random forests and geostatistical mixed-effects models (e.g. Liang et al. 2016). For obtaining information about the status quo at the enterprise or landscape level, inventories are ideal, as this is what they are designed for. Furthermore, forest inventories provide the data for initializing simulation models, for forest planning and scenario analyses.

In the following, we summarize the main reasons why long-term experiments excel by far the information potential of temporary plots and why they are highly important for ecological monitoring and research (Pretzsch et al. 2019). First, only long-term experiments allow to study the long-term structure and growth development at the tree and stand level. They allow revelation of changing ecological processes, legacy and ecological memory effects, and the long-term consequences of ecosystem alteration. In contrast, temporarily observed inventory plots often lack information about the stand history and any intermediate yield before the plot establishment. This lack of information can be only partly remedied by establishing permanent inventory plots.

Second, long-term experiments are necessary for revelation of the cause-effect relationships between various stand treatment options and the response at the tree and stand level. In contrast to inventory plots, experiments usually analyse tree and stand reaction ceteris paribus, i.e. under parity of tree provenance, stand history, site conditions, and other internal and external factors. Temporary plots may indicate correlations but no evidence for casualties. In contrast to experiments, they may vary in many (even unobserved) traits and not only in the analysed factor of interest (Gamfeldt et al. 2013).

Third, long-term experiments usually cover un-thinned or untreated variants, which, for instance, indicate the maximum stand density, unfertilized conditions, etc. and serve as a reference for the density and growth of other treatments (e.g. thinned, mixed, fertilized plots). In addition, long-term experiments often comprise extreme variants that are usually avoided by forest practice and not covered by inventories. For analysing and modelling forest dynamics, however, extreme treatment variants are often even more informative than standard and business-as-usual variants mainly covered by inventories.

Fourth, only long-term surveys provide a full insight into the growth and yield of the remaining and removal stand, i.e. they quantify the total production since stand establishment. This includes also the otherwise not accessible information about the intermediate yield, caused by natural mortality or/and silvicultural treatments.

Fifth, long-term observations often cover a long part of the rotation or even include the subsequent stand generation. In this way, they can reveal effects of changes in environmental conditions at the tree and stand level (Spiecker et al. 1996). As they provide time series of growth and yield data that reach far back in time, they may provide evidence of environmental changes and the human footprint on forest ecosystems. The argument, which similarly long time series can be always obtained by retrospective growth-ring analysis, is only partly true. Retrospective analyses are possible at the individual tree level, but not at all for whole forest stands. This is because only the trees living at the time of sampling can be analysed retrospectively, but not their neighbours and the part of the population, eliminated before due to self-thinning or silvicultural interventions decades before. However, the past development of the population may be important to know its future dynamics (Pretzsch et al. 2021).

5.13.3 *Need for Further Coordination and Standardization of Experimental Design and Set-ups*

The foundation of the International Union of Forest Research Organizations (IUFRO) in 1892 was essential for the coordination and standardization of experimental design and set-ups. The recommendations and definitions for establishment and steering of thinning trials in the early times of IUFRO or its precursor organizations (Verein Deutscher Forstlicher Versuchsanstalten 1873, 1902) had some long-term standardization effects on many kinds of experiments in forests (Becking 1953). However, the objectives and questions of experiments changed and so did the variables and methods to measure. Among others, the measurements were extended to spatial explicit information about coordinates, crown size, stem quality, and vitality at the individual tree level. Natural regeneration, canopy characteristics, as well as variables quantifying the biodiversity, protective functions, recreational, or climate smartness are either added to the variable set of existing experiments. Or they are on the protocol list of new experiments from the beginning on.

Standardizing how to consider mixing proportions would be helpful, as they determine the establishment, steering, and evaluation of experiments (Dirnberger et al. 2017; Halofsky et al. 2018). Standards for height curve, form factor, or diameter growth-diameter functions would harmonize the evaluation of experiments and comparability of their results. Standardization of the result variables of long-term experiments, e.g. regarding merchantable volume, stem volume, stem mass, or total mass, would simplify the common overarching evaluations as realized in some of the studies underlying the book in hand (see Sects. 5.5, 5.6, and 5.7 and Hilmers et al. 2019; Pretzsch et al. 2020a, b; Torresan et al. 2020).

When establishing new experiments, addressing tree species mixing, natural regeneration, or transitioning from age-class forests to continuous cover forestry requires new standards for quantifying and steering mixing portions, stand density, and canopy cover over regeneration or for the horizontal and vertical stand structuring (del Río et al. 2016).

The date of measurement within the year, the frequency and methodology of measurement, and qualitative assessment of stem quality, crown vitality, etc. need some standardization, necessary for later common evaluation.

The silvicultural prescriptions for steering the experiments need common standards, analogous to the early thinning experiments, but more sophisticated as complex forest requires more detailed protocols for quantitative and reproducible, objective experimental steering (kind, strength, and frequency of interferences).

The common standards for evaluation at the stand and tree level simplify the pooling of data. A report of unique essential results will hopefully advance and support the appreciation and support of long-term experiments due to their unique contribution to forest observation, monitoring, and stewardship.

The standard for data storage and exchange will further support this ambition and save a lot of processing, organization, formatting, and time on the long term.

Trans-geographical projects with their international partner groups have the potential to further update the standard for experiments and observation in forest ecosystems from their establishment to the data storage and common evaluations.

5.13.4   Maintenance of Both Unmanaged and Managed Observation Plots

Untreated plots are of special value as reference for natural ecosystem dynamics without direct interference but suitable for revelation of indirect anthropogenic effects. They may reveal and quantify the effects of acid rain (Spiecker et al. 1996), climate change (Pretzsch et al. 2014), and also just local disturbances caused by lowering the groundwater level (Pretzsch and Kölbel 1988) or ozone (Matyssek et al. 2010). The growth and structure in managed forests is often superimposed by management effects in a way that disturbance by management and environmental stress are difficult to separate.

Unmanaged does not mean unmeasured; untreated plots are often accurately measured regarding vitality, growth, mortality, standing, and lying deadwood. Already the early thinning experiments provided un-thinned plots as reference; the inventory of all dropout trees provided information about the total yield over the whole rotation, not available without such concepts.

The reference plots offer information about maximum stand density, self-thinning, natural intra- and interspecific competition processes, and disturbances. They provide information for the derivation of basic relationships (Assmann 2013), model parameterization (Pretzsch et al. 2002), and development of silvicultural prescriptions for both monospecific and especially mixed-species stands (Kelty 1992;

Dieler et al. 2017). Any treatment experiments gain in value if untreated plots are nearby and demonstrate the impact of treatment by visual and quantitative comparison of both. In this network, we used the untreated plots for exploring trends in stand and tree growth that might be superimposed by various kinds of management effects on specifically treated experimental plots or un-specifically managed inventory plots (von Gadow and Kotze 2014). Plots with trees growing solitarily would be of similar interest but are even rarer (Kuehne et al. 2013; Uhl et al. 2015). They would reflect the vitality, growth, allometry, and ageing without competition, which would be of similar interest (for understanding, model parameterization, biomonitoring) as growth under self-thinning and without treatment.

5.13.5 The Relevance and Perspectives of Common Platforms for Forest Research

The use of both long-term experiments as well as temporary plots with increment coring for detecting growth trends reveals the shortcomings of temporal plots for reliable information of growth trends. There are shortcomings of using temporary plots for tree-level evaluations as well as using them for retrospective stand-level growth trend diagnosis.

Trend statements about tree-level growth based on increment cores from temporary plots may be misleading if the tree history is unknown. Sampling in mature stands may lead to biased results, as it means a sampling of survivors that may not represent the mean growth of the population. Notice that a European beech stand may start with a million trees or so and arrive at 50–100 trees per hectare in the mature phase with a mortality rate of 1–7%. The dropout trees may be less vital and lower in growth, and the growth of the remaining trees is higher than the mean (Nehrbass-Ahles et al. 2014). The effects of unknown silvicultural treatment in the past, natural suppression in the understorey, or biotic and abiotic damages (e.g. browsing, acid rain) may have influenced the life history and still have a memory effects of the present and future growth of the trees. By selecting sample trees with long and large crowns indicating permanent dominance and with tree-ring patterns without narrow ring phases by suppression may reduce such biases but cannot completely avoid them.

By sampling on long-term experiments or observation plots, in contrast, it is possible to select representative trees without prior charge by suppression, silvicultural impact, or damages that may interfere with the normally sigmoid growth-age relation.

For the detection of any stand growth trend, information about the growth of trees of various social positions in the stand and about tree removal, mortality, and stand structure in the past is even more important (Torresan et al. 2020). In this platform, we tried to avoid misjudgements by selection of stands un-thinned at least in the near past, by sampling trees for retrospective growth calculation from all biosocial classes over the whole stem diameter range and by assessing tree mortality via stump inventory. In this way, stand growth can be derived retrospectively (Heym et al. 2017, 2018).

All these shortcomings underline the advantages of maintaining a network of long-term experiments for monitoring the vitality and growth of forests (Pretzsch et al. 2019) that after all cover one third of the land surface in Europe and deserve special stewardship, especially in the mountain areas with their manifold ecological, economical, and socio-economic functions and services (Biber et al. 2015; Dieler et al. 2017).

Acknowledgements The authors would like to acknowledge networking support by the COST (European Cooperation in Science and Technology) Action CLIMO (Climate-Smart Forestry in Mountain Regions – CA15226) financially supported by the EU Framework Programme for Research and Innovation HORIZON 2020. This publication is part of a project that has received funding from the European Union's HORIZON 2020 research and innovation programme under the Marie Skłodowska-Curie grant agreement No 778322. Acknowledgements are also due to the European Union for funding the project "Mixed species forest management. Lowering risk, increasing resilience (REFORM)" (# 2816ERA02S under the framework of Sumforest ERA-Net). Further, we would like to thank the Bayerische Staatsforsten (BaySF) for providing the observational plots and to the Bavarian State Ministry of Food, Agriculture, and Forestry for the permanent support of the project W 07 "Long-term experimental plots for forest growth and yield research" (#7831-26625-2017). We also thank the Forest Research Institute, ERTI Sárvár, Hungary, for assistance and for providing observational plots. Research at Polish case studies was additionally supported by the Polish Government MNiSW 2018–2021 matching fund (W117/H2020/2018). In Slovenia, the study was additionally supported by the research programme P4-0059 Forest, forestry, and renewable forest resources. In Bosnia and Hercegovina, collecting data from plots is supported by the Ministry of Civil Affairs of Bosnia and Herzegovina.

References

Anderson-Teixeira KJ, Davies SJ, Bennett AC et al (2015) CTFS-Forest GEO: a worldwide network monitoring forests in an era of global change. Glob Chang Biol 21(2):528–549

Assmann E (2013) The principles of forest yield study: studies in the organic production, structure, increment and yield of forest stands. Elsevier

Bauhus J, Forrester DI, Pretzsch H (2017) From observations to evidence about effects of mixed-species stands. In: Pretzsch H, Forrester DI, Bauhus J (eds) Mixed-species forests. Springer, Berlin, Heidelberg, pp 27–71

Becker A, Bugmann H (2001) Global change and mountains regions – an IGBP initiative for collaborative research. In: Visconti G, Beniston M, Iannorelli ED, Diego B (eds) Global change and protected areas. Kluwer Academic, Dordrecht, pp 3–10

Becking JH (1953) Einige Gesichtspunkte für die Durchführung von vergleichenden Durchforstungsversuchen in gleichaltrigen Beständen. Proceedings of the 11th IUFRO Congress 1953, Rome, pp 580–582

Biber P, Borges JG, Moshammer R et al (2015) How sensitive are ecosystem services in European forest landscapes to silvicultural treatment? Forests 6(5):1666–1695

Blake J (1999) Overcoming the 'value-action gap' in environmental policy: tensions between national policy and local experience. Local Environ 4(3):257–278

Bowditch E, Santopuoli G, Binder F et al (2020) What is Climate-Smart Forestry? A definition from a multinational collaborative process focused on mountain regions of Europe. Ecosyst Serv 43:101113. https://doi.org/10.1016/j.ecoser.2020.101113

Buiteveld J, Vendramin GG, Leonardi S et al (2007) Genetic diversity and differentiation in European beech (Fagus sylvatica L.) stands varying in management history. For Ecol Manag 247:98–106

Carter MR, Gregorich EG (2007) Soil sampling and methods of analysis. CRC press

Cherubini P, Dobbertin M, Innes JL (1998) Potential sampling bias in long-term forest growth trends reconstructed from tree rings: a case study from the Italian Alps. For Ecol Manag 109(1–3):103–118

de Martonne E (1926) Une nouvelle fanction climatologique. L'indice d'aridité. La Meteriologie. Gauthier-Villars, Paris, pp 449–458

del Rio M, Condés S, Pretzsch H (2014) Analyzing size-symmetric vs. size-asymmetric and intra- vs. inter-specific competition in beech (Fagus sylvatica L.) mixed stands. For Ecol Manag 325:90–98. https://doi.org/10.1016/j.foreco.2014.03.047

del Río M, Pretzsch H, Alberdi I et al (2016) Characterization of the structure, dynamics, and productivity of mixed-species stands: review and perspectives. Eur J For Res 135(1):23–49

del Río M, Vergarechea M, Hilmers T et al (2021) Effects of elevation-dependent climate warming on intra- and inter-specific growth synchrony in mixed mountain forests. For Ecol Manag 479. https://doi.org/10.1016/j.foreco.2020.118587

Dieler J, Uhl E, Biber P et al (2017) Effect of forest stand management on species composition, structural diversity, and productivity in the temperate zone of Europe. Eur J For Res 136(4):739–766

Dirnberger G, Sterba H, Condés S et al (2017) Species proportions by area in mixtures of Scots pine (Pinus sylvestris L.) and European beech (Fagus sylvatica L.). Eur J For Res 136(1):171–183

Dorado-Liñán I, Valbuena-Carabaña M, Cañellas I et al (2020) Climate change synchronizes growth and iWUE across species in a temperate-submediterranean mixed oak forest. Front Plant Sci 11:706

EEA (2012) Climate change, impacts and vulnerability in Europe 2012. An indicator-based report, EEA report no 12/2012, European Environment Agency, Copenhagen

Ettinger AK, Ford KR, HilleRisLambers J (2011) Climate determines upper, but not lower, altitudinal range limits of Pacific Northwest conifers. Ecology 92(6):1323–1331

Falk DA, Richards CM, Arlee M et al (2006) Population and ecological genetics. In: Falk DA, Palmer MA, Zedler JB (eds) Restoration ecology. Foundations of restoration ecology. Island Press, Washington, DC, pp 14–41

Ferretti M, Fischer R (2013) Forest monitoring: methods for terrestrial investigations in Europe with an overview of North America and Asia. Vol. 12 of Developments in environmental science. Elsevier, Amsterdam

Forest Europe (2015) Forest Europe C&I set for sustainable forest management (SFM) (https://foresteurope.org/sfm-criteria-indicators/)

Forrester DI, Tachauer IH, Annighoefer P et al (2017) Generalized biomass and leaf area allometric equations for European tree species incorporating stand structure, tree age and climate. For Ecol Manag 396:160–175

Franklin JF (1989) Importance and justification of long-term studies in ecology. In: Likens GE (ed) Long-term studies in ecology. Springer, New York, pp 3–19

Gadow von K (1999) Datengewinnung für Baumhöhenmodelle – permanente und temporäre Versuchsflächen, Intervallflächen. Centralblatt für das gesamte Forstwesen 116(1/2):81–90

Gamfeldt L, Snäll T, Bagchi R et al (2013) Higher levels of multiple ecosystem services are found in forests with more tree species. Nat Commun 4:1340

Gazol A, Camarero JC (2016) Functional diversity enhances silver fir growth resilience to an extreme drought. J Ecol 104(4):1063–1075

Grissino-Mayer HD, Fritts HC (1997) The International Tree-Ring Data Bank: an enhanced global database serving the global scientific community. The Holocene 7(2):235–238

Halofsky JE, Andrews-Key SA, Edwards JE et al (2018) Adapting forest management to climate change: the state of science and applications in Canada and the United States. For Ecol Manag 421:84–97

Hanewinkel M, Kuhn T, Bugmann H et al (2014) Vulnerability of uneven-aged forests to storm damage. Forestry 87(4):525–534

Harmon ME, Pabst RJ (2015) Testing predictions of forest succession using long-term measurements: 100 years of observation in the Oregon Cascades. J Veg Sci 26:722–732

Harris I, Osborn TJ, Jones P et al (2020) Version 4 of the CRU TS monthly high-resolution gridded multivariate climate dataset. Sci Data 7(1):1–18. https://doi.org/10.1038/s41597-020-0453-3

Hernández L, Camarero JJ, Gil-Peregrín E et al (2019) Biotic factors and increasing aridity shape the altitudinal shifts of marginal Pyrenean silver fir populations in Europe. For Ecol Manag 432:558–567. https://doi.org/10.1016/j.foreco.2018.09.037

Heym M, Ruíz-Peinado R, del Río M et al (2017) EuMIXFOR empirical forest mensuration and ring width data from pure and mixed stands of Scots pine (Pinus sylvestris L.) and European beech (Fagus sylvatica L.) through Europe. Ann For Sci 74(3):63

Heym M, Bielak K, Wellhausen K et al (2018) A new method to reconstruct recent tree and stand attributes of temporary research plots: new opportunity to analyse mixed forest stands. In: Gonçalves AC (ed) Conifers. IntechOpen, Rijeka, pp 25–45

Hilmers T, Avdagić A, Bartkowicz L et al (2019) The productivity of mixed mountain forests comprised of Fagus sylvatica, Picea abies, and Abies alba across Europe. Forestry (Lond) 92(5):512–522. https://doi.org/10.1093/forestry/cpz035

Hilmers T, Biber P, Knoke T et al (2020) Assessing transformation scenarios from pure Norway spruce to mixed uneven-aged forests in mountain areas. Eur J For Res. https://doi.org/10.1007/s10342-020-01270-y

Jiang Z, Liu H, Wang H et al (2020) Bedrock geochemistry influences vegetation growth by regulating the regolith water holding capacity. Nat Commun 11:2392. https://doi.org/10.1038/s41467-020-16156-1

Johann K (1993) DESER-Norm 1993. Normen der Sektion Ertragskunde im Deutschen Verband Forstlicher Forschungsanstalten zur Aufbereitung von waldwachstumskundlichen Dauerversuchen. Proc Dt Verb Forstl Forschungsanst, Sek Ertragskd, in Unterreichenbach-Kapfenhardt, pp 96–104

Kangas A, Maltamo M (2006) Forest inventory: methodology and applications. Springer, Dordrecht, London

Kapos V, Rhind J, Edwards M, Price MF, Ravilious C (2000) Developing a map of the world's mountain forests. Forests in sustainable mountain development: a state of knowledge report for 2000. Task Force on Forests in Sustainable Mountain Development, pp 4–19

Kelty MJ (1992) Comparative productivity of monocultures and mixed-species stands. In: The ecology and silviculture of mixed-species forests. Springer, Dordrecht, pp 125–141

Körschens M (2006) The importance of long-term field experiments for soil science and environmental research – a review. Plant Soil Environ 52(Special Issue):1–8

Körner C, Paulsen J, Spehn EM (2011) A definition of mountains and their bioclimatic belts for global comparisons of biodiversity data. Alpine Botany, 121(2):73–78

Kramer K, Degen B, Buschbom J et al (2010) Modelling exploration of the future of European beech (Fagus sylvatica L.) under climate change – range, abundance, genetic diversity and adaptive response. For Ecol Manag 259:2213–2222

Kuehne C, Kublin E, Pyttel P et al (2013) Growth and form of Quercus robur and Fraxinus excelsior respond distinctly different to initial growing space: results from 24-year-old Nelder experiments. J For Res 24(1):1–14

Liang J, Crowther TW, Picard N (2016) Positive biodiversity-productivity relationship predominant in global forests. Science 354(6309):aaf8957. https://doi.org/10.1126/science.aaf8957

Lloret F, Keeling EG, Sala A (2011) Components of tree resilience: effects of successive low-growth episodes in old ponderosa pine forests. Oikos 120(12):1909–1920

Lutz JA (2015) The evolution of long-term data for forestry: large temperate research plots in an era of global change. Northw Sci 89:255–269

Magri D (2008) Patterns of post-glacial spread and the extent of glacial refugia of European beech (Fagus sylvatica). J Biogeogr 35:450–463

Magri D, Vendramin GG, Comps B et al (2006) A new scenario for the Quaternary history of European beech populations: palaeobotanical evidence and genetic consequences. New Phytol 171:199–221

Martín-Benito D, Gea-Izquierdo G, del Río M et al (2008) Long-term trends in dominant-height growth of black pine using dynamic models. For Ecol Manag 256:1230–1238

Matyssek R, Wieser G, Ceulemans R et al (2010) Enhanced ozone strongly reduces carbon sink strength of adult beech (Fagus sylvatica) – resume from the free-air fumigation study at Kranzberg Forest. Environ Pollut 158(8):2527–2532

Nagel J, Spellmann H, Pretzsch H (2012) Zum Informationspotenzial langfristiger forstlicher Versuchsflächen und periodischer Waldinventuren für die waldwachstumskundliche Forschung. Allg For Jagdzeitung 183(5/6):111–116

Nehrbass-Ahles C, Babst F, Klesse S et al (2014) The influence of sampling design on tree-ring-based quantification of forest growth. Glob Chang Biol 20(9):2867–2885

Nogués-Bravo D, Araújoc MB, Erread MP et al (2007) Exposure of global mountain systems to climate warming during the 21st century. Glob Environ Chang 17:420–428

Palmer WC (1965) Meteorological drought, U.S. Weather Bureau, research paper 45. U.S. Department of Commerce, Weather Bureau, Washington, DC

Pennock DJ, van Kessel C (1997) Clear-cut forest harvest impacts on soil quality indicators in the mixedwood forest of Saskatchewan, Canada. Geoderma 75:13–32

Petit RJ, Bialozyt R, Brewer S et al (2001) From spatial patterns of genetic diversity to postglacial migration processes in forest trees. In: Silvertown J, Antonovics J (eds) Integrating ecology and evolution in a spatial context. Blackwell Science, Oxford, pp 295–318

Pfatrisch M (2019) Assessment of climate smartness. Master thesis. Technical University of Munich, Freising, 115p

Porth I, El-Kassaby YA (2014) Assessment of the genetic diversity in forest tree populations using molecular markers. Diversity 6:283–295

Pretzsch H (2009) Forest dynamics, growth and yield. Springer, Berlin, 664 p

Pretzsch H, Biber P (2016) Tree species mixing can increase maximum stand density. Can J For Res 46:1179–1193. https://doi.org/10.1139/cjfr-2015-0413

Pretzsch H, Kölbel M (1988) Einfluß von Grundwasserabsenkungen auf das Wuchsverhalten der Kiefernbestände im Gebiet des Nürnberger Hafens – Ergebnisse ertragskundlicher Untersuchungen auf der Weiserflächenreihe Nürnberg 317. Forstarchiv 59(3):89–96

Pretzsch H, Biber P, Ďurský J (2002) The single tree-based stand simulator SILVA: construction, application and evaluation. For Ecol Manag 162(1):3–21

Pretzsch H, Biber P, Schütze G et al (2014) Forest stand growth dynamics in Central Europe have accelerated since 1870. Nat Commun 5:4967. https://doi.org/10.1038/ncomms5967

Pretzsch H, Biber P, Uhl E et al (2015) Long-term stand dynamics of managed spruce-fir-beech mountain forests in Central Europe: structure, productivity and regeneration success. Forestry 88(4):407–428. https://doi.org/10.1093/forestry/cpv013

Pretzsch H, Schütze G, Biber P (2018) Drought can favour the growth of small in relation to tall trees in mature stands of Norway spruce and European beech. For Ecosyst 5(1):20

Pretzsch H, del Rio M, Biber P et al (2019) Maintenance of long-term experiments for unique insights into forest growth dynamics and trends: review and perspectives. Eur J For Res 138(1):165–185. https://doi.org/10.1007/s10342-018-1151-y

Pretzsch H, Hilmers T, Biber P et al (2020a) Evidence of elevation-specific growth changes of spruce, fir, and beech in European mixed-mountain forests during the last three centuries. Can J For Res 50(7):689–703. https://doi.org/10.1139/cjfr-2019-0368

Pretzsch H, Hilmers T, Uhl E et al (2020b) European beech stem diameter grows better in mixed than in mono-specific stands at the edge of its distribution in mountain forests. Eur J For Res. https://doi.org/10.1007/s10342-020-01319-y

Pretzsch H, del Río M, Giammarchi F et al (2021) Changes of tree and stand growth. Review and implications. IIn: Managing Forest Ecosystems, Vol. 40, Tognetti R, Smith M, Panzacchi P (Eds): Climate-Smart Forestry in Mountain Regions. Springer Nature, Switzerland, AG

Raison RJ, Khanna PK (2011) Possible impacts of climate change on forest soil health, Chapter 12. In: Singh BP et al (eds) Soil health and climate change, Soil biology 29. Springer, Berlin, Heidelberg. https://doi.org/10.1007/978-3-642-20256-8_12

Redman CL, Grove JM, Kuby LH (2004) Integrating social science into the long-term ecological research (LTER) network: social dimensions of ecological change and ecological dimensions of social change. Ecosystems 7(2):161–171

Ruiz-Benito P, Vacchiano G, Lines ER et al (2020) Available and missing data to model impacts of climate change on European forests. Ecol Model 416:108870

Sagarin R, Pauchard A (2010) Observational approaches in ecology open new ground in a changing world. Front Ecol Environ 8(7):379–386

Santopuoli G, Temperli C, Alberdi I et al (2020) Pan-European sustainable forest management indicators for assessing Climate-Smart Forestry in Europe. Can J For Res. https://doi.org/10.1139/cjfr-2020-0166

Seidl R, Albrich K, Erb K et al (2019) What drives the future supply of regulating ecosystem services in a mountain forest landscape? For Ecol Manag 445:37–47. https://doi.org/10.1016/j.foreco.2019.03.047

Spiecker H, Mielikäinen K, Köhl M, Skovsgaard JP (eds) (1996) Growth trends in European forests, Europ For Inst, Res Rep, vol 5. Springer, Heidelberg, 372 p

Tognetti R, Scarascia Mugnozza G, Hofer T (eds) (2017) Mountain watersheds and ecosystem services: balancing multiple demands of forest management in head-watersheds, EFI technical report 101. EFI, Joensuu, 191 p

Tognetti R, Valentini R, Belelli Marchesini L, Gianelle D, Panzacchi P, Marshall JD (2021) Continuous monitoring of tree responses to climate change for smart forestry – a cybernetic web of trees. In: Managing Forest Ecosystems, Vol. 40, Tognetti R, Smith M, Panzacchi P (Eds): Climate-Smart Forestry in Mountain Regions. Springer Nature, Switzerland, AG

Tomppo E, Gschwantner T, Lawrence M et al (2010) National forest inventories. Pathways for common reporting. Springer, Dordrecht

Torresan C, del Río M, Hilmers T et al (2020) Importance of tree species size dominance and heterogeneity on the productivity of spruce-fir-beech mountain forest stands in Europe. For Ecol Manag 457:117716

Torresan C, Luyssaert S, Filippa G, Imangholiloo M, Gaulton R (2021) Remote sensing technologies for assessing climate-smart criteria in mountain forests. In: Managing Forest Ecosystems, Vol. 40, Tognetti R, Smith M, Panzacchi P (Eds): Climate-Smart Forestry in Mountain Regions. Springer Nature, Switzerland, AG

Uhl E, Biber P, Ulbricht M et al (2015) Analysing the effect of stand density and site conditions on structure and growth of oak species using Nelder trials along an environmental gradient: experimental design, evaluation methods, and results. For Ecosyst 2(1):17

van Dam BC (2002) EUROPOP: genetic diversity in river populations of European black poplar for evaluation of biodiversity, conservation strategies, nature development and genetic improvement. In: Dam BC, Bordács S (eds) Genetic diversity in river populations of European black poplar. Proceedings of International Symposium Szekszárd, 16–20 May 2001. Csiszár nyomda, Budapest, pp 15–31

van der Maaten-Theunissen M, van der Maaten E, Bouriaud O (2015) PointRes: An R package to analyze pointer years and components of resilience. Dendrochronologia 35:34–38

van Mantgem PJ, Stephenson NL (2007) Apparent climatically induced increase of tree mortality rates in a temperate forest. Ecol Lett 10:909–916

Verein Deutscher Forstlicher Versuchsanstalten (1873) Anleitung für Durchforstungsversuche. In: Ganghofer von A ed. (1884) Das Forstliche Versuchswesen. Schmid'sche Buchhandlung, Augsburg, vol 2, pp. 247–253

Verein Deutscher Forstlicher Versuchsanstalten (1902) Beratungen der vom Vereine Deutscher Forstlicher Versuchsanstalten eingesetzten Kommission zur Feststellung des neuen Arbeitsplanes für Durchforstungs- und Lichtungsversuche. AFJZ 78:180–184

Versace S, Gianelle D, Frizzera L et al (2019) Prediction of competition indices in a Norway Spruce and Silver Fir-Dominated Forest Using Lidar Data. Remote Sens 11(23):2734

Versace S, Gianelle D, Garfì V et al (2020) Interannual radial growth sensitivity to climatic variations and extreme events in mixed-species and pure forest stands of silver fir and European beech in the Italian Peninsula. Eur J For Res:1–19

Vicente-Serrano SM, Beguería S, López-Moreno JI (2010) A multiscalar drought index sensitive to global warming: the standardized precipitation evapotranspiration index. J Clim 23:1696–1718

von Gadow K, Kotze H (2014) Tree survival and maximum density of planted forests – observations from South African spacing studies. For Ecosyst 1(1):21

Wohner C, Peterseil J, Poursanidis D et al (2019) DEIMS-SDR – a web portal to document research sites and their associated data. Eco Inform 51:15–24

Zeller L, Liang J, Pretzsch H (2018) Tree species richness enhances stand productivity while stand structure can have opposite effects, based on forest inventory data from Germany and the United States of America. For Ecosyst 5(1):4

Chapter 6
Changes of Tree and Stand Growth: Review and Implications

H. Pretzsch, M. del Río, F. Giammarchi, E. Uhl, and R. Tognetti

Abstract In this chapter, we review the current long-term growth trends and short-term growth reaction to single or repeated stress events on tree and stand level in Europe. Based on growth trend analyses, the chapter reveals the strong human footprint on forest ecosystems.

First, we use long-term experiments and increment cores to show change in growth trends within the last centuries. Growth reactions are caused by deposition and climate change rather than by silvicultural measures. Second, we look closer on regional-specific deviations from the general trend. Climate change, drought events, acid rain and O_3 are causing regional-specific growth reaction patterns. Third, we assess stress events and the resilience and resistance of monospecific and mixed stands against biotic and abiotic stress in view of the ongoing growth trends.

H. Pretzsch (✉)
Forest Growth and Yield Science, TUM School of Life Sciences Weihenstephan, Technical University of Munich, Freising, Germany
e-mail: hans.pretzsch@tum.de

M. del Río
INIA, Forest Research Centre, Madrid, Spain

iuFOR, Sustainable Forest Management Research Institute, University of Valladolid & INIA, Valladolid, Spain
e-mail: delrio@inia.es

F. Giammarchi
Faculty of Science and Technology, Free University of Bolzano-Bozen, Bolzano, Italy
e-mail: francesco.giammarchi@unibz.it

E. Uhl
Forest Growth and Yield Science, TUM School of Life Sciences Weihenstephan, Technical University of Munich, Freising, Germany

Bavarian State Institute of Forestry (LWF), Freising, Germany
e-mail: enno.uhl@tum.de

R. Tognetti
Dipartimento di Agricoltura, Ambiente e Alimenti, Università degli Studi del Molise, Campobasso, Italy
e-mail: tognetti@unimol.it

© The Author(s) 2022
R. Tognetti et al. (eds.), *Climate-Smart Forestry in Mountain Regions*, Managing Forest Ecosystems 40, https://doi.org/10.1007/978-3-030-80767-2_6

The revealed tree and stand growth behaviours are highly relevant, as any changes of forest growth and structure have strong impacts on the provision of goods and ecosystem services. The results underline the importance of biomonitoring and suggest counteracting measures by forest planning, adaptation of silvicultural guidelines for existing forest and innovative design of future forests stands.

6.1 Introduction: The Information Potential of Tree and Stand Growth Trajectories

At first sight, the reports about the status quo of forest health, vitality and productivity in Europe seem contradictory; on the one hand messages about growth acceleration (Kauppi et al. 2014), on the other hand forest dieback due to stress factors (Allen et al. 2015). The situation seems similarly contrary at the global level, with accelerating growth in the northern latitudes and growth decline in the Mediterranean and dry continental zones. At higher altitudes, the findings are similar to northern latitudes; some tree species in some regions seem to benefit from the changing environmental conditions, and other species in various regions suffer. In terms of canopy cover, European and North American forests have experienced a history of recovery (Nabuurs et al. 2013; Zhu et al. 2018). Climate change and land use changes interfere with forest regrowth, changing the temporal trajectories of biomass recovery and the age structure of forest stands. These forests, although continuing to sequester a significant amount of carbon, already show signs of growth saturation, which may limit their sink potential in the decades to come, as it will gradually saturate as trees age. Information on carbon uptake in regrowth forests derives from modelling exercises (Pugh et al. 2019). Sink saturation occurs due to a decrease in the assimilation component or an increase in the emission component for a given period of time. Global terrestrial carbon sink in forests is largely driven by tree physiological responses to increasing CO_2, N deposition, air pollution, evaporative demand, land use changes (vegetation shift, forest regrowth) and changes in forest cover (forest fire, forest dieback).

Certainly, the growth reaction pattern of a species depends on the respective site conditions and the specific environmental changes. When growing at its ecological optimum, i.e. in the centre of its realised ecological niche, a species will be less affected by environmental changes than when growing at the edge or even beyond its natural range. In Central Europe, a dieback of Norway spruce under drought in the lowlands beyond its natural range is not surprising. As it is not adapted to the increasing warmth and drought, the environmental conditions simply develop away from its ecological niche. Similarly, a growth acceleration of European beech in the mountain zone of the Alps and in its northern range area is not surprising as here the environmental conditions may improve the growing conditions of beech as they change towards its optimal growing conditions. A general increasing trend of basal area increment was observed for European beech throughout the twentieth century, across the Italian Peninsula, with the exception of southernmost populations (Tognetti et al. 2014), which was accompanied by a continuous enhancement in isotope-derived intrinsic water-use efficiency.

In addition to the interplay between initial site conditions and large-scale environmental change, regional conditions may influence the growth behaviour. Examples for this are recovery from litter raking (Gimmi et al. 2013), soil acidification by deposition and over-exploitation (Lundström et al. 2003; Hoegberg et al. 2006), lowering of groundwater level (Pretzsch and Kölbel 1988), ozone exposure (Matyssek et al. 2010) or eutrophication by N deposition (Hofmann et al. 1990; Pretzsch et al. 2014a).

Yet, shifts of species' relative abundance in mixed-species forests might strongly affect regional carbon balances, which warrants the establishment/maintenance of long-term experiments to measure and monitor growth dynamics and trends in representative forests of major biogeographic areas (Pretzsch et al. 2019a). Water retention experiments, fertilizing experiments and other long-term experiments with different silvicultural treatments can reveal, for given site conditions and species setups, how the effects of environmental changes can be modulated or mitigated by management activities. They may contribute to answering whether tree species mixing, stand density reduction or choice of provenances can mitigate the effects of climate change, extending the present limit of forest carbon sink. Trees and forest stands may have the ability to acclimate to stress and to recover. Morphological acclimation and epigenetic adaptation to stress are far from being sufficiently understood. Due to the site dependency of growth behaviour, silvicultural recipes that are useful on one site may be inappropriate for counteracting environmental changes on other sites.

An important basis for fact-finding on growth reactions to external factors are long-term observations by, e.g. national inventory plots, environmental monitoring networks and networks of long-term experiments, covering extended environmental gradients. For the study at hand, we tapped these databases:

(i) To provide the theoretical basis for better understanding the spatial-temporal pattern of forest growth behaviour
(ii) To review and give overview of the growth trends and events of tree and stands in order to derive measures for silvicultural adaptation
(iii) To outline the process of acclimation to stress, adaptation and recovery
(iv) To discuss the implications of the revealed growth behaviour for environmental monitoring, forest ecology and management

6.2 Theoretical Considerations on Growth Changes: Effects of Site Conditions and Species Identity

6.2.1 Standard of Comparison

The unimodal relationship between environmental conditions and growth (or any other indicator of fitness) is helpful for understanding both the site-specific and non-linear growth reactions of monocultures in case of disturbances (Tognetti et al. 2019) and the growth-stabilizing effect of tree species mixtures (Bauhus et al.

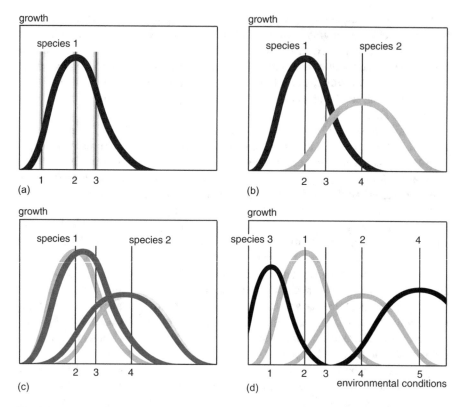

Fig. 6.1 Growth of various tree species depending on their realised ecological niche and prevailing site conditions in schematic representation. The vertical lines represent specific site conditions:
- (**a**) Unimodal relationship between site conditions and growth of species 1. Any changes of site conditions (grey bands left and right to site conditions 1, 2 and 3) have stronger non-linear effects on sites 1 and 3 than on site 2
- (**b**) Species 2 with a niche different from species 1 may compensate for growth losses when site conditions change from 2 to 3 or from 3 to 4
- (**c**) Tree species interactions in mixed stands can extend (facilitation) or reduce (competition) the ecological niches of tree species (grey lines)
- (**d**) By combining various tree species with different niches in mixtures (above numbers), stand growth under climate change (below numbers) may be stabilised

2017). In Figure 6.1a, the environmental conditions change along the x-axis (for simplicity one dimensional); growth is high in the centre of the real niche and decreases if temperature, water and nutrient supply become scarce or excessive. Growth is rather stable under site condition 2 (see middle vertical line), as it is close to the saddle of the unimodal curve, where temporal variation of the resources supply (grey bands left and right of site 2) hardly changes the growth. However, on sites represented by the steep left or right branch of the niche, changes may have strong non-linear positive or negative effects (grey bands left and right of the sites 1 and 3, left and right vertical lines). This is why climate warming by 2 °C may have hardly any effect on temperate sites but strong positive and negative effects on cold and warm sites, respectively.

A combination of two species with differing niches, as shown in Figure 6.1b, may stabilise stand growth if site conditions change. In this case, species 2 may compensate for growth losses of species 1, if the site conditions shift from 2 to 3 or to 4. Combining two species in a stand can also trigger competition reduction and facilitation (e.g. by hydraulic redistribution through hydraulic lift, atmospheric nitrogen fixation by symbionts) that may extend their real niche (grey unimodal curves in the background, Fig. 6.1c) and improve their growth stability (Fig. 6.1c).

Multispecies forest stands, as represented by their niche composition in Figure 6.1d, may be inferior in growth under stable environmental conditions as some species are growing suboptimally. However, if the environmental conditions change, e.g. gradually to a warmer and drier climate, or episodically caused by drought years, this may change. In this case an assemblage of various tree species with different niches can mean a risk distribution and stabilisation of stand productivity, although one or the other tree species may decline or even drop out.

6.2.2 Long- and Short-Term Deviations from Normality

Useful references for detection of any trend or event of the tree or stand development are basic growth and yield curves as shown in Figure 6.2a. Under constant environmental conditions, the growth has a unimodal development over time, and the yield, as the integral of the growth, has a sigmoid shape. With progressing size, tree and stand growth declines, and the yield curve has an asymptotic course. Marziliano et al. (2019) observed that the increase in tree size, more than senescence, explained the reduction in height increment in older silver fir trees. Certainly, the growth and yield curves of diameter, basal area and volume differ in rhythm (slope, point of inflexion, turning point), and they are also species specific and depend, among others, on site conditions, but the unimodal pattern and sigmoid shape remain as their general characteristics and can be used as reference for detecting abnormal developments in case of environmental changes.

Based on the unimodal growth curve (black) as a standard of comparison, any acceleration or deceleration of growth by environmental changes (Fig. 6.2b) with improved growing conditions may be interrupted by episodic stress events (Fig. 6.2c), or normal growing conditions terminated by stress events with short- or long-lasting growth reduction (Fig. 6.2d) can be revealed and quantified.

6.3 Empirical Evidence of Growth Trends and Events

6.3.1 Overarching Growth Trends in the Lowlands of Europe

Shortly after the acid rain phenomena from 1970 to 1980, Kenk et al. (1991) detected a site index improvement by up to 7 metres of dominant height at age 100 according to the yield table by Assmann and Franz (1963) for Norway spruce stands on poor

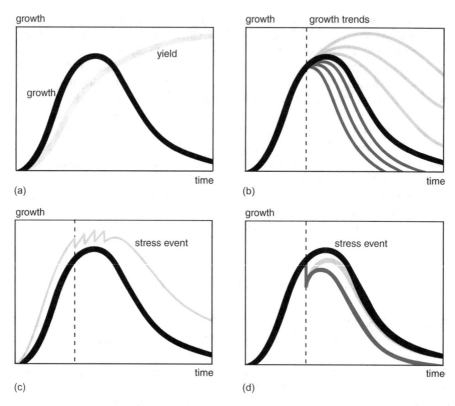

Fig. 6.2 Modification of the normal unimodal growth curve (black) of forest trees or even-aged forest stands by changing environmental conditions. (**a**) Unimodal curve of tree and stand growth (integrated yield in grey) under normal, constant and undisturbed growing conditions. (**b**) Acceleration (green) or deceleration (red) of growth caused by environmental changes. (**c**) Improved growing conditions, interrupted by episodic stress events (green). (**d**) Normal growing conditions, halted by stress events with subsequent low (red) and high (green) resilient reaction

and medium sites. Among others, Spiecker et al. (1996), Bontemps et al. (2012), Kauppi et al. (2014) and Pretzsch et al. (2014a) revealed that forest stand growth dynamics in most parts of Europe have strongly accelerated in recent decades.

Pretzsch et al. (2019a) compiled a data set of several hundred long-term experiments from Austria, Denmark, France, Germany, Great Britain, Poland, Spain, Sweden and Switzerland for closer analyses of the current growth trends in European lowlands. Figure 6.3 shows the total stand volume production of 577 fully stocked and unthinned or just moderately thinned long-term experiments of Norway spruce, Scots pine, European beech and common and sessile oak in comparison to the yield tables for Norway spruce by Wiedemann (1936/1942), Scots pine by Wiedemann (1943), European beech by Schober (1967) and sessile oak by Jüttner (1955). The yield classes I and IV of these tables represent the development of the respective species growing in the early twentieth century under optimal and poor conditions, respectively.

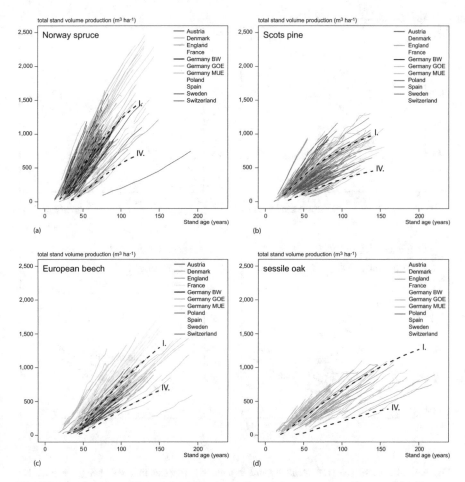

Fig. 6.3 Development of the total volume production on long-term experiments of (**a**) Norway spruce (*n* = 202), (**b**) Scots pine (*n* = 189), (**c**) European beech (*n* = 97) and (**d**) sessile oak (*n* = 89) in Europe established between 1848 and 2010, according to Pretzsch et al. (2019a). The observed trajectories strongly exceed the spectrum of common yield tables (yield classes I and IV) for Norway spruce by Wiedemann (1936/1942), Scots pine by Wiedemann (1943), European beech by Schober (1967) and sessile oak by Jüttner (1955). The German experimental plots were provided by the Forest Research Station of Baden-Württemberg (BW), the Research Station of Lower Saxony in Goettingen (GOE) and the Chair for Forest Growth and Yield Science at TUM in München (MUE)

The same data set also revealed significant relocations of the age trajectories of total stand volume production and related variables when considering the year of stand establishment. Figure 6.4 shows, representatively for Scots pine, how the total stand volume production, standing volume and absolute and relative cumulative volume, achieved at a given age, changed during the last 150 years. Similar trends were found for other major species. The total stand production and standing stock in a mature Scots pine stand is reached 50 years earlier today than for stands that

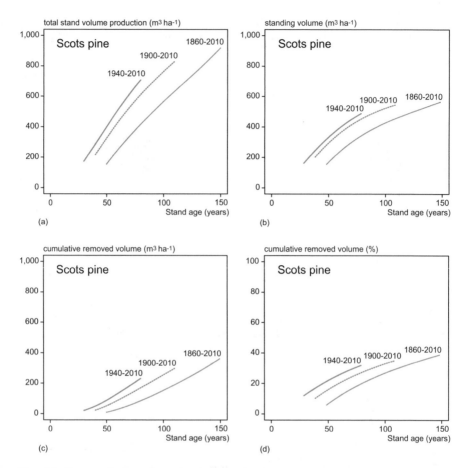

Fig. 6.4 Change of (**a**) total stand volume production, (**b**) standing volume, (**c**) cumulative removed volume and (**d**) percentage of removed (mortality, thinning) volume of fully stocked, unthinned or just moderately thinned long-term experiments of Scots pine established in 1860, 1900 and 1940, according to Pretzsch et al. (2019a)

were established 100 years earlier. As a consequence, the intermediate yield (i.e. the cumulative removed volume) is 200 m³ ha⁻¹ at the age of 75 years at present, while it was just 75 m³ ha⁻¹ for stands that were established 100 years earlier. This reflects an increase by 150%. The cumulative removed volume is the total production minus standing volume; the percentage of removed volume results from (total production − standing volume)/total production × 100.

For a subsample of the plots in Germany, Pretzsch et al. (2018) showed that wood density decreased by 8–12% since 1900. Even if stand and trees grew much faster in terms of wood volume, stand biomass increment increased by 9–24 percentage points less compared to volume increment. This does not at all cancel the remarkable volume growth acceleration in the past 100 years, but it slightly reduces the findings based on stand volume records. The decreasing wood density goes along with an increased early wood fraction and points to the observed extension of the growing season and fertilisation effect of dry deposition as the main causes of the detected changes in growth trends.

The growth trends shown in Figures 6.3 and 6.4 indicate changes in growth conditions in terms of rising temperature, extension of the growing season, rising atmospheric CO_2 level, nitrogen deposition and abandonment of nutrient-exporting treatments such as litter raking (Pretzsch et al. 2014a). In other regions, environmental changes can be, of course, also detrimental for growth rates and slow down stand dynamics. A recent study about tree growth in forests and urban areas revealed that the beneficial effects of climate change can turn into growth decrease and losses in regions with limited water and nutrient availability (Pretzsch et al. 2017). Again, without unmanaged long-term experiments, the knowledge of growth trends and their causes would be strongly biased or blurred. Nevertheless, Keenan et al. (2016), using global carbon budget estimates, ground, atmospheric and satellite observations, and multiple global vegetation models, reported increases in the terrestrial sink during the past decade. This was associated with the effects of rising atmospheric CO_2, though many other factors may influence the carbon cycle at the local scale. Indeed, partitioning of photosynthates can be largely influenced by environmental factors, stand age and forest management.

6.3.2 Growth Trends in High-Elevation Forest Ecosystems

High-altitude forest ecosystems often provide invaluable ecosystem services, e.g. protection against avalanches, landslides, rockfall or flooding (Bebi et al. 2001). Due to their prevailing limitation by low temperatures and short growing season, especially mixed mountain forest ecosystems at higher elevations are expected to be strongly affected by climate warming (Piao et al. 2011; Vayreda et al. 2012; Ruiz-Benito et al. 2014) similarly to forest ecosystems in the northern latitudes. Knowledge of any growth trends or structural changes of mountain forests may enable forest management to adapt and stabilise these stands and thus avoid decline of productivity and other forest ecosystem services.

Hilmers et al. (2019) and Pretzsch et al. (2020b) showed, for mixed mountain forest of Norway spruce (*Picea abies* (L.) Karst.), silver fir (*Abies alba* Mill.) and European beech (*Fagus sylvatica* L.), how environmental changes can modify the growth trend of tree species, depending on the altitude. The revealed altitudinal-related growth trends were found in European mixed mountain forests but probably show reaction patterns that may be characteristic for forest ecosystems at higher elevations under environmental change.

Pretzsch et al. (2020b) sampled increment cores from 1721 Norway spruces, silver firs and European beeches on 28 long-term experimental plots in mixed-mountain forest, in seven European countries from Bulgaria to Switzerland. The plots were located between 621 and 1569 m a.s.l., having an annual temperature range between 2.9 and 8.2 °C and an annual precipitation of 794–2767 mm yr^{-1}.

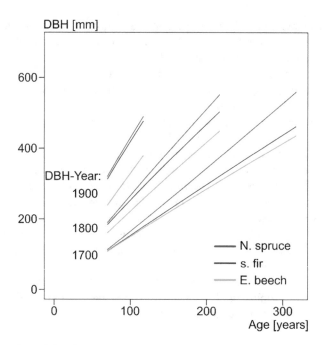

Fig. 6.5 Overview of the changes in the level and steepness of the growth curves of Norway spruce, silver fir and European beech during the last three centuries, according to Pretzsch et al. (2020b). For all the three species, the steepness of the curve increased the strongest in the case of European beech and silver fir and the least in the case of Norway spruce

The tree ring series revealed an increase of both the level and the slope of the diameter-age relationship for all the three species during the last 300 years, as shown in Figure 6.5. The trend of steepening the diameter-age relationship from year 1700 to the present is strongest for silver fir and European beech and the least pronounced for Norway spruce. In the past (1700), the size growth of Norway spruce was far ahead of silver fir and European beech, whereas, in the two recent centuries, the slopes of the growth curves of silver fir and European beech increased, indicating a trend to a more similar growth vigour.

For Norway spruce, the change in growth trend was not dependent on altitude (Fig. 6.6a), but, for silver fir (not shown) and especially for European beech (Fig. 6.6b), we found an altitude-dependent behaviour. In the past (DBH-year 1700), the growth of silver fir and European beech was the highest at low altitudes and the lowest at high elevations. Both species changed this pattern from 1700 until present. Trees with DBH-year 1900 are growing better at high elevations and less at lower altitudes.

This spatio-temporal pattern suggests temperature increase as the main factor for significant changes in the growth and interspecific competition at the expense of Norway spruce in mixed mountain forests. The long-term growth trajectories of Norway spruce in relation to silver fir and European beech hint at a relative advantage of fir and beech at the expense of spruce. The relative inferior growth trend of spruce in relation to fir and beech corresponds to a loss of fitness. On the long run,

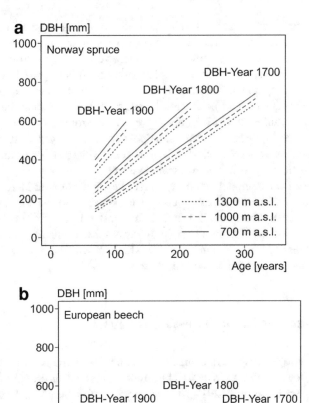

Fig. 6.6 Altitude-dependent changes of the growth of (**a**) Norway spruce and European beech (**b**) in the last 300 years, according to Pretzsch et al. (2020b). (**a**) In the case of Norway spruce, the growth was the highest at lower elevations and decreased with altitude in the past (DBH-year 1700 and 1800) as well as at present. (**b**) In the case of European beech, the growth was also the highest at lower elevations and decreased with altitude in the past. During the last 200 years, the relationships turned to a superior growth at higher altitudes compared to lower elevations

this may reduce the competitive strength and success of Norway spruce in stand development and natural regeneration.

Similar altitude-dependent temporal changes were found when analysing intraspecific synchrony in tree growth response to inter-annual fluctuations during the last century (del Río et al. 2020). Synchrony within beeches decreased with climate warming at high altitudes, reflecting a lower dependency of climate conditions in the last decades. By contrast, Norway spruce synchrony maintained constant at high altitudes but increased remarkably at lower altitudes, indicating that climate warming is exerting a stronger control on spruce growth at these altitudes.

At lower altitudes, Norway spruce is endangered especially by drought stress and subsequent bark beetle infestation; at higher altitudes, it is impaired by the growth acceleration of competing silver fir and European beech in mixed mountain stands. The climate change-induced promotion of silver fir and European beech will, in absence of silvicultural activities, gradually replace the role that Norway spruce had in the past in the high-montane and subalpine zone. At lower altitudes, Norway spruce will be limited to cold sites in hollow relief where silver fir and European beech, but not Norway spruce, suffer from late frost.

Regardless of the competitive ability of Norway spruce under increasing disturbance regimes, a decrease in wood quality can be expected for this species in the years to come (e.g. wood density). Being the most relevant timber for softwood products in Central and Northern Europe, Norway spruce calls for new silvicultural strategies addressing climate change issues, which include the right choice of provenances and degree of stand density regulation.

6.3.3 Stress Events and Low-Growth Years

Many studies use the long-term course of the annual stem diameter growth of trees for analysing low-growth years caused by stress events such as drought, late frost or insect injury (Schweingruber 2012). The studies are often based on dominant trees, in order to keep the effects of competition by neighbours as small as possible (Pretzsch et al. 2020a). The normal level of the sampled individuals is assumed as standard for comparison, and any slump of growth in low-growth years is quantified by the ratio between the reduced and normal level. This results in information about relative growth losses. Zang et al. (2011) revealed, by this method, that the ranking between primary tree species in Central Europe regarding growth losses is Norway spruce, silver fir, Douglas fir, European beech, Scots pine and sessile oak. By comparing the growth reactions in the given years in monospecific stands with neighbouring mixed-species ones, Pretzsch et al. (2013), Grossiord et al. (2014), Thurm et al. (2016) and Ammer (2019) could show that, in some cases, interspecific neighbourhood can reduce species-specific drought stress effects.

Drought events in Central Europe, among others, in 1976, 2003 and 2015, triggered many studies about the effects of episodic drought on the growth and mortality of forest tree species (Ciais et al. 2005; Bréda et al. 2006; Allen et al. 2015). These findings suggest that tree species especially growing at or beyond the border of their natural range, such as Norway spruce or European larch (*Larix decidua* Mill.) in Central Europe, can show severe growth losses and mortality (Kölling et al. 2009; Lévesque et al. 2013). Scots pine (*Pinus sylvestris* L.) and sessile oak (*Quercus petraea* L.) often serve as examples for rather drought-tolerant species (Walentowski et al. 2007; Zang et al. 2011, 2012), more suitable for forestry in Central Europe under climate change towards warm and dry conditions. In order to mitigate drought, silviculture aims at the selection of well acclimated species and provenances (Atzmon et al. 2004; Arend et al. 2011; Zang et al. 2011), at reducing stand density (D'Amato et al. 2013; Sohn et al. 2016; Bottero et al. 2017), at modifying the kind of thinning (Rodríguez-Calcerrada et al. 2011; Gebhardt et al. 2014; Pretzsch et al. 2018) or at favouring tree species mixing. The latter, however, is not rated effective for drought mitigation in general (Grossiord 2019). Indeed, in forest stands, species diversity is not always positively related to drought resistance. Conte et al. (2018) observed that stem radial growth and isotope-derived intrinsic water-use efficiency were generally higher in pure than in mixed stands of European beech and Scots pine.

Exemplarily, Figure 6.7 shows species-specific stress reactions caused by the drought year 1976 quantified in relation to the mean growth level in the 3-year-period 1973–1975 before the drought stress (reference line = 1.0). The study was based on tree ring measurement on cores from increment chronologies from 559 trees of Norway spruce, European beech and sessile oak in South Germany, with half of them sampled in monospecific stands and the other half in mixed stands (Pretzsch et al. 2013).

For quantifying the resistance, recovery and resilience, indices introduced by Lloret et al. (2011) were applied, allowing for retracing the tree's growth reaction on the episodic drought stress in the years 1976 and 2003. The following general reaction patterns, visualised in Figure 6.7, were found. In pure stands, spruce had the lowest resistance but the best recovery. Oak and beech were more resistant but recovered less pronounced and thus were less resilient. In mixture, spruce and oak performed like in pure stands, but beech was significantly more resistant and resilient than in monocultures. Especially when mixed with oak, beech was facilitated. We hypothesise that the revealed water stress release of beech emerges in mixture because of the asynchronous stress reaction pattern of beech and oak and a facilitation of beech by hydraulic lift of water by oak. A potential positive contribution of species

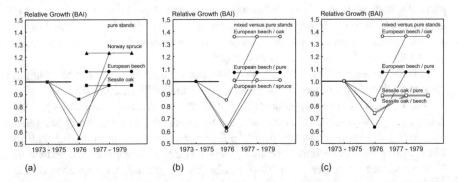

(a)   (b)   (c)

Fig. 6.7 Species-specific stress reactions caused by the drought year 1976 shown in relation to the mean growth level in the 3-year-period 1973–1975 before the drought stress (reference line = 1.0) (according to Pretzsch et al. 2013). (**a**) Norway spruce, European beech and sessile oak in pure stands. (**b**) European beech in pure and mixed stands. (**c**) European beech and sessile in pure and mixed stands. The courses represent the growth in the dry year 1976 and in the recovery period (periodical mean of 1977–1979) in relationship to the growth in the reference period (periodical mean of 1973–1975)

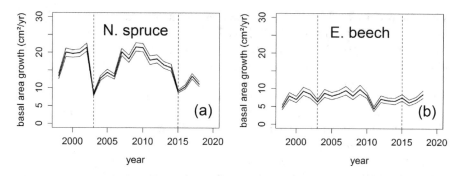

Fig. 6.8 Mean (± SE) annual basal area increment (± SE) of all (**a**) Norway spruce (*n* = 268) and (**b**) European beech (*n* = 141) from 1998 to 2018. Trees facing rainfall exclusion since 2014 were excluded (tree numbers refer to the year 2003)

with a deep root system (e.g. oak) towards those with shallow roots (e.g. beech) can be hypothesised in mixed forests subjected to drought stress (Zapater et al. 2011).

As another example, we show the growth reactions of Norway spruce and European beech to natural episodic and experimentally extended drought in mature monospecific and mixed-species stands of Norway spruce and European beech in the Kranzberg Forest (Fig. 6.8). From the annual diameter growth records since 1998 based on girth tapes, Norway spruce and European beech both reflect the episodic drought events in 2003 and 2015. The courses of the annual basal area growth (± SE) from 1998 to 2018 is visualised in Figure 6.8. The long-term trend in this period is slightly decreasing for Norway spruce and rather parallel to the x-axis for European beech. This long-term trend, however, is interrupted by slumps of the annual growth in 2003 and 2015, especially in the case of Norway spruce. European beech was much more resistant to the drought years. Norway spruce shows a strong growth reduction in the drought years 2003 and 2015, while European beech shows just a slight growth reduction in 2003 and even an increase in 2015. It is obvious that the growth of Norway spruce is severely reduced by about 50–60% compared to the growth before the drought period. European beech trees do not exceed half of the growth losses and react much less to drought.

In order to show intra- and inter-species-specific response patterns to drought, we also compared the behaviour in the drought years 2003 and 2015 with the 3-year periods before and after the events (Fig. 6.9). Here, we visualise the results just for 2003, as they were similar concerning the relationships between the species and concerning the intra- and interspecific differences, but more pronounced than in 2015. Figure 6.9a underpins the much stronger effect of drought on the growth of Norway spruce compared with European beech, in general, without considering their intra- and interspecific neighbourhoods.

Interestingly, Norway spruce suffered 10–20% less under drought when growing in the neighbourhood of European beech (see Fig. 6.9b, mean and SE lines for sb). The growth losses were stronger in the intraspecific neighbourhood of Norway spruce in 2003. The tree growth of the group with intraspecific competition (group

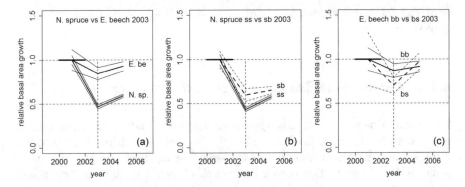

Fig. 6.9 Visualisation of the growth resistance and resilience to the 2003 drought event based on the annual basal area increment (± SE) (according to Pretzsch et al. 2020a). The pre-drought growth in the period 2000–2002 is set to 1.0 (1.0 line). The growth in the drought year 2003 and in the post-drought period 2004–2006 was sketched in relation to this reference level. (**a**) On average, the growth of Norway spruce dropped steeply and recovered slowly; European beech was hardly affected by the 2003 drought. (**b**) When Norway spruce grew in the interspecific neighbourhood with European beech (sb, broken lines), it was less affected by drought compared with intraspecific constellations (ss). (**c**) When European beech grew in the interspecific neighbourhood with Norway spruce (bs, broken lines), it was more affected by drought compared with intraspecific constellations (bb)

ss) was significantly lower than the growth in the group with interspecific competition (group sb).

European beech behaved differently (Fig. 6.9c). European beech growth declined significantly stronger under drought in the interspecific neighbourhood but recovered quickly (see Fig. 6.9c, mean and SE lines for bs). In contrast, European beech growing in the neighbourhood of beech trees was much less affected by drought (mean and SE lines for bb). In dry years, Norway spruce seemingly benefits from interspecific neighbourhoods. Similar growth-stabilizing interspecific interactions were reported by del Río et al. (2014).

The results of such studies indicate growth reactions of selected sampled trees. The diagnosis of growth reactions and growth losses at the stand level can show a different picture. Small trees may react differently to stress and partially compensate the growth losses of tall trees in low-growth years (Pretzsch et al. 2018). In addition to the growth of individual trees, stand growth is determined by stand density, mortality and existing natural regeneration that can strongly modify the growth under drought stress (Allen et al. 2015; Pretzsch et al. 2015; Sohn et al. 2016). Stress may also modify height growth, form factor or allometry between root and shoot; all of these aspects are not considered by relative growth analyses based on increment cores of individual trees at breast height (Rötzer et al. 2017).

In summary, the dendrochronological analyses of growth oscillation in drought years may serve as a bioindication and evidence of stress by environmental factors (Dobbertin 2005; Allen et al. 2010), whereas the growth analyses at the stand level provide integrated information of growth reactions that are relevant on ecosystem level and for management and planning purposes. Both are useful indicators of climate-smart forestry.

6.3.4 Vulnerability Related to High Productivity Level

Compared to the past, forest stand growth has accelerated during the last decades in many areas of Europe (Spiecker et al. 1996; Kauppi et al. 2014; Pretzsch et al. 2014a). Although proceeding on a raised growth level, stands are encountering a series of drought years. But, even if tree and stand growth by this may be reduced below the present level, they still exceed the past level, represented, for instance, by the yield tables. The negative deflection from the currently increased level is naturally interpreted as stress exposure. However, at the same time, the absolute level is still higher than in the past. This may evoke contradictory assessments regarding forest health and vitality.

For showing this concurrency of long-term upward trend and episodic stress events, we present results from an ongoing study of monospecific and mixed-species stands (triplets) of Norway spruce, Scots pine, European beech and sessile oak across Europe (Pretzsch et al. 2020c). Figure 6.10 depicts the stand volume growth of monospecific stands of the respective species in Germany in comparison with the corresponding yield tables by Wiedemann (1943), Jüttner (1955), Assmann and Franz (1963) and Schober (1967).

In the case of Norway spruce (Fig. 6.10a), the observed (oscillating course) and expected growths (unimodal growth curves) were the closest to each other. In the case of European beech (Fig. 6.10b, d), the observed growth strongly deviated from the yield table predictions. Most of the stands showed an increased growth level and an increasing growth trend since one or two decades. Recent drought years just cause an oscillation on a luxury hypertrophy level and only occasionally cause a slump of the growth trajectories below the yield tables used as standard for comparison. Also, Scots pine (Fig. 6.10c, e) strongly exceeded the level of the yield tables and strongly increased in growth. In the last few years, the upward trend is, from time to time, interrupted by low-growth years. The most positive deviations from the yield tables and upward trends were found for the stand growth of sessile oak (Fig. 6.10f); the drought years in 2003 and 2015 had hardly any effects on the annual course of growth of sessile oak. The growth of the mixed-species stands also proceeded above the yield tables and showed an increasing trend (not shown). But, compared with the monospecific stands, the inter-annual oscillation in mixed-species stands was lower. So, species interactions not only stabilise growth at tree level (Fig. 6.9) but also contribute to greater temporal stability at the stand level (del Rio et al. 2017).

On average, the growth of the analysed stands of spruce, pine, beech and oak was 48% above the historic level represented by the yield tables (Pretzsch et al. 2020c); this reflects a considerable change of forest growth and potential for wood utilisation compared to the previous century. Growth loss due to drought was highest in

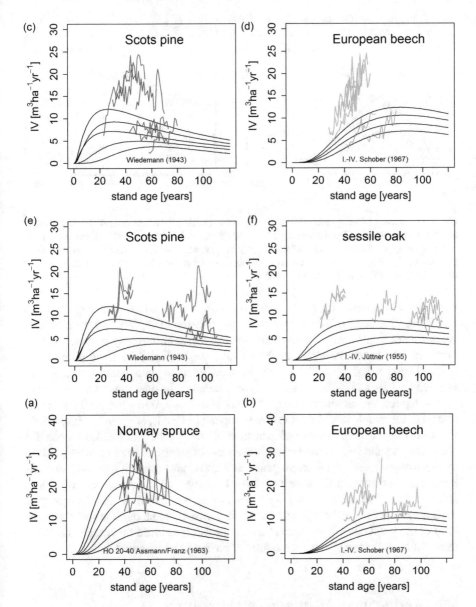

Fig. 6.10 Courses of the absolute annual volume growth (merchantable wood with diameter 7 cm at the smaller end including bark) of monospecific stands of the triplets of (**a** and **b**) Norway spruce and European beech, (**c** and **d**) Scots pine and European beech and (**e** and **f**) Scots pine and sessile oak in the period 1997–2018 compared with the courses of the yield tables for moderate thinning by Wiedemann (1943), Jüttner (1955), Assmann and Franz (1963) and Schober (1967)

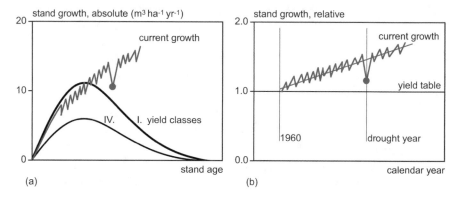

Fig. 6.11 Growth of traditional yield table vs. current reality in schematic (**a**) absolute and (**b**) relative representation. Growth predicted by yield tables (yield classes I and IV), current growth trend and low-growth events caused by drought years (growth decrease marked by filled circle). In the marked drought year, the stand growth would lie under the level of the upward trend but still above the level of the yield tables (1.0 line)

Norway spruce stands, lowest in sessile oak stands and medium in European beech and Scots pine stands. Mixed-species stands performed slightly better in drought years but not significantly. The current growth courses of the analysed stands can be understood as a new standard. In the recent drought years, the stand growth was reduced on average by 28%; in the most favoured or unfavoured cases, it was reduced by 19% and 37%, respectively, referenced at the new standard.

The aggregated picture applying for sites with medium to good growing conditions, i.e. for many sites in the Central European lowlands, is given in Figure 6.11. Figure 6.11a shows the course of growth predicted by the yield tables (curves for yield classes I and IV), the current annual course of growth (sawtooth curve) and the low-growth collapse in a drought year (minimum value marked by a dot) in schematic representation at the absolute scale. Figure 6.11b shows the same relationships at a relative scale. Both graphs reveal that the growth in the drought year declines below the level of the current upward growth trend, but it is still above the level of the yield table. This means that the growth is reduced by drought stress but may still lie considerably above the growth predicted by the yield tables.

6.4 Acclimation, Adaptation and Recovery

6.4.1 Acclimation

Trees and stands may physiologically or morphologically acclimate to drought stress (Cinnirella et al. 2002; Lapenis et al. 2005; Reich et al. 2016). Here, we show an example of trees' acclimation to extended drought stress (Fig. 6.12). In a rainfall exclusion experiment, Goisser et al. (2016) and Pretzsch et al. (2020a) analysed the

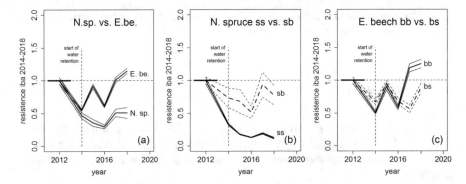

Fig. 6.12 Visualisation of the resistance to the 2014–2018 rainfall exclusion based on the single-tree annual basal area increment (± SE). Pre-drought growth in the period 2011–2013 is set to 1.0 (1.0 line, solid black); the growth during years of the rainfall exclusion is related to the pre-drought level. (**a**) On average, growth of Norway spruce dropped steeply and recovered slowly; growth of European beech dropped less and even increased above the level of the pre-drought period after 4 years. (**b**) When Norway spruce grew in the interspecific neighbourhood with European beech (sb, broken lines), it was less affected (20–30%) during the rainfall exclusion in the first years and recovered remarkably compared with spruce trees in intraspecific constellations (ss). (**c**) When European beech grew in the interspecific neighbourhood with Norway spruce (bs, broken lines), it performed better in the first years of drought but then relapsed the growth of beech trees in intra-specific neighbourhoods (bb)

courses of the annual basal area increments within a 5-year period of water exclusion over all summer periods. Figure 6.12a shows the behaviour of Norway spruce and European beech in monospecific and mixed stands. Both species strongly reduced their growth in the first year after the rainfall exclusion. Norway spruce continued to decrease but stabilised in 2016–2018. European beech stabilised even earlier and rearrived at the base level in 2018. Since 2015, Norway spruce grew on average significantly less than European beech on the treatment plots. This underpins the species-specific drought tolerance reviewed by Niinemets and Valladares (2006). The asymptotic growth trend of Norway spruce and the upward growth trend of European beech after multi-year drought can be interpreted as acclimation.

Especially in the case of Norway spruce, the neighbourhood seems to matter for the growth behaviour imposed by experimentally extended drought stress (Fig. 6.12b). Norway spruce growing in the neighbourhood of spruce trees displayed a much stronger growth decrease than those in vicinity to European beech. In the case of European beech, both trees in inter- and intraspecific neighbourhood behave similarly in the first 3 years (Fig. 6.13c). Since 2016, we found significant differences between the two groups, i.e. beech trees in intraspecific competition outgrew those neighboured by spruce trees.

In summary, the experimentally induced drought from 2014 to 2018 caused a strong growth reduction in the first year, followed by a slight acclimation to the dry conditions. European beech acclimatised and recovered better than Norway spruce; Norway spruce acclimatised better in the neighbourhood of beech trees. For Norway spruce tree mortality was fivefold, while for European beech it is similar to the long-term mortality rate on the untreated plots.

6.4.2 Adaptation

The intraspecific genotypic and phenotypic variability plays an important role in terms of adaptation. Figure 6.13a illustrates how a high genetic diversity can contribute to adaptation, as each genotype shows a distinct optimum curve to climate conditions.

Similarly, there can be substantial differences in expression of the optimum curve among different provenances of a species (Benito Garzón et al. 2011) through adaptation to varying climate conditions (Aitken et al. 2008).

The Scots pine provenance trial Schwabach 304, established in 1927, provides an example for provenance-dependent behaviour of growth (Fig. 6.14a). This experiment revealed that the production of the provenance "Bamberg" was inferior until an age of 50 years compared to the provenances "Schwabach" and "Unterfranken". The provenance Bamberg seems to be better suitable for the changing growing conditions since the last century. However, it turned to the most productive provenance, possibly indicating higher adaptation to the specific site conditions. In the case of the Douglas fir provenance trial Kösching 95 (established in 1961, first survey 1961, last survey 2015), all provenances performed very similarly during the juvenile stand development. With increasing age, however, the difference in total stand growth between the weakest and strongest performing provenance becomes a remarkable 500 m³ ha⁻¹ (Fig. 6.14b). This trial also includes plots of Norway spruce,

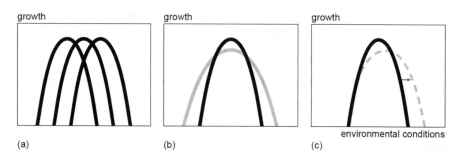

Fig. 6.13 Growth of different genotypes depending on their real ecological niche in schematic representation (niche reduced to one axis for simplifying the representation). (**a**) Unimodal relationship between site conditions for different genotypes. A greater genetic diversity can contribute to a better adaptation to climate change. (**b**) Growth of two species with different phenotypic plasticity. A higher phenotypic plasticity can reduce the impact of climate change and favour "adaptation" in a broad sense of the term. (**c**) Shift in species growth response along the ecological niche caused by epigenetic preconditions. Different growth responses to environmental conditions by epigenetic changes can be inherited in the new generation and contribute to tree species adaptation

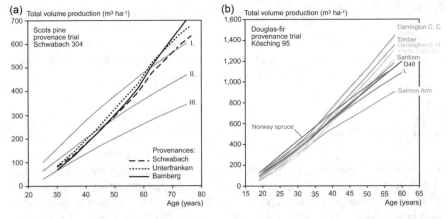

Fig. 6.14 Long-term observations can reveal changes in the performance ranking of different spe-cies' provenances as well as deviations of their growth and yield characteristics from the prediction by yield tables. (**a**) On the Scots pine provenance trial Schwabach 304, initially low-performing provenances finally have superior total production. After 20 years, all the three provenances sur-pass the level of the yield table. (**b**) Douglas fir provenance trial Kösching 95, where some Douglas fir provenances significantly excel the other and the total production of the Norway spruce refer-ence plots in advanced age. The yield tables for Scots pine by Wiedemann (1943, yield classes I–III) and for Douglas fir by Bergel (1985, yield class I) and Norway spruce by Assmann and Franz (1963, O 40) were used as reference

which performs similarly to Douglas fir initially but lags behind most of the Douglas fir provenances at advanced ages. The deviations from the reference yield tables are negligible in the beginning but accumulate strongly with increasing stand age.

The long-term changes in ranking and trend, depicted in Figure 6.14, underpin that the choice of silvicultural options (e.g. provenance, thinning and tree species) should not be based just on early tests but on long-term observation. Real-time series of observations cannot simply be replaced by artificial time series, which try to derive the development over age by measuring and combining stands of different ages on the same site.

Generally, a higher genetic diversity implies more possibilities of adaptation to changing conditions, but species adaptation through selection during the regenera-tion phase is a long-lasting process. Therefore, other aspects, such as species phe-notypic plasticity to environmental conditions or the above-mentioned ability of acclimation, are important for adaptation to climate change, in a broader perception of the term "adaptation". More plastic species or provenances are expected to resist better to climate changes in large parts of their distribution areas (Fig. 6.13b). However, plasticity is not always linked to genetic diversity (Mutke et al. 2010); therefore, it does not always guarantee good performance.

Epigenetic changes can also have important effects on species adaptability (Fig. 6.13c). Current research scrutinises to what extent climate change can cause heritable epigenetic changes and adaptation of trees beyond the genetic adaptation by sexual reproduction (Franks Hoffmann 2012; Verhoeven et al. 2016). That environmentally induced epigenetic changes may be inherited by future generations is of special relevance for long-lived organisms, such as trees (Bräutigam et al. 2013). Rico et al. (2014) and Shanker et al. (2014) report on epigenetic changes under drought stress, and Bossdorf et al. (2008) discussed how to consider such high important aspects in future experiments in ecology.

6.4.3 Recovery

The recovery capacity after cessation of disturbance is an important characteristic of tree growth to better understand growth reaction patterns under changing environment. Beyond the already revised growth response to extreme annual events through the analysis of resistance, recovery and resilience indices by Lloret et al. (2011), here, we present two examples of growth response to longer disturbances, such as SO₂ pollution and O₃ exposure, and the recovery afterwards.

Growth Reactions of Norway Spruce and Silver Fir to SO₂ Pollutants The stem diameter growth trajectories of Norway spruce and silver fir in Fig. 6.15 show the stress reaction and recovery caused by SO₂ pollutants (1950–1985) and dry years (1976, 2003). The trajectories are based on 118 trees per species sampled by increment boring in 22 mixed-species stands across Germany. The stands cover a wide

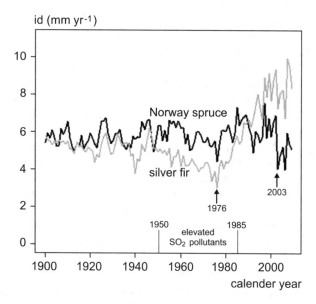

Fig. 6.15 Average course of the stem diameter increment (id mm yr⁻¹) of about 100-year-old silver fir ($n = 118$) and Norway spruce ($n = 118$), in Bavaria and Lower Saxony. The graph shows the period 1950–1985 with increased SO₂ pollution, the species-specific growth trend after 1985 with reduced SO₂ pollution and the drought years 1976 and 2003 (Uhl et al. 2013)

range of site conditions, within and beyond the natural distribution of both species, in Bavaria and Lower Saxony. In Bavaria, the tree age of Norway spruce and silver fir was 110 and 114 years, respectively, and in Lower Saxony it was 72 and 88 years, respectively. Since 1900, both species first show a rather similar and steady level of growth. Silver fir then shows a growth depression from 1950 to 1980 and a growth recovery afterwards. The growth of Norway spruce is superior to silver fir in the period 1950–1980 and turns to be inferior after 1980.

The growth of silver fir was strongly reduced by the high SO_2 pollutants in the period 1950–1980 (Uhl et al. 2013). After reduction of the SO_2 load in the atmosphere by installation of filters in power stations (Joos 2006; Elling et al. 2009), the stem diameter growth of silver fir increased continuously in the last 20–30 years despite the advancing age. The finding that the growth of silver fir dropped less and recovered faster than the growth of Norway spruce in drought years means that the vitality changed in favour of silver fir in the last decades. Consequently, in the very widespread mixed-species stands of Norway spruce, silver fir and European beech in European mountain regions, silver fir gained in competitive strength and productivity compared to the other species (Pretzsch et al. 2015; Bosela et al. 2018; Hilmers et al. 2019).

Growth Reactions to Chronic O_3 Stress and Recovery After Temporal O_3 Exposure Increased O_3 concentration in the air can have significantly detrimental effects on the health and growth of trees in urban areas and forests (Matyssek und Sandermann 2003). O_3 fumigation experiments show that ozone exposure can reduce the photosynthesis (Botkin et al. 1972; Karnosky et al. 2007; Wittig et al. 2009), the carbon assimilation (Chappelka and Samuelson 1998; Sitch et al. 2007), the growth (Wipfler et al. 2005; Pretzsch und Schütze 2018) and allometry of trees (Dickson et al. 2001; Grantz et al. 2006; Pretzsch et al. 2010). The effect of O_3 on growth can be analysed by tree ring analyses and comparison of exposed trees with controls (Pretzsch und Schütze 2018).

The free-air ozone fumigation experiment in the Kranzberg Forest showed how 80-year-old Norway spruce and European beech trees reacted to double-ambient O_3 exposure (2000–2007) and also how they behaved after the stop of the temporary exposure (Matyssek et al. 2010; Häberle et al. 2012). During the phase with double-ambient O_3 exposure (2 × O_3), the annual basal area increments of Norway spruce and European beech decreased by 24% and 32%, respectively (Fig. 6.16). In the period 2008–2016, after the stop of the exposure, both species recovered. The basal area increments of the previously O_3-fumigated trees recovered and raised by 14% and 24% above the level of the control trees (1 × O_3). So, they nearly caught up with the control trees regarding diameter growth. However, the resistance and resilience of the previously fumigated trees to drought and late frost were lower compared with the control trees. This underlines that measures for pollution control may result in a quick recovery from chronic O_3 stress and emphasises the relevance of O_3 control for the health and the carbon sink function of trees and forests (Matyssek et al. 2010).

Tree diameter increment

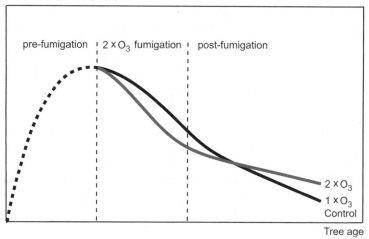

Fig. 6.16 Pattern of growth responses of mature Norway spruce and European beech to temporary $2 \times O_3$ ozone fumigation, according to Pretzsch et al. (2018). Free-air twice-ambient ozone fumigation ($2 \times O_3$) reduced diameter increment of mature Norway spruce and European beech within the $2 \times O_3$ fumigation period from 2000 to 2007. In the post-fumigation period, growth of the former fumigated trees can recover and even exceed the growth of the control trees. $1 \times O_3$ means ambient ozone concentration and represents the control tree (black line); $2 \times O_3$ means twice-ambient ozone-fumigated trees (red line)

6.5 Discussion: Implications for Environmental Monitoring, Forest Ecology and Management

The reviewed tree and stand growth behaviour by reasons of changing environments is highly relevant as any change of forest growth and structure has strong impacts on the provision of ecosystem goods and services. In the following, we discuss the implications for environmental monitoring, forest ecology and management.

6.5.1 Environmental Monitoring

As tree and stand growth strongly depends on environmental conditions, their observation delivers a unique contribution to environmental monitoring. Thus, the tree and stand growth measured on long-term experiments and the tree growth rates retrospectively derived from increment cores and stem discs can provide evidence of environmental changes (Schweingruber 2012). Growth records from trees or stands without active forest management are of special value, as their growth courses are not superimposed by silvicultural management effects. Long-term observation helps to quantify the human footprint on nature (Pretzsch et al. 2014a).

In temperate Europe, there is a patchwork of forests with positive, neutral and negative growth reactions to the ongoing change of environmental conditions. In regions with sufficient water supply, growth strongly accelerated due to favourable effects of global change, such as nitrogen deposition, extension of the growing season (McMahon et al. 2010) and rise of the CO_2 concentration (Ewald et al. 2004). In drier regions, forest growth is reduced by drought and accompanied biotic calamities (Martin-StPaul et al. 2013). In fact, many forests show growth slumps in dry years but are currently growing at higher levels than before, requiring a long-term reference for impact assessment (Fig. 6.11).

The demonstration of forest growth and decline and the practical consequences on long-term experiments can motivate measures for environmental care and pollution control (Matyssek and Sandermann 2003; Uhl et al. 2013). The growth trends and events on environmental conditions can also be used for assessing the climate smartness of various forest and stand types and contribute to developing indicators of stability and climate smartness for monitoring and decision-making purposes (Bowditch et al. 2020; Santopuoli et al. 2021).

6.5.2   Forest Ecology

The accelerated tree growth and forest aging and the projected sink saturation require conformance of all associated organisms, including humans. Plants and animals inhabiting these habitats depend on specific phases in stand development and structure; faster growth means interference in species living conditions and demands for higher mobility. In other areas, growth decline may call in question the future of tree species and the associated plants and animals. In the lowlands, we find horizontal mosaics and, in the mountains, elevation zones with changing growth, changing competitive strength of the species and a transition to species with ecological niches better suitable for both the continuously and abruptly changing growth conditions.

If we interpret tree growth trends as an indication of fitness (Dobbertin 2005), we may consider that some species, such as Norway spruce and Scots pine, especially when cultivated beyond their natural range in dry lowland areas, lose competitiveness. In many cases their growth is still on a rather high level but with a decreasing tendency, especially caused by drought years that became more frequent in the last decades.

Long-term observation of tree growth shows that trees can recover from episodic and chronic stress and that they are also able to acclimatise to a certain degree to, e.g. drought stress (Pretzsch et al. 2020a). Future research will show the potential of acclimatisation and of adaptation over successive generations. Natural regeneration may pave the way to a continuous adaptation by natural selection of the fittest. Permanent plots surveyed with a combination of high resolution and coarse size will be required to track forest demography in time and space (e.g. tree mortality, age structure) and understand scale and pattern of forest development (e.g. species diversity, vegetation shift).

The shown growth trends suggest new experiments in monospecific and mixed-species stands, in order to better understand the effects of environmental changes on forest resilience. Of importance are, e.g. experiments analysing the provenance selection and adaptation to climate change, the impacts of tree species mixing on growth stability, the potential for acclimation and adaptation of tree species, the effects of various silvicultural measures (e.g. thinning), the interactive consequences of disturbances and the effects of nutrient addition on stress resistance, resilience and recovery.

6.5.3   Forest Management

Management options and decisions regarding the choice of species and provenance, stand regeneration, spacing, thinning and rotation length depend on information about the current tree and stand growth. Susceptibility of presently used species to biotic or abiotic damages will determine the selection of new provenances and foreign species (Atzmon et al. 2004; Arend et al. 2011; Zang et al. 2011). By early stand regeneration and transition to natural regeneration, the stand acclimation and adaptation to drought may be promoted (Aasamaa et al. 2004). Spacing and thinning may be intensified in order to use the accelerated growth or to reduce stand density and drought stress (Sohn et al. 2016). Long-term experiments show, in many regions, an increase of the maximum stand density; this means that keeping stands at the traditional density levels may cause severe growth losses as the optimal density will be under matched. The accelerated growth may suggest a shortening of the rotation period (Pretzsch et al. 2014a).

All these measures depend on information about growth trends and short-term growth reactions to adjust forest management and maintain sustainability, in particular, under changing environmental conditions. More rapid tree growth can result in earlier harvest threshold diameters and rotation due to increased stand productivity, which can raise the annual harvest. Recent growth trends allow foresters to maintain much higher growing stocks. However, strong thinning, which uses past conditions as a guideline, might reduce stand density, such that the actual growth potential is not fully realised. In addition, a shortened rotation period can mean reduced risk in terms of forest damage, including windthrow, bark beetle infestation and/or snow breakage. On the other hand, at sites with decreasing growth trends and decay symptoms, urgent adaptation strategies are required, such as species diversification, density regulation, promotion of natural regeneration or even, in some cases, species or provenance assisted migration.

For a specific stand mean tree size, standing stock and mortality rate can be achieved one or more decades earlier. This leaves age-based experience values, widely used yield tables and other models and many traditional management guidelines to become obsolete. Trade-offs exist between shortening rotation period to maximise carbon sequestration and lengthening tree growth to reduce biodiversity loss. Growth records may further serve for updating forest growth models not only for monocultures but also for more complex mixed-species stands. Such models should provide forest management with appropriate growth and yield information for calculation of the annual sustainable harvest. The underlying experimental and observation plots may serve as demonstration and training plots on how to manage forest under environmental changes and as silvicultural assistance to the transition from high-risk monocultures to close-to-nature forest systems, towards climate-smart forestry.

6.6 The Importance of Long-Term Experiments for Fact-Finding

As most of the results presented in this chapter are based on long-term growth and yield experimental plots, we finally discuss the importance of long-term experiments for fact-finding. The founding fathers of forest yield science, such as Danckelmann and Schwappach, were convinced that trees and forests require long-term observation, and for this purpose they established the first forest research stations, beginning in 1870 (Ganghofer 1881; Milnik 1999; Landesforstanstalt Eberswalde 2001). Most of our scientific knowledge of tree and stand dynamics, and practical decision, is based on long-term experiments. However, long-term experiments are often under scrutiny, endangered and sacrificed for cost reduction. The reasons are as follows: forest areas with long-term experiments have to be left out from regular forest operations, long-term experiments are costly and waiting for some results more than a couple of years hardly fits into these fast-moving times.

A misleading argument for discontinuance is that forest inventories established in the last decades at national or enterprise level render long-term experiments superfluous. Long-term experiments and temporary plots are rather complementary, more than redundant (Nagel et al. 2012). Long-term experiments provide far-backward-reaching time series of growth and yield for selected sites; they also include extreme variants, such as unthinned stands or solitary growing conditions, which are hardly represented by inventories. Unthinned plots can be important as reference (maximum density, site-specific productivity), and solitary plots are relevant for understanding tree behaviour (aging, wood quality) under strong thinning and the trade-off between tree and stand level performance. Temporary plots and forest inventories, in contrast, include mainly routinely managed stands (mean stand density, common silvicultural treatment) and provide a representative overview of the current growth behaviour.

Forest inventories can provide large-scale representative data, and the information potential can be exploited by big data methods (e.g. Liang et al. 2016). However, inventories may hardly provide information about the stand history and intermediate yield; a lack that can be partly remedied by permanent inventory plots, when they get measured repeatedly and gradually, provide longer time series. Long-term experiments, in contrast, can reveal the cause-effect relationship of various treatment options on tree and stand behaviour, as they are established under controlled continuous (ceteris paribus) conditions. Inventory plots indicate correlations but provide no evidence for casualties, as they can vary in many traits beyond the factor in question. Long-term experiments strive for general rules, laws and understanding, while inventory plots aim at regional information for practical purposes. Simulation models can make use of both long-term plots for developing the thinning, mortality and regeneration algorithms and inventory data for calibration of the site-growth relationship, model initialisation and risk assessment.

Large-scale and long-term monitoring plots may be useful to characterise ecological processes (succession, outbreak, windthrow, etc.), thus providing models with data from heterogeneous conditions and including temporal variability of climate. Monitoring of large plots helps to emerge the effect of processes that are spatially autocorrelated (dispersal, competition, facilitation or mortality). Monitoring of permanent plots helps intercept the impact of phenomena that are temporally discrete (e.g. extreme events). During the careers of research topics, such as impact of acid rain, nitrogen deposition or climate change or global change on forests, repeated observation on permanently marked plots provided information far beyond the purpose they originally were established for. They represent stands with known history regarding establishment, silvicultural treatment and disturbances. They offer time series of stand development for biomonitoring, development of silvicultural treatment, modelling and demonstration and training.

Current course of growth may reveal site- and species-specific reaction to various disturbances and may contribute to a less emotional but more objectified discussion of the human influence on tree growth and forest dynamics. While the public debate about forest ecology and health changes its focus typically minute by minute, long-term experiments provide a differentiated and consolidated information base about where and to what extent not only forest growth is influenced by humans but also that pollution and climate control can help.

Acknowledgements The authors would like to acknowledge networking support by the COST (European Cooperation in Science and Technology) Action CLIMO (Climate-Smart Forestry in Mountain Regions – CA15226), which received funding under the EU Framework Programme for Research and Innovation HORIZON 2020.

The first author wishes to thank the German Science Foundation (Deutsche Forschungsgesellschaft) for funding the projects "Structure and dynamics of mixed-species stands of Scots pine and European beech compared with monospecific stands. Analysis along an ecological gradient through Europe" (# DFG PR 292/15-1) and "From near-death back to life: Mixed stands of spruce and beech under drought stress and stress recovery. From pattern to process" (# DFG PR 292/22-1). We would also like to thank the Bavarian State Ministry for Environment and Consumer Protection for funding the project "Pine (*Pinus sylvestris*) and beech (*Fagus sylvatica*) in mixed stands: Suitable partners to ensure productivity on dry sites in times of climate change (KROOF II)" (# GZ: TKP01KPB-73853) and the Bavarian State Ministry for Nutrition, Agriculture and Forestry for funding the project "W047" (# GZ: 7831-28160-2018). We further thank the

Bayerische Staatsforsten (BaySF) for supporting the establishment of the plots and the Bavarian State Ministry for Nutrition, Agriculture and Forestry for permanent support of the project W007 "Long-term experimental plots for forest growth and yield research" (# 7831-26625-2020). Thanks to all the project partners of the mentioned projects for providing data on tree and stand growth across Europe. Thanks also goes to anonymous reviewers for their constructive criticism.

References

Aasamaa K, Söber A, Hartung W, Niinemets Ü (2004) Drought acclimation of two deciduous tree species of different layers in a temperate forest canopy. Trees 18(1):93–101

Aitken SN, Yeaman S, Holliday JA et al (2008) Adaptation, migration or extirpation: climate change outcomes for tree populations. Evol Appl 1(1):95–111

Allen CD, Macalady AK, Chenchouni H et al (2010) A global overview of drought and heat-induced tree mortality reveals emerging climate change risks for forests. For Ecol Manag 259(4):660–684

Allen CD, Breshears DD, McDowell NG (2015) On underestimation of global vulnerability to tree mortality and forest die-off from hotter drought in the Anthropocene. Ecosphere 6(8):1–55

Ammer C (2019) Diversity and forest productivity in a changing climate. New Phytol 221(1):50–66

Arend M, Kuster T, Günthardt-Goerg MS, Dobbertin M (2011) Provenance-specific growth responses to drought and air warming in three European oak species (*Quercus robur*, *Q. petraea* and *Q. pubescens*). Tree Physiol 31(3):287–297

Assmann E, Franz F (1963) Vorläufige Fichten-Ertragstafel für Bayern. Forstl Forschungsanst München, Inst Ertragskd, München, 104 p

Atzmon N, Moshe Y, Schiller G (2004) Eco-physiological response to severe drought in *Pinus halepensis* Mill. trees of two provenances. Plant Ecol 171(1–2):15–22

Bauhus J, Forrester DI, Gardiner B et al (2017) Ecological stability of mixed-species forests. In: Mixed-species forests. Springer, Berlin, Heidelberg, pp 337–382

Bebi P, Kienast F, Schönenberger W (2001) Assessing structures in mountain forests as a basis for investigating the forests' dynamics and protective function. For Ecol Manag 145(1/2):3–14

Benito Garzón M, Alía R, Robson TM et al (2011) Intra-specific variability and plasticity influence potential tree species distributions under climate change. Glob Ecol Biogeogr 20:766–778

Bergel D (1985) Douglasien-Ertragstafel für Nordwestdeutschland. Niedersächs Forstl Versuchsanst, Abt Waldwachstum, Göttingen, p 72

Bontemps JD, Hervé JC, Duplat P et al (2012) Shifts in the height-related competitiveness of tree species following recent climate warming and implications for tree community composition: the case of common beech and sessile oak as predominant broadleaved species in Europe. Oikos 121:1287–1299

Bosela M, Lukac M, Castagneri D et al (2018) Contrasting effects of environmental change on the radial growth of co-occurring beech and fir trees across Europe. Sci Total Environ 615:1460–1469

Bossdorf O, Richards CL, Pigliucci M (2008) Epigenetics for ecologists. Ecol Lett 11(2):106–115

Botkin DB, Janak JF, Wallis JR (1972) Some ecological consequences of a computer model of forest growth. J Ecol 60:849–872

Bottero A, D'Amato AW, Palik BJ et al (2017) Density-dependent vulnerability of forest ecosystems to drought. J Appl Ecol 54(6):1605–1614

Bowditch E, Santopuoli G, Binder F et al (2020) What is Climate Smart Forestry? A definition from a multinational collaborative process focused on mountain regions of Europe. Ecosyst Serv 43:101113. https://doi.org/10.10/j.ecoser.2020.101113

Bräutigam K, Vining KJ, Lafon-Placette C et al (2013) Epigenetic regulation of adaptive responses of forest tree species to the environment. Ecol Evol 3(2):399–415

Bréda N, Huc R, Granier A et al (2006) Temperate forest trees and stands under severe drought: a review of eco-physiological responses, adaptation processes and long-term consequences. Ann For Sci 63(6):625–644

Chappelka AH, Samuelson LJ (1998) Ambient ozone effects on forest trees of the eastern United States: a review. New Phytol 139(1):91–108

Ciais P, Reichstein M, Viovy N et al (2005) Europe-wide reduction in primary productivity caused by the heat and drought in 2003. Nature 437(7058):529

Cinnirella S, Magnani F, Saracino A et al (2002) Response of a mature *Pinus laricio* plantation to a three-year restriction of water supply: structural and functional acclimation to drought. Tree Physiol 22(1):21–30

Conte E, Lombardi F, Battipaglia G et al (2018) Growth dynamics, climate sensitivity and water use efficiency in pure *vs.* mixed pine and beech stands in Trentino (Italy). For Ecol Manag 409:707–718

D'Amato AW, Bradford JB, Fraver S et al (2013) Effects of thinning on drought vulnerability and climate response in north temperate forest ecosystems. Ecol Appl 23(8):1735–1742

del Río M, Schütze G, Pretzsch H (2014) Temporal variation of competition and facilitation in mixed species forests in Central Europe. Plant Biol 16(1):166–176

del Rio M, Pretzsch H, Ruiz-Peinado R et al (2017) Species interactions increase the temporal stability of community productivity in *Pinus sylvestris*-Fagus sylvatica mixtures across Europe. J Ecol 105(4):1032–1043

del Río M, Vergarechea M, Hilmers T et al (2020) Effects of elevation-dependent climate warming on intra- and inter-specific growth synchrony in mixed mountain forests. For Ecol Manag 479:10. https://doi.org/10.1016/j.foreco.2020.118587

Dickson RE, Coleman MD, Pechter P et al (2001) Growth and crown architecture of two aspen genotypes exposed to interacting ozone and carbon dioxide. Environ Pollut 115(3):319–334

Dobbertin M (2005) Tree growth as indicator of tree vitality and of tree reaction to environmental stress: a review. Eur J For Res 124(4):319–333

Elling W, Dittmar C, Pfaffelmoser K et al (2009) Dendroecological assessment of the complex causes of decline and recovery of the growth of silver fir (Abies alba Mill.) in southern Germany. For Ecol Manag 257(4):1175–1187

Ewald J, Felbermeier B, von Gilsa H et al (2004) Zur Zukunft der Buche (*Fagus sylvatica* L.) in Mitteleuropa. Eur J For Res 123:45–51

Franks SJ, Hoffmann AA (2012) Genetics of climate change adaptation. Annu Rev Genet 46:185–208

Ganghofer von A (1881) Das Forstliche Versuchswesen, Band I. Augsburg, 1881, p 505

Gebhardt T, Häberle KH, Matyssek R et al (2014) The more, the better? Water relations of Norway spruce stands after progressive thinning. Agric For Meteoral 197:235–243

Gimmi U, Poulter B, Wolf A, Portner H et al (2013) Soil carbon pools in Swiss forests show legacy effects from historic forest litter raking. Landsc Ecol 28(5):835–846

Goisser M, Geppert U, Rötzer T et al (2016) Does belowground interaction with *Fagus sylvatica* increase drought susceptibility of photosynthesis and stem growth in *Picea abies*? For Ecol Manag 375:268–278

Grantz DA, Gunn S, Vu HB (2006) O3 impacts on plant development: a meta-analysis of root/ shoot allocation and growth. Plant Cell Environ 29(7):1193–1209

Grossiord C (2019) Having the right neighbors: how tree species diversity modulates drought impacts on forests. New Phytol. https://doi.org/10.1111/nph.15667

Grossiord C, Granier A, Ratcliffe S et al (2014) Tree diversity does not always improve resistance of forest ecosystems to drought. PNAS 111(41):14812–14815

Häberle KH, Weigt R, Nikolova PS et al (2012) Case study "Kranzberger Forst": growth and defence in European beech (*Fagus sylvatica* L.) and Norway Spruce (*Picea abies* (L.) Karst). In: Matyssek R et al (eds) Growth and defence in plants, Ecological studies 220. Springer, Berlin, Heidelberg, pp 243–271. https://doi.org/10.1007/978-3-642-30645-7_11

Hilmers T, Avdagić A, Bartkowicz L et al (2019) The productivity of mixed mountain forests comprised of Fagus sylvatica, Picea abies, and Abies alba across Europe. Forestry. https://doi.org/10.1093/forestry/cpz035

Hoegberg P, Fan H, Quist M et al (2006) Tree growth and soil acidification in response to 30 years of experimental nitrogen loading on boreal forest. Glob Change Biol 12(3):489–499

Hofmann G, Heinsdorf D, Krauss HH (1990) Wirkung atmogener Stickstoffeinträge auf Produktivität und Stabilität von Kiefern-Forstökosystemen. Beiträge für die Forstwirtschaft 24(2):59–73

Joos F (2006) Sekundäre Maßnahmen der Abgasreinigung. In: Technische Verbrennung: Verbrennungstechnik, Verbrennungsmodellierung, Emissionen. Springer, Berlin, Heidelberg, pp 711–723

Jüttner O (1955) Eichenertragstafeln. In: Schober R (ed) (1971) Ertragstafeln der wichtigsten Baumarten. JD Sauerländer's Verlag, Frankfurt am Main, pp 12-25, 134–138

Karnosky DF, Werner H, Holopainen T et al (2007) Free-air exposure systems to scale up ozone research to mature trees. Plant Biol 9(02):181–190

Kauppi PE, Posch M, Pirinen P (2014) Large impacts of climatic warming on growth of boreal forests since 1960. PLoS One 9(11):e111340

Keenan TF, Prentice IC, Canadell JG et al (2016) Recent pause in the growth rate of atmospheric CO2 due to enhanced terrestrial carbon uptake. Nat Commun 7:1–9

Kenk G, Spiecker H, Diener G (1991) Referenzdaten zum Waldwachstum, KfK-PEF 82. Kernforschungszentrum, Karlsruhe, p 59

Kölling C, Knoke T, Schall P et al (2009) Überlegungen zum Risiko des Fichtenanbaus in Deutschland vor dem Hintergrund des Klimawandels. Forstarchiv 80(2):42–54

Landesforstanstalt Eberswalde (2001) Adam Schwappach: Ein Forstwissenschaftler und sein Erbe. Nimrod Verlag, Hanstedt, p 448

Lapenis A, Shvidenko A, Shepaschenko D et al (2005) Acclimation of Russian forests to recent changes in climate. Glob Change Biol 11(12):2090–2102

Lévesque M, Saurer M, Siegwolf R et al (2013) Drought response of five conifer species under contrasting water availability suggests high vulnerability of Norway spruce and European larch. Glob Change Biol 19(10):3184–3199

Liang J, Crowther TW, Picard N et al (2016) Positive biodiversity-productivity relationship predominant in global forests. Science 354(6309):1–12

Lloret F, Keeling EG, Sala A (2011) Components of tree resilience: effects of successive low-growth episodes in old ponderosa pine forests. Oikos 120(12):1909–1920

Lundström US, Bain DC, Taylor AF et al (2003) Effects of acidification and its mitigation with lime and wood ash on forest soil processes: a review. Water Air Soil Pollut Focus 3(4):5–28

Martin-StPaul NK, Limousin JM, Vogt-Schilb H et al (2013) The temporal response to drought in a Mediterranean evergreen tree: comparing a regional precipitation gradient and a throughfall exclusion experiment. Glob Change Biol 19(8):2413–2426

Marziliano P, Tognetti R, Lombardi F (2019) Is tree age or tree size reducing height increment in Abies alba Mill. at its southernmost distribution limit? Ann For Sci 76:17

Matyssek R, Sandermann H (2003) Impact of ozone on trees: an ecophysiological perspective. In: Progress in botany. Springer, Berlin, Heidelberg, pp 349–404

Matyssek R, Wieser G, Ceulemans R et al (2010) Enhanced ozone strongly reduces carbon sink strength of adult beech (Fagus sylvatica) – resume from the free-air fumigation study at Kranzberg Forest. Environ Pollut 158(8):2527–2532

McMahon SM, Parker GG, Miller DR (2010) Evidence for a recent increase in forest growth. Proc Natl Acad Sci USA 107(8):3611–3615

Milnik A (1999) Bernhard Danckelmann. Leben und Leistungen eines Forstmannes. Nimrod Verlag, Suderburg, p 352

Mutke S, Gordo J, Chambel MR et al (2010) Phenotypic plasticity is stronger than adaptative differentiation among Mediterranean stone pine provenances. For Syst 19:354

Nabuurs GJ et al (2013) First signs of carbon sink saturation in European forest biomass. Nat Clim Chang 3:792–796

Nagel J, Spellmann H, Pretzsch H (2012) Zum Informationspotenzial langfristiger forstlicher Versuchsflächen und periodischer Waldinventuren für die waldwachstumskundliche Forschung. Allg For Jagdztg 183(5/6):111–116

Niinemets Ü, Valladares F (2006) Tolerance to shade, drought, and waterlogging of temperate northern Hemisphere trees and shrubs. Ecol Monogr 76(4):521–547

Piao S, Cui M, Chen A et al (2011) Altitude and temperature dependence of change in the spring vegetation green-up date from 1982 to 2006 in the Qinghai-Xizang Plateau. Agric For Meteorol 151(12):1599–1608

Pretzsch H, Kölbel M (1988) Einfluß von Grundwasserabsenkungen auf das Wuchsverhalten der Kiefernbestände im Gebiet des Nürnberger Hafens – Ergebnisse ertragskundlicher Untersuchungen auf der Weiserflächenreihe Nürnberg 317. Forstarchiv 59(3):89–96

Pretzsch H, Schütze G (2018) Growth recovery of mature Norway spruce and European beech from chronic O3 stress. Eur J For Res 137(2):251–263

Pretzsch H, Dieler J, Matyssek R et al (2010) Tree and stand growth of mature Norway spruce and European beech under long-term ozone fumigation. Environ Pollut 158(4):1061–1070

Pretzsch H, Schütze G, Uhl E (2013) Resistance of European tree species to drought stress in mixed versus pure forests: evidence of stress release by inter-specific facilitation. Plant Biol 15:483–495

Pretzsch H, Biber P, Schütze G et al (2014a) Forest stand growth dynamics in Central Europe have accelerated since 1870. Nat Commun 5(4957):1–10

Pretzsch H, Rötzer T, Matyssek R et al (2014b) Mixed Norway spruce (Picea abies [L] Karst) and European beech (Fagus sylvatica [L]) stands under drought: from reaction pattern to mechanism. Trees 28(5):1305–1321

Pretzsch H, Biber P, Uhl E et al (2015) Long-term stand dynamics of managed spruce-fir-beech mountain forests in Central Europe: structure, productivity and regeneration success. Forestry 88(4):407–428

Pretzsch H, Biber P, Uhl E et al (2017) Climate change accelerates growth of urban trees in metropolises worldwide. Sci Rep 7:10. https://doi.org/10.1038/s41598-017-14831-w

Pretzsch H, Schütze G, Biber P (2018) Drought can favour the growth of small in relation to tall trees in mature stands of Norway spruce and European beech. For Ecosyst 5(1):20. https://doi.org/10.1186/s40663-018-0139-x

Pretzsch H, del Rio M, Biber P et al (2019a) Maintenance of long-term experiments for unique insights into forest growth dynamics and trends: review and perspectives. Eur J For Res 138(1):165–185

Pretzsch H, Steckel H, Heym M et al (2019b) Stand growth and structure of mixed-species and monospecific stands of Scots pine (Pinus sylvestris L) and oak (Quercus robur L, Quercus petraea (MATT) LIEBL) analysed along a productivity gradient through Europe. Eur J For Res. https://doi.org/10.1007/s10342-019-01233-y

Pretzsch H, Grams T, Häberle KH et al (2020a) Growth and mortality of Norway spruce and European beech in monospecific and mixed-species stands under natural episodic and experimentally extended drought. Results of the KROOF rainfall exclusion experiment. Trees Struct Funct. https://doi.org/10.1007/s00468-020-01973-0

Pretzsch H, Hilmers T, Biber P, Avdagić A et al (2020b) Evidence of elevation-specific growth changes of spruce, fir and beech in European mixed-mountain forests during the last three centuries. Can J For Res. https://doi.org/10.1139/cjfr-2019-0368

Pretzsch H, Ammer C, Wolff B et al (2020c) Zuwachsniveau, Zuwachstrend und episodische Zuwachseinbrüche. Ein zusammenfassendes Bild vom aktuellen Zuwachsgang in Rein- und Mischbeständen aus Fichte, Kiefer, Buche und Eiche. Allg For Jagdztg 192:1

Pugh TAM et al (2019) Role of forest regrowth in global carbon sink dynamics. Proc Natl Acad Sci USA 116:4382–4387

Reich PB, Sendall KM, Stefanski A et al (2016) Boreal and temperate trees show strong acclimation of respiration to warming. Nature 531(7596):633

Rico L, Ogaya R, Barbeta A et al (2014) Changes in DNA methylation fingerprint of Quercus ilex trees in response to experimental field drought simulating projected climate change. Plant Biol 16(2):419–427

Rodríguez-Calcerrada J, Pérez-Ramos IM, Ourcival JM et al (2011) Is selective thinning an adequate practice for adapting Quercus ilex coppices to climate change? Ann For Sci 68(3):575

Rötzer T, Biber P, Moser A et al (2017) Stem and root diameter growth of European beech and Norway spruce under extreme drought. For Ecol Manag 406:184–195

Ruiz-Benito P, Madrigal-Gonzalez J, Ratcliffe S et al (2014) Stand structure and recent climate change constrain stand basal area change in European forests: a comparison across boreal, temperate, and Mediterranean biomes. Ecosystems 17(8):1439–1454

Santopuoli G, Temperli C, Alberdi I et al (2021) Pan-European sustainable forest management indicators for assessing Climate-Smart Forestry in Europe. Can J For Res. https://doi.org/10.1139/cjfr-2020-0166

Schober R (1967) Buchen-Ertragstafel für mäßige und starke Durchforstung. In: Schober R (1972) Die Rotbuche 1971. Schr Forstl Fak Univ Göttingen u Niedersächs Forstl Versuchsanst 43/44, JD Sauerländer's Verlag, Frankfurt am Main, p. 333

Schweingruber FH (2012) Tree rings: basics and applications of dendrochronology. Springer, Berlin, Heidelberg

Shanker AK, Maheswari M, Yadav SK et al (2014) Drought stress responses in crops. Funct Integr Genomics 14(1):11–22

Sitch S, Cox PM, Collins WJ et al (2007) Indirect radiative forcing of climate change through ozone effects on the land-carbon sink. Nature 448(7155):791–794

Sohn JA, Saha S, Bauhus J (2016) Potential of forest thinning to mitigate drought stress: a meta-analysis. For Ecol Manag 380:261–273

Spiecker H, Mielikäinen K, Köhl M, Skovsgaard JP (eds) (1996) Growth trends in European forests, Europ For Inst, Res Rep 5. Springer, Berlin, Heidelberg, p 372

Thurm EA, Uhl E, Pretzsch H (2016) Mixture reduces climate sensitivity of Douglas-fir stem growth. For Ecol Manag 376:205–220

Tognetti R, Lombardi F, Lasserre B et al (2014) Tree-ring stable isotopes reveal twentieth-century increases in water-use efficiency of Fagus sylvatica and Nothofagus spp. in Italian and Chilean mountains. PLoS One 9:e113136

Tognetti R, Lasserre B, Di Febbraro M et al (2019) Modeling regional drought-stress indices for beech forests in Mediterranean mountains based on tree-ring data. Agric For Meteorol 265:110–120

Uhl E, Ammer C, Spellmann H et al (2013) Zuwachstrend und Stressresilienz von Tanne und Fichte im Vergleich. Allg For Jagdztg 184(11/12):278–292

Vayreda J, Martinez-Vilalta J, Gracia M et al (2012) Recent climate changes interact with stand structure and management to determine changes in tree carbon stocks in Spanish forests. Glob Change Biol 18(3):1028–1041

Verhoeven KJ, Vonholdt BM, Sork VL (2016) Epigenetics in ecology and evolution: what we know and what we need to know. Mol Ecol 25(8):1631–1638

Walentowski H, Kölling C, Ewald J (2007) Die Waldkiefer – bereit für den Klimawandel? LWF Wissen 57:37–46

Wiedemann E (1936/1942) Die Fichte 1936, vol 248. Verlag M & H Schaper, Hannover

Wiedemann E (1943) Kiefern-Ertragstafel für mäßige Durchforstung, starke Durchforstung und Lichtung. In: Wiedemann E (1948) Die Kiefer 1948. Verlag M & H Schaper, Hannover, p 337

Wipfler P, Seifert T, Heerdt C et al (2005) Growth of adult Norway Spruce (Picea abies [L.] Karst.) and European beech (Fagus sylvatica [L.]) under free-air ozone fumigation. Plant Biol 7(6):611–618

Wittig VE, Ainsworth EA, Naidu SL et al (2009) Quantifying the impact of current and future tropospheric ozone on tree biomass, growth, physiology and biochemistry: a quantitative meta-analysis. Glob Change Biol 15(2):396–424

Zang C, Rothe A, Weis W et al (2011) Zur Baumarteneignung bei Klimawandel: Ableitung der Trockenstress-Anfälligkeit wichtiger Waldbaumarten aus Jahrringbreiten. Environ Sci Pol 14:100–110

Zang C, Pretzsch H, Rothe A (2012) Size-dependent responses to summer drought in Scots pine, Norway spruce and common oak. Trees 26(2):557–569

Zapater M, Hossann C, Bréda N et al (2011) Evidence of hydraulic lift in a young beech and oak mixed forest using 18O soil water labelling. Trees 25:885–894

Zhu K, Zhang J, Niu S et al (2018) Limits to growth of forest biomass carbon sink under climate change. Nat Commun 9:2709

Chapter 7
Modelling Future Growth of Mountain Forests Under Changing Environments

Michal Bosela, Katarína Merganičová, Chiara Torresan, Paolo Cherubini, Marek Fabrika, Berthold Heinze, Maria Höhn, Milica Kašanin-Grubin, Matija Klopčič, Ilona Mészáros, Maciej Pach, Katarina Strelcová, Christian Temperli, Giustino Tonon, Hans Pretzsch, and Roberto Tognetti

Abstract Models to predict the effects of different silvicultural treatments on future forest development are the best available tools to demonstrate and test possible climate-smart pathways of mountain forestry. This chapter reviews the state of the art in modelling approaches to predict the future growth of European mountain forests under changing environmental and management conditions. Growth models, both mechanistic and empirical, which are currently available to predict forest

The author Giustino Tonon died prior to the publication of this chapter.

M. Bosela (✉)
Faculty of Forestry, Technical University in Zvolen, Zvolen, Slovakia

National Forest Centre, Zvolen, Slovakia
e-mail: ybosela@tuzvo.sk

K. Merganičová
Faculty of Forestry and Wood Sciences, Czech University of Life Sciences Prague, Praha, Suchdol, Czech Republic

Department of Biodiversity of Ecosystems and Landscape, Slovak Academy of Sciences, Institute of Landscape Ecology, Nitra, Slovak Republic
e-mail: k.merganicova@forim.sk

C. Torresan
Institute of BioEconomy – National Research Council of Italy, San Michele all'Adige, TN, Italy
e-mail: chiara.torresan@ibe.cnr.it

P. Cherubini
Swiss Federal Research Institute WSL, Birmensdorf, Switzerland

Department of Forest and Conservation Sciences, Faculty of Forestry, University of British Columbia, Vancouver, BC, Canada
e-mail: paolo.cherubini@wsl.ch

M. Fabrika · K. Strelcová
Faculty of Forestry, Technical University in Zvolen, Zvolen, Slovakia
e-mail: fabrika@tuzvo.sk; strelcova@tuzvo.sk

© The Author(s) 2022   223
R. Tognetti et al. (eds.), *Climate-Smart Forestry in Mountain Regions*, Managing
Forest Ecosystems 40, https://doi.org/10.1007/978-3-030-80767-2_7

growth are reviewed. The chapter also discusses the potential of integrating the effects of genetic origin, species mixture and new silvicultural prescriptions on biomass production into the growth models. The potential of growth simulations to quantify indicators of climate-smart forestry (CSF) is evaluated as well. We conclude that available forest growth models largely differ from each other in many ways, and so they provide a large range of future growth estimates. However, the fast development of computing capacity allows and will allow a wide range of growth simulations and multi-model averaging to produce robust estimates. Still, great attention is required to evaluate the performance of the models. Remote sensing measurements will allow the use of growth models across ecological gradients.

B. Heinze
Department of Forest Genetics, BFW Austrian Federal Research Centre for Forests, Vienna, Austria
e-mail: berthold.heinze@bfw.gv.at

M. Höhn
Department of Botany, Budai Campus, Hungarian University of Agriculture and Life Sciences, Budapest, Hungary
e-mail: hohn.maria@uni-mate.hu

M. Kašanin-Grubin
University of Belgrade, Institute for Chemistry, Technology and Metallurgy, Beograd, Serbia
e-mail: mkasaningrubin@chem.bg.ac.rs

M. Klopčič
Biotechnical Faculty, Department of Forestry and Renewable Forest Resources, University of Ljubljana, Ljubljana, Slovenia
e-mail: matija.klopcic@bf.uni-lj.si

I. Mészáros
Faculty of Science and Technology, Department of Botany, University of Debrecen, Debrecen, Hungary
e-mail: immeszaros@unideb.hu

M. Pach
Department of Ecology and Silviculture, Faculty of Forestry, University of Agriculture in Krakow, Krakow, Poland
e-mail: rlpach@cyf-kr.edu.pl

C. Temperli
Swiss Federal Institute for Forest, Snow and Landscape Research WSL, Birmensdorf, Switzerland
e-mail: christian.temperli@wsl.ch

G. Tonon (deceased)

H. Pretzsch
Chair for Forest Growth and Yield Science, TUM School of Life Sciences in Freising Weihenstephan, Technical University of Munich, Freising, Germany
e-mail: Hans.Pretzsch@tum.de

R. Tognetti
Dipartimento Agricoltura, Ambiente e Alimenti, Università degli Studi del Molise, Campobasso, Italy

Centro di Ricerca per le Aree Interne e gli Appennini (ArIA), Università degli Studi del Molise, Campobasso, Italy
e-mail: tognetti@posta.unimol.it

Acronyms

CSF	Climate-smart forestry
C	Carbon
CO_2	Carbon dioxide
GHG	Greenhouse gas
GCM	Global climate model
RCM	Regional climate models
IPCC	Intergovernmental Panel on Climate Change
SRES	Special Report on Emissions Scenarios
RCP	Representative Concentration Pathway
AR5	Fifth Assessment Report
YM	Yield model
EM	Empirical model
PM	Mechanistic models
ES	Growth simulator
SES	Hybrid (semi-empirical) model
SFM	Sustainable forest management
TRW	Tree-ring width
VS	Vaganov-Shashkin
LDM	Landscape dynamics or forest landscape model
NFI	National forest inventory
EFDM	European Forest Dynamics Model
EFISCEN	European Forest Information Scenario model
ICP Forests	International Co-operative Programme on Assessment and Monitoring of Air Pollution Effects on Forests
EC	Eddy covariance
NEP	Net ecosystem productivity
MODIS	Moderate Resolution Imaging Spectroradiometer
GPP	Gross primary productivity
NPP	Net primary productivity
LUE	Light-use efficiency
APAR	Absorbed photosynthetically active radiation
NDVI	Normalised difference vegetation index
PAR	Photosynthetically active radiation
ER	Ecosystem respiration
LiDAR	Light detection and ranging
SAR	Synthetic aperture radar
EVI	Enhanced vegetation index
NDWI	Normalized difference water index
WDRVI	Wide dynamic range vegetation index
PI	Phenology index
LAI	Leaf area index
DA	Data assimilation

ALS	Airborne laser scanning
TRW	Tree ring width
ITRDB	International Tree-Ring Data Bank
NCEI	National Centers for Environmental Information

7.1 Introduction

Globally, the forest sector plays a crucial role in climate change mitigation because forests store a significant amount of carbon (C) and absorb around 30% of the annual anthropogenic global carbon dioxide (CO_2) emission. For example, Pan et al. (2011) estimated a total forest sink of 2.4 ± 0.4 petagrams of C per year (Pg C year^{-1}) globally from 1990 until 2007. However, in the same study, the authors estimated a C source of 1.3 ± 0.7 Pg C year^{-1} due to land-use change in the tropical forests.

Climate change imposes direct effects on forest ecosystems through increasing the concentration of atmospheric CO_2 or change in temperature and precipitation (Keenan et al. 2013). Individual organisms living in forest ecosystems respond to climate change in different ways. If their adaptation to new environmental conditions is successful, forest ecosystems continue to provide ecosystem services specific to the type of forest ecosystem, and, by storing C, they can significantly aid in mitigating the impacts of climate change too (Fig. 7.1). However, signs of C saturation in European (Nabuurs et al. 2013) and tropical (Hubau et al. 2020) forests indicate that forests cannot infinitely absorb CO_2. Moreover, trees can adapt to new conditions by reducing their biomass production, which may, in turn, lessen the mitigation effect (Sperry et al. 2019).

Forestry actions that lead to a reduction in greenhouse gas (GHG) emissions and maximise carbon sequestration are considered climate-smart (Nabuurs et al. 2018; Yousefpour et al. 2018). The recently developed comprehensive definition of climate-smart forestry suggests that it should enable forest practitioners to

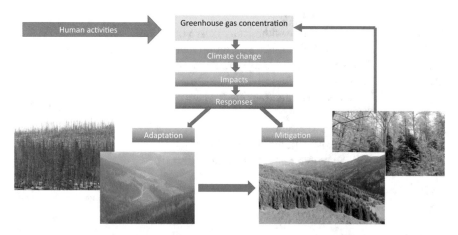

Fig. 7.1 Scheme describing the links between climate change and forests and the role of forests in mitigation of climate change impacts

transform management targets, allowing forests to adapt to and mitigate climate-induced changes while delivering other services to the society (Bowditch et al. 2020).

Mountain forests (for a definition see, e.g., Kapos et al. 2000), are particularly affected by environmental changes, because they are energy and temperature limited, and potentially exposed to warming-induced drought stress (Albrich et al. 2020). In addition, mountain forests are further exposed to and affected by large-scale wind disturbances, frequently followed by outbreaks of pests and fungal diseases (Seidl et al. 2014). Therefore, forest policy decision-makers and forest managers need to be advised by science about the potential and vulnerability of different tree species under predicted climate change.

Predictions by growth models are the best available source of information to optimise forest management and to assess the potential adaptation of the forests to climate change and the mitigation of climate change by the forests. Forest growth models have the potential to test many different variants of forest management, including various species compositions and silvicultural systems, from stand to regional or landscape levels (Fontes et al. 2010; Kramer et al. 2010; Merganič et al. 2020). Example applications of growth models include prediction of future yields, exploration of silvicultural options, preparing resource forecasts, providing insights into stand dynamics, etc. Growth models are generally classified into (i) empirical and (ii) process-based (also known as mechanistic) models. Empirical models are based on empirical equations (regression functions) describing particular relationships without knowing the causal mechanism of the complex system (Fontes et al. 2010; Mäkelä et al. 2012). In contrast, process-based models are based on a theoretical understanding of relevant ecological processes.

Currently, not all available models can test different management variants relevant for ecosystem management and ecopolitical decisions. In the following sections, we review the potential of various models to test the effects of climate change on the growth (the difference in standing volume between the beginning and end of a specified period of time) and productivity (the potential amount of wood produced by the forest within a specified time period, usually rotation period) of mountain forests and their potential to continue to be or become climate-smart.

7.2 Prediction of Future Climate Conditions

To obtain consistent predictions of future tree and forest growth, reliable past climate data as well as predictions of future climate in specific spatial and temporal resolutions must be provided as input into growth models and simulators. To obtain past climate data, spatially interpolated databases at varying spatial and temporal resolutions have been developed (Harris et al. 2014; Moreno and Hasenauer 2016; Cornes et al. 2018). The continental and global databases of past climate data are often the products of spatial interpolation of instrumental time series from climate stations. Therefore, the precision of the interpolated data depends on the density of climate station data provided by individual countries. In the following section, we briefly review the existing approaches to model and predict climate conditions.

7.2.1 Climate Models

Climate models are numerical representations of the Earth's climate system based on global patterns of physical processes, including chemical and biological components of the climate system, simulating the transfer of energy and materials through the system. Currently, there are a variety of models available from simple, simulating only a certain process in the atmosphere, to complex, simulating many processes of the climate system.

Global climate models (GCMs) are general circulation models, which were developed on general principles of fluid dynamics and thermodynamics (Stute et al. 2001). A crucial limitation of global models for their use in ecological modelling is the coarse spatial resolution. Therefore, regional climate models (RCMs) were developed by downscaling GCMs to the region of interest. The more recently developed RCMs have provided a tool to characterise past and future climates at various spatial scales (Rummukainen 2010).

7.2.2 Climate Change Scenarios

The intensifying greenhouse effect leads to global warming and to change in other climate characteristics on the Earth. The most serious consequences are changes in general atmospheric circulation, shifting in frontal and climate zones and the high speed of climate change, exceeding all previous climate changes at least tenfold. This is what scientists have learned from the mathematical modelling of the Earth climate system, where critical physical and chemical processes in the atmosphere and the oceans and physical processes associated with the cryosphere, biosphere and lithosphere were considered (IPCC 2014).

In 2000, the Intergovernmental Panel on Climate Change (IPCC) issued its Special Report on Emissions Scenarios (SRES) and introduced four scenario families to describe a range of possible future climate conditions. Each scenario (A1, A2, B1 and B2) was based on a complex relationship between the socioeconomic forces driving greenhouse gases and aerosol emissions (Nakicenovic et al. 2000). The SRES scenarios have been in use for more than a decade.

In 2009, a new set of scenarios was developed based on the concentration of greenhouse gases in the atmosphere in 2100 (Moss et al. 2010). These scenarios are known as Representative Concentration Pathways (RCPs). Each RCP indicates the amount of radiative forcing, expressed in watts per square metre, that would result from greenhouse gases in the atmosphere in 2100. These four RCPs were used for climate modelling in the IPCC Fifth Assessment Report (AR5) (IPCC 2014): RCP2.6 with radiative forcing peaking at approximately 3 W m^{-2}, RCP4.5 at 4.5 W m^{-2}, RCP6.0 at 6 W m^{-2} and RCP8.5 peaking at 8.5 W m^{-2}, being the most pessimistic scenario at the time.

As in the case of SRES, the GCM/RCM is used to derive data under different RCP scenarios (Jacob et al. 2014). The simulations from the climate models are then used as input to growth models, in some cases with a preceding statistical downscaling to account for topographic effects at a scale below 10 km and to match the grain size of forest models (Temperli et al. 2012; Seidl et al. 2019).

7.3 Simulating Future Forest Growth in the Context of CSF

Forest growth models are used to predict the development of trees, stands and forest ecosystems in the near or distant future, under various scenarios. Forest modelling science has developed from simple empirical yield models (YM), based on either single-time inventories or repeated empirical measurements and regression equations, to more complex empirical models (EM) and dendroclimatic models (DM) and to mechanistic models (PM), which describe physiological mechanisms and processes to predict forest growth. More complex empirical growth simulators (ES) and hybrid (semi-empirical) models (SES), which combine empirical regression equations with physiological processes, are better placed to be used to simulate future forest development than the simple YM and EM, because they often directly include growth sensitivity to climate. A range of growth models available include whole landscape or biome models, stand models, diameter distribution and size class models and individual-tree models (Burkhart and Tomé 2012a). The classification of forest growth models was presented in many studies (Porté and Bartelink 2002; Mäkelä et al. 2012; Fabrika and Pretzsch 2013; Fabrika et al. 2019). Growth models can be classified according to their ability to account for inter- and intraspecific competition and according to the sensitivity of simulated tree/stand growth to climate variation (Fig. 7.2). Tree-level (individual tree or gap/patch) ecophysiological models (the rightmost dark green box in Fig. 7.2) are believed to be most suited for simulations of forest development, because they combine causal effects of climate change and inter- and intraspecific competition (Rötzer et al. 2010; Seidl et al. 2012).

Simulations of forest development with forest growth models require input data, according to the spatial scale for which the prediction of future forest development is required (Fig. 7.3). Input data sources are reviewed in Sect. 7.4 in more detail.

Until recently, forest growth models (mainly ES) were used to predict biomass production and to test the effects of different management approaches and climate change scenarios. However, increasing requirements for a variety of ecosystem services as well as for sustainable forest management have raised the demands on models to expand the spectrum of outputs (Mäkelä et al. 2012; Temperli et al. 2020).

To allow assessment of CSF with modelling approaches, forest growth models must be able to simulate forest stand development under varying forest management alternatives (e.g. different silvicultural treatments) and policy strategies (e.g.

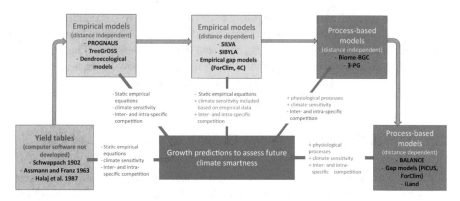

Fig. 7.2 Classification of forest growth models in the context of climate-smart forestry with some examples of existing models or groups of models for each class (in bold black letters). The brown arrow denotes increasing details on ecosystem processes implemented in models

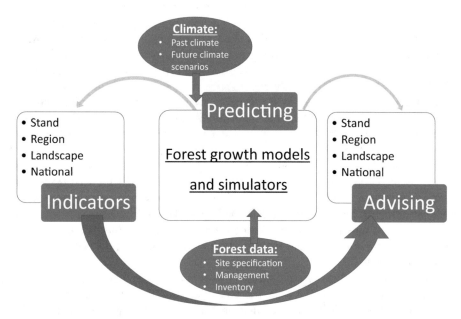

Fig. 7.3 Scheme of the process of predicting future growth and development of forest ecosystems and advising at different spatial scales

emission scenarios). Models should be able to simulate direct and indirect effects of different silvicultural treatments not only on tree growth rates (representing their potential to store carbon) but also on wood quality (the potential of timber to be used for substitution of carbon-intensive materials or fossil fuels) (e.g. Mäkelä et al. 2010).

Bowditch et al. (2020) selected a set of CSF indicators, by combining the pan-European indicators for sustainable forest management (SFM) (FOREST EUROPE

2015) with the ecosystem services defined by the European Environment Agency in the Common International Classification of Ecosystem Services (Haines-Young and Potschin 2018), to assess if the applied CSF practices are on track to meet the goals of forest adaptation and mitigation to climate change. Annex 7.1 resumes the ability of example growth models to address the indicators of CSF based on Bowditch et al. (2020). The models listed in the table represent individual groups following the classification provided above (Fig. 7.3).

7.3.1   Empirical Growth Models

Empirical models use correlation relationships translated into a set of regression equations to simulate tree and/or stand growth. Empirical models include YM, DM and ES.

7.3.1.1   Yield Models

YMs, experiencing more than 250 years of history, are the oldest models in forestry science and practice. They "predict" forest development over the rotation period or longer and are usually based on long-term monitoring or permanent research plots. They are based on regression functions derived from the empirical data and are often presented in the form of handy yield tables (which summarises expected yield tabulated by measurable stand characteristics, such as age, site index and stand density) to enable their use in forestry practice. Pretzsch (2009) and Fabrika and Pretzsch (2013) provided a comprehensive review of yield tables developed since the eighteenth century.

These models mainly rely on the classical assumption of the stationarity of site conditions (Vanclay and Skovsgaard 1997; Skovsgaard and Vanclay 2008) and thus are not capable of predicting forest growth under changing environmental conditions. They predict stand characteristics, such as stand height and diameter, standing volume (merchantable), stand density, etc., and do not consider inter-tree and interspecific competition when used to simulate the growth under different forest management. They use a species site index (top or mean stand height at a standard age, e.g. 100 years) based on height-age curves to consider site potentials to produce wood. In mixed forests, they are used to predict the growth of individual species and their predictions are subsequently combined to a stand level. An exception is the YM developed for mixed forests (Christmann 1949).

These models are, in many cases, well suited to estimate the current amount of wood in forests based on a few measurements but not to predict the future growth of tree species under scenarios of various environmental changes.

7.3.1.2 Empirical Growth Simulators

ES have the second longest tradition in forest modelling after YM. Therefore, in the recent decades, these models began to be intensively used to forecast the development of the forest and to study the impact of changed conditions (environmental, economic, socio-economic) on the growth and structure of the forests (Sodtke et al. 2004). Compared to YMs, the range of conditions for ES application has been largely expanded. Their applications are no longer limited to monospecific and even-aged forests. ESs can model forest stands of various species, age spatial structures. Their ability to account for these forest characteristics depends on the spatial detail of modelling (population, class/cohort or individual) and how the model accounts for the spatial changes in environmental conditions due to inter-tree competition for resources (Fabrika and Pretzsch 2013). Based on the competition for resources, we recognise distance-independent models (Sterba 1995; Nagel 1996) and distance-dependent models (Hasenauer 1994; Pretzsch et al. 2002) (Fig. 7.3). Distance-independent models are biogroup-, ecosystem- or landscape-scale models, whereas distance-dependent models simulate individual trees and thus require spatial coordinates of trees in the stand. The emergence and development of ES made it possible to address the impact of different forest management on forest growth and structure (e.g. thinning) and thus eliminated the limitations of YM, i.e. their applicability only for a few methods of forest management. The range of applicable forest management variants is increasing from population through class/cohort to individual-tree models and from distance-independent to distance-dependent models (Fabrika et al. 2019). This advance in the modelling has opened the space for the use of models, if they are, at the same time, sensitive to climate, to assess the impact of climate change on forest growth and structure (Hlásny et al. 2011). However, an additional limitation related to the response of the increment in tree size to environmental conditions needed to be further addressed in EMs. First EMs used the phytocentric method to quantify site quality that affects the growth of trees and stands (Nagel 1999). However, the static nature of the site index (discussed in Sect. 7.3.1.1) is at odds with the principle of the forest's response to climate change. Therefore, models with a geocentric method (Kahn 1994) have been developed. The geocentric approach considers a direct response of the tree/stand growth to climatic and soil characteristics, for example, expressed by a direct regression model (Monserud and Sterba 1996) or by means of a dose-effect function (Kahn 1994). The link between environmental conditions and diameter/height/volume increment is ensured through empirical (statistically derived) relationships. Such a modification of the models allows their use to assess the impact of climate change on forest growth and structure. Although the introduction of the geocentric approach has expanded the range of ES applications for environmental studies, the very nature of empirical models still limits their use. They cannot be used outside the range of environmental conditions for which they were developed. Therefore, to assess the climate smartness of forest management, statistical relationships should be replaced with causal relationships, which represents the shift from empirical to process-based (mechanistic) models.

7.3.1.3   Dendroecological Models

Dendroecology refers to the use of annual tree rings and dendrochronological techniques to address questions in ecology (Fritts and Swetnam 1989). Tree rings are the products of multiple processes related to the energy, water, carbon and nutrient budget (Babst et al. 2014a). They represent a part of the above-ground carbon accumulation. Dendroecological models establish relationships between tree-ring measurements and environmental factors (Fritts 2001). Originally, dendrochronologists used the relationships between tree-ring formation and climate variance as a proxy to estimate climate variability in a distant past for which weather measurements did not exist. Recently, tree rings have been increasingly used by forest scientists to investigate and model the effects of climate on tree growth and to build empirical models to predict the future growth of forest tree species (Girardin et al. 2008; Chen et al. 2010).

Although tree rings represent only radial stem growth at a particular stem height, tree-ring data-based estimation of above-ground carbon sequestration has been found to be coherent with the net ecosystem productivity measured using eddy covariance techniques (Babst et al. 2014b). However, the contrasting results found in other studies dealing with tree-ring data (Rocha et al. 2006) suggest that there is often a problem with scaling from a tree stem to a forest ecosystem because of sampling bias and stand dynamics (Cherubini et al. 1998; Nehrbass-Ahles et al. 2014).

The developed models were often based on linear relationships between mean tree-ring width (TRW) chronologies and climate variables (Cook and Kairiukstis 1990; Fritts 2001; Dorado-Liñán et al. 2019). They did not consider changes in the relationships over time due to changes in the environment other than climate variation (Guiot et al. 2014). Such empirically-based models should not be used to predict the growth outside the period and the range of site characteristics for which they were developed. Therefore, process-based dendroecological models have been developed to reproduce the daily cellular development (Wilson and Howard 1968; Rauscher et al. 1990; Fritts et al. 1991; Tolwinski-Ward et al. 2011). The first dendroecological process-based model was the TRACH model (Fritts et al. 1991). More recently, the Vaganov-Shashkin (VS) model of tree-ring formation was developed (Vaganov et al. 2006). The VS model and its simplified version called the VS-light model (Tolwinski-Ward et al. 2011), which uses daily climatic input variables and more than 30 parameters for simulating secondary growth of xylem and anatomical features of annual rings, are now frequently used in dendroecological studies (e.g. Sánchez-Salguero et al. 2017).

Although some studies indicated the potential of empirical models developed from tree-ring data for predicting the future growth of forest trees (Dorado-Liñán et al. 2019), process-based models are currently preferred over EMs. However, the crucial role of tree-ring data is to inform vegetation models about long-term forest growth variability and disturbance regime from local to global scales (Babst et al. 2014a). Moreover, process-based models, in general, should be compared against regional and stand-level tree-ring data in shorter periods to avoid potentially biased estimations of net primary productivity by mechanistic models.

7.3.2 Process-Based Growth Models

Unlike empirical model and dendroecological process-based models that focus only on tree-ring formation, PMs simulate physiological processes in the whole plant (photosynthesis, respiration, allocation, mortality, transpiration, translocation and nutrient uptake) and their interactions with processes in the atmosphere and soil. The models relate plant carbon budgets to environmental drivers, climatic variables and/or biogeochemical processes (Battaglia and Sands 1998). This enables PMs to simulate forest responses to changing environmental conditions (e.g. future climate change).

However, our understanding of individual processes differs, some being understood better than others. For example, carbon allocation, which has a critical role in forest adaptation to environmental changes, is often simplified due to insufficient knowledge on driving mechanisms (Merganičová et al. 2019). The other crucial uncertainties in PMs include mortality and regeneration (Mäkelä et al. 2000; Bugmann et al. 2019).

In contrast to EMs, PMs usually work at a finer temporal resolution, starting from less than a minute (Fontes et al. 2010; Pretzsch et al. 2015). Only a few process models, such as 3-PG, FORMIND or TRAGIC, use a coarser scale than 1 day (Hauhs et al. 1995; Köhler and Huth 1998; Forrester and Tang 2016). Many process-based models use different temporal resolutions for simulating different processes, e.g. carbon allocation is frequently simulated at a coarser scale than photosynthesis (Merganičová et al. 2019).

To incorporate physiological processes as realistically as possible, PMs use many physiological parameters as well as input stand and environmental variables. Environmental variables often include solar radiation, temperature, precipitation, wind speed and direction, vapour pressure deficit, nitrogen deposition, CO_2 content in the atmosphere and available soil water content measured at fine temporal resolutions. Long-term, cost-effective and highly instrumented monitoring plots may provide such data, enabling key forest indicators to be modelled. Discussion on highly instrumented experimental plots is presented in Chaps. 10 and 16 of this book (respectively, Tognetti et al. 2021; Pappas et al. 2021). Several models are more simplistic and use only some of these variables and at a coarser temporal resolution (e.g. 3-PG, Landsberg and Waring 1997), whereas others are more complex requiring most of the variables at a finer scale (e.g. ANAFORE, Deckmyn et al. 2008, or FORCLIM, Bugmann 1996). The models range from biome-scale (e.g. Biome-BGC, Thornton et al. 2005) to individual-tree models (e.g. BALANCE, Rötzer et al. 2010).

Landscape dynamics or forest landscape models (LDM) are another group of PMs. The LDMs are based on the interaction of spatial patterns and ecological processes at various spatio-temporal scales. They usually simulate forest dynamics at a site scale (up to 300 ha) and landscape processes at a larger scale (He 2008; Shifley et al. 2017). For example, the process-based model of forest landscape dynamics iLand (Seidl et al. 2012) simulates forest landscape dynamics via modelling spatially explicit resource availability at the landscape scale and integrating local resource competition and physiological resource use. Moreover, it simulates spatial and temporal interactions of vegetation and disturbance agents, which place this model among the most complex models from the point of simulated landscape dynamics. Another widely used LDM in Europe is LandClim (Schumacher et al. 2004). It basically uses a gap-model approach to simulate forest dynamics in individual grid cells of 25×25 m of a landscape wide up to 50 km^2 and accounts for spatial processes, such as wind, fire disturbance, bark beetle, seed dispersal and forest management. Recent applications include estimates of adaptive management effects on forest ecosystem service provision (Temperli et al. 2012), reconstruction of species range shifts in the Holocene (Henne et al. 2013), analyses of disturbance interaction with climate change (Temperli et al. 2015) and biodiversity (Schuler et al. 2019). The FLM TreeMig is also raster based but can be applied from the watershed to the continental scale (Lischke et al. 2006). While it primarily focuses on tree migration under climate drivers (Meier et al. 2012; Scherrer et al. 2020), it has also been applied to assess avalanche-forest interactions (Zurbriggen et al. 2014). Remote sensing technologies may contribute with spatially explicit time series of vegetation traits to estimate temporal changes in CSF indicators at the landscape scale as well as to serve as input for models. Benefits and challenges of remote sensing for monitoring forest ecosystems are presented in Chaps. 11 and 16 of this book (respectively, Torresan et al. 2021; Pappas et al. 2021).

Reliability of growth predictions using PMs depends on various factors, including the spatio-temporal scale of the predictions, level of details available to calibrate and validate the models, etc. The scale and detail of various types of forest monitoring data strongly influence the reliability of simulations. In this regard, new approaches include, for example, model-data fusion with Bayesian inference, which have the potential to strongly reduce the prediction biases and increase their reliability (Trotsiuk et al. 2020).

PMs are thus well placed to address the CSF and support decisions in adaptation and mitigation strategies, because they consider species sensitivity to environmental conditions via physiological processes. PMs can be used to test different scenarios of future environmental conditions. However, there are some components that still need to be developed or improved in PMs to predict future forest growth and landscape dynamics more realistically. These include, for example, intra- and interspecific competition/facilitation, tree mortality, deadwood, natural regeneration and carbon allocation to different tree components. In particular, below-ground carbon allocation needs to be further validated in most of the PMs and analysed with greater accuracy, since it can strongly affect the ecosystem response to climate change. Management is often not simulated by PMs in detail – particularly in the group of models that do not account for the inter-tree competitive interactions in the stand. In such cases, simplified rules need to be applied to test the impact of different management scenarios, e.g. the proportion of biomass extracted (Merganičová et al. 2005).

7.3.3 Considering Environmental Conditions in Growth Models

In the case of YMs, environmental conditions are indirectly considered by the site index. Site index usually refers to the mean (Halaj and Petráš 1998) or dominant (Burkhart and Tomé 2012b) stand height at a standard age (e.g. 100 years). However, YMs consider that site index is a temporarily static parameter that represents site conditions at the time of data collection. Once used for predictions outside the conditions and region of empirical data, small to large systematic errors can be expected, and reliability of predictions is strongly limited. Recently, advanced methods were proposed to develop dynamic site index models (Socha and Tymińska-Czabańska 2019), which consider the changes of site index due to changing environmental conditions.

In ESs and SESs, site conditions affect tree or forest growth potential defined by a growth function (Burkhart and Tomé 2012c) using a modifier based on, for example, ecological site classification (Pretzsch et al. 2002). Among the ESs that apply this growth reduction approach are SILVA (Pretzsch et al. 2002) and SIBYLA (Fabrika 2005). Other ESs use explicit empirical relationships between climate predictors (temperature, precipitation, drought indices) and growth, regeneration and mortality processes to simulate forest development (Stadelmann et al. 2019; Zell et al. 2019).

Unlike YMs and ESs, PMs simulate physiological mechanisms that are directly affected by environmental conditions. PMs are thus more reliable and better suited to simulate future growth of forests under alternative climate change scenarios under the assumption that processes are correctly described, whereas ESs are confined to the climatic space that is represented by the data they have been parameterised with.

7.3.4 Integrating the Effects of Species Mixture into Growth Models

Recently, a strong research activity with the aim to explore how species interactions influence wood and biomass production (Pretzsch and Schütze 2009, 2015; Rötzer et al. 2009; Pretzsch et al. 2010; Jucker et al. 2014; Toïgo et al. 2015) and how to improve forest multifunctionality (van der Plas et al. 2016) has been ongoing. According to Pretzsch et al. (2015) and Bravo et al. (2019), one of the following four principles can be used to predict the growth of mixed-species forests: (1) by applying weighted means of monocultures, (2) multipliers, (3) species-specific growing space competition indices or (4) process-based representation of mixing effects. Most common single-species YM can be used to predict growth in mixed-species forests by simple weighted means of species growth predicted in monocultures. This approach does not consider interspecific competition, and thus, it is less suitable to simulate the future growth of mixed-species forests. Individual-based empirical growth simulators often include inter-tree and interspecific interactions by means of various distance-dependent competition indices (Pretzsch et al. 2002, 2015; Fabrika 2005). The use of empirically based multipliers or more advanced competition indices used in ES often assumes that interactions do not change over time. However, a few competition indices use tree dimensions in the calculation and thus consider competition changes over time. PMs have the potential to overcome the shortcomings of ES by modelling species interactions in a mechanistic way. PMs differ in which processes are considered to be affected by species-mixing effects, i.e. radiation, water, phenology, nutrients and structure (Pretzsch et al. 2015). However, only a few PMs and LDMs consider species interactions in most of the processes to simulate the growth of mixed forests more realistically (Rötzer et al. 2010; Seidl et al. 2012; Temperli et al. 2012; Forrester and Tang 2016; Huber et al. 2018). Also, in this case, it is challenging to better understand how the species composition affects carbon allocation within the tree and among the trees belonging to different species and social positions, and its comprehension would dramatically improve the prediction ability of PMs.

7.3.5 Integrating Silvicultural Prescriptions and the Induced Treatment Responses into Growth Models

Forest management, specifically silvicultural treatments applied over the rotation period, can modify species composition and canopy structure, which in turn can influence forest response to environmental change, including direct effects of warming and drying or other disturbances caused by various factors, such as wind, snow, game and ice (Seidl et al. 2011; Mausolf et al. 2018). The effects of silvicultural interventions and past natural and human-induced perturbations should be correctly considered in forest growth simulation studies, especially in the case of intensively managed European forests (Spiecker 2003; Fontes et al. 2010). Silvicultural techniques influence not only the productivity (and so the carbon sequestration) of the forest stand but also carbon allocation among the tree and stand components, forest vertical and horizontal structure, crown morphology, forest stability and vitality, which alter the resistance of forest to various types of disturbances (Noormets et al. 2015). For example, "heavy thinning from below" applied in some European countries removes all suppressed trees and keeps dominant trees that are all directly exposed to macroclimate (Bosela et al. 2016c). On the other hand, vertically more diversified forests after "thinning from above" may be more resistant under predicted future climate change conditions. Applied silvicultural treatments, including different regeneration methods, have a significant role in creating more complex forests that are expected to be more resistant and resilient to changes in environmental conditions and natural disturbances (O'Hara 2006; Puettmann 2011; O'Hara and

Ramage 2013; Lafond et al. 2014; Fahey et al. 2018). Forest microclimate, altered by different silvicultural treatments, will probably have a crucial role in buffering extreme weather events in the future (Zellweger et al. 2020).

Several models consider silvicultural treatments as a very important component of future growth predictions (Fontes et al. 2010; Fabrika et al. 2018). The most common management intervention employed in growth models is thinning, which can vary by type, intensity and timing. Different types of thinning are implemented mostly in individual-tree process-based models, functional-structural plant models, distance-dependent and distance-independent empirical tree models, tree and cohort gap models or distribution stand models (Fabrika et al. 2018). Other management prescriptions rather rarely applied in growth models are early stand treatments (weeding, cleaning), fertilisation often combined with thinning and harvesting (Weiskittel et al. 2011).

Up to now, less than one-third of all existing growth models consider species-mixing effects and can be directly applied to mixed forests (Pretzsch et al. 2015). The present tendency in silvicultural prescriptions to convert monospecific to multispecies stands or establish new mixed forests is very much favoured, considering matching species composition to site conditions, and requires the development and implementation of appropriate silvicultural strategies for mixtures into growth models (Bravo et al. 2019).

Despite the importance of silvicultural treatments on stand productivity and climate sensitivity of tree species, forest growth models are not always capable of simulating the effects at individual tree level. This limited capacity is because some of the models operate at a stand or biome level and thus are unable to consider inter- and intraspecific competition or facilitation (Jucker et al. 2014). Stand- and biome-scale distance-independent models can simulate the growth under varying stand densities (Horemans et al. 2016), whereas individual-tree distance-dependent ESs/SESs and PMs are well placed to simulate the growth of individual trees and forest stands under different silvicultural treatments and/or forest disturbances (Pretzsch et al. 2002; Fabrika 2005; Seidl et al. 2012; Mina et al. 2017). Other constraints of the wider application of silvicultural methods in growth models include the quality and quantity of experimental data available and appropriate determination of temporal resolution (Weiskittel et al. 2011).

Nevertheless, much effort needs to be invested to improve the existing forest growth models to include an entire portfolio of silvicultural strategies and forest management that would address global climate change (D'Amato et al. 2011).

7.3.6 Effects of Genetic Structure on Forest Growth

Postglacial migrations have altered the genetic diversity of organisms (Hewitt 2004). Evidence has suggested that the populations in refugial areas are typically genetically more diverse and the allelic richness may gradually decline along the migration routes (Hewitt 2000; Petit et al. 2003). A certain level of genetic diversity is required to allow populations to adapt to changing conditions (Howe et al. 2003). A recent study showed the impact of postglacial migration on genetic diversity of European silver fir, *Abies alba* Mill. (Liepelt et al. 2009), which might have had strong effects on the growth and climate responses of the species (Bosela et al. 2016a). Climate-driven natural selection also leads to local adaptation if the climate remains static over at least one tree generation. It may be questioned whether this was ever the case in any tree species' history since the last glaciation in Europe. Tree populations usually exhibit moderate to strong local adaptation; however, fast environmental change may cause local populations experience conditions to which they are not yet adapted (Howe et al. 2003; Wang et al. 2010). Therefore, the higher the genetic diversity is at the population level, the more chance for populations to adapt to the changing environmental conditions (Howe et al. 2003). Consequently, recommendations have stressed the importance of high genetic variability of forest plant material for uncertain futures (Eriksson et al. 1993; Yousefpour et al. 2017). Strong selection, especially among seedlings, would play a crucial role for local selection in natural forests, and varying adaptation effects would recur and act differently in time and space (at different locations and on each tree generation). How far such effects also come to bear on plant material raised under optimal conditions in nurseries and planted under growth-promoting forest management measures remains an open question (Namkoong 1998). Co-occurring tree species can develop quite different adaptive strategies under identical environmental conditions. Contrasting genecological patterns reported for spruce and fir (strong climate-related differentiation in spruce vs. modest differentiation in fir) suggested that spruce can be considered an adaptive specialist while fir is more an adaptive generalist (Frank et al. 2017).

Strong latitudinal clines in the bud burst of tree species (Kramer et al. 2015), which depends on critical temperature sums specific to the climate a provenance is adapted to, and the effects of genetic diversity on tree growth (Bosela et al. 2016a) suggest (successful) genetic adaptations to local environmental conditions in the standing tree generation. However, the bud burst response of, e.g. European beech (*Fagus sylvatica* L.) to temperature sums proved to be plastic (Kramer et al. 2017), which further complicates the evaluation of the issue. Despite that the effects of intraspecific genetic variability on the responses to local climate conditions were ascertained (Neale and Wheeler 2019), still only a very few forest growth models address this aspect (Kramer et al. 2015; Berzaghi et al. 2020). It is important to stress that strongest selection/adaptation effects are experienced by tree populations in the seedling stage (when individuals are most vulnerable to extreme conditions) or in the event of catastrophic disturbances that kill the less resilient individuals. Ignoring the above effects in predicting future adaptive responses of tree species may under- or overestimate the potential of species under changing environmental conditions. Moreover, phenotypic plasticity and adaptive capacity of tree species may be significantly modified by epigenetic variation (Bräutigam et al. 2013).

7.4 Source of Data to Parameterise, Calibrate and Validate Growth Models

In this section, we review different sources of forest data, including national and stand-wise forest inventories, long-term research plots, eddy covariance system, dendrochronological networks (e.g. ITRB), climate and soil databases and remotely sensed data that can be acquired or are available to be used for forest growth models.

7.4.1 National Forest Inventory

It was as early as in the 1910s and 1920s when the European Nordic countries, namely, Norway, Sweden and Finland, launched the first sample-based national forest inventories (NFIs) as a response to the increasing importance of forests and wood for their economy (Vidal et al. 2016a). However, sample-based inventories were not initiated in the rest of Europe until after World War II. Since then, their importance has increased, and the country-scale inventories were launched in France (1958), Austria, Spain, Portugal and Greece (1960s), followed by Switzerland, Italy, Germany (1980s) and other countries. Nowadays, almost all countries in Europe conduct their NFI (Vidal et al. 2016a).

NFIs represent the main source of information about the state and changes of wood resources primarily at a national scale, but in some cases also at a regional scale. However, varying sampling designs among the European countries due to varying policy needs in the past limited the use of European NFIs for international reporting. Recent international activities were successful in harmonising the outputs of the European NFIs at European scale (Vidal et al. 2008, 2016a, b; Bosela et al. 2016b; Fischer et al. 2016; Gschwantner et al. 2016; Alberdi et al. 2020).

NFIs are valuable sources of data for parameterisation and calibration of forest growth models and simulators, because they often provide repeated measurements, representative of a region, landscape and country (McCullagh et al. 2017). They cover a broad range of site conditions, where tree species grow. Data from NFIs have been successfully used to calibrate and validate empirical growth simulators (Fabrika 2005; McCullagh et al. 2017). In Switzerland, the NFI-based forest management scenario model MASSIMO is, among others, successfully used to simulate future harvesting potentials, forest-related carbon budgets and forest reference levels used in greenhouse gas reporting (Stadelmann et al. 2019). In Germany, the NFI-based forest model WEHAM is used to evaluate the sustainability of potential future forest policy scenarios (Seintsch et al. 2017). Similarly, the Câldis system was recently used to evaluate climate-smart management scenarios in terms of standing biomass and carbon as well as soil carbon based on the data of the Austrian NFI (Jandl et al. 2018). NFI data were also used to parameterise or calibrate physiological forest development models (van Oijen et al. 2013; Gutsch et al. 2018; Minunno et al. 2019).

Recently, NFI data of 23 European countries have been used to prepare future projections of the forest growing stock, above-ground carbon and harvesting until 2040 (Vauhkonen et al. 2019). The European Forest Dynamics Model (EFDM) was parameterised using NFI data, and future development of forest resources was simulated under business-as-usual forest management. Further, the large-scale European Forest Information Scenario model (EFISCEN) (Schelhaas et al. 2007) uses data of European NFIs and has been applied to evaluate the development of forest resources in the future under various management scenarios (Verkerk et al. 2011). These modelling activities suggest that European NFIs are suitable to serve the increasing information demands from national to international levels. As they are statistically sound and sufficiently cover the European forest area, NFIs can become the main source of data to aid in sustaining the resilience and climate smartness of the European mountain forests. However, the shortcoming of NFI data is that stand history is not known, and extreme densities and treatments are often insufficiently represented, although these are of special importance for model parameterisation and evaluation. The strength and limitation of NFI-derived CSF indicators, as well as an example of their application in two case studies, are presented in Chap. 4 of this book (Temperli et al. 2021).

7.4.2 Stand-Wise Forest Inventory

Stand-wise inventory, or inventory by compartments, is the assessment of wood resources of the forest stand defined as "geographically contiguous parcels of land whose site type and growing stock is homogenous" (Koivuniemi and Korhonen 2006). The first stand-wise forest inventories were often local and conducted by timber producers to estimate timber resources (Tomppo et al. 2010). For Central and Eastern European countries (especially the former socialistic countries with centrally planned economy), it has been typical to collect forest data at a stand level for management planning. State administration used these data to strictly regulate the use of forest resources at stand and forest district levels. In many Central-Eastern European countries (including the Czech Republic, Slovakia, Romania, Poland, Slovenia, etc.), stand-level inventories continue to be the main source of data for strategic management planning and regulation, despite the fact that the countries have already launched their sample-based NFIs. However, stand-wise inventory data are not available in most European countries. Moreover, stand-wise inventories are often conducted as surveys of forest managers and include expert assessments of forest characteristics, such as species composition, of unknown or limited precision, which strongly limits their use for scientific investigations (Grīnvalds 2014). Stand-wise inventory data are also limited in the spectrum of provided variables. They often only include mean stand variables, such as mean stand diameter and height, growing stock and stand density. The stand (compartment) area largely varies and often changes over time, which also limits the use of these data for temporal studies.

7.4.3 Long-Term Research and Monitoring Plots

Long-term research or monitoring plots are a unique source of data that can be used to either build empirical models or to calibrate and validate available growth models and simulators (Pretzsch et al. 2014; Pretzsch 2020). Compared to tree-ring data, long-term plots usually include mortality data and thus provide information on the true development of forest stands. The spectrum of variables assessed and measured in the plots depends on the aims of monitoring but is often reduced to dendrometric characteristics, which limits their use for assessing the indicators of SFM and CSF. Long-term monitoring plots often include only simple diameter and height measurements, because volume stock estimation was the main purpose of establishing such plots in the past, and thus bring uncertainty when scaling to estimations of biomass and carbon stocks and fluxes. However, the European network of forest condition monitoring plots (International Co-operative Programme on Assessment and Monitoring of Air Pollution Effects on Forests, ICP Forests) provide an example of the long-term monitoring plots (from national to European scales) that go beyond the measurement of basic dendrometric characteristics (Michel et al. 2019). The biggest advantages of long-term monitoring plots are the long time period they cover (often several decades and in few cases more than 100 years) and the known management treatments. In an actual example, such data were used for estimating parameters of the process-based 3-PG model (Landsberg and Waring 1997) in conjunction with other data sources such as NFI data (Trotsiuk et al. 2020). Zell (2018) used data from long-term experimental forest management plots to parameterise an empirical climate-sensitive stand development model that includes an empirical management module. Thanks to long-term data, the capacity of forest gap-models to simulate accurate forest management prescriptions has greatly increased over the past decade (Rasche et al. 2011; Mina et al. 2017). A shortcoming of long-term plots is their uneven spatial distribution, covering only a small portion of the range of site conditions and only a few tree species. These include mostly productive sites and commercially interesting tree species, for studying the growth of which long-term plots have historically been set up. Moreover, monitoring plots that span across centuries are scarce and missing for most regions. However, establishing long-term monitoring plots across the range of site conditions is crucial to calibrate and validate growth simulators under changing climatic conditions and to support climate-smart forest management decision-making (Thrippleton et al. 2020). In Chap. 5 (Pretzsch et al. 2021), the design of a smart network of observational forest plots across European mountain regions is described, and a discussion on their relevance for monitoring growth patterns in monospecific European beech and mixed-species stands of Norway spruce, European beech and silver fir is provided.

7.4.4   Eddy Covariance Measurements

The eddy covariance (EC) technique is an atmospheric measurement technique based on measuring vertical turbulent fluxes within atmospheric boundary layers. It is one of the most appropriate ways to measure local turbulent fluxes of CO_2 (Wang et al. 2009). The technique is used to estimate seasonal fluctuations in carbon exchange between the forest and the atmosphere (Baldocchi 2003). This technique has been successfully used to estimate the net ecosystem productivity (NEP). EC measurements are often used for calibration and validation of NEP estimated by growth simulators (Kramer et al. 2002; Mo et al. 2008; Meyer et al. 2018). However, using the EC technique (but not only EC) includes two potential sources of uncertainty: measurement error and representativeness error (Lasslop et al. 2008; Youhua et al. 2016). Measurement error can be minimised by, for example, the calibration of the instruments. However, representativeness error depends on surface roughness and thermal stability, which further depends on the vegetation heterogeneity (Youhua et al. 2016). EC measurements are more accurate when the atmospheric conditions are steady, the terrain is flat and the surrounding vegetation is homogeneous (Baldocchi 2003). Hence, in the mountainous areas, i.e. in highly complex terrain, and in forests strongly affected by natural disturbances (e.g. fire, diseases, insect infestation), the precision of EC estimates of NEP strongly decreases. The more complicated orography and vegetation heterogeneity were likely the reason for different findings from nearly no link to the high correlation between biometric data and EC measurements (Rocha et al. 2006; Zweifel et al. 2010; Babst et al. 2014b). The distance between the forest under study and the nearest EC tower is another factor affecting the coherence between biometric and EC data (Babst et al. 2014b). A recent study based on a 5 km × 5 km gridded EC measurements revealed large variability in the representativeness of single EC towers to estimate NEP (Youhua et al. 2016).

7.4.5   Remote and Proximal Sensing

Remotely sensed data, such as Landsat or Moderate Resolution Imaging Spectroradiometer (MODIS) satellite imagery, are increasingly used to estimate gross or net primary productivity (GPP, NPP, Neumann et al. 2016) and NEP or to derive vegetation indices further used in large-scale ecological studies, including the characterisation of forest disturbance regimes (Yuan et al. 2010; Jin and Eklundh 2014; Liang et al. 2015; Hart et al. 2017; Liu et al. 2018; Yang et al. 2020). Light-use efficiency (LUE), defined as the amount of carbon produced per unit of absorbed photosynthetically active radiation (APAR), has been successfully used to quantify the dynamics in GPP. The models to estimate LUE, and therefore GPP, are usually based on normalised difference vegetation index (NDVI), photosynthetically active radiation (PAR), fraction of PAR, air temperature, moisture and other environmental conditions (Yuan et al. 2010). There are, however, various definitions of LUE used in developing LUE models, which have implications for the estimation of forest productivity (Gitelson and Gamon 2015). Ecosystem respiration (ER) is an essential component of water and energy budgets and is used to estimate NPP of forest ecosystems by its subtracting from GPP. It is, however, the most difficult component to estimate because of the heterogeneity of the landscape, soil properties and topography, among other factors (Yuan et al. 2010; Zhang et al. 2016). NDVI is a remotely sensed vegetation index frequently used to assess leaf phenology or changes in the canopy due to disturbances, such as bark beetle outbreak or wind storms (Jönsson et al. 2009; Meddens et al. 2013; Jin and Eklundh 2014). Although it is popular because of its robustness against noise, in some forest types, the index is too sensitive to snow cover and much less sensitive to growth of close-canopy forests (Jönsson et al. 2009). Canopy nitrogen content and chlorophyll light-absorbance variables, used as indices to nutrient cycling and maximum photosynthetic capacity, can be estimated using both aerial and satellite optical hyperspectral imagery. Variables, such as above-ground tree height and vertical and horizontal distribution of tree crowns, used for the model parameterisation, can be computed using light detection and ranging (LiDAR) data and interferometric synthetic aperture radar (SAR). Other vegetation indices used in ecological studies include the enhanced vegetation index (EVI), normalized difference water index (NDWI), wide dynamic range vegetation index (WDRVI), phenology index (PI) and leaf area index (LAI). To obtain the remotely sensed data, the MODIS instrument aboard the Terra satellite is often used. MODIS is viewing the entire Earth's surface every 1 to 2 days and acquires data in 36 spectral bands, or groups of wavelengths. The spatial resolution of MODIS images is 250 m (bands 1–2), 500 m (bands 3–7) and 1000 m (bands 8–36). To increase the spatial resolution of MODIS-derived indices, a combination of MODIS and Landsat time series (available at finer 30 m resolution) provides a solution (Yang et al. 2020). Another promising remote sensing-based data sources are the Sentinel-1 and Sentinel-2 SAR data that can be collected independently from daylight or weather conditions and were recently used for rapid detection of windthrows (Frampton et al. 2013; Rüetschi et al. 2019).

There are various models able to utilise remote sensing data in different ways, but the satellite-driven version of the 3-PG model (Physiological Principles in Predicting Growth), developed by Landsberg and Waring 1997 and Waring et al. 2010 is probably the most known and used. The physiological variables used in the model can be estimated from remote-sensing measurements of factors that influence those variables. In the model, GPP is ultimately a function of the APAR and the canopy quantum use efficiency.

It is worth mentioning that recent data assimilation (DA) techniques have been used to estimate forest stand data by sequentially combining remote sensing-based estimates of forest variables with predictions from growth models (Nyström et al. 2015). DA provides a way of blending the monitoring properties of remotely sensed data with the predictive and explanatory abilities of forest growth models (Huang et al. 2019). Input to the data assimilation may be canopy height models, obtained from airborne laser scanning (ALS) data or from image matching of digital aerial images at different time points during the growth season. With this approach, the prior forecast is updated to the posterior forecast when a new estimate is considered. This kind of approach needs modification of the existing growth models that would allow data assimilation and also requires the possibility to interrupt the model simulations before the end and use remote sensing data to update specific characteristics.

7.4.6   Tree-Ring Time Series

Annually resolved TRW series represent a valuable source of information on past growth dynamics of individual trees and forests (Babst et al. 2018, 2019; Klesse et al. 2018). Over the past century, TRW data have been collected across the globe for many different purposes, and a high portion of these data has been archived in the International Tree-Ring Data Bank (ITRDB) managed by Paleoclimatology Team of National Centers for Environmental Information and the World Data System for Paleoclimatology. ITRDB now includes TRW series from over 4000 sites and six continents. Tree-ring networks were frequently used to reconstruct past climate as well as to investigate responses of forest trees to variation in environmental characteristics to assess species vulnerability to changing conditions (Babst et al. 2013). Recently, TRW data have been successfully applied to predict future forest growth and climate responses (Charney et al. 2016; Dorado-Liñán et al. 2019). TRW data have also been used to reconstruct regimes of windthrow, bark beetle, storm and other disturbance regimes (Veblen et al. 1994; Svoboda et al. 2014). Explaining relationships between climate and disturbance dynamics (Hart et al. 2014) forms the basis to parameterise models of disturbance dynamics under climate change (Temperli et al. 2015; Thom et al. 2017). However, using TRW data for detecting long-term growth trends and species climate responses must follow purpose-oriented sampling designs (Nehrbass-Ahles et al. 2014) and appropriate detrending methods (Peters et al. 2015) to minimise potential prediction biases (Klesse et al. 2018).

7.5 Conclusions and Perspectives

Expected rapid climate change will likely challenge the adaptation capacity of many forest ecosystems. Forest growth models represent a promising tool to predict the effects of different climate change scenarios on the growth of individual trees and forest stands as well as the future distribution of forest tree species under changing conditions and thus to support forest managers and policymakers in developing long-term strategies. Available forest growth models largely differ from each other in many ways due to which they provide a large range of future growth estimates. A multi-model averaging technique has been found a good way to avoid biased estimates of single models due to shortcomings of individual modelling approaches (Picard et al. 2012; Hlásny et al. 2014; Dormann et al. 2018). Although modelling the relationships between forest production and future climate is complex and intrinsically uncertain, forest growth models may help to guide climate-smart strategies aimed at overcoming mitigation, adaptation and production gaps. For example, synergies and trade-offs between biodiversity conservation and timber production can be assessed, and user-friendly interactive decision support tools can be developed, ensuring that all stakeholders envisage the risks of adapting their management strategies to changes in climate and society and anticipate the consequences of environmental disturbances.

Past constraints that limited the capacity to model forest dynamics, such as the availability of data for model calibration and validation, the computing capacity, the model applicability to real-world problems and the ability to integrate biological, social and economic drivers of change, have become less restrictive. For this, the role of models for predicting forest growth and yield under changing environments is now central in applied decision-making. For that, to ensure their role, great attention is required to evaluate the performance, to expand the driver of changes and to incorporate variables as input social and economic trends and needs.

Acknowledgements We would like to thank Laura Dobor, Sonja Vospernik and Thomas Ledermann for their valuable information regarding the ability of iLand and PROGNAUS models to address the indicators of climate-smart forestry. Michal Bošeľa received support from the Slovak Research and Development Agency via the grants No. APVV-15-0265 and APVV-19-0183. Katarína Merganičová was supported by the grant "EVA4.0", No. CZ.02.1.01/0.0/0.0/16_01 9/0000803, financed by OP Research, Development and Education, and the project "Scientific support of climate change adaptation in agriculture and mitigation of soil degradation" (ITMS2014 + 313011 W580), supported by the Integrated Infrastructure Operational Programme funded by the European Regional Development Fund. Ilona Mészáros was supported by grant No. NKFI-125652 from the National Research, Development, and Innovation Office, Hungary. In memory of Giustino Tonon, a co-author of this chapter, who had passed away prior to the publication of the book.

Appendix

Annex 7.1 Identification of ability of selected growth models (representing the model groups defined in Fig. 7.3) to address the indicators of climate-smart forestry (indicator list based on (Bowditch et al. 2020))

Indicator	Specific indicator	Yield tables	PROGNAUS	SIBYLA	Biome-BGC	iLand
1.1 Forest area		Not provided	Input and output	Not provided	Not provided	Input and output
1.2 Growing stock		Output	Input and output	Output	Derived from output	Output
1.3 Age structure and/or diameter distribution	Age structure	Input and output in the form of mean stand age	May be obtained as output	Input and output	Stand age as a parameter driving the start of normal simulation	Input and output
	Diameter distribution	Not internally considered	Input and output	Input and output	Not considered	Input and output
1.4 Carbon stock	Carbon stock in forest biomass	Not provided, may be derived from growing stock	Not provided, may be derived from growing stock	Output	Output	Output
	Carbon stock in forest soils	Not provided	Not provided	Not provided	Output	Output
	Carbon stock in harvested wood products	Not provided, may be derived from model output if coupled with assortment model	Not provided, may be derived from model output if coupled with assortment model	Can be derived from model output	Not provided, may be derived from model output if coupled with assortment model	Not provided, may be derived from model output if coupled with assortment model
2.1 Deposition of air pollutants	Carbon dioxide	Not considered	Not considered	Input	Input	Input
	Nitrogen deposition	Not considered	Not considered	Input	Input	Input
	Sulphur deposition	Not considered	Not considered	Not considered	Not considered	Not considered
	Other (CO, O_3, …)	Not considered	Not considered	Not considered	Not considered	Not considered

(continued)

Annex 7.1 (continued)

Indicator	Specific indicator	Yield tables	PROGNAUS	SIBYLA	Biome-BGC	iLand
2.2 Soil condition	pH	Not considered	Not considered	Not considered	Not considered	Not considered
	CEC = cation exchange capacity	Not considered	Not considered	Not considered	Not considered	Not considered
	C/N	Not considered	Not considered	Not considered	Can be derived from model output	Input
	Organic C	Not considered	Not considered	Not considered	Output	Output
	Base saturation (percentage of CEC occupied by bases (Ca^{2+}, Mg^{2+}, K^+ and Na^+))	Not considered	Not considered	Not considered	Not considered	Not considered
	Soil type	Not considered	Input	Not considered	Not considered	Not considered
	Soil fertility	Not considered	Not considered	Input	Not considered	Not considered
	Soil moisture	Not considered	Input	Input	Output	Output
	Site index	Input	Not considered	Input	Not considered	Not considered
	Growth region	Input	Input	Input	Not applied	Not applied
2.3 Defoliation		Not considered	Not considered	Not considered	Not considered	Not considered

2.4 Forest damage	Wind	Not considered	Not considered	Output	Not internally considered, can be approximated by modifying mortality rate	Input of windstorm characteristics (length, severity) and output/ simulated spatially and explicitly at landscape scale
	Fire	Not considered	Not considered	Output	Input (annual fire mortality rate) and output	Input and output
	Bark beetle	Not considered	Not considered	Output	Not internally considered, can be approximated by modifying mortality rate	Output
	Defoliators	Not considered	Not considered	Output	Not considered	Not considered
3.1 Increment and fellings	Annual increment in volume, biomass or carbon stock	Output	Output	Output	Output	Can be derived from model output
	Annual felling of wood volume, biomass or carbon stock	Output	Output	Output	Output	Output
3.2 Roundwood	Quantity of roundwood	Output from assortment yield tables	Output	Output	Not considered, may be derived from model output if coupled with assortment tables	Can be derived from model output
	Market value of roundwood	Not considered, may be derived from assortment yield tables	Can be derived from model output	Output	Not considered	Can be derived from model output

(continued)

Annex 7.1 (continued)

Indicator	Specific indicator	Yield tables	PROGNAUS	SIBYLA	Biome-BGC	iLand
3.5 Forests under management plans	Forests under FMP	Not considered	Input and output if provided by a user	Not considered	Not considered	Input
4.1 Tree species composition	Number of tree species	Not considered by individual models	Input and output	Input and output	Not internally considered	Input and output
4.2 Regeneration		Not considered	Output	Input and output	Not considered	Input and output
4.3 Naturalness		Not considered	Can be derived from model output	Not considered but can be derived from model output	Not internally considered, but the degree of naturalness (not area) can be considered as input information affecting simulation set-up (past management specification)	Not considered but can be derived from model output
4.4 Introduced tree species		Not considered	Not considered	Not considered	Not considered	Input (seeds or planting) and output
4.5 Deadwood	Volume of deadwood	Not considered	Output	Output	Can be derived from model output	Output
	Carbon stock in deadwood	Not considered	Can be derived from model output	Can be derived from model output	Output	Output
4.6 Genetic resources		Not considered	Not considered	Not considered	Not considered	Not considered
4.7 Landscape pattern		Not considered	Not considered	Not considered	Not considered	Not considered but can be derived from model output
4.8 Threatened forest species		Not considered	Not considered	Not considered	Not considered	Not considered

4.9 Protected forests		Not considered	Input and output if provided by a user	Not considered	Not considered	Input
5.1 Protective forests – soil and water		Not considered	Input and output if provided by a user	Not considered	Not considered	Not considered
6.7 Wood consumption		Not considered	Not considered	Not considered	Not considered	Not considered
6.8 Trade in wood		Not considered	Not considered	Not considered	Not considered	Not considered
6.9 Energy from wood resources		Not considered	Not considered	Not considered	Not considered	Not considered
6.10 Accessibility for recreation		Not considered	Input and output if provided by a user	Not considered	Not considered	Not considered but can be derived from model output
Management system		Predefined	Input	Input	Input	Input
Slenderness coefficient		Can be derived from mean stand values	Output	Output	Not considered	Input, can be derived from output as well
Vertical distribution of tree crowns		Not considered	Output	Output	Model input – simplified two-layer structure	Included in model simulations but without any output information
Horizontal distribution of tree crowns	Trees per hectare, basal area per hectare	Output	Output	Output	Not considered	Input and output
Crown area, tree crown diameter	Crown area, tree crown diameter	Not provided	Output	Output	Not considered	Included in model simulations but without any output information

References

Alberdi I, Bender S, Riedel T et al (2020) Assessing forest availability for wood supply in Europe. For Policy Econ 111:102032. https://doi.org/10.1016/j.forpol.2019.102032

Albrich K, Rammer W, Seidl R (2020) Climate change causes critical transitions and irreversible alterations of mountain forests. Glob Chang Biol. https://doi.org/10.1111/gcb.15118

Babst F, Poulter B, Trouet V et al (2013) Site- and species-specific responses of forest growth to climate across the European continent. Glob Ecol Biogeogr 22:706–717. https://doi.org/10.1111/geb.12023

Babst F, Alexander MR, Szejner P et al (2014a) A tree-ring perspective on the terrestrial carbon cycle. Oecologia 176:307–322. https://doi.org/10.1007/s00442-014-3031-6

Babst F, Bouriaud O, Papale D et al (2014b) Above-ground woody carbon sequestration measured from tree rings is coherent with net ecosystem productivity at five eddy-covariance sites. New Phytol 201:1289–1303. https://doi.org/10.1111/nph.12589

Babst F, Bodesheim P, Charney N et al (2018) When tree rings go global: challenges and opportunities for retro- and prospective insight. Quat Sci Rev 197:1–20. https://doi.org/10.1016/j.quascirev.2018.07.009

Babst F, Bouriaud O, Poulter B et al (2019) Twentieth century redistribution in climatic drivers of global tree growth. Sci Adv 5:eaat4313. https://doi.org/10.1126/sciadv.aat4313

Baldocchi DD (2003) Assessing the eddy covariance technique for evaluating carbon dioxide exchange rates of ecosystems: past, present and future. Glob Chang Biol 9:479–492. https://doi.org/10.1046/j.1365-2486.2003.00629.x

Battaglia M, Sands PJ (1998) Process-based forest productivity models and their application in forest management. For Ecol Manag 102:13–32. https://doi.org/10.1016/S0378-1127(97)00112-6

Berzaghi F, Wright IJ, Kramer K et al (2020) Towards a new generation of trait-flexible vegetation models. Trends Ecol Evol 35:191–205. https://doi.org/10.1016/j.tree.2019.11.006

Bosela M, Popa I, Gömöry D et al (2016a) Effects of postglacial phylogeny and genetic diversity on the growth variability and climate sensitivity of European silver fir. J Ecol 104:716–724. https://doi.org/10.1111/1365-2745.12561

Bosela M, Redmond J, Kučera M et al (2016b) Stem quality assessment in European National Forest Inventories: an opportunity for harmonised reporting? Ann For Sci 73. https://doi.org/10.1007/s13595-015-0503-8

Bosela M, Štefančík I, Petráš R, Vacek S (2016c) The effects of climate warming on the growth of European beech forests depend critically on thinning strategy and site productivity. Agric For Meteorol 222:21–31. https://doi.org/10.1016/j.agrformet.2016.03.005

Bowditch E, Santopuoli G, Binder F et al (2020) What is climate-smart forestry? A definition from a multinational collaborative process focused on mountain regions of Europe. Ecosyst Serv 43:101113. https://doi.org/10.1016/j.ecoser.2020.101113

Bräutigam K, Vining KJ, Lafon-Placette C et al (2013) Epigenetic regulation of adaptive responses of forest tree species to the environment. Ecol Evol 3:399–415. https://doi.org/10.1002/ece3.461

Bravo F, Fabrika M, Ammer C et al (2019) Modelling approaches for mixed forests dynamics prognosis. Research gaps and opportunities. For Syst 28:eR002. https://doi.org/10.5424/fs/2019281-14342

Bugmann HKM (1996) A simplified Forest model to study species composition along climate gradients. Ecology 77:2055–2074. https://doi.org/10.2307/2265700

Bugmann H, Seidl R, Hartig F et al (2019) Tree mortality submodels drive simulated long-term forest dynamics: assessing 15 models from the stand to global scale. Ecosphere 10:e02616. https://doi.org/10.1002/ecs2.2616

Burkhart HE, Tomé M (2012a) Modeling forest stand development. In: Burkhart HE, Tomé M (eds) Modeling Forest trees and stands. Springer, Dordrecht, pp 233–244

Burkhart HE, Tomé M (2012b) Evaluating site quality. In: Burkhart HE, Tomé M (eds) Modeling forest trees and stands. Springer, Dordrecht, pp 131–173

7 Modelling Future Growth of Mountain Forests Under Changing Environments 253

Burkhart HE, Tomé M (2012c) Growth functions. In: Burkhart HE, Tomé M (eds) Modeling forest trees and stands. Springer, Dordrecht, pp 111–130

Charney ND, Babst F, Poulter B et al (2016) Observed forest sensitivity to climate implies large changes in 21st century North American forest growth. Ecol Lett 19:1119–1128. https://doi.org/10.1111/ele.12650

Chen PY, Welsh C, Hamann A (2010) Geographic variation in growth response of Douglas-fir to interannual climate variability and projected climate change. Glob Chang Biol 16:3374–3385. https://doi.org/10.1111/j.1365-2486.2010.02166.x

Cherubini P, Dobbertin M, Innes JL (1998) Potential sampling bias in long-term forest growth trends reconstructed from tree rings: a case study from the Italian Alps. For Ecol Manag 109:103–118. https://doi.org/10.1016/S0378-1127(98)00242-4

Christmann (1949) Ertragstafel für Kiefern-Fichten-Mischbestand. In: Ertragstafeln der wichtigsten Holzarten bei verschiedener Durchforstung sowie einiger Mischbestandsformen. Schaper, Hannover, p 100

Cook E, Kairiukstis L (1990) Methods of dendrochronology: applications in the environmental sciences. Springer, Berlin

Cornes R, van der Schrier G, van der Besselaar EJM, Jones PD (2018) An ensemble version of the E-OBS temperature and precipitation datasets. J Geophys Res Atmos 123:9391–9409. https://doi.org/10.1029/2017JD028200

D'Amato AW, Bradford JB, Fraver S, Palik BJ (2011) Forest management for mitigation and adaptation to climate change: insights from long-term silviculture experiments. For Ecol Manag 262:803–816. https://doi.org/10.1016/j.foreco.2011.05.014

Deckmyn G, Verbeeck H, Op de Beeck M et al (2008) ANAFORE: a stand-scale process-based forest model that includes wood tissue development and labile carbon storage in trees. Ecol Model 215:345–368. https://doi.org/10.1016/j.ecolmodel.2008.04.007

Dorado-Liñán I, Piovesan G, Martínez-Sancho E et al (2019) Geographical adaptation prevails over species-specific determinism in trees' vulnerability to climate change at Mediterranean rear-edge forests. Glob Chang Biol 25:1296–1314. https://doi.org/10.1111/gcb.14544

Dormann CF, Calabrese JM, Guillera-Arroita G et al (2018) Model averaging in ecology: a review of Bayesian, information-theoretic, and tactical approaches for predictive inference. Ecol Monogr 88:485–504. https://doi.org/10.1002/ecm.1309

Eriksson G, Namkoong G, Roberds JH (1993) Dynamic gene conservation for uncertain futures. For Ecol Manag 62:15–37. https://doi.org/10.1016/0378-1127(93)90039-P

Fabrika M (2005) Simulátor biodynamiky lesa SIBYLA, koncepcia, konštrukcia a programové riešenie. Technical University in Zvolen

Fabrika M, Pretzsch H (2013) Forest ecosystem analysis and modelling, 1st edn. Technical University in Zvolen, Zvolen

Fabrika M, Pretzsch H, Bravo F (2018) Models for mixed forests BT – dynamics, silviculture and management of mixed forests. In: Bravo-Oviedo A, Pretzsch H, del Río M (eds) . Springer, Cham, pp 343–380

Fabrika M, Valent P, Merganicova K (2019) Forest modelling and visualisation – state of the art and perspectives. Cent Eur For J 66:147–165. https://doi.org/10.2478/forj-2019-0018

Fahey RT, Alveshere BC, Burton JI et al (2018) Shifting conceptions of complexity in forest management and silviculture. For Ecol Manag 421:59–71. https://doi.org/10.1016/j.foreco.2018.01.011

Fischer C, Gasparini P, Nylander M et al (2016) Joining criteria for harmonizing European Forest available for wood supply estimates. Case studies from National Forest Inventories. Forests 7:104. https://doi.org/10.3390/f7050104

Fontes L, Bontemps J-D, Bugmann H et al (2010) Models for supporting forest management in a changing environment. For Syst 3:8. https://doi.org/10.5424/fs/201019s-9315

FOREST EUROPE (2015) State of Europe's Forests:2015

Forrester DI, Tang X (2016) Analysing the spatial and temporal dynamics of species interactions in mixed-species forests and the effects of stand density using the 3-PG model. Ecol Model 319:233–254. https://doi.org/10.1016/j.ecolmodel.2015.07.010

Frampton WJ, Dash J, Watmough G, Milton EJ (2013) Evaluating the capabilities of Sentinel-2 for quantitative estimation of biophysical variables in vegetation. ISPRS J Photogramm Remote Sens 82:83–92. https://doi.org/10.1016/j.isprsjprs.2013.04.007

Frank A, Sperisen C, Howe GT et al (2017) Distinct genecological patterns in seedlings of Norway spruce and silver fir from a mountainous landscape. Ecology 98:211–227. https://doi.org/10.1002/ecy.1632

Fritts HC (2001) Tree rings and climate. The Blackburn Press, New York/San Francisco

Fritts HC, Swetnam TW (1989) Dendroecology: a tool for evaluating variations in past and present forest environments. Academic

Fritts HC, Vaganov EA, Sviderskaya IV, Shashkin AV (1991) Climatic variation and tree-ring structure in conifers: empirical and mechanistic models of tree-ring width, number of cells, cell size, cell-wall thickness and wood density. Clim Res 1:97–116

Girardin MP, Raulier F, Bernier PY, Tardif JC (2008) Response of tree growth to a changing climate in boreal Central Canada: a comparison of empirical, process-based, and hybrid modelling approaches. Ecol Model 213:209–228. https://doi.org/10.1016/j.ecolmodel.2007.12.010

Gitelson AA, Gamon JA (2015) The need for a common basis for defining light-use efficiency: implications for productivity estimation. Remote Sens Environ 156:196–201. https://doi.org/10.1016/j.rse.2014.09.017

Grīnvalds A (2014) The accuracy of standwise forest inventory in mature stands. Proc Latv Univ Agric 32:1–8. https://doi.org/10.2478/plua-2014-0007

Gschwantner T, Lanz A, Vidal C et al (2016) Comparison of methods used in European National Forest Inventories for the estimation of volume increment: towards harmonisation. Ann For Sci:73. https://doi.org/10.1007/s13595-016-0554-5

Guiot J, Boucher E, Gea-Izquierdo G (2014) Process models and model-data fusion in dendroecology. Front Ecol Evol 2:52

Gutsch M, Lasch-Born P, Kollas C et al (2018) Balancing trade-offs between ecosystem services in Germany's forests under climate change. Environ Res Lett 13:45012. https://doi.org/10.1088/1748-9326/aab4e5

Haines-Young R, Potschin MB (2018) Common International Classification of Ecosystem Services (CICES) V5.1

Halaj J, Petráš R (1998) Rastové tabuľky hlavných drevín [Growth tables of the main tree species]. SAP – Slovak Academic Press, Bratislava

Harris I, Jones PD, Osborn TJ, Lister DH (2014) Updated high-resolution grids of monthly climatic observations – the CRU TS3.10 Dataset. Int J Climatol 34:623–642. https://doi.org/10.1002/joc.3711

Hart SJ, Veblen TT, Eisenhart KS et al (2014) Drought induces spruce beetle (Dendroctonus rufipennis) outbreaks across northwestern Colorado. Ecology 95:930–939

Hart SJ, Veblen TT, Schneider D, Molotch NP (2017) Summer and winter drought drive the initiation and spread of spruce beetle outbreak. Ecology 98:2698–2707. https://doi.org/10.1002/ecy.1963

Hasenauer H (1994) Ein Einzelbaumsimulator für ungleichaltrige Fichten-Kieferen- und Buchen-Fichtenmischbestände. Forstliche Schriftenreihe Universität für Bodenkultur, Wien, Band 8

Hauhs M, Kastner-Maresch A, Rost-Siebert K (1995) A model relating forest growth to ecosystem-scale budgets of energy and nutrients. Ecol Model 83:229–243. https://doi.org/10.1016/0304-3800(95)00011-Z

He HS (2008) Forest landscape models: definitions, characterization, and classification. For Ecol Manag 254:484–498. https://doi.org/10.1016/j.foreco.2007.08.022

Henne PD, Elkin C, Colombaroli D et al (2013) Impacts of changing climate and land use on vegetation dynamics in a Mediterranean ecosystem: insights from paleoecology and dynamic modeling. Landsc Ecol 28:819–833. https://doi.org/10.1007/s10980-012-9782-8

Hewitt G (2000) The genetic legacy of the quaternary ice ages. Nature 405:907–913. https://doi.org/10.1038/35016000

Hewitt GM (2004) Genetic consequences of climatic oscillations in the quaternary. Philos Trans R Soc Lond Ser B Biol Sci 359:183–195. https://doi.org/10.1098/rstb.2003.1388

Hlásny T, Barcza Z, Fabrika M et al (2011) Climate change impacts on growth and carbon balance of forests in Central Europe. Clim Res 47:219–236. https://doi.org/10.3354/cr01024

Hlásny T, Barcza Z, Barka I et al (2014) Future carbon cycle in mountain spruce forests of Central Europe: modelling framework and ecological inferences. For Ecol Manag 328:55–68. https://doi.org/10.1016/j.foreco.2014.04.038

Horemans JA, Bosela M, Dobor L et al (2016) Variance decomposition of predictions of stem biomass increment for European beech: contribution of selected sources of uncertainty. For Ecol Manag:361. https://doi.org/10.1016/j.foreco.2015.10.048

Howe GT, Aitken SN, Neale DB et al (2003) From genotype to phenotype: unraveling the complexities of cold adaptation in forest trees. Can J Bot 81:1247–1266

Huang J, Gómez-Dans JL, Huang H et al (2019) Assimilation of remote sensing into crop growth models: current status and perspectives. Agric For Meteorol 276–277:107609. https://doi.org/10.1016/j.agrformet.2019.06.008

Hubau W, Lewis SL, Phillips OL et al (2020) Asynchronous carbon sink saturation in African and Amazonian tropical forests. Nature 579:80–87. https://doi.org/10.1038/s41586-020-2035-0

Huber N, Bugmann H, Lafond V (2018) Global sensitivity analysis of a dynamic vegetation model: model sensitivity depends on successional time, climate and competitive interactions. Ecol Model 368:377–390. https://doi.org/10.1016/j.ecolmodel.2017.12.013

IPCC (2014) Climate change 2014: synthesis report. Contribution of Working Groups I, II and III to the Fifth Assessment Report of the Intergovernmental Panel on Climate Change [Core Writing Team, Pachauri RK, Meyer LA (eds)]

Jacob D, Petersen J, Eggert B et al (2014) EURO-CORDEX: new high-resolution climate change projections for European impact research. Reg Environ Chang 14:563–578. https://doi.org/10.1007/s10113-013-0499-2

Jandl R, Ledermann T, Kindermann G et al (2018) Strategies for climate-smart forest management in Austria. Forest 9

Jin H, Eklundh L (2014) A physically based vegetation index for improved monitoring of plant phenology. Remote Sens Environ 152:512–525. https://doi.org/10.1016/j.rse.2014.07.010

Jönsson AM, Appelberg G, Harding S, Bärring L (2009) Spatio-temporal impact of climate change on the activity and voltinism of the spruce bark beetle, Ips typographus. Glob Chang Biol 15:486–499. https://doi.org/10.1111/j.1365-2486.2008.01742.x

Jucker T, Bouriaud O, Avacaritei D, Coomes DA (2014) Stabilizing effects of diversity on aboveground wood production in forest ecosystems: linking patterns and processes. Ecol Lett 17:1560–1569. https://doi.org/10.1111/ele.12382

Kahn M (1994) Modellierung der Höhenentwicklung ausgewählter Baumarten in Abhängigkeit vom Standort. Forstliche Forschungsber. München, vol 141

Kapos V, Rhind J, Edwards M et al (2000) Developing a map of the world's mountain forests. In: Price MF, Butt N (eds) Forests in sustainable mountain development: a state-of knowledge report for 2000. CAB International, Wallingford, pp 4–19

Keenan TF, Hollinger DY, Bohrer G et al (2013) Increase in forest water-use efficiency as atmospheric carbon dioxide concentrations rise. Nature 499:324–327. https://doi.org/10.1038/nature12291

Klesse S, DeRose RJ, Guiterman CH et al (2018) Sampling bias overestimates climate change impacts on forest growth in the southwestern United States. Nat Commun 9:1–9. https://doi.org/10.1038/s41467-018-07800-y

Köhler P, Huth A (1998) The effects of tree species grouping in tropical rainforest modelling: simulations with the individual-based model Formind. Ecol Model 109:301–321. https://doi.org/10.1016/S0304-3800(98)00066-0

Koivuniemi J, Korhonen K (2006) Inventory by compartments. In: Kangas A, Maltamo M (eds) Forest inventory: methodology and applications. Springer Dordrecht, pp. 271–278

Kramer K, Leinonen I, Bartelink HH et al (2002) Evaluation of six process-based forest growth models using eddy-covariance measurements of CO2 and H2O fluxes at six forest sites in Europe. Glob Chang Biol 8:213–230. https://doi.org/10.1046/j.1365-2486.2002.00471.x

Kramer K, Degen B, Buschbom J et al (2010) Modelling exploration of the future of European beech (Fagus sylvatica L.) under climate change-range, abundance, genetic diversity and adaptive response. For Ecol Manag 259:2213–2222. https://doi.org/10.1016/j.foreco.2009.12.023

Kramer K, van der Werf B, Schelhaas M-J (2015) Bring in the genes: genetic-ecophysiological modeling of the adaptive response of trees to environmental change. With application to the annual cycle. Front Plant Sci 5:1–10. https://doi.org/10.3389/fpls.2014.00742

Kramer K, Ducousso A, Gömöry D et al (2017) Chilling and forcing requirements for foliage bud burst of European beech (Fagus sylvatica L.) differ between provenances and are phenotypically plastic. Agric For Meteorol 234–235:172–181. https://doi.org/10.1016/j.agrformet.2016.12.002

Lafond V, Lagarrigues G, Cordonnier T, Courbaud B (2014) Uneven-aged management options to promote forest resilience for climate change adaptation: effects of group selection and harvesting intensity. Ann For Sci 71:173–186. https://doi.org/10.1007/s13595-013-0291-y

Landsberg JJ, Waring RH (1997) A generalised model of forest productivity using simplified concepts of radiation-use efficiency, carbon balance and partitioning. For Ecol Manag 95:209–228. https://doi.org/10.1016/S0378-1127(97)00026-1

Lasslop G, Reichstein M, Kattge J, Papale D (2008) Influences of observation errors in eddy flux data on inverse model parameter estimation. Biogeosciences 5:1311–1324. https://doi.org/10.5194/bg-5-1311-2008

Liang L, Di L, Zhang L et al (2015) Estimation of crop LAI using hyperspectral vegetation indices and a hybrid inversion method. Remote Sens Environ 165:123–134. https://doi.org/10.1016/j.rse.2015.04.032

Liepelt S, Cheddadi R, de Beaulieu JL et al (2009) Postglacial range expansion and its genetic imprints in Abies alba (Mill.) – a synthesis from palaeobotanic and genetic data. Rev Palaeobot Palynol 153:139–149. https://doi.org/10.1016/j.revpalbo.2008.07.007

Lischke H, Zimmermann NE, Bolliger J et al (2006) TreeMig: a forest-landscape model for simulating spatio-temporal patterns from stand to landscape scale. Ecol Model 199:409–420. https://doi.org/10.1016/j.ecolmodel.2005.11.046

Liu Q, Fu YH, Liu Y et al (2018) Simulating the onset of spring vegetation growth across the Northern Hemisphere. Glob Chang Biol 24:1342–1356. https://doi.org/10.1111/gcb.13954

Mäkelä A, Landsberg J, Ek AR et al (2000) Process-based models for forest ecosystem management: current state of the art and challenges for practical implementation. Tree Physiol 20:289–298. https://doi.org/10.1093/treephys/20.5-6.289

Mäkelä A, Grace DG et al (2010) Simulating wood quality in forest management models. For Syst 19:48–68. https://doi.org/10.5424/fs/201019S-9314

Mäkelä A, del Río M, Hynynen J et al (2012) Using stand-scale forest models for estimating indicators of sustainable forest management. For Ecol Manag 285:164–178. https://doi.org/10.1016/j.foreco.2012.07.041

Mausolf K, Wilm P, Härdtle W et al (2018) Higher drought sensitivity of radial growth of European beech in managed than in unmanaged forests. Sci Total Environ 642:1201–1208. https://doi.org/10.1016/j.scitotenv.2018.06.065

McCullagh A, Black K, Nieuwenhuis M (2017) Evaluation of tree and stand-level growth models using national forest inventory data. Eur J For Res 136:251–258. https://doi.org/10.1007/s10342-017-1025-8

Meddens AJH, Hicke JA, Vierling LA, Hudak AT (2013) Evaluating methods to detect bark beetle-caused tree mortality using single-date and multi-date Landsat imagery. Remote Sens Environ 132:49–58. https://doi.org/10.1016/j.rse.2013.01.002

Meier ES, Lischke H, Schmatz DR, Zimmermann NE (2012) Climate, competition and connectivity affect future migration and ranges of European trees. Glob Ecol Biogeogr 21:164–178. https://doi.org/10.1111/j.1466-8238.2011.00669.x

Merganič J, Merganičová K, Výbošťok J et al (2020) Searching for Pareto fronts for forest stand wind stability by incorporating timber and biodiversity values. Forest 11

Merganičová K, Pietsch SA, Hasenauer H (2005) Testing mechanistic modeling to assess impacts of biomass removal. For Ecol Manag 207:37–57. https://doi.org/10.1016/j.foreco.2004.10.017

Merganičová K, Merganič J, Lehtonen A et al (2019) Forest carbon allocation modelling under climate change. Tree Physiol 39:1937–1960. https://doi.org/10.1093/treephys/tpz105

Meyer G, Black TA, Jassal RS et al (2018) Simulation of net ecosystem productivity of a lodgepole pine forest after mountain pine beetle attack using a modified version of 3-PG. For Ecol Manag 412:41–52. https://doi.org/10.1016/j.foreco.2018.01.034

Michel A, Prescher A-K, Schwärzel K (2019) Forest condition in Europe: 2019 technical report of ICP forests. Report under the UNECE Convention on Long-range Transboundary Air Pollution (Air Convention). BFW-Dokumentation 27/2019, Vienna, Austria

Mina M, Bugmann H, Klopcic M, Cailleret M (2017) Accurate modeling of harvesting is key for projecting future forest dynamics: a case study in the Slovenian mountains. Reg Environ Chang 17:49–64. https://doi.org/10.1007/s10113-015-0902-2

Minunno F, Peltoniemi M, Härkönen S et al (2019) Bayesian calibration of a carbon balance model PREBAS using data from permanent growth experiments and national forest inventory. For Ecol Manag 440:208–257. https://doi.org/10.1016/j.foreco.2019.02.041

Mo X, Chen JM, Ju W, Black TA (2008) Optimization of ecosystem model parameters through assimilating eddy covariance flux data with an ensemble Kalman filter. Ecol Model 217:157–173. https://doi.org/10.1016/j.ecolmodel.2008.06.021

Monserud RA, Sterba H (1996) A basal area increment model for individual trees growing in even- and uneven-aged forest stands in Austria. For Ecol Manag 80:57–80. https://doi.org/10.1016/0378-1127(95)03638-5

Moreno A, Hasenauer H (2016) Spatial downscaling of European climate data. Int J Climatol 36:1444–1458. https://doi.org/10.1002/joc.4436

Moss RH, Edmonds JA, Hibbard KA et al (2010) The next generation of scenarios for climate change research and assessment. Nature 463:747–756. https://doi.org/10.1038/nature08823

Nabuurs G-J, Lindner M, Verkerk PJ et al (2013) First signs of carbon sink saturation in European forest biomass. Nat Clim Chang 3:792–796. https://doi.org/10.1038/nclimate1853

Nabuurs GJ, Arets EJMM, Schelhaas MJ (2018) Understanding the implications of the EU-LULUCF regulation for the wood supply from EU forests to the EU 07 Agricultural and Veterinary Sciences 0705 Forestry Sciences Georgii Alexandrov. Carbon Balance Manag 13:18. https://doi.org/10.1186/s13021-018-0107-3

Nagel J (1996) Anwendungsprogramm zur Bestandesbewertung und zur Prognose der Bestandesentwicklung. Forst und Holz 3:76–78

Nagel J (1999) Konzeptionelle Überlegungen zum schrittweisen Aufbau eines waldwachstum-skundlichen Simulationssystems für Nordwestdeutschland. Schriften aus der Forstlichen Fakultät der Universität Göttingen und der Niedersächsischen Forstlichen Versuchsanstalt. J. D. Sauerländer's verlag, Frankfurt am Main

Nakicenovic N, Davidson O, Davis G et al (2000) Special report on emissions scenarios: a special report of the Working Group III of the Intergovernmental Panel on Climate Change

Namkoong G (1998) Forest genetics and conservation in Europe. In: Turok J, Palmberg-Lerche C, Skroppa T, Ouedraogo AS (eds) Conservation of forest genetic resources in Europe. Proceedings of the European Forest Genetic Resources Workshop, 21 November 1995. International Plant Genetic Resources Institute, Sopron, Hingary, pp. 3–10

Neale DB, Wheeler N (2019) The conifers: genomes, variation and evolution. Springer International Publishing

Nehrbass-Ahles C, Babst F, Klesse S et al (2014) The influence of sampling design on tree-ring-based quantification of forest growth. Glob Chang Biol 20:2867–2885. https://doi.org/10.1111/gcb.12599

Neumann M, Moreno A, Thurnher C et al (2016) Creating a regional MODIS satellite-driven net primary production dataset for European forests. Remote Sens 8

Noormets A, Epron D, Domec JC et al (2015) Effects of forest management on productivity and carbon sequestration: a review and hypothesis. For Ecol Manag 355:124–140. https://doi.org/10.1016/j.foreco.2015.05.019

Nyström M, Lindgren N, Wallerman J et al (2015) Data assimilation in forest inventory: first empirical results. Forests 6:4540–4557

O'Hara KL (2006) Multiaged forest stands for protection forests: concepts and applications. For Snow Landsc Res 80:45–55

O'Hara KL, Ramage BS (2013) Silviculture in an uncertain world: utilizing multi-aged management systems to integrate disturbance†. For An Int J For Res 86:401–410. https://doi.org/10.1093/forestry/cpt012

Pan Y, Birdsey RA, Fang J et al (2011) A large and persistent carbon sink in the world's forests. Science 333:988–993. https://doi.org/10.1126/science.1201609

Pappas C, Bélanger N, Bergeron Y, et al (2021) Smartforests Canada - A network of monitoring plots for forest management under environmental change. In: Managing Forest Ecosystems, Vol. 40, Tognetti R, Smith M, Panzacchi P (Eds): Climate-Smart Forestry in Mountain Regions. Springer Nature, Switzerland, AG

Peters RL, Groenendijk P, Vlam M, Zuidema PA (2015) Detecting long-term growth trends using tree rings: a critical evaluation of methods. Glob Chang Biol 21:2040–2054. https://doi.org/10.1111/gcb.12826

Petit RJ, Aguinagalde I, de Beaulieu J-L et al (2003) Glacial refugia: hotspots but not melting pots of genetic diversity. Science (80-) 300:1563–1565. https://doi.org/10.1126/science.1083264

Picard N, Henry M, Mortier F et al (2012) Using Bayesian model averaging to predict tree aboveground biomass in tropical moist forests. For Sci 58:15–23. https://doi.org/10.5849/forsci.10-083

Porté A, Bartelink HH (2002) Modelling mixed forest growth: a review of models for forest management. Ecol Model 150:141–188. https://doi.org/10.1016/S0304-3800(01)00476-8

Pretzsch H (2009) Forest Dynamics, Growth and Yield. From Measurement to Model. Springer-Verlag Berlin Heidelberg, 664 pp. https://doi.org/10.1007/978-3-540-88307-4

Pretzsch H (2020) The course of tree growth. Theory and reality. For Ecol Manag 478:118508. https://doi.org/10.1016/j.foreco.2020.118508

Pretzsch H, Schütze G (2009) Transgressive overyielding in mixed compared with pure stands of Norway spruce and European beech in Central Europe: evidence on stand level and explanation on individual tree level. Eur J For Res 128:183–204. https://doi.org/10.1007/s10342-008-0215-9

Pretzsch H, Schütze G (2015) Effect of tree species mixing on the size structure, density, and yield of forest stands. Eur J For Res. https://doi.org/10.1007/s10342-015-0913-z

Pretzsch H, Biber P, Ďurský J (2002) The single tree-based stand simulator SILVA: construction, application and evaluation. For Ecol Manag 162:3–21. https://doi.org/10.1016/S0378-1127(02)00047-6

Pretzsch H, Block J, Dieler J et al (2010) Comparison between the productivity of pure and mixed stands of Norway spruce and European beech along an ecological gradient. Ann For Sci 67:712–712. https://doi.org/10.1051/forest/2010037

Pretzsch H, Biber P, Schütze G et al (2014) Forest stand growth dynamics in Central Europe has accelerated since 1870. Nat Commun 5:4967. https://doi.org/10.1038/ncomms5967

Pretzsch H, Forrester DI, Rötzer T (2015) Representation of species mixing in forest growth models: a review and perspective. Ecol Model 313:276–292. https://doi.org/10.1016/j.ecolmodel.2015.06.044

Pretzsch H, Hilmers T, Uhl E, et al (2021) Efficacy of trans-geographic observational network design for revelation of growth pattern in mountain forests across Europe. In: Managing Forest Ecosystems, Vol. 40, Tognetti R, Smith M, Panzacchi P (Eds): Climate-Smart Forestry in Mountain Regions. Springer Nature, Switzerland, AG

Puettmann KJ (2011) Silvicultural challenges and options in the context of global change: "simple" fixes and opportunities for new management approaches. J For 109:321–331. https://doi.org/10.1093/jof/109.6.321

Rasche L, Fahse L, Zingg A, Bugmann H (2011) Getting a virtual forester fit for the challenge of climatic change. J Appl Ecol 48:1174–1186. https://doi.org/10.1111/j.1365-2664.2011.02014.x

Rauscher HM, Isebrands JG, Host GE et al (1990) ECOPHYS: an ecophysiological growth process model for juvenile poplar. Tree Physiol 7:255–281. https://doi.org/10.1093/treephys/7.1-2-3-4.255

Rocha A, Goulden M, Dunn A, Wofsy S (2006) On linking interannual tree ring variability with observations of whole-forest CO2 flux. Glob Chang Biol 12:1378–1389. https://doi.org/10.1111/j.1365-2486.2006.01179.x

Rötzer T, Seifert T, Pretzsch H (2009) Modelling above and below ground carbon dynamics in a mixed beech and spruce stand influenced by climate. Eur J For Res 128:171–182. https://doi.org/10.1007/s10342-008-0213-y

Rötzer T, Leuchner M, Nunn AJ (2010) Simulating stand climate, phenology, and photosynthesis of a forest stand with a process-based growth model. Int J Biometeorol 54:449–464. https://doi.org/10.1007/s00484-009-0298-0

Rüetschi M, Small D, Waser LT (2019) Rapid detection of windthrows using Sentinel-1 C-band SAR data. Remote Sens 11

Rummukainen M (2010) State-of-the-art with regional. Clim Chang 1:82–96. https://doi.org/10.1002/wcc.008

Sánchez-Salguero R, Camarero JJ, Gutiérrez E et al (2017) Assessing forest vulnerability to climate warming using a process-based model of tree growth: bad prospects for rear-edges. Glob Chang Biol:2705–2719. https://doi.org/10.1111/gcb.13541

Schelhaas MJ, Eggers J, Lindner M et al (2007) Model documentation for the European Forest Information Scenario model (EFISCEN 3.1.3). Alterra, 268, Centrum Ecosystemen,

Scherrer D, Vitasse Y, Guisan A et al (2020) Competition and demography rather than dispersal limitation slow down upward shifts of trees' upper elevation limits in the Alps. J Ecol. https://doi.org/10.1111/1365-2745.13451

Schuler LJ, Bugmann H, Petter G, Snell RS (2019) How multiple and interacting disturbances shape tree diversity in European mountain landscapes. Landsc Ecol 34:1279–1294. https://doi.org/10.1007/s10980-019-00838-3

Schumacher S, Bugmann H, Mladenoff DJ (2004) Improving the formulation of tree growth and succession in a spatially explicit landscape model. Ecol Model 180:175–194. https://doi.org/10.1016/j.ecolmodel.2003.12.055

Seidl R, Schelhaas MJ, Lexer MJ (2011) Unraveling the drivers of intensifying forest disturbance regimes in Europe. Glob Chang Biol 17:2842–2852. https://doi.org/10.1111/j.1365-2486.2011.02452.x

Seidl R, Rammer W, Scheller RM, Spies TA (2012) An individual-based process model to simulate landscape-scale forest ecosystem dynamics. Ecol Model 231:87–100. https://doi.org/10.1016/j.ecolmodel.2012.02.015

Seidl R, Schelhaas M-J, Rammer W, Verkerk PJ (2014) Increasing forest disturbances in Europe and their impact on carbon storage. Nat Clim Chang 4:806–810. https://doi.org/10.1038/nclimate2318

Seidl R, Albrich K, Erb K et al (2019) What drives the future supply of regulating ecosystem services in a mountain forest landscape? For Ecol Manag 445:37–47. https://doi.org/10.1016/j.foreco.2019.03.047

Seintsch B, Döring P, Dunger K et al (2017) Das WEHAM-Szenarien Verbundforschungsprojekt. AFZ/Der Wald 72:10–13

Shifley SR, He HS, Lischke H et al (2017) The past and future of modeling forest dynamics: from growth and yield curves to forest landscape models. Landsc Ecol 32:1307–1325. https://doi.org/10.1007/s10980-017-0540-9

Skovsgaard JP, Vanclay JK (2008) Forest site productivity: a review of the evolution of dendrometric concepts for even-aged stands. Forestry 81:13–31. https://doi.org/10.1093/forestry/cpm041

Socha J, Tymińska-Czabańska L (2019) A method for the development of dynamic site index models using height–age data from temporal sample plots. Forest 10

Sodtke R, Schmidt M, Fabrika M et al (2004) Anwendung und Einsatz von Einzelbaummodellen als Komponenten von entscheidungsunterstützenden Systemen für die strategische Forstbetriebsplannung. Forstarchiv 75:51–64

Sperry JS, Venturas MD, Todd HN et al (2019) The impact of rising CO2 and acclimation on the response of US forests to global warming. Proc Natl Acad Sci 116:25734–25744. https://doi.org/10.1073/pnas.1913072116

Spiecker H (2003) Silvicultural management in maintaining biodiversity and resistance of forests in Europe—temperate zone. J Environ Manag 67:55–65. https://doi.org/10.1016/S0301-4797(02)00188-3

Stadelmann G, Temperli C, Rohner B et al (2019) Presenting MASSIMO: a management scenario simulation model to project growth, harvests and carbon dynamics of Swiss forests. Forest 10

Sterba H (1995) PROGNAUS – ein absandsunabhängiger Wachstumssimulator für ungleichaltrige Mischbestände. In: DVFF – Sektion Ertragskunde. Joachimstahl, pp 173–183

Stute M, Clement A, Lohmann G (2001) Global climate models: past, present, and future. Proc Natl Acad Sci U S A 98:10529–10530. https://doi.org/10.1073/pnas.191366098

Svoboda M, Janda P, Bače R et al (2014) Landscape-level variability in historical disturbance in primary Picea abies mountain forests of the Eastern Carpathians, Romania. J Veg Sci 25:386–401. https://doi.org/10.1111/jvs.12109

Temperli C, Bugmann H, Elkin C (2012) Adaptive management for competing forest goods and services under climate change. Ecol Appl 22:2065–2077. https://doi.org/10.1890/12-0210.1

Temperli C, Veblen TT, Hart SJ et al (2015) Interactions among spruce beetle disturbance, climate change and forest dynamics captured by a forest landscape model. Ecosphere 6:art231. https://doi.org/10.1890/ES15-00394.1

Temperli C, Blattert C, Stadelmann G et al (2020) Trade-offs between ecosystem service provision and the predisposition to disturbances: a NFI-based scenario analysis. For Ecosyst 7:27. https://doi.org/10.1186/s40663-020-00236-1

Temperli C, Santopuoli G, Bottero A, et al (2021) National Forest Inventory data to evaluate Climate-Smart Forestry. In: Managing Forest Ecosystems, Vol. 40, Tognetti R, Smith M, Panzacchi P (Eds): Climate-Smart Forestry in Mountain Regions. Springer Nature, Switzerland, AG

Thom D, Rammer W, Seidl R (2017) The impact of future forest dynamics on climate: interactive effects of changing vegetation and disturbance regimes. Ecol Monogr 87:665–684. https://doi.org/10.1002/ecm.1272

Thornton PE, Running SW, Hunt ER (2005) Biome-BGC: terrestrial ecosystem process model, Version 4.1.1

Thrippleton T, Lüscher F, Bugmann H (2020) Climate change impacts across a large forest enterprise in the Northern Pre-Alps: dynamic forest modelling as a tool for decision support. Eur J For Res 139:483–498. https://doi.org/10.1007/s10342-020-01263-x

Tognetti R, Valentini R, Belelli Marchesini L, Gianelle D, Panzacchi P, Marshall JD (2021) Continuous monitoring of tree responses to climate change for smart forestry – a cybernetic web of trees. In: Managing Forest Ecosystems, Vol. 40, Tognetti R, Smith M, Panzacchi P (Eds): Climate-Smart Forestry in Mountain Regions. Springer Nature, Switzerland, AG

Toïgo M, Vallet P, Perot T et al (2015) Overyielding in mixed forests decreases with site productivity. J Ecol 103:502–512. https://doi.org/10.1111/1365-2745.12353

Tolwinski-Ward SE, Evans MN, Hughes MK, Anchukaitis KJ (2011) An efficient forward model of the climate controls on interannual variation in tree-ring width. Clim Dyn 36:2419–2439. https://doi.org/10.1007/s00382-010-0945-5

Tomppo E, Gschwantner T, Lawrence M, McRoberts RE (2010) National forest inventories – pathways for common reporting. Springer

Torresan C, Luyssaert S, Filippa G, Imangholiloo M, Gaulton R (2021) Remote sensing technologies for assessing climate-smart criteria in mountain forests. In: Managing Forest Ecosystems, Vol. 40, Tognetti R, Smith M, Panzacchi P (Eds): Climate-Smart Forestry in Mountain Regions. Springer Nature, Switzerland, AG

Trotsiuk V, Hartig F, Cailleret M et al (2020) Assessing the response of forest productivity to climate extremes in Switzerland using model–data fusion. Glob Chang Biol 26:2463–2476. https://doi.org/10.1111/gcb.15011

Vaganov EA, Hughes MK, Shashkin AV (2006) Growth dynamics of conifer tree rings: images of past and future environments. Springer, Berlin/Heidelberg

van der Plas F, Manning P, Allan E et al (2016) Jack-of-all-trades effects drive biodiversity-ecosystem multifunctionality relationships in European forests. Nat Commun 7:11109. https://doi.org/10.1038/ncomms11109

van Oijen M, Reyer C, Bohn FJ et al (2013) Bayesian calibration, comparison and averaging of six forest models, using data from Scots pine stands across Europe. For Ecol Manag 289:255–268. https://doi.org/10.1016/j.foreco.2012.09.043

Vanclay JK, Skovsgaard JP (1997) Evaluating forest growth models. Ecol Model 98:1–12. https://doi.org/10.1016/S0304-3800(96)01932-1

Vauhkonen J, Berger A, Gschwantner T et al (2019) Harmonised projections of future forest resources in Europe. Ann For Sci 76:79. https://doi.org/10.1007/s13595-019-0863-6

Veblen TT, Hadley KS, Nel EM et al (1994) Disturbance regime and disturbance interactions in a Rocky Mountain Subalpine Forest. J Ecol 82:125–135. https://doi.org/10.2307/2261392

Verkerk PJ, Anttila P, Eggers J et al (2011) The realisable potential supply of woody biomass from forests in the European Union. For Ecol Manag 261:2007–2015. https://doi.org/10.1016/j.foreco.2011.02.027

Vidal C, Lanz A, Tomppo E et al (2008) Establishing forest inventory reference definitions for forest and growing stock: a study towards common reporting. Silva Fenn 42:247–266. https://doi.org/10.14214/sf.255

Vidal C, Alberdi I, Hernández L, Redmond J (2016a) National Forest Inventories: assessment of wood availability and use, 1st edn. Springer International Publishing, Cham

Vidal C, Alberdi I, Redmond J et al (2016b) The role of European National Forest Inventories for international forestry reporting. Ann For Sci. https://doi.org/10.1007/s13595-016-0545-6

Wang Y-P, Trudinger CM, Enting IG (2009) A review of applications of model–data fusion to studies of terrestrial carbon fluxes at different scales. Agric For Meteorol 149:1829–1842. https://doi.org/10.1016/j.agrformet.2009.07.009

Wang T, O'Neill GA, Aitken SN (2010) Integrating environmental and genetic effects to predict responses of tree populations to climate. Ecol Appl 20:153–163. https://doi.org/10.1890/08-2257.1

Waring RH, Coops NC, Landsberg JJ (2010) Improving predictions of forest growth using the 3-PGS model with observations made by remote sensing. For Ecol Manag 259:1722–1729. https://doi.org/10.1016/j.foreco.2009.05.036

Weiskittel A, Hann D, Kershaw J, Vanclay J (2011) Forest growth and yield modeling

Wilson B, Howard R (1968) A computer model for cambial activity. For Sci 14:77–90

Yang Y, Anderson M, Gao F et al (2020) Investigating impacts of drought and disturbance on evapotranspiration over a forested landscape in North Carolina, USA using high spatiotemporal resolution remotely sensed data. Remote Sens Environ 238:111018. https://doi.org/10.1016/j.rse.2018.12.017

Youhua R, Li X, Sun R et al (2016) Spatial representativeness and uncertainty of eddy covariance carbon flux measurements for upscaling net ecosystem productivity to the grid scale. Agric For Meteorol 230–231:114–127

Yousefpour R, Temperli C, Jacobsen JB et al (2017) A framework for modeling adaptive forest management and decision making under climate change. Ecol Soc 22. https://doi.org/10.5751/ES-09614-220440

Yousefpour R, Augustynczik ALD, Reyer CPO et al (2018) Realizing mitigation efficiency of European commercial forests by climate smart forestry. Sci Rep 8:1–11. https://doi.org/10.1038/s41598-017-18778-w

Yuan W, Liu S, Yu G et al (2010) Global estimates of evapotranspiration and gross primary production based on MODIS and global meteorology data. Remote Sens Environ 114:1416–1431. https://doi.org/10.1016/j.rse.2010.01.022

Zell J (2018) Climate sensitive tree growth functions and the role of transformations. Forest 9

Zell J, Rohner B, Thürig E, Stadelmann G (2019) Modeling ingrowth for empirical forest prediction systems. For Ecol Manag 433:771–779. https://doi.org/10.1016/j.foreco.2018.11.052

Zellweger F, De Frenne P, Lenoir J et al (2020) Forest microclimate dynamics drive plant responses to warming. Science (80-) 368:772–775. https://doi.org/10.1126/science.aba6880

Zhang K, Kimball JS, Running SW (2016) A review of remote sensing based actual evapotranspiration estimation. WIREs Water 3:834–853. https://doi.org/10.1002/wat2.1168

Zurbriggen N, Nabel JEMS, Teich M et al (2014) Explicit avalanche-forest feedback simulations improve the performance of a coupled avalanche-forest model. Ecol Complex 17:56–66. https://doi.org/10.1016/j.ecocom.2013.09.002

Zweifel R, Eugster W, Etzold S et al (2010) Link between continuous stem radius changes and net ecosystem productivity of a subalpine Norway spruce forest in the Swiss Alps. New Phytol 187:819–830. https://doi.org/10.1111/j.1469-8137.2010.03301.x

Chapter 8
Climate-Smart Silviculture in Mountain Regions

Maciej Pach, Kamil Bielak, Andrej Bončina, Lluís Coll, Maria Höhn,
Milica Kašanin-Grubin, Jerzy Lesiński, Hans Pretzsch, Jerzy Skrzyszewski,
Peter Spathelf, Giustino Tonon, Andrew Weatherall, and Tzvetan Zlatanov

Abstract Mountain forests in Europe have to face recently speeding-up phenomena related to climate change, reflected not only by the increases in the mean global temperature but also by frequent extreme events, that can cause a lot of various damages threatening forest stability. The crucial task of management is to adapt forests to environmental uncertainties using various strategies that should be undertaken to enhance forest resistance and resilience, as well as to maintain forest biodiversity and provision of ecosystem services at requested levels. Forests can play an important role in the mitigation of climate change. The stand features that increase forest climate smartness could be improved by applying appropriate silvicultural measures, which are powerful tools to modify forests. The chapter provides information on the importance of selected stand features in the face of climate change and silvicultural prescriptions on stand level focusing to achieve the required level of climate smartness. The selection of silvicultural prescriptions should be also supported by the application of simulation models. The sets of the various treatments and management alternatives should be an inherent part of adaptive forest management that is a leading approach in changing environmental conditions.

M. Pach (✉) · J. Skrzyszewski
Department of Ecology and Silviculture, Faculty of Forestry, University of Agriculture in Krakow, Krakow, Poland
e-mail: rlpach@cyf-kr.edu.pl; rlskrzys@cyf-kr.edu.pl

K. Bielak
Department of Silviculture, Institute of Forest Sciences, Warsaw University of Life Sciences, Warsaw, Poland
e-mail: kamil.bielak@wl.sggw.pl

A. Bončina
Department of Forestry and Renewable Forest Resources, Biotechnical Faculty, University of Ljubljana, Ljubljana, Slovenia
e-mail: andrej.boncina@bf.uni-lj.si

© The Author(s) 2022
R. Tognetti et al. (eds.), *Climate-Smart Forestry in Mountain Regions*, Managing Forest Ecosystems 40, https://doi.org/10.1007/978-3-030-80767-2_8

8.1   Introduction

European forests in mountain regions are particularly vulnerable to the impact of climate change that could endanger the provision of ecosystem services. Hanewinkel et al. (2013) showed that the expected loss of value of European forest lands due to the decline of economically valuable species, in the absence of effective counter-measures, varies between 14% and 50% by 2100, depending on the interest rate and climate scenario applied. Adaptive forest management can address environmental uncertainties with strategies that enhance forest resistance and resilience, maintain forest biodiversity, and provide ecosystem services at requested levels. Various types of adaptation can be distinguished (Locatelli et al. 2010; Yousefpour et al. 2017; Lindner et al. 2020): (1) anticipatory or proactive adaptation, which takes place before the impacts of climate change are observed, (2) reactive adaptation,

L. Coll
Department of Agriculture and Forest Engineering, School of Agrifood and Forestry Science and Engineering, University of Lleida, Lleida, Spain
e-mail: lluis.coll@udl.cat

M. Höhn
Department of Botany, Budai Campus, Hungarian University of Agriculture and Life Sciences, Budapest, Hungary
e-mail: hohn.maria@uni-mate.hu

M. Kašanin-Grubin
Institute for Chemistry, Technology and Metallurgy, University of Belgrade, Belgrade, Serbia
e-mail: mkasaningrubin@chem.bg.ac.rs

J. Lesiński
Department of Forest Biodiversity, Faculty of Forestry, University of Agriculture in Krakow, Krakow, Poland
e-mail: jerzy.lesinski@urk.edu.pl

H. Pretzsch
Chair for Forest Growth and Yield Science, TUM School of Life Sciences in Freising Weihenstephan, Technical University of Munich, Freising, Germany
e-mail: hans.pretzsch@tum.de

P. Spathelf
Faculty of Forest and Environment, Eberswalde University for Sustainable Development, Eberswalde, Germany
e-mail: Peter.Spathelf@hnee.de

G. Tonon
Faculty of Science and Technology, Free University of Bolzano/Bozen, Bolzano, Italy
e-mail: giustno.tonon@unibz.it

A. Weatherall
National School of Forestry, University of Cumbria, Cumbria, UK
e-mail: andrew.weatherall@cumbria.ac.uk

T. Zlatanov
Institute of Biodiversity and Ecosystem Research, Bulgarian Academy of Sciences, Sofia, Bulgaria

which takes place after impacts of climate change have been observed, and (3) autonomous or spontaneous adaptation that does not constitute a conscious response to climatic stimuli, but is triggered by ecological changes in natural systems and by market or welfare changes in human systems. The selection of the adaptation strategy should be based on a profound analysis of environmental and socioeconomic circumstances at a local and regional level and requires planning and implementation of forward-looking adaptation measures considering projected climate change (Lindner et al. 2020).

Climate-Smart Forestry (CSF) defined as "*sustainable adaptive forest management and governance to protect and enhance the potential of a forest to adapt to, and mitigate climate change*" (Bowditch et al. 2020) can be characterized by selected criteria and indicators originating from sustainable forest management (SFM) indicators (Santopuoli et al. 2021). The stand features that increase forest smartness could be improved by silvicultural measure (e.g., horizontal and vertical spatial structure, mixed species composition, deadwood amount, etc.). This chapter presents possible silvicultural measures for CSF with analysis via simulation models to evaluate their reliability.

8.2 Risks to Forests Induced by Climate Change

Mountain forests are considered to be particularly vulnerable to the effects of climate warming as temperature determines the upper limit of the altitudinal range for plant communities (Lenoir et al. 2008). Most studies conducted in mountain areas predict an upward shift of forest communities in response to temperature increases (Guisan et al. 1998). However, other factors (i.e., land-use changes, disturbances, biotic interactions) also modulate these responses (Martín-Alcón et al. 2010; Ameztegui and Coll 2013; Ameztegui et al. 2016), which can lead to unforeseen dynamics such as downslope displacements (see Bodin et al. 2013).

The increasing occurrence of extreme drought and heat events is at the origin of many declining forests and tree mortality episodes worldwide (Allen et al. 2010; Martínez-Vilalta et al. 2012; Margalef-Marrase et al. 2020) and mountain forests are not an exception (Galiano et al. 2010; Linares and Camarero 2012). Climate warming is predicted to intensify the disturbance regimes to which these systems are exposed (Seidl et al. 2017). For example, in Mediterranean mountains, the combined effect of fuel flammability increases and fuel accumulation associated to land-abandonment is expected to have a high impact on fire risk (Pausas 2004) compromising the local persistence of some populations that do not present adaptive mechanisms to such events (Vilà-Cabrera et al. 2012). In temperate and boreal areas, warming is also expected to intensify the frequency and severity of windstorms events (Seidl et al. 2014), insect outbreaks (Weed et al. 2013; Biedermann et al. 2019), and pathogen attacks (Sturrock et al. 2011) and to modulate the interactions among different disturbances (Temperli et al. 2013; Seidl and Rammer 2017).

The intensification of disturbance regimes is particularly important in mountain areas where the occurrence of these events (e.g., extensive bark beetle attacks, crown fires) was rare in the past. Recent catastrophic events, such as the *Vaia* storm (that caused in October 2018 damages of millions of cubic meters in northern Italy) or the unprecedented outbreaks of bark beetles in central Europe, point to the need of implementing effective monitoring strategies and designing managing regimes accounting for increasing risks.

Large-scale natural disturbances usually are followed by salvage logging: the main aim of it is to reduce economic losses. Besides, sanitary and aesthetic reasons are of some importance with this respect, too. On the other hand, the salvage logging practices indicate its strong impact on the functioning of the forest ecosystem, such as ecosystem restoration due to deterioration of the regenerative capacity of forests (Pons et al. 2020) and to threats to biodiversity conservation (Thorn et al. 2018). In order to maintain populations of the saproxylic species, Lonsdale et al. (2008) strongly suggest reducing salvage logging intensity in damaged tree-stands.

8.3   Indicators that Could Be Modified by Silvicultural Measures at Stand Level (Silvicultural Indicators)

Criteria and Indicators (C&I) of CSF originated from C&I of Sustainable Forest Management (Forest Europe 2015; Bowditch et al. 2020; Santopuoli et al. 2021) may refer to the stand, landscape, or even regional/national level. In this chapter, we are focusing on "silvicultural indicators" of CSF, which are manageable by silviculture measures at the stand level. Their evaluation is based on classification of indicators, presented by Bowditch et al. (2020) (Table 8.1).

8.4   Silvicultural Treatments Improving Stand Adaptation

8.4.1   *Forest Area (Afforestation)*

In the last decades, European mountains have undergone important forest expansion processes associated with the abandonment of traditional agrosilvopastoral activities (Kozak 2003; Gehrig-Fasel et al. 2007; Ameztegui et al. 2010). These processes include the encroachment of woody vegetation in areas previously occupied by cultures or pastures, and the densification of pre-existing forest stands. The rate of forest expansion is not homogeneous and depends on several factors operating and different spatiotemporal scales such as the browsing pressure (Coop and Givnish 2007), physiographic factors (Poyatos et al. 2003), or local socioeconomic conditions (Dirnböck et al. 2003; Ameztegui et al. 2016), among others. The ecological consequences of these processes differ. The progressive regression of abandoned land is leading to a homogenization of the landscape, and to the loss of

Table 8.1 Criteria and indicators (Bowditch et al. 2020) considered as those that can be shaped by silvicultural treatments within adaptation and mitigation strategy at the stand level

Strategy	Criteria	Indicator	Label	Description
Adaptation	Forest resources and global carbon cycles	Forest area	Forest area	Area of forest and other wooded lands, classified by forest type and by availability for wood supply, and share of forest and other wooded lands in total land area.
		Age structure and/or diameter distribution	Forest structure	Age structure and/or diameter distribution of forest and on other wooded lands, classified by availability for wood supply.
	Forest health and vitality	Soil condition	Soil condition	Chemical soil properties (pH, CEC, C/N, organic C, base saturation) in forest and on other wooded lands related to soil acidity and eutrophication, classified by main soil types.
		Forest damage	Forest damage	Forest and other wooded lands with damage, classified by primary damaging agent (abiotic, biotic, and human induced).
	Productive functions of forests	Increment and felling	Increment/felling	A balance between net annual increment and annual felling of wood in forest available for wood supply.
	Forest biological diversity	Tree species composition	Diversity	Area of forest and other wooded lands, classified by the number of tree species occurring.
		Regeneration	Regeneration	Total forest area by stand origin and area of annual forest regeneration and expansion.
		Naturalness	Naturalness	Area of forest and other wooded lands by the class of naturalness ("undisturbed by man," "seminatural," or "plantations").
		Introduced tree species	New species	Area of forest and other wooded lands dominated by introduced tree species.
		Deadwood	Deadwood	The volume of standing deadwood and of lying deadwood in forest and on other wooded lands.
		Genetic resources	Genetic resources	Area managed for conservation and utilization of forest tree genetic resources (*in situ* and *ex situ* genetic conservation) and area managed for seed production.
		Threatened forest species	Threatened species	Number of threatened forest species, classified according to IUCN Red List categories to the total number of forest species.
	Protective function (soil and water)	Protective forests – soil, water, and other ecosystem functions, and infrastructures	Protective forests	Area of forest and other wooded lands designated to prevent soil erosion, preserve water resources, maintain other protective functions, protect infrastructure and managed natural resources against natural hazards.
	New indicators	Slenderness coefficient	Slenderness	The ratio of total tree height to stem diameter outside bark at 1.3 m above ground level.
		Vertical distribution of tree crowns	Vertical crowns	Distribution of tree crowns in the vertical space. It can be measured in terms of layers (one, two, multiple), or in terms of the ratio between tree height and crown length.
		Horizontal distribution of tree crowns	Horizontal crowns	Canopy space-filling and can be expressed in measure of the density of tree crowns, such as crown area, tree crown diameter. It can be also expressed in measure of the density of trees, such as trees per hectare, basal area per hectare (in this case, the horizontal distribution refers to the tree).

Table 8.1 (continued)

Strategy	Criteria	Indicator	Label	Description
Mitigation	Forest resources and global carbon cycles	Growing stock	Growing stock	Growing stock in forest and on other wooded lands, classified by forest type and by availability for wood supply.
		Carbon stock	Carbon stock	Carbon stock and carbon stock changes in forest biomass, forest soils, and in harvested wood products.
	Productive functions of forests	Roundwood	Roundwood	Quantity and market value of roundwood.

mosaic-type structure, which is important for maintaining high biodiversity (Edwards 2005). The increase of stand density causes both higher fuel accumulations in the stands and higher competition for growing resources among the individuals, thus increasing the vulnerability of these systems to wildfires and drought (Nocentini and Coll 2013). Forest expansion in the upper parts of catchments can induce significant streamflow reductions in semiarid regions (Gallart and Llorens 2004).

Reforestation programs took place in mountain areas of many European countries during the twentieth century. The primary objective of these actions was to avoid soil degradation and regulate the hydrological conditions of watersheds (Mansourian et al. 2005). Conifer species were mainly used due to its pioneer character and ability to establish in difficult environmental conditions (Ceballos 1960). Unfortunately, management after afforestation was not adequately conducted and, at present, they show excessive densification, growth stagnation, and generalized poor health status (Pausas et al. 2004). The current management of these stands (some of which are rather aged) represents a big challenge for forest practitioners due to the location (often in inaccessible areas) and their primary protective role (Brang et al. 2006).

8.4.2 Structure of Forest Stands (Age and Diameter Distribution, Vertical and Horizontal Distribution of Tree Crowns)

Age structure, diameter distribution, and vertical and horizontal distribution of tree crowns are closely interrelated. Structural diversity in forests encompasses different age cohorts, size classes of trees and the spatial arrangement of different patches of tree groups, and structural elements, such as large living and dead trees, coarse woody debris or seed-producing tree clusters on a stand level. These stand legacies provide essential ecosystem processes (e.g., seed dispersal, nutrient translocation) and preserve genetic information in the phase of an ecosystem's recovery after disturbance. They are important elements in the reorganization loop of the adaptive cycle (Drever et al. 2006; Bauhus et al. 2009). Furthermore, stand legacies enhance faunal species richness, for example, as antagonist species, which reduce forest vulnerability.

The multiaged stands with structural diversity have the potential to increase both the resistance and resilience to various-scale forest disturbances (improve response diversity of a forest) (Elmqvist et al. 2003; Brang et al. 2013; O'Hara and Ramage 2013; Spathelf et al. 2018) and also productivity (Torresan et al. 2020). Such structural diversity in a forest can be achieved using several ways during stand management.

Thinning may become increasingly important for adaptation in many forest types, reducing stand density and increasing the individual stability and stress resistance of the remaining best crop trees in the stand (Misson et al. 2003;

Rodríguez-Calcerrada et al. 2011; Sohn et al. 2013; Spathelf et al. 2018). The application of the selected method and type of thinning must be compatible with the silvicultural objectives. Among the various thinning methods, there are those as selective (Schädelin), classical differentiation thinning, interfering with all layers of the stand, and free or variable density thinning, belonging to the crown or all-layer thinning type, that contribute to the increased structural diversity in the stand (Leibundgut 1982; Schütz 1987; Helms 1998; Schütz 2001b; Spiecker 2004; O'Hara et al. 2012; Silva et al. 2018). All types of thinnings, besides the improvement of timber quality, can help to create a diversity of age classes; decrease the water, nutrient, and light competition; increase individual tree resistance to biotic and abiotic factors; and, in some cases, encourage a wider range of species, which is a way of reducing and dispersing of silvicultural risk (Silva et al. 2018). Thinning, especially accomplished in medium-aged and/or older stands, may create conditions for the establishment of natural regeneration of the same or different species, thereby introducing new young age-classes of trees into the stand. Such vertically structured stands are more resilient after disturbance, since advanced regeneration is going to be quickly released (Brang et al. 2013).

Diameter and age structural diversity of forest stands is also associated with the occurrence and severity of natural disturbances; for example, a study from the Julian Alps showed that occurrence of windthrow disturbances in forest stands is negatively related to the volume of small-diameter trees (<30 cm in diameter at breast height), and positively with the volume of medium- (30–50 cm in diameter at breast height) and large diameter trees (>50 cm in diameter at breast height), while a large amount of small-diameter trees (<30 cm in diameter at breast height) increased the likelihood for snow breakage occurrence (Klopčič et al. 2009). The integration of various-scale disturbances into forest management could be the way of achieving a multiaged, multilayer, and multispecies forests that can fulfil multiple purposes. A wide range of measures to promote uneven-aged stands and structural diversity including emulation of disturbances, planning salvage operations, and variable treatment intervals or intensities is presented by O'Hara and Ramage (2013). However, many of these measures can be applied on forest (a group of stands) or landscape scale leading to high structural diversity that reduces the probability of stand-replacing disturbances.

Forest stability, vitality, and resilience can be also enhanced through silvicultural activities making the best use of natural structures and processes. This includes proactive steering of natural successions instead of passive waiting for natural processes to occur (Silva et al. 2018; Lindner et al. 2020). These processes can supplement the structural diversity in terms of species composition, and vertical and horizontal stand diversification.

The application of some silvicultural systems is one of the possible measures leading to the formation of structurally diversified forests. At present, slightly less than 70% of forests in Europe are reported as even-aged, whereas uneven-aged forests appear to be the main forest type in South-West Europe (Forest Europe 2015). But this does not mean that all even-aged stands should be converted. The long-term process of transition from even-aged to uneven-aged stand could be performed only in those places where site, stand, and socioeconomic conditions allow its realization and where it is advisable. Several methods to achieve uneven-aged, structurally diversified stands composed of shade-tolerant tree species exist including silvicultural systems (irregular shelterwood system and its variations, selection system), thinning (selection with intervention in all stand layers), and other methods combining different felling schedule with various methods of natural and artificial regeneration (Schütz 2001b; Nyland 2003; Pretzsch 2019; Hilmers et al. 2020). The transformation from regular to irregular stands can be accomplished in the present stand with the sequence of differentiation thinning or in the following stand generation depending on the stability and irregularity of a stand (Schütz 2001b). The implementation of the methods depends on the silvicultural objectives, species composition and stability of existing stand, and site conditions. Irregular shelterwood system and its many variations are characterized by the greatest potential and versatility to shape uneven-aged forests composed of various tree species of functional traits (Puettmann et al. 2009; Raymond et al. 2009; D'Amato et al. 2011; Lussier and Meek 2014; Raymond and Bédard 2017).

8.4.3 Soil Condition

Forest cover is strongly influenced by soil productivity, which is partially governed by climate, but more significantly by bedrock composition and erosion rate (Hahm et al. 2014; Milodowski et al. 2015; Wolf et al. 2016). Forest soil productivity is crucial for sustainable forest management and is a function of soil potential properties and soil conditions. Soil potential properties are the ones that are not easily altered such as soil depth, stoniness, the content of organic matter, texture, porosity, clay mineralogy, while soil conditions can be altered more easily and are represented by soil thickness, porosity, and soil organic matter (Poff 1996). Soil depth, as a basic criterion of soil classifications, represents the depth from the surface of the soil to the parent material. Soil porosity is a combined volume of solids and pores filled with air and/or water. The size and interconnection of pores determine water infiltration and retention, gas exchange, biological activity, and rootable soil volume, thus representing an extremely important link in soil productivity.

Forest soils hold a substantial portion of terrestrial carbon and any alterations in carbon cycling are significant for forest productivity and ecosystem services (James and Harrison 2016). Change in quality or quantity of soil organic matter caused by climate change is probably one of the most important factors affecting forest soils (Raison and Khanna 2011), since soil organic matter, together with nitrogen and phosphorous, is one of the principal components of soils and has a crucial role in several biological, chemical, and physical properties (James and Harrison 2016). At large scale, the variability of soil organic carbon is mostly governed by climate, while on a local scale, it depends on forest management practices, type of bedrock, soil properties, and topography (Conforti et al. 2016).

Bedrock has a significant role in vegetation growth by regulating physical and chemical properties of soils (Hamh et al. 2014) and has a substantial influence on soil erosion processes (Milodowski et al. 2015). It is the source of mineral nutrients and influences soil texture characteristics controlling the water and nutrient retention capacity, but can also present a supply of heavy metals that have an adverse influence on plant growth (Jiang et al. 2020). Soil texture in forest soils determines soil water and aeration, both important for tree growth and microbial processes (Gomez-Guarrerro and Doane 2018). Soil degradation includes higher bulk density and lower hydraulic conductivity and extensive nutrient losses in soils (Hajabbasi et al. 1997). Loss of porosity leads to infiltration reduction, loss of soil volume, and enhances soil erosion. However, these effects might be happening at the same time and causal (Poff 1996). Similar is true for textural properties. Soils with high silt, low clay, and low organic matter content are generally considered to be more erodible. However, this is not straightforward and particle size distribution has to be considered in relation to other soil physical and chemical properties (Wischmeier and Mannering 1969).

Altieri et al. (2018) have experimentally shown that soil erosion is not a substantial problem in well-managed forests and minimal values of soil loss were reported in areas with high canopy cover and biomass. However, some authors indicate that new silvicultural treatments should be planned with care, since, as established by earlier studies, loss of forest cover, either due to deforestation or climate change, can impose a serious problem with long-term consequences. If the topsoil layer is disturbed due to natural or human-induced causes, such as wildfire, harvesting, and prescription burning, erosion rates can substantially rise. Relationship between soil disturbance and soil productivity is a complex interconnection among soil physical properties, nutrient cycling, and climate. The disturbance effect depends on soil local characteristics and microclimate, so mitigation solutions have to be site-specific (Elliot et al. 1996).

8.4.4   Forest Damages

Forest disturbances are, in many cases, inseparably related to climate change (Dale et al. 2000, 2001; Reyer et al. 2017; Seidl et al. 2017). Disturbances, human-induced and naturally caused mostly by wind, insects, fungi, fires, droughts, heatwaves, and their interactions, shape the forest ecosystems in terms of species composition, structure, and processes. Proactive disturbance-risk management should encompass adaptive silvicultural measures, being a part of the adaptive forest management, which enable using some strategies to counteract the effects of climate change resulting in forest disturbances. These possible actions should be undertaken considering uncertainties about climate change impacts on forests and their reactions (Lindner et al. 2014). The potential silvicultural disturbance-prevention measures include (1) the use of more climate-adapted tree species or their genotypes (Kauppi et al. 2018; Thurm et al. 2018), the introduction of economic alternatives to main species (Deuffic et al. 2020), management to facilitate the establishment of species outside of historical natural ranges and genomics-based assisted migration in refor-estation (transformation) (Hagerman and Pelai 2018); (2) application of more diver-sified species composition of forests (mixtures of conifers with broadleaves, shade-tolerant with intolerant species, more drought-resistant with less-resistant species) involving also conversion from single-species to multispecies stands where site conditions permit (Kerr et al. 2010; Jactel et al. 2017), this allows to distribute silvicultural risk to many tree species in a stand; (3) managing for and/or increasing resilience (Hagerman and Pelai 2018); (4) more frequent and intensive thinnings (selective or differentiation) and shorter and/or diversified rotation length (Jactel et al. 2009; Silva et al. 2018; Deuffic et al. 2020); (5) shaping the diversified age structure of forests (uneven-aged/selection structure) (Schütz 2002; O'Hara 2014; Deuffic et al. 2020); (6) increasing stand stability and decreasing stand density (Knoke et al. 2008; Deuffic et al. 2020); (7) fire-smart landscape management tech-niques (Kauppi et al. 2018; Lindner et al. 2020). In addition, the realization of the concepts of close-to-nature silviculture (Schütz 1999; Brang et al. 2014; O'Hara 2014) and/or continuous cover forestry (Mason et al. 1999; Pukkala and Gadov 2012) seems to enhance adaptation to climate change and, to some extent, mitigate its effects on forests (Fig. 8.1).

Fig. 8.1 Forest damages caused by the windstorm Xynthia (2010) in la Val d'Aran (NE, Spain). (Photo: Álvaro Aunós)

8.4.5 *Increment and Felling*

In the case of close-to-nature mountain mixed and uneven-aged forests, comprised of species combination such as silver fir (*Abies alba* Mill.), Norway spruce (*Picea abies* (L.) Karst.) and European beech (*Fagus sylvatica* L.), all silvicultural operations attempt to achieve growth sustainability from one cutting cycle to the next and continuous forest cover for preventing soil erosion. To this end, the single and/or group selection (plenter) system can be used in mountain regions across Europe (Schütz 2001a). In the selection forest, the mixture of trees of different sizes (diameter at breast height and height), ages, and species, growing together in a small area (<0.1 ha) (Schütz 2001a; Bončina 2011a; O'Hara 2014), results in much more steady course of growth, in comparison to one species dominated even-aged stands, at both tree and stand level (Fig. 8.2). The higher resilience of stand growth to silviculturally induced density reductions in mixed, uneven-aged mountain forests can be observed as in this case, trees in the medium and lower canopy layers can compensate (buffer) for losses in the upper layer and vice versa (Mitscherlich 1952; Assmann 1970).

In the structural stable mountain forests (Fig. 8.3), the equilibrium state is achieved when standing volume remains relatively constant from one cutting cycle to the next; in the other words, the harvest volume equals increment. Therefore, the value of periodic volume increment may serve as an additional important parameter to consider, when regulating the long-term development of mixed, uneven-aged

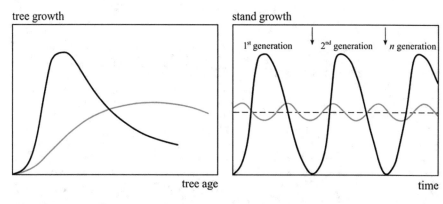

Fig. 8.2 Tree level and stand level growth pattern in two contrasting silvicultural systems: simple even-aged forests (black line) and complex uneven-aged forests (grey line), managed by a clearcutting and selection system, respectively. *In the first case, the growth (e.g., tree diameter left and stand volume increment right) follows the unimodal curve and changes more rapidly (up and down) over the time with a clear pick during the optimal developmental phase, to decreases, however, to zero at the time of the initiation of subsequent forest generation by clearcuttings or shelterwood cuttings with short regeneration period (10–20 years). In the selection forest, the combination of trees of many sizes, age classes, and species buffers the changes in the growth pattern and thus a steadier increment course can be observed.* (Adapted from Schütz (2001a) and Pretzsch et al. (2015) (after Assmann 1970) in case of a tree and stand level, respectively)

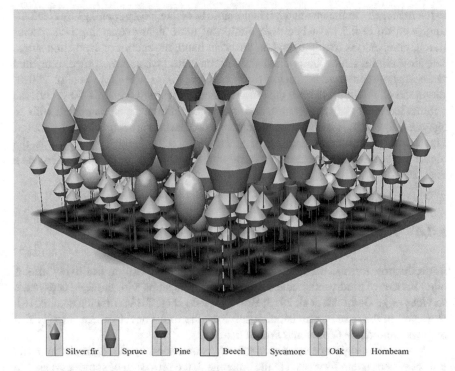

Silver fir   Spruce   Pine   Beech   Sycamore   Oak   Hornbeam

Fig. 8.3 The vertical and horizontal structural stable close-to-nature mountain mixed and uneven-aged stand, comprised of silver fir (Abies alba Mill.), European beech (Fagus sylvatica L.) and Norway spruce (Picea abies (L.) Karst.), as well as other minor tree species, managed by selection (plenter) system on the long-term experimental plot in the Zagnansk Forest District (Poland)

mountain forests. Schütz (2001a) described the plenter structure as being maintained through continuous control of the growing stock (standing volume); a growing stock over the equilibrium would lead to reduced regeneration and recruitment into smaller diameter classes, whereas levels of growing stock below the equilibrium would reduce total increment and the quality of trees as well as to overpromote the natural regeneration and recruitment rate. However, in some cases such as the current stand diameter structure deviates significantly from the equilibrium curve (e.g., when stands previously were managed by uniform or irregular shelterwood cuttings), the transformation by applying heavy structural differentiation thinning is recommended, to reduce mainly the density in the middle diameter at breast height classes and, therefore, as a consequence, also the stand productivity.

Finally, if the proportion of more light-demanding tree species in the stand is required, the irregular shelterwood system, emulating the natural gap dynamic pattern, would be also recommended in the scope of climate-smart silviculture. The main difference between selection and irregular shelterwood systems lies in an emphasis on individual trees in the former case versu. cohorts of trees in the latter (Schütz et al. 2016). Moreover, the irregular shelterwood system gives a free hand to the manager. On the one hand, it is possible to create larger openings by clearcutting (e.g., up to 0.5 ha) when regeneration of more light-demanding tree species (larch, pine, oak) is required, and on the other hand, the shelter or group and single selection cuttings may be applied for intermediate (sycamore, spruce, elm) and more shade-tolerant tree species (lime and beech) as well (cf. Schütz 2001a; Raymond et al. 2009). During the long regeneration period (e.g., 40–50 years), the volume increment of regeneration and overstorey add up. Thus, compared with an age-class forest, the total volume increment's oscillations are much less distinct (cf. Fig. 8.2). The more multilayered a stand and the more horizontally diverse it is, the higher is its growth resilience to natural and silvicultural interferences (Pretzsch et al. 2015).

8.4.6 Tree Species Composition

In production forests, the diversity of tree species composition positively affects other indices of biodiversity and shows close relationships with multiple ecosystem services (e.g., Gamfeldt et al. 2013; Bravo-Oviedo et al. 2014; Almeida et al. 2018). However, in the case of provisioning services, the forest development stage is also of great importance (Zeller and Pretzsch 2019).

None of the single tree species in a forest is able to safeguard a provision of the full set of ecosystem services. On the other hand, the provision of some services can be impossible, since they might be negatively correlated with each other. Therefore, in order of satisfying the society demands regarding multiple ecosystem services the production forests should be managed considering use of the various tree species. No doubt that tree species diversity positively influences ecosystem functioning, but in some cases, probably the effect of species identity is stronger compared with diversity (Nadrowski et al. 2010). If the dominating tree species is badly chosen, then changing it back to the former one might reverse the negative outcomes for biodiversity, production, and recreational values, as well as on stand vulnerability to a wind, frost, and drought damage, as well as on risks of pathogen or insect outbreak (Felton et al. 2019).

Tree species diversity of temperate mountain forests is much lower if compared to the tree species diversity in forests of lower vegetation belts (i.e., planar-hilly, sub-montane). Therefore, management options regarding tree species are much broader in lower areas. For the adaptation of mountain forests to climate change, it is highly important (1) to maintain/increase genetic variation in the species, (2) to increase structural diversity (Brang et al. 2014), and (3) to assure that all "natural" tree species of mountain forests are present in a forest stand. However, quite often, European mountain forests contain mainly three tree species – Norway spruce, silver fir, and European beech. Especially, recruitment of silver fir into these forests is often restricted or even prevented due to browsing pressure (e.g., Ficko et al. 2011; Simončič et al. 2019), which noticeably decreases adaptation capacity of forests to climate change. The additional characteristics if compared to the forests in lower elevations are that in mountain forests, many minor tree species (e.g., *Sorbus aucu-paria*, *Salix* sp.) cannot compete or are economically less important for forest managers/owners, while some other with possible high economic value (i.e., *Ulmus glabra*, *Fraxinus excelsior*) suffer from diseases. Therefore, tree species from genus *Acer*, *Larix*, and *Pinus* gain higher importance for the adaptation of mountain forests to climate change.

There is increasing evidence that tree species mixtures positively influence forest functionality. Forest stands with tree species with different functional traits enhance forest fitness in the face of climate change as they include different "strategies" concerning plant establishment and competitiveness (Jactel et al. 2017). Moreover, in many cases, species-rich forests with high functional diversity are more productive than less diverse forests (Pretzsch et al. 2010). In stands where light demanding and shade-tolerant, canopy and understory or deep-rooting, and shallow-rooting species are combined, resources such as light, water, and nutrients can be spatially and temporarily used differently and thus more efficiently. Such forests are more resistant to various abiotic disturbance events, such as drought, fires, or storms (Bravo-Oviedo et al. 2014; Knoke et al. 2008; Schütz et al. 2006; Spellmann et al. 2011; Lebourgeois et al. 2013), and more resilient once a disturbance has occurred (Jactel et al. 2009, 2017). With an increasing number of functionally different species, the probability increases that some of these species can resist external disturbances or changing environmental conditions (i.e., the ecological insurance concept, according to Yachi and Loreau 1999). Examples are the bark beetle *Ips typographus* that attacks Norway spruce (*Picea abies*), but not broad-leaved species or silver fir (*Abies alba*) (Wermelinger 2004), or the ash dieback (*Hymenoscyphus pseudoalbidus*) affecting exclusively *Fraxinus excelsior* (Kjær et al. 2012). In addition to functional diversity, the redundancy of species increases the probability that one species can take over the role of another species that does not survive (Walker et al. 1999; Messier et al. 2019).

It is assumed that in the future also mountain forest ecosystems are severely affected by water shortage (Collin 2020). The admixture of broadleaved tree species in conifer stands can have positive effects on soil water availability, thus reducing water stress for the trees. There is evidence that interception losses are higher in pure conifer stands with Scots pine and Norway spruce compared to broadleaved or mixed stands with European beech (Barbier et al. 2009; Berger et al. 2009). In a study in northeastern Germany, Müller (2009) analyzed seepage rates in mixtures of Scots pine with European beech compared to pure Scots pine stands. The higher seepage in mixed stands is due to reduced interception losses and a higher stemflow on broadleaved trees compared to pine. Moreover, in pure (pine) stands, dense ground vegetation layers of the grass *Calamagrostis* lead to a further reduction of the soil water content (Müller and Bolte 2009).

Other complementarity effects with respect to water supply of species mixtures are reported, such as hydraulic redistribution or the different stomatal behavior of the trees. Thus, water availability in mixed stands can be positively influenced, although many effects are observed at the drier end of the gradients and are not yet quantified (Grossiord et al. 2014; Bauhus et al. 2017).

8.4.7   Regeneration

One of the most important practices to increase species richness is the choice of regeneration cut or the silvicultural system, respectively. Here, the future species composition of the forest can actively be changed by replacing tree species and/or tree individuals sensitive to climate change with trees of native or introduced species and/or species' provenances that are potentially better adapted to future climate conditions (called active adaptation; Martín-Alcón et al. 2016; Bolte et al. 2009). Examples for this strategy are the ongoing conversion of pure Norway spruce stands into mixed stands or silvicultural measures aiming at replacing species such as Norway spruce by other species of comparable economic value (e.g., Douglas fir). In Germany, the Bavarian State Department of Environment, Health and Consumer Protection published a regional climate program in November 2007 that includes an example for the application of the "active adaptation" concepts on species level. It is planned to convert about 200,000 ha of pure Norway spruce forests by 2030 in areas where a high risk of drought damage is assumed to less sensitive mixed forests, predominantly with European beech and oak (Stmelf 2018).

The concomitant natural establishment of diverse species can be controlled by creating large variations in light conditions, allowing both light-demanding and shade-tolerant species to regenerate (e.g., group selection or irregular shelterwood in combination with strip cuts). In young growth originating from natural regeneration, enrichment planting is a valuable practice to introduce additional species. Once young trees are established, species richness can be maintained by appropriate tending measures, such as precommercial thinning or thinning. Especially rare species or species with low competitiveness, in particular, if they are adapted to a warmer and drier climate, have to be released in this case (Brang et al. 2008). Finally, the successful establishment of species-rich stands depends very much on the control of ungulates (Gill 1992; Götmark et al. 2005; Ameztegui and Coll 2015). To achieve the optimal adaptive effect of species mixtures, large monospecific patches should be avoided as well as very intimate mixtures, which usually require high tending investments. Forest conversion encompasses the use of the many native (mostly broadleaved) tree species, and selected exotic tree species, respectively. From tree species trials and recent dendroecological analyses, we know that rare native species and non-native tree species can increase forest resilience in a landscape (Kunz et al. 2018; Vitali et al. 2017). Furthermore, provenances of native tree species from warmer regions of the species' distribution range could enrich forest diversity in mountainous regions. Especially these rear-edge populations (Hampe and Petit 2005) of native species often show desired adaptation traits, such as higher drought stress tolerance compared to provenances from the core distribution area of a species.

In the early stage of conversion, an interesting option to diversify forest stands is currently discussed in Germany. Some authors have recommended integrating early successional species, which seem to be more adapted to the drier site conditions into regular stand management in the tree species portfolio (Lüpke 2009). Early succes-sional species quickly cover bare regeneration sites. They recover nutrients, which otherwise would likely be lost and are valuable elements for enhancing biodiversity (nurse crops). Moreover, there is growing evidence to abandon the practice of a merely local provenance choice. Assisted migration, the planned translocation of provenances and species beyond their natural occurrence range, has the potential to ensure that provenances or species are adapted enough to cope with the future warmer climate in the final stage of their development cycle. At the same time, stands must be robust enough to get along with still harsh climatic conditions in the establishment phase when they are young plants. Already well-developed recom-mendations for assisted migration transfer distances have been developed for Douglas fir in the Northwest USA (Sáenz-Romero et al. 2016).

8.4.8   Naturalness

The concept of naturalness has been broadly used in forestry. The naturalness of forest stands can be assessed by different indicators (e.g., Brumelis et al. 2011; Winter 2012); tree species composition is one of the most important. The natural-ness of tree species composition for a given forest site is estimated by comparison between current tree species composition and the natural tree species composition of forest stand, which is a part of potential natural vegetation. Due to climate change, natural vegetation may change over several decades (e.g., Hickler et al. 2012).

The analyses of forest stands in the Dinaric mountain areas (Bončina et al. 2017) showed that the alteration of the natural tree species composition of forest stands is primarily the result of forest management and past land-use, conditioned either by topography or accessibility of forests. The portion of Norway spruce increased due to past forest management. A higher level of alteration of natural tree species com-position of mountain forests significantly increases the susceptibility of forest stands to natural disturbances – mainly windthrows and insect outbreaks. Therefore, sanitary felling can be a few times larger than in stands with natural tree species composition (e.g., Pasztor et al. 2015; Bončina et al. 2017).

In general, for the introduction of new, exotic tree species and provenances, it is suggested to follow the order: (1) species that are already adapted on a larger scale in the planting region and tested non-autochthonous provenances, then (2) new spe-cies with knowledge on their behavior but no adaptation yet, and finally (3) com-pletely new species (Spathelf and Bolte 2020). Currently, only a few forest owners have started to plant nonnative tree species other than Douglas fir, red oak, and grand fir on a larger scale. Nevertheless, around 10–20 "new" species are in the search of forest research institutes across Europe. Existing trials with non-native species are currently evaluated and new trials established (de Avila and Albrecht 2017; Brang et al. 2016; Metzger et al. 2012).

8.4.9 Introduced Tree Species

Tree species have been deliberately moved by humans for as long as humans have been cultivating land for food, fuel, and fiber. Indeed, it is highly probable that tree species were inadvertently moved by our hunter-gatherer ancestors, just as other primates do today (Chapman 1989). Long before the development of countries with borders and associated concepts, such as nativeness, immigration, introduction, or invasiveness, humans travelled and traded and the only considerations were what thrived where, and what value it had as a product. For trees, the main considerations would have been fruit and nut production, foliage and bark for animal fodder, firewood potential, and use as a building material.

As concepts of forest management developed, the choice of tree species for timber production has become increasingly sophisticated. It may even be that tree species selection was the first conscious silvicultural decision? Originally it is probable that some native tree species were preferred by foresters; for example, oak has long been advocated in Britain for ship-building (Evelyn 1664; Fisher 1763). Where one tree species is favored, others inevitably decline in abundance. In the past, when there was no concept of genetic diversity or origin, although native tree species might have been selected for planting, the seed itself might have been introduced from another country.

It is difficult to trace the widespread use of nonnative (also known as exotic or introduced) species in plantation forestry, but it is always likely to have been most prevalent in countries like the UK with low native tree species richness. For example, at least one introduced tree species, European larch (*Larix decidua*), has been being planted in the UK since the mid eighteenth century when medals and cash prizes were awarded to those who planted most trees by the Royal Society for the Encouragement of the Arts. In their first full transactions published in 1783, the summary of the activities since the inauguration in 1754 showed that a sum of 50 £ had been paid alongside the award of 45 gold and 14 silver medals for the "encouragement of planting to raise Timber" from a list of trees, including oak, but also larch (Anon 1783). Larch would have been included on the list, because softwood timber was considered best for ship's masts and the UK has only three native conifers, juniper (*Juniperus communis*), yew (*Taxus baccata*), and Scots pine (*Pinus sylvestris*), of which only one, Scots pine, can grow sufficiently straight and tall to be used as a ship's mast.

In the early nineteenth century, the great plant hunters, such as David Douglas, sent new coniferous tree species back to Britain, from the Pacific North-West, where the climate is similar to the Atlantic North-West of Europe. It was soon noted how fast and straight these species, particularly Sitka spruce (*Picea sitchensis*) and, of course, Douglas fir (*Pseudotsuga menziesii*) (Savill 1991) grew. Consequently, they were soon widely planted, not only in the UK but on suitable sites throughout northwestern Europe. Sitka spruce now comprises more than a quarter of all forest trees in the UK (Forestry Statistics 2019).

In recent years, the recognition of the role of trees, woods, and forests in combating climate breakdown has led to an appreciation that introduced tree species may have a role to play in climate mitigation. Specifically, if introduced trees grow faster than native species, they are considered to sequester carbon faster in the forest via net photosynthesis. This means that they have faster biomass accumulation to provide a carbon substitution benefit from the forest sooner (as wood fuel, or by replacing building materials, such as concrete and steel that have higher carbon footprints). Whether faster-growing trees or more frequent harvests of biomass provide better climate mitigation as timber density, not just volume needs to be quantified, carbon transfer via roots into the mycorrhizae and soil needs to be measured and the effects of more frequent harvests on soil disturbance need to be included.

The concept of planting introduced tree species as a silvicultural treatment for improving forest stand adaptation to climate breakdown is novel. It has been recognized that long-lived, slow to reproduce, heavy seeded plants, including many native tree species, are unable to rapidly adapt to climate change by moving or adapting, so increased tree mortality and associated forest dieback is projected to occur in many regions over the twenty-first century (Field et al. 2014). Consequently, although the native tree species in a given country may be adapted to survive the current pests, diseases, and abiotic threats they face, they may not be resistant and resilient to future threats. As tree species in certain locations may be adapted to climatic conditions that are similar to the ones predicted to be faced in others, it can be argued that to maintain a forest structure, for commercial timber production and other ecosystem services, but also as a habitat/ecosystem for biodiversity, the introduction of tree species likely to thrive in the future climate is justified (Forestry Commission 2020).

The novel argument for the introduced tree species, as a silvicultural treatment to help forests adapt to climate breakdown and thus maintain the delivery of ecosystem services, including climate mitigation, is controversial. However, a CSF approach means putting climate adaptation and mitigation first among multiple sustainable forest management objectives and all options need to be considered (Bowditch et al. 2020). Research into the effectiveness of this approach is needed and indicators need to be developed to guide if, how, and where this is viable. For example, in commercial plantation forestry with introduced tree species, it is not a great issue to introduce others. However, in the current plantations of native species, it may need more careful consideration. In our most pristine native woodlands, introduced tree species may be viewed as too damaging to their integrity.

8.4.10 Deadwood

In forest ecosystems, deadwood influences the nutrient and water cycling, humus formation, carbon storage, fire frequency, natural regeneration and represents a crucial component of forest ecosystems for maintaining and improving forest biodiversity. Decaying wood, such as logs, snags and stumps, as well as rot holes, dead limbs and roots, heart rot and hollowing in living ancient or veteran trees, all of them are habitats for the specific species of fungi, flora, and fauna (Humphrey et al.

2004). Thus, the deadwood volume narrowly meant as the coarse woody debris (CWD), that is, logs and snags, has been selected as the main Pan-European SFM indicator regarding biodiversity, and it is also one of 15 main indicators of biodiversity as proposed by European Environmental Agency (Humphrey et al. 2004; Merganičová et al. 2012)., However, the ancient and veteran trees in all forests also are of key importance for rare and threatened saproxylic species, but, unfortunately, they arenot used for biodiversity monitoring (Humphrey et al. 2004).

Deadwood constitutes habitats for many species of cryptogams, such as bryophytes, lichens, and fungi (Humphrey et al. 2002; Lonsdale et al. 2008; Stokland and Larson 2011; Persiani et al. 2015; Preikša et al. 2015), invertebrates like saproxylic beetles (Martikainen et al. 2000; Franc 2007; Müller et al. 2008; Lassauce et al. 2011), as well as amphibians, birds, and mammals (Merganičová et al. 2012).

Wood-decaying fungi are essential for the functioning of forest ecosystems. They provide habitat for many other deadwood-dependent organisms and enable the regeneration of forests. There are plenty of examples of enhanced survival of seedlings of various forest tree species (mainly conifers) occurring on decaying deadwood (Lonsdale et al. 2008). To support the decaying fungi species of varying requirements, a wide range of CWD of different sizes and stages of decay is necessary (Lonsdale et al. 2008).

All types of deadwood are a substrate for the development of rare cryptogam species. The intermediate decay stages are extremely important for fungi, while bryophytes or lichens do not show such a clear preference. The highest number of cryptogam species is found on the deadwood of Common ash, English oak, and Norway spruce, while deadwood of other tree species hosts less than half cryptogam species (Preikša et al. 2015).

Throughout Europe, saproxylic beetle species have been identified as the most threatened community of invertebrates (Davies et al. 2008). Species richness in saproxylic beetles has a significant positive correlation with the main deadwood variables (Martikainen et al. 2000). It depends not only on deadwood amount, but also on other microhabitat factors, such as the richness of wood-inhabiting fungi, and, for the threatened saproxylic beetles – on the frequency of *Fomes fomentarius* (Müller et al. 2008).

Natural variation of deadwood niches – including decay stages, snag sizes, tree cavities, and wood-decaying fungi species – must be maintained to efficiently preserve the whole saproxylic beetle fauna. To better assess the quantitative relationships between deadwood and biodiversity of saproxylic beetles, apart from the deadwood volume, deadwood type or decay stage should also be considered (Lassauce et al. 2011).

Pieces of evidence have shown that climate change will speed up tree growth and accumulation ending up in a higher stock of deadwood available *in situ* (Mazziotta et al. 2014). However, due to increased decomposition rates, the time the deadwood stock is available for deadwood-associated species will diminish and the carbon stored in deadwood will return to the carbon cycle faster (Büntgen et al. 2019). Disturbances from fire, insects, and pathogens, in particular, are likely to increase in a warming world (Seidl et al. 2017), which could markedly modify the distribution of deadwood across the forested landscapes in time and space. Under such circumstances, it is going to become increasingly challengeable to manage deadwood in a sustainable way. Some authors recommend that the structure and dynamics of old-growth forests are used as a reference system for managed forests (Jandl et al. 2019). Based on modelling results, it was found that continuous cover forestry, based on emulating natural disturbances and leaving 10% of stands uncut with no deadwood extraction, will result in greater dendrobiotic birds habitat quality per unit of current volume increment under climate change (ARANGE 2020). However, it is not clear whether the carbon sink function will decrease or even stop when the forests get into a steady-state of carbon sequestration in biomass and soil organic matter and of carbon loss due to decomposition of deadwood debris and soil organic matter (Desai et al. 2005; Pukkala 2017). In all cases, forest owners should be flexible and prepared to diversify the silvicultural systems across forested landscapes. They will need to follow natural disturbances in a way that will guarantee the presence of enough deadwood to adequately address the various trade-offs between wood biomass production for carbon sequestration, on the one hand, and forest protective functions, on the other.

Forest management should mimic the natural stand dynamics, increasing the number of dead trees and the diversification of the vertical and horizontal tree layers, considering the good potential for restoring and increasing the diversity of saproxylic communities and their associated ecological functions. For monitoring the ecological sustainability of forest management, we must focus on threatened species (Müller et al. 2008).

For strategies to increase deadwood amount in managed forests, the best results will be achieved in areas close to existing reserves or other important habitats (Müller et al. 2008). Research into deadwood dynamics carried out in unmanaged forest ecosystems (Christensen et al. 2005; Persiani et al. 2015) has proved useful as a reference tool to implement rehabilitation criteria in sustainable management, to maintain and increase biodiversity and other ecosystem services provided by managed forests.

Sanitary cuttings, carried out mainly to avoid outbreaks of insect pest populations or to reduce risk of forest fires, are another measures leading to severe restriction of the capacity of managed forest ecosystems to provide habitats for saproxylic species (Humphrey et al. 2004). Since healthy, resistant, and resilient managed forest should, partly, consist of diseased or injured trees (Szwagrzyk 2020), their retention until natural death would allow accumulation of deadwood of various types and sizes, and representing all tree species that grow in a forest, that would create niches for all deadwood depending species.

To improve the status of the deadwood-depending organisms, the managed forests should maintain a long-term continuous provision of greater amounts of dead and decaying wood microhabitats that deadwood-depending organisms require for their survival (Christensen et al. 2005; Davies et al. 2008; Lonsdale et al. 2008). However, no simple deadwood stocking recommendations can be applied, due to the inherent complexity of all the stand, site, and management factors that drive deadwood dynamics (Persiani et al. 2015).

8.4.11 Genetic Resources

Since the advent of the population genetic studies based on molecular markers, it has been postulated that long-term survival and adaptation of species and populations to the changing environment strongly depends on the high genetic variation accumulated in the gene stock of populations over historical times (van Dam 2002). Natural forest tree populations tend to maintain a high level of genetic diversity along with the distribution range because of the high outcrossing rate, the long life span of individuals that besides preserving their highly heterogeneous genomes can fix beneficial alleles for a longer period (Petit and Hampe 2006). Several acting forces on population-level, however, can shape the uniform distribution of the genetic variation especially at the range periphery of the species where gene flow usually decreases or the environment reaches the tolerance limit of populations. Genetic drift and inbreeding acting at the range margins can cause differentiation by changing the frequency distribution of alleles and selecting population-specific alleles (Hampe and Petit 2005). These selected alleles might be beneficial in the local adaptation on the range margins, but can be also harmful forcing populations to counteract against their fixation (balancing selection). Differences between central versus peripheral populations and the role of the beneficial alleles helping population adaptation have been much discussed in different studies (Gibson et al. 2009; Logana et al. 2019).

Natural or human-induced fragmentation in species' distribution area can increase the effect of marginality and can cause isolation within the species' range, not only at the range periphery. Fragmentation increases genetic divergence and will promote the overall structuring of populations. If gene flow becomes limited among the fragmented sites, the long-lasting drift and inbreeding end up in pauperization of the gene stock causing a higher rate of homozygous individuals. Homozygote excess usually produces limited resilience and lower fitness, impeding population adaptation to the changing environment (Mátyás 2002; Allendorf et al. 2013). All these processes are strongly affecting populations in the time of the ongoing climate change that has an unpredictable impact on the structure of the ecosystems and populations therein.

Forest trees having a large genome and preserving a considerable amount of genetic variation are expected to have high resilience. Moreover, the high phenotypic plasticity allows them to withstand even large environmental fluctuations during their lifetime. However, researchers have expressed their concerns also. Tree species with long generation time due to the long-life span of individuals might be unable to react to the fast changes experienced due to the ongoing climate change. These events are too rapid relative to the tree's age and populations may not have adequate time to adapt or to disperse and colonize the newly available habitats. As many species are unlikely to migrate fast enough to track the rapidly changing climate in the future, their standing adaptive variation will likely play an increasingly important role in their response (Jump and Peñuelas 2005; Mátyás and Kramer 2016).

Studies have shown that the effect of climate change acts strongly at the species' ecological limits, as tolerance to climatic extremes is genetically determined. For example, European beech populations at the lower xeric limit of the species' distribution are more exposed to the impacts of climate change; hence, the decline of population in these territories has been anticipated (Mátyás and Kramer 2016). In turn, more recent studies have shown that beech, currently dominating lower elevations in mountain sites, has a high potential to advance to higher altitudes, where it can perform better in mixed stands than in monospecific stands (Pretzsch et al. 2020c).

Since the first utilization of forests ecosystems, humans intentionally or unconsciously have altered the gene pool of forest tree species (Buiteveld et al. 2007). The decrease in forest area size, habitat degradation, change in species composition, forest plantations, or tree breeding, all these, have influenced natural gene stock of forest communities. Most European forests are also affected by historical forest management and despite the intention of the last decades, to preserve sustainable silviculture, the long history in forestry has left strong imprints in the genetic makeup of forest tree species.

Genetic studies in forest tree populations using a large stock of genetic markers make it possible to reveal all these historical processes, to evaluate the composition and quality of the forest gene stock, and foresight future ability of communities to adapt to changing environmental conditions.

To mitigate the effects of the changing climate in forest ecosystems and to conduct CSF primarily, it is important to have deep insights into the genetic constitution of species and their populations. High levels of "standing genetic diversity" in populations is a prerequisite for species to face fast environmental changes as selection and fixation of new adaptive mutations take a comparatively longer time. In contrast to new adaptive mutations, standing variation most probably has already passed through a "selective filter" and might have been formerly tested by selection in past environments (Barrett and Schluter 2008). Selection of the new alleles in tree species with their long-life span needs comparatively more time; thus, lack of adaptation may end in the decline of the functional traits.

Former case studies on the phenotypic variation of forest trees, experiments in provenance trials beginning from the early 1970s, and many common garden experiments provided a large source of data helping in understanding the adaptive behavior of species. Although these were not designed to monitor the effects of climate change, they still provide insights into the aspects of the genetic variation and of the adaptive response to species to the acting environmental forces. A more recent study by use of field trials and modelling tools tried to determine the extent to which four widespread forest tree species in Europe (Norway spruce, Scots pine, European beech and sessile oak) may be affected by the climatic change (Mátyás and Kramer 2016).

To explore the impacts of environmental change on the adaptive potential of trees, functional phenotypic traits need to be assigned to allele composition. This is not always unequivocal as for many traits, there is still limited knowledge, likely because they are regulated by multilocus systems. Thus, a genome-wide scanning of the changes in population genetic diversity based on neutral markers represents another but a more conservative approach (Kramer et al. 2010). However, an increasing number of projects mapping complete genomes of mountain forest tree species (Mosca et al. 2019) and the development of gene-specific primers makes it already possible to identify nucleotide diversity in genes and candidates responsible for the adaptive variation.

To grant forest tree populations the ability to keep an adequate level of genetic diversity, to maintain viability, and to support long-term evolutionary potential, genetic aspects should be embedded in the forest management (Buiteveld et al. 2007). Forest ecosystems will only persist if genetic variation and allele composition of trees are maintained at a high level and this especially holds in view of the environmental changes. Therefore, studies on the genetic makeup of species and populations should be performed before starting the planning of any forest management activity. Moreover, selection and conservation of multiple genetic resources should be also in the focus.

A case study on the genetic variation of pure beech stands along species' distribution range was initiated within CLIMO (COST Action, CA15226). The novelty of this work is the coupling of the genetic data with other empirical measurements within the considered study plots. Dominant trees from 12 study plots were subjected to molecular evaluation based on six nuclear microsatellite markers. The overall high genetic variation of the stands was correlated to local climate variables. Among the genetic indices, the number of alleles and Shannon genetic diversity were shown to be highly correlated with daily temperature and the frequency of frost days (Höhn et al. 2021).

8.4.12 Threatened Forest Species

Because of the extensive studies that had been carried out for years, species diversity turned out to be the easiest aspect to implement among the main biodiversity components (Kraus and Krumm 2013). Species diversity remains a keystone, and the loss of species is the most recognizable form of biodiversity decline.

European Red List of Trees identifies those species that are threatened with extinction at the European level to inform about actions needed to improve their conservation status (Rivers et al. 2019). The list summarizes the results for the assessment of all known native European trees, a total of 454 species, of which 265 (over 58%) are endemic to continental Europe. In common with vascular plants (Bilz et al. 2011), some of the highest levels of endemism are found in the main mountain chains, such as the Alps, Pyrenees, Carpathians, Apennines, Dinaric Mts., and others. The mountain areas also represent the richest centers in Europe (Rivers et al. 2019).

Overall, 191 (42%) European tree species have been assessed as having a high risk of extinction, that is, assessed as critically endangered, endangered, or vulnerable. A further 13 tree species are assessed as almost meeting the criteria for a threatened category, while 216 of them are considered of not being of conservation concern (Rivers et al. 2019).

Each country may produce its national red lists (most often containing more species than in the European red list), which assess the risk of species' extinction within the country borders. The species considered of not being under threat in the European scale locally can be seen as critically endangered or vulnerable. Thus, several countries have developed management or action plans for various species and have legislation in place to protect certain species legally. Some examples of successful initiatives include the Regional Programme of Conservation and Restitution of *Sorbus torminalis* in Poland (Zwierzyński and Bednorz 2012), and the *Zelkova* global action plan (Kozlowski and Gratzfeld 2013).

Forest management can negatively affect several taxa by reducing the number of natural attributes, such as microhabitats bearing trees, deadwood quantity, and its diversity (Lafond et al. 2015). However, the opposite happens: for example, in European beech forests, no loss in vascular plant species restricted to forests occurred over the past 250 years despite forest management (Schulze et al. 2016). On the other hand, some field studies highlight a positive effect of forest management on the diversity of such taxa, as bird species related to open spaces and species of the understory vegetation (Lafond et al. 2015).

In uneven-aged forests, large tree retention has a positive effect on several structure and biodiversity indicators. In particular, it seems efficient to compensate for the negative impacts of increased harvesting intensity by limiting the decrease in tree sizes diversity. Through the analysis of compensating effects, it has been revealed the existence of possible ecological intensification pathways, that is, the possibility to increase management intensity while maintaining biodiversity through the promotion of nature-based management principles, that is, gap creation and retention measures (Lafond et al. 2015).

Nowadays, climate change represents an important driver of species extinction, the importance of which is increasing, when acting in synergy with habitat destruction and fragmentation. Climate change will cause many tree species to lose a part of their current habitat but will also enable them to colonize new habitats (e.g., del Río et al. 2018). As natural migration is often slow, forest managers are examining the option of assisted migration (Martín-Alcón et al. 2016; Gömöry et al. 2020). A main premise of SFM is that silviculture based on patterns and processes found in old-growth forests will maintain the provision of important habitats for biodiversity (Schütz et al. 2016). Although in some parts of Europe is implemented "close-to-nature management" (Schütz 1999; Brang et al. 2014; O'Hara 2016), the resulting forests lack the diversity in composition and structure of forest ecosystems that are driven by natural succession and dynamics so characteristic to old-growth forests (Kraus and Krumm 2013).

8.4.13 Protective Forests (Soil, Water, and Other Ecosystem Functions)

Protective forests as one of SFM indicators are presented here in the broader context, that is, including all FRA 2020 categories as described in Global Forest Resources Assessment (FAO 2020) for European countries, except Russian Federation and Turkey. Thus, apart from protective forests primarily designated for protection of soil and water, Table 8.2 contains data on areas primarily designated for other forest ecosystem functions, such as conservation of biodiversity, social services, and of multiple use. Such an approach enables a general understanding of the role of forest in maintaining also other forest functions.

Any undisturbed forest has a capacity to effectively sustain the maintenance of biological diversity and protection of soil and water, since these capacities/functions are integrated with natural processes ongoing in forest ecosystems. Forest functions as such, reflect a forest ecosystem capacity of sustaining expected people's demands, while forest ecosystem services account for the provision of requested benefits (Lesiński 2012). These benefits are of the basic value for peoples' welfare.

Forests primarily designated for various management objectives (FAO 2020), with exception of production, occur in as many as 30 out of 36 countries and they all together cover 42.55% of the total forest area in Europe (Table 8.2). The representativeness of the above FRA 2020 categories in individual countries and their average share in the total forest area in Europe are as follows:

– Multiple use: 18 countries; area – 19.50%
– Protection of soil and water: 21 countries; area 10.70%
– Conservation of biodiversity: 26 countries; area 10.03%
– Social services: 19 countries; area 2.32%

In 12 countries' data in the matter, production has not been at all mentioned as one of primarily designated forest management objectives, and in six of them, forests have not been designated for any other objective, either. The latter have been described either as other, none, or unknown (Table 8.2).

When it comes to distinguishing forest functions, two main approaches might be considered – the zoning approach and the integration approach (Bončina 2011b). The zoning approach takes place mainly in scarcely populated countries/regions, large forest areas, or mountain ranges (Lesiński 2012). Protective forests are typical examples of such an approach. In Europe, they cover large areas in the mountains of Switzerland, Austria, Romania, Moldova, and Norway (Table 8.2). The integrated approach suits much better in densely populated countries/regions (Lesiński 2012). Integration approach results in multifunctionality that is achieved by combining various objectives at a larger scale. France, Belgium, Netherlands, and Spain are very good examples of implementing such an approach (Table 8.2).

Table 8.2 Primary designated forest management objectives 2020 expressed by the percent of the total forest area in the country

Country	Production (%)	Protection of soil and water (%)	Conservation of biodiversity (%)	Social services (%)	Multiple use (%)	Other/none/ unknown (%)	Total forest area (1000 ha)
France	–	–	–	–	100.00	–	17,253.00
Belgium	–	–	1.80	–	98.20	–	689.30
Netherlands	0.82	–	39.31	–	59.87	–	369.50
Poland	–	20.29	10.29	10.77	58.65	–	9483.00
Slovakia	22.88	18.23	2.33	8.44	48.14	–	1925.90
Spain	19.14	17.63	22.18	–	41.05	–	18,572.17
Slovenia	47.38	19.17	0.90	2.05	30.50	–	1237.83
Bulgaria	38.53	9.38	18.08	5.78	28.23	–	3893.00
Iceland	33.30	29.97	0.33	12.85	23.38	0.17	51.35
Moldova	–	58.09	–	19.63	22.28	–	386.50
Denmark	80.23	–	5.08	–	14.69	–	628.52
Croatia	68.80	12.57	2.87	1.41	14.35	–	1924.98
Sweden	70.00	–	18.30	–	11.70	–	27,980.00
Lithuania	71.66	6.13	9.40	2.68	10.13	–	2201.00
Norway	46.55	37.84	6.02	0.16	9.43	0.21	12,180.00
Estonia	72.85	4.55	13.63	–	8.97	–	2438.01
Serbia	65.65	21.96	6.67	2.93	2.13	0.66	2722.65
Portugal	–	–	–	–	0.10	99.90	3312.00
Austria	61.50	37.27	–	1.23	–	–	3899.15
Albania	79.60	16.48	3.92	–	–	–	789.00
Latvia	76.31	6.41	15.26	2.02	–	–	3410.79
Czech Rep.	73.78	9.45	9.10	5.28	–	2.39	2677.09
Belarus	49.91	16.31	15.56	14.71	–	3.51	8767.60
Hungary	49.55	9.28	24.69	0.95	–	5.35	2053.01

(continued)

Table 8.2 (continued)

Country	Production (%)	Protection of soil and water (%)	Conservation of biodiversity (%)	Social services (%)	Multiple use (%)	Other/none/ unknown (%)	Total forest area (1000 ha)
Switzerland	27.51	54.60	8.43	2.65	–	6.81	1269.11
Ukraine	37.18	25.87	14.55	14.96	–	7.46	9690.00
Romania	42.71	36.30	4.63	–	–	16.36	6929.05
Ireland	44.38	–	5.55	0.07	–	50.00	782.02
Finland	–	–	12.64	2.36	–	89.00	22,409.00
Italy	3.43	–	0.97	–	–	95.60	9566.13
Germany	–	–	–	–	–	100.00	11,419.00
UK	–	–	–	–	–	100.00	3194.06
Greece	–	–	–	–	–	100.00	3901.80
Bosnia & Herzegovina	–	–	–	–	–	100.00	2187.91
Northern Macedonia	–	–	–	–	–	100.00	1001.49
Luxembourg	–	–	–	–	–	100.00	88.70
Total – area	59,101.37	21,476.84	20,147.61	4664.84	39,132.60	56,161.58	200,684.84
Total – %	29.45	10.70	10.03	2.32	19.50	28.00	100.00

Order of countries by Multiple use (descending) and by Other/None/Unknown (ascending)
Source: Reports for individual European countries; FAO (2020)

Protective forests are forests that have as their primary function to prevent soil erosion, to preserve water resources and protect infrastructure, people and/or their assets against the impact of natural hazards. The main tool used by these forests are standing trees, which act as obstacles to downslope mass movements such as rock falls, snow avalanches, erosion, landslides, debris flows, and floods. The protective effect of these forests is ensured only if the silvicultural system used and any natural disturbances that occur leave a sufficient amount of forest cover (Brang et al. 2006). Thus, the continuous cover approach maintaining uneven-aged tree-stands seems to be the best solution (Mason et al. 1999; Pukkala and Gadov 2012).

8.4.14 Slenderness Coefficient

The slenderness coefficient (H/D ratio) is commonly used in studies of the resilience of trees and forest stands to the destructive activity of wind and snow (Abetz 1976; Burschel and Huss 1997). This is a synthetic indicator describing the shape of the tree trunk (the stem taper, the opposite of slenderness). The slenderness coefficient is calculated by dividing the height (H) by the diameter at breast height (D). The higher the diameter at breast height of a tree with the same height (lower H/D values), the stronger the force necessary to bend the trunk. A low slenderness coefficient is found for a longer crown, lower center of mass, and better-developed root system due to the large space for growth. Free growth of the crown facilitates the increase of growth of the diameter at breast height and the increase of the stem taper (Petty and Worrell 1981). Expansion of the space for the growth of the crown also limits their asymmetric development (Petty and Swain 1985; Valinger et al. 1994), with trees with regular crowns less susceptible to leaning or swinging (e.g., due to an asymmetric snow load). A torsional moment is also observed less frequently. The tree roots in a loose stand tend to be better anchored in the soil, leading to increased resilience to tree damages (Nielsen 1995). Peltola et al. (1997) explain that with a low slenderness coefficient, snow is dislodged from the tree crowns by gusts of wind, and these two damaging factors do not accumulate.

In some of the publications, the authors use the slenderness coefficient as a measure of tree/stand stability. They formulate conclusions and assessments concerning the rules of tending stands, assuming that lower H/D values will result in higher resistance to the wind (snow) (Wang et al. 1998; Wilson and Oliver 2000; Castedo-Dorado et al. 2009; Vacchiano et al. 2013; Meng et al. 2017). A different methodological approach is the creation of models allowing the prediction of the fact or probability of wind damage emergence. In these models, the slenderness coefficient is one of the many variables. An analysis of efficiency, measures of fit, prediction, and classification capacity of these models indicates that the slenderness coefficient, as a single variable, is frequently of relatively limited value as an explanatory variable. The studies by Pukkala et al. (2016) show the slenderness coefficient as useful, but only in interaction with specific basal area and diameter at breast height values. In the work by Martín-Alcón et al. (2010), the slenderness coefficient influenced the prediction of the percentage of damaged trees in a stand, but only when divided by the basal area. Slenderness alone is not a good indicator of tree stand stability, one also requires a factor indicating mutual tree support within a forest stand (Schütz et al. 2006; Schelhaas et al. 2007; Martín-Alcón et al. 2010). The work by Albrecht et al. (2012) concludes that a variable tree height should be used in conjunction with H/D. The usability of the slenderness coefficient as a predictive measure was also critically assessed in studies covering historic data. Oliveira (1987) and Schütz et al. (2006) also do not recommend the usage of the H/D ratio as a single variable to predict resistance to wind damage. In the work of Díaz-Yáñez et al. (2017), the best prediction models use the slenderness coefficient to clarify the probability of damage at less than 10%. Tree height covered the majority of the probability of damage emergence.

Due to the common usage of the slenderness coefficient as the measure (indicator) of resistance against wind damage, many authors indicate its desired/critical values. In Germany, Abetz (1987) recommended a slenderness factor of ca. 80 as suitable for Norway spruce. Burschel and Huss (1997) suggest the following value scale for coniferous species: very unstable (H/D > 100), unstable (H/D 80–100), stable (H/D < 80), alone trees (H/D < 45). The values given by other authors are similar (Johann 1981; Cremer et al. 1982; Rottmann 1986; Becquey and Riou-Nivert 1987; Lohmander and Helles 1987; Peltola et al. 1997; Wilson and Oliver 2000). Summarizing their research, Wilson and Oliver (2000) conclude that there is no single H/D value guaranteeing stability. It depends on the wind strength. We are not able to eliminate the hazard of wind, and a lower H/D value leads to a reduction in value or production costs (pruning). Skrzyszewski (1993) concluded that promotion of trees (European spruce) with a low slenderness coefficient (below 80) during thinning leads to the emergence of tree stands with trees that have branches extending low over the ground, with a large share of ingrown knots and showing a very tapered stem and broad rings, resulting in low mechanical strength and reduced wood durability in coniferous species. Safe strategies are sparser planting spacing and/or high-intensity cuttings at a young age (sapling and pole stage), and in neglected forest stands – very low-intensity cuttings during many entries. An alternative may be growing all-age (selection) stands (Dobbertin 2002; Griess and Knoke 2011; Jaworski 2013; Hanewinkel et al. 2014; Pukkala et al. 2016).

The slenderness coefficient is the result (effect, derivative) of a specific vertical structure of a stand, its age (height), and density as well as the silvicultural treatments executed in the past. The quoted publications indicate that the slenderness coefficient should be analyzed in association with other variables (site conditions, soil humidity, exposure, slope, altitude and age as well as the height of neighboring tree stands).

8.5 Silvicultural Treatments Improving Stand Mitigation

8.5.1 Growing Stock

Keeping the average growing stock low is mentioned as one of the principles for climate change adaptation in forest management (Brang et al. 2014). This might be especially true in the case when forest fires cause stand damages, which is typical for Mediterranean forests. However, the high growing stock of forest stands can increase damage susceptibility also in mountain areas, especially in the case of wind throws or insect attacks, and much less for ice or snow breaks (Jalkanen and Mattila 2000; Klopčič et al. 2009).

At present, the average growing stock in European forests is 163 m³/ha, ranging from 10 m³/ha in Iceland to 352 m³/ha in Switzerland (Forest Europe 2015). During the last decades, the rate of annual felling to wood volume increment has been relatively stable and remained under 80% for most countries in Europe. Such a ratio of wood utilization allows the forest stock to increase being a right action within mitigation of climate change. However, the ratio of harvesting to increment is assumed to increase in the nearest future since the demand for woody biomass as a renewable energy source is expected to increase (Forest Europe 2015). On the other hand, the higher growing stock as well the age of forest stands may increase the occurrence and the severity of insect outbreaks in forests prone to the pest damages (Pasztor et al. 2015). The crucial issue is to keep the average value of growing stock on a certain level allowing to control trade-offs between climate change mitigation actions focused on growing stock increase (increase carbon stores) and adaptation potentials focused on wood harvesting enabling to create multilayer and uneven-aged stands (D'Amato et al. 2011). For example, Pretzsch et al. (2014) and Schütz (2003) suggest to strive to achieve the average growing stock amounted to 300–400 m³/ha on fertile site conditions and 250–350 m³/ha in medium site conditions in mountain mixed-species forests.

8.5.2 Carbon Stock (Soil)

Soils are the largest carbon pools of most of the mountain forest ecosystems and are more stable and less exposed to sudden fluctuations than trees (Scharlemann et al. 2014; Achat et al. 2015). Nevertheless, forest management can affect soil organic carbon (SOC) stock and accumulation rate (Mayer et al. 2020; Tonon et al. 2011). The extent to which different silvicultural systems, harvest intensities (percentage of original biomass), frequency, modality (tree stem or whole tree collection), and the degree of mechanization might impact SOC is still debated due to the huge number of external factors (climate, species composition, litter quality, type of soil, and relative clay content) involved in the ecosystem response (Hoover 2011). In this context, although conflicting results can be found in literature, often linked to the different timescale of the experiments, some general indications can be provided to maximize the SOC stock of mountain forests. First, all forest management strategies with a positive effect on the ecosystem's net primary productivity have to be considered as putative strategies to increase SOC accumulation through the increased production of leaf, root, and woody litter. An example is a conversion of pure to mixed forest in central Europe (Aguirre et al. 2019; Pretzsch et al. 2020b; Torresan et al. 2020) with evident positive effects in terms of productivity and resilience to climate changes. Recent evidence shows that intensive forest harvest, based on the collection of tree stems and logging residues, has a deleterious effect on SOC content resulting in a consistent and lasting loss of carbon that affects both the surface and mineral soil layers, suggesting that residues management is one of crucial aspect to take into account in the context of the CSF (Achat et al. 2015). On the other hand, it has been shown that traditional forest harvest has a temporary effect limited to the forest floor without any serious impact on the mineral deeper soil layers (Achat et al. 2015; Mayer et al. 2020).

The transition from the traditional age class forestry to the continuous cover forestry toward the conversion of even-aged to irregular or uneven-aged forests is a further indication that emerged from the literature as a potential option with a positive long-term effect on soil and forest carbon storage (Seidl et al. 2008). This structural change can be reached through the application of selective harvesting, such as single-tree selection or small-group selection cutting. However, this transition is not possible everywhere as it requires local technical skills, suitable climate conditions to support natural regeneration under the typical uneven-aged forest microclimate, and a flexible timber market. In agreement with the ECCP-Working Group on Forest Sink (2003), 30 million hectares of forests at the European level could be converted to continuous cover forestry with an important carbon gain at the regional level. Where social, technical, or ecological reasons make the conversion to the continuous cover forestry inapplicable, the elongation of the rotation period of even-aged forest can be considered as a possible strategy to promote carbon sequestration. Indeed, the time that the system needs to recover to the pre-disturbance SOC content can largely vary according to climate, soil conditions, and magnitude of the perturbation. Therefore, the rotation period in even-aged forestry should be at least longer than the recovery time. An additional point supporting the elongation of the rotation period is the evidence that forest ecosystems are far to be carbon-saturated, since several old-growth forests are still accumulating carbon in the different compartment, soil included, at different altitudinal and latitudinal belts with an important accumulation rate (Gunn et al. 2014; Schrumpf et al. 2014; Vedrova et al. 2018; Badalamenti et al. 2019; Keeton 2019).

8.5.3 Roundwood (Timber Products)

One of the main effective measures to mitigate climate change is to retrieve carbon from the atmosphere and store it as long as possible in roundwood as a timber product resulting from the stand management expected by society (Kauppi et al. 2018). The roundwood timber can be used for many purposes in building construction and human everyday life. The best climate change mitigation effect is achieved when the wood is converted into long-lived products and where the same wood unit is used in several, successive product cycles (Jandl et al. 2018). Wood production belongs to the component of the CSF concept concerning mitigation related to sustainably increasing forest productivity and income and relies on using wood resources sustainably to substitute nonrenewable, carbon-intensive materials (Nabuurs et al. 2017; Verkerk et al. 2020). To produce the best quality and sufficient amount of different assortments of roundwood timber, many different optimal silvicultural measures depending on the species composition, site conditions, and societal expectations can be implemented. The most important silvicultural treatments are thinnings, both precommercial and commercial, which systematically applied in the stand can lead to the rotation age with high-quality crop trees. Some economic analyses have shown that properly made precommercial thinning often is the most rewarding long-term investment that can be made during the silvicultural prescription throughout a stand's life (Smith et al. 1997). The various methods of commercial thinning focused on the production of best timber products at the end of rotation are undertaken in Europe. The common thinning rule is the selective promotion of the best trees (taking into account vitality, social position, quality, stability) in the stand using the procedure based on Assmann concept (1961) of maintaining stand basal area allowing to maintain the growth at the level of at least 95% of the maximum periodic annual volume increment. Other models of thinnings are mostly adapted to the main tree species in the stand and their selection is also determined by the final production objective (target timber assortments) of stand management. Here we are presenting some examples of particular methods applied in stands consisting of main tree species: *Pinus sylvestris* stands – thinning leading to construction timber and/or valuable wood (Burschel and Huss 1997); *Picea abies* stands – the selection of thinning procedure based on the stand age, method of stand formation, origin, and stability (Burschel and Huss 1997); Abetz's model (1975) of thinning with a selection of a certain number of future best trees; *Abies alba* stands – selection of the best crop trees and selective thinning (Korpel 1975), differentiation thinning leading to the forest with selection structure (Schütz 1989; Korpel and Saniga 1993); *Quercus* stands – three methods to produce veneer, valuable, and sawn timber (Burschel and Huss 1997); *Fagus sylvatica* stands – selective thinning (Burschel and Huss 1997), qualitative group thinning (Kato and Mülder 1983; Röhring et al. 2006), thinning procedures by Freist and Altherr

(Burschel and Huss 1997). However, most of them were developed in the period when climate change was not an important issue, so they need to be modified according to the current state of knowledge and predicted climate change effect on timber production.

8.6 Application of Simulation Models for Development, Testing, and Improving Silvicultural Prescriptions

8.6.1 The Role of Models in Forest Science and Practice

Under changing environmental conditions, models become particularly important as they may provide information about the efficacy regarding the mitigation and adaptation potential of new silvicultural measures, which are so far not at all or not sufficiently covered by experiments.

For instance, experiments addressing the effects of thinning in mixed stands or the transition from even-aged monocultures to uneven-aged mixed stands are still rare; so, the potential of mixing and transitioning on the resistance, resilience, or recovery under drought stress may be explored by model scenarios.

The main role of models is the integration of existing knowledge for better understanding and regulation of well-analyzed ecosystems under stable conditions; however, in the view of unknown future development, their contribution to exploring future management strategies by scenario analyses gains an increasing importance. Ecophysiologically based models derive the system behavior from mechanistic relationships but often lack an interface to silvicultural management. Management models statistically reflect the tree stand growth observed on experimental and inventory plots, cover the practical relevant dendrometric variables but are less flexible regarding changing growing conditions. By deriving relationships between growth and water supply with ecophysiological models, statistically modelling such relationships and integrating them into management models pros and cons of both model approaches may be combined as shown by Schwaiger et al. (2018a, b).

Silvicultural prescriptions can serve as guidelines for tree and stand regulation. Their development starts with a participative target development (Pretzsch et al. 2008). Their quantitative development is based on expert knowledge, existing experimental plots, and inventory data and is also often based on scenario analyses with models. Models may reveal which treatment variants match the best with the ecosystem functions and services aimed at by the defined target state. In this case, models are used as virtual experiments for testing the long-term consequences of various treatment options. By comparing the model outcome with the target, the most promising and stable variants may be selected and formulated as easy-to-apply rules for use in forest practice.

8.6.2   *Models as a Substitute for Missing Experiments*

Strong stand density reductions, thinning in mixtures, transitioning from even-aged monocultures to uneven-aged mixed stands, gap cuts, or inner edges may be caused by silvicultural measures of mitigation or adaptation to drought and climate change. The effects of such measures on the long-term are unclear, as respective long-term experiments are hardly available (Pretzsch et al. 2020a).

However, spatially explicit individual tree models, parameterized based on long-term experiments with distant-dependent prognosis algorithms, are suitable for assessing the reactions of such silvicultural measures. Especially models that are based on basic rules such as allometric principles, self-thinning lines, density competition relationships, or dose–response functions for quantification of the relationships between growth and environmental conditions may be helpful. They interpolate or extrapolate beyond the range of stand and structure conditions covered by the parameterization data sets.

Starting with defined stands and environmental conditions, various treatment options may be tested by simulation runs. Link functions to various ecosystem services enable the assessment of the long-term consequences of various treatment options and their match with the target settings by management. The treatment options that meet the best the various criteria of the defined target are an appropriate basis for further development of practical guidelines.

Certainly, additional aspects, for example, of forest utilization, wood quality, and ecological and socioeconomic impact, need to be considered; however, the effects of the silvicultural measures on the natural production represent an essential pillar and step in the process of guideline development.

8.6.3   *Models as Decision Support in the Case of an Unclear Future Development*

In the last few decades, forest science and forestry were faced with the environmental impact on forest ecosystems such as acid rain, increasing atmospheric ozone concentration, and eutrophic deposition as well as climate change. There is hardly any previous experience from experiments or monitoring on how forestry may mitigate or adapt to such environmental changes. Chamber experiments are restricted to small trees and field experiments are costly and very long-lasting; they are important but not sufficient to quickly provide forest management with recommendations for decision making under environmental stress.

Model predictions and model scenarios are often the only alternatives for getting decision support, but their results should be interpreted and applied with due care for the following reasons: Mechanistic models that are based on general relationships between environmental conditions and ecophysiological processes. As they are hardly parameterized for the special stress conditions, they may provide false estimations of the stress reactions, acclimation, and adaptation reactions of forest ecosystems (Bréda et al. 2006). Statistical models that apply the "space for time" approach and derive the temporal reactions on different environmental conditions from the plant's reactions along spatial gradients may underestimate the time trees need to adapt to or recover from exposition to stress events. Individual tree models that estimate the tree's growth response on tree size and competition within the stands may be parametrized for even-aged stands where most trees have a similar history. If their functions are used to predict tree growth in mixtures, in heterogeneous stands or regeneration phases with gaps and edge trees, there may be false estimations as the trees are no longer growing similarly but have much more diverse courses of growth than in monospecific and even-aged stands. Thus, their growth courses can hardly be predicted just from their present size and competition. More information about past development, for example, about the inner and outer traits, may be necessary for appropriate predictions. Improved models are on the way (Sievänen et al. 2000; Pretzsch et al. 2002; Rötzer et al. 2012) and may be used for silvicultural scenario development for mitigation and adaptation measures (Hilmers et al. 2020).

8.6.4   Model Scenarios to Fathom Out the Potential of Adapting Forest Stands to Climate Change by Silvicultural Measures

Model scenario calculations may help to better adapt forest ecosystems to environmental changes and stress. We see mainly five lines of models' applications for the development of climate-smart silvicultural measures.

(i)   Models, mainly ecophysiologically based models that consider the water cycle, may contribute to developing thinning strategies for mitigation and adaptation to drought stress (Rötzer et al. 2012).

(ii) Models may contribute to test the climate smartness of so far not well-known neglected domestic and introduced tree species. Of special interest is whether species, for example, such as sorb tree (*Sorbus domestica*) or Douglas-fir (*Pseudotsuga menziesii*) or Turkey oak (*Quercus cerris*), may be suitable substitutes for more drought-susceptible tree species.

(iii) Tree species mixing by mixing from the beginning on or underplanting may improve the stand growth stability due to risk distribution, improved recovery, or resistance to drought due to interspecific facilitation (Forrester and Bauhus 2016).

(iv) As there are still only a very few long-term experiments on transitioning from even-aged monocultures to uneven-aged mixed stands (Pretzsch 2019), models may support the derivation of suitable scenarios concerning multiple ecosystem services. Of special interest are the effects of a transition to continuous cover forestry regarding the reduction of damages and growth losses by wind and storm, bark beetle, and drought stress.

(v) Reduction of the rotation length or introduction of midrotation mixture may reduce risks and increase productivity due to adaptation to the species-specific growth rhythms and damage susceptibilities and temporary tree species mixing.

In all cases, model scenarios may serve as a substitute or completion of information derived from experiments or observations.

8.6.5 Example of the Application of Models for the Development of Silvicultural Guidelines

Since early medieval times, mixed mountain forests of Norway spruce, silver fir, and European beech have been replaced by Norway spruce monocultures for wood supply of the salt works in Southern Germany (Hilmers et al. 2020). The resulting nonnatural and unstable secondary Norway spruce monocultures suffered various damages by snow and wind in the last decades. In addition, they are strongly affected by drought stress events and continuous climate warming (Pretzsch et al. 2020b).

To increase the stability of these historically destabilized and presently endangered forests, alternative management concepts are being intensively discussed and introduced. A promising option to restore the stability of these ecosystems is their transformation from pure Norway spruce stands into site-appropriate, sustainable, and stable mixed mountain forests of mainly Norway spruce, silver fir, European beech, and sycamore maple.

However, there are hardly any practical examples for such silvicultural transformations. As way out we used the stand growth simulator SILVA to develop and assess the results of various transformation scenarios. We also analyzed any trade-offs between the pros and cons of various scenarios and the resulting success criteria (Hilmers et al. 2020). Here we show the results of the test of seven different transformation scenarios (e.g., slit-coupes, shelterwood and gap-coupes, strip clear-cutting, do-nothing). We report their impact on the five evaluation criteria forest stand growth, economical results, carbon store, stand stability, and biodiversity.

Some of the results are visualized in Fig. 8.4. In essence, we found out that the scenarios applying gap or slit-coupes resulted in the most beneficial overall utility values regarding the five evaluation criteria. We could also show the best way to transform destabilized forests to sustainable and stable ecosystems. In this way, guidelines for restoration and transformation can be developed, compared, and ranked regarding their achievement, even if respective model stands are scarce or missing. A precondition is the availability of simulation tools that are based on basic principles of tree and stand dynamics and thus can realistically simulate also treatment options and development scenarios that were not used for the model parameterization (Pretzsch et al. 2008).

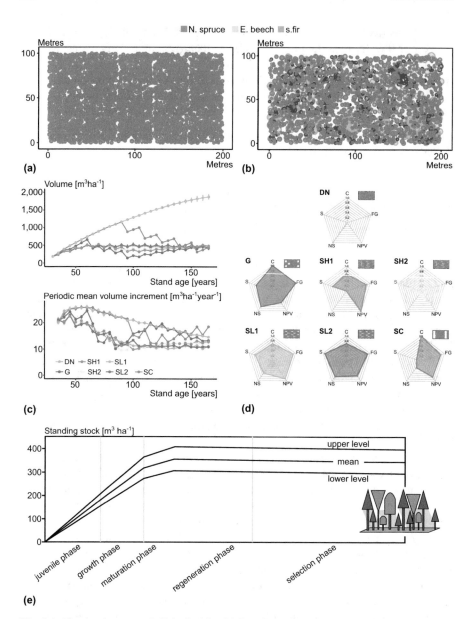

Fig. 8.4 The development of silvicultural guidelines by model scenario calculations comprises the (**a** and **b**) initial state and aim of the modelling exercise, (**c**) definition of the standing stock for the model scenarios (above) and the resulting growth (below), (**d**) the multicriteria overview of the long-term consequences of various options regarding ecological, economical, and socioeconomic criteria, (**e**) simplified silvicultural guidelines for practical application. (After Hilmers et al. 2020) **a and b visualizes the objective, that is, the transition from even-aged monospecific Norway spruce stands to uneven-aged mixed-species stands of mainly Norway spruce, silver fir, and European beech**

8.6.6   From Models for Regulation and Optimization to Guidelines for Silvicultural Steering

We showed that model-based scenario analysis may contribute to deriving an appropriate silvicultural option. For this purpose, various treatment variants may be implemented algorithmically in a growth model, for example, various tree number-mean tree height guidelines (N-h-curves) for steering the stand development over time. For each of $i = 1 \ldots n$ treatment options, the stand development can be simulated, and the result may be compared with the target state. The very treatment options that show the best approximation of the stand to the defined target may be of special interest as a suitable guideline for practical application. The derivation is often based on a combination of a normative, experimental, and simulation approach and also includes ecological and socioeconomic criteria.

The involvement of a scenario analysis requires a quantitative formulation of a set of treatment options and finally provides a quantitatively based treatment option that might be used for target-oriented silvicultural steering in practice.

This model-based derivation of a silvicultural treatment guideline represents a regulation process (Fig. 8.5). By definition (Berg and Kuhlmann 1993), regulation means that a development is controlled by the initial conditions, the time exogen variables and that the rules can be modified by a feedback between the current development and the applied rules (closed-loop-control).

The application of this once derived curve for stand treatment represents steering. Steering means that system development is controlled just by the initial conditions, the time exogen variables, and static rules (open-loop control). In this case, once derived and prescribed rules (e.g., guideline curve or threshold), provide the setpoint for stand characteristic (e.g., density, mixing proportion, number of future crop trees) to which the stand is adjusted by silvicultural interventions in defined intervals. In the case of steering, there is no feedback between the stand development and the once fixed guideline. The application of a given silvicultural treatment in forest practice, that is, the thinning of a stand based on a defined N-h curve, represents steering.

Fig. 8.4 (continued)

c shows the stand volume of the remaining stand in $m^3\ ha^{-1}$ (above) and the mean stand periodic annual volume increment in $m^3\ ha^{-1}\ year^{-1}$ (below) resulting from 30 simulations of the following variants: DN, do-nothing scenario; G, Gap-coupes with the planting of beech and fir; SH1, shelterwood-coupes with natural regeneration; SH2, shelterwood-coupes with the planting of fir and beech; SL1, slit-coupes with natural regeneration; SL2, slit-coupes with the planting of fir and beech; SC, strip clear-cutting with natural regeneration

d shows radar charts of the evaluation of multiple criteria. FG, forest growth; NPV, net present value; C, carbon sequestration; S, stability; NS, number of species. The scaled results of the respective factors of each criterion are shown. Results were scaled between 0 and 1. Results evaluated with 1 represent the best scenario in comparison to the other scenarios. Categories rated 0 show the worst scenario. DN, do-nothing scenario; G, Gap-coupes with the planting of beech and fir; SH1, shelterwood-coupes with natural regeneration; SH2, shelterwood-coupes with the planting of fir and beech; SL1, slit-coupes with natural regeneration; SL2, slit-coupes with the planting of fir and beech; SC, strip clear-cutting with natural regeneration

e represents the main aspect of the derived silvicultural guideline, the target standing stock for stands growing under different site conditions

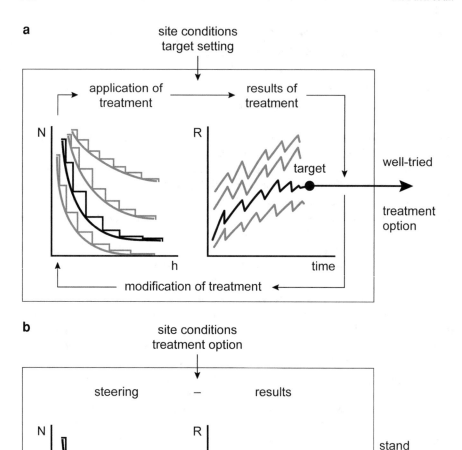

Fig. 8.5 Regulation (**a**) vs. steering (**b**) of systems after Berg and Kuhlmann (1993). (**a**) The effects of silvicultural treatments on stand development may be simulated by a model. The stand development can be compared with the management objective, and the repeated modification of the treatment and simulation generates a set of scenarios with some of them appropriate for reaching a defined management objective. The feedback between the simulated stand development and the treatment option makes this process a regulation. (**b**) A derived silvicultural treatment may be applied as a silviculture guideline for stand management. As there is no feedback between stand development and guideline characteristics, this process is called steering

References

Abetz P (1975) Eine Entscheidungshilfe für die Durchforstung von Fichtenbeständen. Allg Forstzeitschr 30(33/34)

Abetz P (1976) Beiträge zum Baumwachstum. Der h/d-Wert – mehr als ein Schlankheitsgrad! Forst-u. Holzwirt 31 (19)

Abetz P (1987) Why the crop tree aligned thinning system (ZB-Df) increases the stability and productivity of stands. In: Knutell H (ed) Development of thinning systems to reduce stand damages. Proceedings of IUFRO Group S1.05–05, June 1987, Sweden. Department of Operational Efficiency, Faculty of Forestry, Swedish Unversity of Agricultural Sciences, Garpenberg, pp 35–42

Achat DL, Fortin M, Landmann G et al (2015) Forest soil carbon is threatened by intensive biomass harvesting. Sci Rep 5

Aguirre A, del Río M, Condés S (2019) Productivity estimations for monospecific and mixed pine forests along the Iberian Peninsula aridity gradient. Forests 10(5):430

Albrecht A, Hanewinkel M, Bauhus J, Kohnle U (2012) How does silviculture affect storm damage in forests of south-western Germany? Results from empirical modeling based on long-term observations. Eur J For Res 131:229–247

Allen CD, Macalady AK, Chenchouni H et al (2010) A global overview of drought and heat-induced tree mortality reveals emerging climate change risks for forests. For Ecol Manag 259(4):660–684

Allendorf FW, Luikart G, Aitken SN (2013) Conservation and the Genetics of populations. Wiley-Blackwell

Almeida I, Rösch C, Saha S (2018) Comparison of ecosystem services from mixed and monospecific forests in Southwest Germany: a survey on public perception. Forests 9(10):627

Altieri V, De Franco S, Lombardi F, Marziliano PA, Menguzzato G, Porto P (2018) The role of silvicultural systems and forest types in preventing soil erosion processes in mountain forests: a methodological approach using cesium-137 measurements. J Soils Sedim 18:3378–3387

Ameztegui A, Coll L (2013) Unraveling the role of light and biotic interactions on seedling performance of four Pyrenean species along environmental gradients. For Ecol Manag 303:25–34

Ameztegui A, Coll L (2015) Herbivory and seedling establishment in Pyrenean forests: influence of micro- and meso-habitat factors on browsing pressure. For Ecol Manag 342:103–111

Ameztegui A, Brotons L, Coll L (2010) Land-use changes as major drivers of Mountain pine (*Pinus uncinata* Ram.) expansion in the Pyrenees. Glob Ecol Biogeogr 19(5):632–641

Ameztegui A, Coll L, Brotons L, Ninot JM (2016) Land-use legacies rather than climate change are driving the recent upward shift of the mountain treeline in the Pyrenees. Glob Ecol Biogeogr 25(3):267–273

Anon (1783) Summary abstracts of the rewards restowed by the society, from the institution in 1754, to 1782, inclusive. Transactions of the Society, Instituted at London, for the Encouragement of Arts, Manufactures, and Commerce 1:1–62

ARANGE (2020) Advanced Multifunctional Forest Management in European Mountain Ranges (ARANGE). Project within the European commission's 7th framework program, grant agreement number 289437. http://www.arange-project.eu/. Accessed 03 June 2020

Assmann E (1961) Waldertragskunde. Organische Produktion, Struktur, Zuwachs und Ertrag von Waldbeständen. BLV Verlagsgesellschaft München Bonn Wien 490 pp

Assmann E (1970) The principles of forest yield study. Pergamon Press 506 pp

Badalamenti E, Battipaglia G, Gristina L et al (2019) Carbon stock increases up to old-growth forest along a secondary succession in Mediterranean island ecosystems. PLoS One 14

Barbier S, Balandier P, Gosselin F (2009) Influence of several tree traits on rainfall partitioning in temperate and boreal forests: a review. Ann For Sci 66(602)

Barrett RHD, Schluter D (2008) Adaptation from standing genetic variation. Trends Ecol Evol 23(1):38–44

Bauhus J, Puettmann K, Messier C (2009) Silviculture for old-growth attributes. For Ecol Manag 258:525–537

Bauhus J, Forrester DI, Pretzsch H (2017) From observations to evidence. About effects of mixed-species stands. In: Pretzsch H, Forrester DI, Bauhus J (eds) Mixed-species forests. Ecology and management. Springer, Dordrecht, pp 27–71

Becquey J, Riou-Nivert P (1987) L'existence de zones de stabilité des peuplements. Conséquences sur la gestion. Rev For Fr 39:323–334

Berg E, Kuhlmann F (1993) Systemanalyse und Simulation. Ulmer Stuttgart, pp 269–271

Berger TW, Inselsbacher E, Mutsch F (2009) Nutrient cycling and soil leaching in eighteen pure and mixed stands of beech (Fagus sylvatica) and spruce (Picea abies). For Ecol Manag 258:2578–2592

Biedermann PHW, Müller J, Grégoire J-C et al (2019) Bark beetle population dynamics in the Anthropocene: challenges and solutions. Trends Ecol Evol 34:914–924

Bilz M, Kell SP, Maxted N, Lansdown RV (2011) European red list of vascular plants. Publications Office of the European Union, Luxembourg

Bodin J, Badeau V, Bruno E, Cluzeau C, Moisselin J-M, Walther G-R, Dupouey J-L (2013) Shifts of forest species along an elevational gradient in Southeast France: climate change or stand maturation? J Veg Sci 24:269–283

Bolte A, Ammer C, Löf M, Madsen P, Nabuurs GJ, Schall P, Spathelf P, Rock J (2009) Adaptive forest management in central Europe: climate change impacts, strategies, and integrative concept. Scand J For Res 24(6):471–480

Bončina A (2011a) History, current status and future prospects of uneven-aged forest management in the Dinaric region: an overview. Forestry 84:467–478

Bončina A (2011b) Conceptual approaches to integrate nature conservation into forest management: a Central European perspective. Int For Rev 13(1):13–22

Bončina A, Klopčič M, Simončič T, Dakskobler I, Ficko A, Rozman A (2017) A general framework to describe the alteration of natural tree species composition as an indicator of forest naturalness. Ecol Indic 77:194–204

Bowditch E, Santopuoli G, Binder F et al (2020) What is Climate-Smart Forestry? A definition from a multinational collaborative process focused on mountain regions of Europe. Ecosyst Serv 43:101113

Brang P, Schoenenberger W, Frehner M, Schwitter R, Thormann J-J, Wasser B (2006) Management of protection forests in the European Alps: an overview. For Snow Landsc Res 80:23–44

Brang P, Bugmann H, Bürgi A, Mühlethaler U, Rigling A, Schwitter R (2008) Klimawandel als-waldbauliche Herausforderung. Schweiz Z Forstwes 159:362–373

Brang P, Breznikar A, Hanewinkel M, Jandl R, Maier B (2013) Managing Alpine forests in a changing climate. In: Cerbu GA, Hanewinkel M, Gerosa G, Jandl R (eds) Management strategies to adapt Alpine space forests to climate change risks, pp 369–383

Brang P, Spathelf P, Larsen JB et al (2014) Suitability of close-to-nature silviculture for adapting temperate European forests to climate change. Forestry 87(4):492–503

Brang P, Pluess AR, Bürgi A, Born J (2016) Potenzial von Gastbaumarten bei der Anpassung an den Klimawandel. In: Pluess AR, Augustin S, Brang P (eds) Wald im Klimawandel. Grundlagen für Adaptationsstrategien. Bundesamt für Umwelt BAFU, Bern; Eidg. Forschungsanstalt WSL, Birmensdorf; Haupt. Bern, Stuttgart, pp 385–406

Bravo-Oviedo A, Pretzsch H, Ammer C et al (2014) European mixed forests: definition and perspectives. For Syst 23(3):518–533

Bréda N, Huc R, Granier A, Dreyer E (2006) Temperate forest trees and stands under severe drought: a review of ecophysiological responses, adaptation processes and long-term consequences. Ann For Sci 63(6):625–644

Brumelis G, Jonsson BG, Kouki J, Kuuluvainen T, Shorohova E (2011) Forest naturalness in northern Europe: perspectives on processes, structures and species diversity. Silva Fenn 45:807–821

Buiteveld J, Vendramin GG, Leonardi S, Kamer K, Geburek T (2007) Genetic diversity and differentiation in European beech (*Fagus sylvatica* L.) stands varying in management history. For Ecol Manag 247:98–106

Büntgen U, Krusic P, Piermattei A et al (2019) Limited capacity of tree growth to mitigate the global greenhouse effect under predicted warming. Nat Commun 10(1):2171

Burschel P, Huss J (1997) Grundriss des Waldbaus. Berlin Parey Buchverlag

Castedo-Dorado F, Crecente-Campo F, Álvarez-Álvarez P, Barrio-Anta M (2009) Development of a stand density management diagram for radiata pine stands including assessment of stand stability. Forestry 82(1):1–16

Ceballos I (1960) Repoblación forestal española en los últimos años (1940–1960). Estudios geográficos 21(81):497–507

Chapman CA (1989) Primate seed dispersal: the fate of dispersed seeds Biotropica. 21(2):148–154

Christensen M, Hahn K, Mountford EP et al (2005) Deadwood in European beech (*Fagus sylvatica*) forest reserves. For Ecol Manag 210:267–282

Collin S (2020) Drought in the forest. www.waldwissen.net. Accessed 3 Mar 2021

Conforti M, Luca F, Scarciglia F, Matteucci G, Buttafuoco G (2016) Soil carbon stock in relation to soil properties and landscape position in a forest ecosystem of southern Italy (Calabria region). Catena 144:23–33

Coop JD, Givnish TJ (2007) Spatial and temporal patterns of recent forest encroachment in montane grasslands of the Valles Caldera, New Mexico, USA. J Biogeogr 34:914–927

Cremer KW, Borough CJ, McKinnel FH, Carter PR (1982) Effects of stocking and thinning on wind damage in plantations. NZ J For Sci 12:244–268

D'Amato A, Bradford JB, Fraver S, Palik BJ (2011) Forest management for mitigation and adaptation to climate change: insights from long-term silviculture experiments. For Ecol Manag 262:803–816

Dale VH, Joyce LA, McNulty S, Neilson RP (2000) The interplay between climate change, forests, and disturbances. Sci Total Environ 262:201–204

Dale VH, Joyce LA, Mcnulty S et al (2001) Climate change and forest disturbances. Bioscience 51(9):723–734

Davies O, Haufe J, Pommerening A (2008) Silvicultural principles of continuous cover forestry – a guide to best practice. Forestry Commission Wales, UK, 111 pp

De Avila A, Albrecht A (2017) Alternative Baumarten im Klimawandel: Artensteckbriefe – eine Stoffsammlung. Forstliche Versuchs- und Forschungsanstalt Baden-Württemberg, Freiburg, 122 p

del Río S, Álvarez-Esteban R, Cano E, Pinto-Gomes C, Penas A (2018) Potential impacts of climate change on habitat suitability of Fagus sylvatica L. forests in Spain. Plant Biosyst 152:1205–1213

Desai AP, Bolstad P, Cook B, Davis K, Carey E (2005) Comparing net ecosystem exchange of carbon dioxide between old growth and mature forest in upper Midwest, USA. Agric For Meteorol 128:33–55

Deuffic P, Garms M, He J, Brahic E, Yang H, Mayer M (2020) Forest dieback, a tangible proof of climate change? A cross-comparison of forest stakeholders' perceptions and strategies in the mountain forests of Europe and China. Environ Manag 66:858–872

Díaz-Yáñez O, Mola-Yudego B, Ramón González-Olabarria JR, Pukkala T (2017) How does forest composition and structure affect the stability against wind and snow? For Ecol Manag 401:215–222

Dirnböck T, Dullinger S, Grabherr G (2003) A regional impact assessment of climate and land-use change on alpine vegetation. J Biogeogr 30:401–417

Dobbertin M (2002) Influence of stand structure and site factors on wind damage comparing the storms Vivian and Lothar. For Snow Landsc Res 77:187–205

Drever CR, Peterson G, Messier C, Bergeron Y, Flannigan M (2006) Can forest management based on natural disturbance maintain ecological resilience? Can J For Res 36:2285–2299

ECCP-Working Group on Forest Sinks (2003) Conclusions and recommendations regarding forest related sinks and climate change mitigation. Tech. Rep., EC-DG Environment

Edwards ME (2005) Landscape history and biodiversity conservation in the uplands of Norway and Britain: comparisons and contradictions. In: Thompson DBA, Price MF, Galbraith CA (eds) Mountains of Northern Europe: conservation, management, people and nature. TSO Scotland, Edinburgh

Elliot WJ, Page-Dumroese D, Robichaud PR (1996) The Effects of forest management on erosion and soil productivity. An invited paper presented at the Symposium on Soil Quality and Erosion Interaction sponsored by The Soil and Water Conservation Society of America held at Keystone, Colorado July 7th, 1996

Elmqvist T, Thomas E, Carl F, Magnus N, Garry P, Jan B, Brian W, Jon N (2003) Response diversity, ecosystem change, and resilience. Front Ecol Environ 1:488

Evelyn J (1664) Sylva, or a discourse of forest-trees and the propagation of timber in his majesty's dominions. John Martyn, London, 120 pp

FAO (2020) Global forest resources assessment 2020. Main Report, Rome, 184 pp. ISBN:978-92-5-132974-0

Felton A, Petersson L, Nilsson O et al (2019) The tree species matters: Biodiversity and ecosystem service implications of replacing Scots pine production stands with Norway spruce. Ambio 15 pp

Ficko A, Poljanec A, Bončina A (2011) Do changes in spatial distribution, structure and abundance of silver fir (*Abies alba* Mill.) indicate its decline? For Ecol Manag 261:844–854

Field CB, Barros VR, Mach KJ, Mastrandrea MD, van Aalst M, Adger WN, Arent DJ, Barnett J et al (2014) Technical summary. In: Field CB et al (eds) Climate change 2014: impacts, adaptation, and vulnerability. Part A: Global and sectoral aspects. Contribution of Working Group II to the Fifth Assessment Report of the Intergovernmental Panel on Climate Change. Cambridge University Press, Cambridge/New York, pp 35–94

Fisher R (1763) Heart of oak, the British bulwark. J. Johnson, London, 97 pp

Forest Europe (2015) State of Europe's forests 2015, Ministerial Conference on Protection of Forests in Europe, FOREST EUROPE, Liaison Unit Madrid 312 p

Forestry Commission (2020) Managing England's Woodlands in a climate emergency. Forestry Commission, 16 pp

Forestry Statistics (2019) Forest research. Edinburgh, UK

Forrester DI, Bauhus J (2016) A review of processes behind diversity – productivity relationships in forests. Curr For Rep 2(1):45–61

Franc N (2007) Standing or downed dead trees – does it matter for saproxylic beetles in temperate oak-rich forest? Can J For Res 37:2494–2507

Galiano L, Martínez-Vilalta J, Lloret F (2010) Drought-induced multifactor decline of Scots pine in the Pyrenees and potential vegetation change by the expansion of co-occurring oak species. Ecosystems 13:978–991

Gallart F, Llorens P (2004) Catchment management under environmental change: impact of land cover change on water resources. Water Int 28:334–340

Gamfeldt L, Snall T, Bagchi R et al (2013) Higher levels of multiple ecosystem services are found in forests with more tree species. Nat Commun 4:1340

Gehrig-Fasel J, Guisan A, Zimmermann NE (2007) Tree line shifts in the Swiss Alps: climate change or land abandonment? J Veg Sci 18:571–582

Gibson SY, van der Marel RC, Starzomski BM (2009) Climate change and conservation of leading-edge peripheral populations. Conserv Biol 23(6):1369–1373

Gill RMA (1992) A review of damage by mammals in north temperate forests. 1. Deer. Forestry 65:145–169

Gomez-Guarrerro A, Doane T (2018) The response of forest ecosystems to climate change. Chapter 7 in Developments of soil science. 35:185–206

Gömöry D, Krajmerová D, Hrivnák M, Longauer R (2020) Assisted migration vs. close-to-nature forestry: what are the prospects for tree populations under climate change? Cent Eur For J 66:63–70

Götmark F, Berglund A, Wiklander K (2005) Browsing damage on broadleaved trees in semi-natural temperate forest in Sweden, with a focus on oak regeneration. Scand J For Res 20:223–234

Griess VC, Knoke T (2011) Growth performance, windthrow, and insects: meta-analyses of parameters influencing performance of mixed-species stands in boreal and northern temperate biomes. Can J For Res 41:1141–1159

Grossiord C, Granier A, Ratcliffe S et al (2014) Tree diversity does not always improve resistance of forest ecosystems to drought. PNAS 111(41):14812–14815

Guisan A, Theurillat JP, Kienast F (1998) Predicting the potential distribution of plant species in an alpine environment. J Veg Sci 9:65–74

Gunn JS, Ducey MJ, Whitman AA (2014) Late-successional and old-growth forest carbon temporal dynamics in the Northern Forest (Northeastern USA). For Ecol Manag 312:40–46

Hagerman SM, Pelai R (2018) Responding to climate change in forest management: two decades of recommendations. Front Ecol Environ 16:579–587

Hahm WJ, Riebe CS, Lukens CE, Araki S (2014) Bedrock composition regulates mountain ecosystems and landscape evolution. Proc Natl Acad Sci U S A 111:3338–3343

Hajabbasi MA, Jalalian A, Karimzadeh HR (1997) Deforestation effects on soil physical and chemical properties, Lordegan, Iran. Plant Soil 190:301–308

Hampe A, Petit RJ (2005) Conserving biodiversity under climate change: the rear edge matters. Ecol Lett 8:461–467

Hanewinkel M, Cullmann DA, Schelhaas M-J, Nabuurs G-J, Zimmermann NE (2013) Climate change may cause severe loss in the economic value of European forest land. Nat Clim Chang 3:203–207

Hanewinkel M, Kuhn T, Bugmann H, Lanz A, Brang P (2014) Vulnerability of uneven-aged forests to storm damage. Forestry 87:525–534

Helms JA (1998) The dictionary of forestry. Society of American Foresters, Bethesda, 125 p

Hickler T, Vohland K, Feehan J et al (2012) Projecting the future distribution of European potential natural vegetation zones with a generalized, tree species-based dynamic vegetation model. Glob Ecol Biogeogr 21:50–63

Hilmers T, Biber P, Knoke T, Pretzsch H (2020) Assessing transformation scenarios from pure Norway spruce to mixed uneven-aged forests in mountain areas. Eur J For Res 139:567–584

Höhn M, Major E, Avdagic A, Bielak K, Bosela M, Coll L, Dinca L, Giammarchi F, Ibrahimspahic A, Mataruga M, Pach M, Uhn E, Zlatanov T, Cseke K, Kovács Z, Palla B, Ladanyi M, Heinze B (2021) Local characteristics of the standing genetic diversity of European beech with high within-region differentiation at the eastern part of the range. Can J For Res. https://doi.org/10.1139/cjfr-2020-0413

Hoover CM (2011) Management impacts on forest floor and soil organic carbon in Northern temperate forests of the US. Carbon Balance Manag 6

Humphrey JW, Davey S, Peace AJ, Ferris R, Harding K (2002) Lichens and bryophyte communities of planted and semi-natural forests in Britain: the influence of site type, stand structure and deadwood. Biol Conserv 107:165–180

Humphrey JW, Sippola AL, Lempérière G, Dodelin B, Alexander KNA, Butler JE (2004) Deadwood as an indicator of biodiversity in European forests: from theory to operational guidance. In: Marchetti M (ed) Monitoring and indicators of forest biodiversity in Europe – from ideas to operationality, vol 51. EFI Proceedings, pp 193–206

Jactel H, Nicoll BC, Branco M et al (2009) The influences of forest stand management on biotic and abiotic risks of damage. Ann For Sci 66(7):701–701

Jactel H, Bauhus J, Boberg J et al (2017) Tree diversity drives forest stand resistance to natural disturbances. Curr For Rep 3(3):223–243

Jalkanen A, Mattila U (2000) Logistic regression models for wind and snow damage in northern Finland based on the National Forest Inventory data. For Ecol Manag 135:315–330

James J, Harrison R (2016) The effect of harvest on forest soil carbon: a meta-analysis. Forests 7(12):308

Jandl R, Ledermann T, Kindermann G, Freudenschuss A, Gschwantner T, Weiss P (2018) Strategies for climate-smart forest management in Austria. Forests 9:592

Jandl R, Spathelf P, Bolte A, Prescott C (2019) Forest adaptation to climate change – is non-management an option? Ann For Sci 76:48

Jaworski A (2013) Hodowla lasu. Pielęgnowanie lasu. PWRiL, Warszawa, p 359

Jiang Z, Liu H, Wang H et al (2020) Bedrock geochemistry influences vegetation growth by regulating the regolith water holding capacity. Nat Commun 11:2392

Johann K (1981) Nicht Schnee, sondern falsche Bestandsbehandlung versacht Katastrophen. Allg Forstztg 92:163–171

Jump AS, Penuelas J (2005) Running to stand still: adaptation and the response of plants to rapid climate change. Ecol Lett 8:1010–1020

Kato F, Mülder D (1983) Qualitative Gruppendurchforstung der Buche. Allg Forst-u Jagdztg 154:8

Kauppi P, Hanewinkel M, Lundmark L, Nabuurs GJ, Peltola H, Trasobares A, Hetemäki L (2018) Climate smart forestry in Europe. European Forest Institute

Keeton WS (2019) Source or sink? Carbon dynamics in eastern old-growth forests and their role in climate change mitigation. In: Anonymous ecology and recovery of eastern old-growth forests, pp 267–288

Kerr G, Morgan G, Blyth J, Stokes V (2010) Transformation from even-aged plantations to an irregular forest: the world's longest running trial area at Glentress, Scotland. Forestry 83:329–344

Kjær ED, McKinney LV, Nielsen LR, Hansen LN, Hansen JK (2012) Adaptive potential of ash (*Fraxinus excelsior*) populations against the novel emerging pathogen Hymenoscyphus pseudoalbidus. Evol Appl 5:219–228

Klopčič M, Poljanec A, Gartner A, Bončina A (2009) Factors related to natural disturbances in Mountain Norway Spruce (*Picea abies*) forests in the Julian Alps. Ecoscience 16:48–57

Knoke T, Ammer C, Stimm B, Mosandl R (2008) Admixing broadleaved to coniferous tree species: a review on yield, ecological stability and economics. Eur J For Res 127:89–101

Korpel Š (1975) Zásady pestovanie v porostach s trvalým zastúpenim Jedle. In: Pestovanie a ochrana jedle. Zvolen, VŠLD

Korpel Š, Saniga M (1993) Výberný hospodàrsky spôsob. Matice Lesnicka Pisek, Praha

Kozak J (2003) Forest cover change in the Western Carpathians in the past 180 Years. Mt Res Dev 23:369–375

Kozlowski G, Gratzfeld J (2013) Zelkova – an ancient tree. Global status and conservation action. Natural History Museum Fribourg

Kramer K, Degen B, Buschbom J, Hickler T, Thuiller W, Sykes MT, Winter W (2010) Modelling exploration of the future of European beech (Fagus sylvatica L.) under climate change – Range, abundance, genetic diversity and adaptive response. For Ecol Manag 259:2213–2222

Kraus D, Krumm F (eds) (2013) Integrative approaches as an opportunity for the conservation of forest biodiversity. EFI 284 pp

Kunz J, Löffler G, Bauhus J (2018) Minor European broadleaved tree species are more drought-tolerant than Fagus sylvatica but not more tolerant than Quercus petraea. For Ecol Manag 414:15–27

Lafond V, Cordonier T, Courbaud B (2015) Reconciling biodiversity conservation and timber production in mixed uneven-aged mountain forests: identification of ecological intensification pathways. Environ Manag 56:1118–1133

Lassauce A, Paillet Y, Jactel H, Bouget C (2011) Deadwood as a surrogate for forest biodiversity: meta-analysis of correlations between deadwood volume and species richness of saproxylic organisms. Ecol Indic 11:1027–1039

Lebourgeois F, Gomez N, Pinto P, Mérian P (2013) Mixed stands reduce *Abies alba* tree ring sensitivity to summer drought in the Vosges mountains, western Europe. For Ecol Manag 303:61–71

Leibundgut H (1982) Über die Anzahl Auslesenbäume bei der Auslesedurchforstung. Schweiz Z Forstwes 133:115–119

Lenoir J, Gégout JC, Marquet PA, de Ruffray P, Brisse H (2008) A significant upward shift in plant species optimum elevation during the 20th century. Science 320:1768–1771

Lesiński J (2012) Forest functions, goods and services: an attempt to integrate various approaches. In: Chubinsky M (ed) Proceedings of the IUFRO-EFI-ICFFI conference "Ecosystem design for multiple services – with an emphasis on Eurasian Boreal Forests", held in St. Petersburg on 9–11 November, 2011. ICFFI News, 1, 14:22–38

Linares JC, Camarero JJ (2012) Growth patterns and sensitivity to climate predict silver fir decline in the Spanish Pyrenees. Eur J For Res 131:1001–1012

Lindner M, Fitzgerald JB, Zimmermann NE et al (2014) Climate change and European forests: what do we know, what are the uncertainties, and what are the implications for forest management? J Environ Manag 146:69–83

Lindner M, Schwarz M, Spathelf P, de Koning JHC, Jandl R, Viszlai I, Vančo M (2020) Adaptation to climate change in sustainable forest management in Europe. FOREST EUROPE, Liaison Unit Bratislava, Zvolen

Locatelli B, Brockhaus M, Buck A et al (2010) Forests and adaptation to climate change: challenges and opportunities. In: Mery G et al (eds) Forests and society – responding to global drivers of change, IUFRO World Series, vol 25, pp 21–42

Logana SA, Phuekvilaia P, Sandersona R, Wolffa K (2019) Reproductive and population genetic characteristics of leading-edge and central populations of two temperate forest tree species and implications for range expansion. For Ecol Manag 433:475–486

Lohmander P, Helles F (1987) Windthrow probability as a function of stand characteristics and shelter. Scand J For Res 2:227–238

Lonsdale D, Pautasso M, Holdenrieder O (2008) Wood-decaying fungi in the forest: conservation needs and management options. Eur J For Res 127:1–22

Lussier J-M, Meek P (2014) Managing heterogeneous stands using a multiple-treatment irregular shelterwood method. J For 112(3):287–295

Mansourian S, Vallauri D, Dudley N (eds) (2005) Forest restoration in landscapes: beyond planting trees. Springer, New York

Margalef-Marrase J, Pérez-Navarro MA, Lloret F (2020) Relationship between heatwave-induced forest die-off and climatic suitability in multiple tree species. Glob Chang Biol. https://doi.org/10.1111/gcb.15042

Martikainen P, Siitonen J, Punttila P, Kaila L, Rauh J (2000) Species richness of Coleoptera inmature managed and old-growth boreal forests in southern Finland. Biol Conserv 94(2):199–209

Martín-Alcón S, González-Olabarria JR, Coll L (2010) Wind and snow damage in the Pyrenees pine forests: effect of stand attributes and location. Silva Fenn 44(3):399–410

Martín-Alcón S, Coll L, Ameztegui A (2016) Diversifying sub-Mediterranean pinewoods with oak species in a context of assisted migration: responses to local climate and light environment. Appl Veg Sci 19:254–267

Martínez-Vilalta J, Lloret F, Breshears DD (2012) Drought-induced forest decline: causes, scope and implications. Biol Lett 8:689–691

Mason B, Kerr G, Simpson J (1999) What is continuous cover forestry? Forestry Commission Information Note 29

Mátyás CS (2002) Erdészeti és természetvédelmi genetika (Forest and nature conservation genetics). Mezőgazda kiadó, Budapest. (in Hungarian)

Mátyás CS, Kramer K (2016) Reference to the document: Climate change affects forest genetic resources: consequences for adaptive management. ForGer Brief. www.fp7-forger.eu

Mayer M, Prescott CE, Abaker WEA et al (2020) Influence of forest management activities on soil organic carbon stocks: a knowledge synthesis. For Ecol Manag 466:118127

Mazziotta A, Mönkkönen M, Strandman H, Routa J, Tikkanen O-P, Kellomäki S (2014) Modeling the effects of climate change and management on the dead wood dynamics in boreal forest plantations. Eur J For Res 133(3):405–421

Meng J, Bai Y, Zeng W, Wu Ma W (2017) A management tool for reducing the potential risk of windthrow for coastal Casuarina equisetifolia L. stands on Hainan Island, China. Eur J For Res 136:543–554

Merganičová K, Merganič J, Svoboda M, Bače R, Šebeň V (2012) Chapter 4: Deadwood in forest ecosystems. In: Blanco JA, Lo YH (eds) Forest ecosystems – more than just trees, pp 81–108

Messier C, Bauhus J, Doyon F, Maure F, Sousa-Silva R, Nolet P, Mina M, Aquilé N, Fortin M-J, Puettmann K (2019) The functional complex network approach to foster forest resilience to global changes. For Ecosyst 6(21)

Metzger HG, Schirmer R, Konnert M (2012) Neue fremdländische Baumarten im Anbautest. AFZ-Der Wald 67(5):32–34

Milodowski DT, Mudd SM, Mithard ETA (2015) Erosion rates as a potential bottom-up control of forest structural characteristics in the Sierra Nevada Mountains. Ecology 96(1):31–38

Misson L, Nicault A, Guiot J (2003) Effects of different thinning intensities on drought response in Norway spruce (*Picea abies*[L.] Karst.). For Ecol Manag 183:47–60

Mitscherlich G (1952) Der Tannen-Fichten-(Buchen)-Plenterwald. Schriften der Badischen Forstlichen Versuchsanstalt Freiburg im Breisgau 8:1–42

Mosca E, Cruz F, Gómez-Garrido J, Bianco L et al (2019) A reference genome sequence for the European Silver Fir (*Abies alba* Mill.): a community-generated genomic resource. G3-Genes Genomes, Genetics 9(7):2039–2049

Müller J (2009) Forestry and water budget of the lowlands in northeast Germany: consequences for the choice of tree species and for forest management. J Water Land Dev 13a:133–148

Müller J, Bolte A (2009) The use of lysimeters in forest hydrology research in northeast Germany. Landbauforsch 59(1):1–10

Müller J, Bussler H, Kneib T (2008) Saproxylic beetle assemblages related to silvicultural management intensity and stand structures in a beech forest in Southern Germany. J Insect Conserv 12:107–124

Nabuurs G-J, Delacote P, Ellison D, Hanewinkel M, Hetemäki L, Lindner M (2017) By 2050 the mitigation effects of EU forests could nearly double through climate smart forestry. Forests 8:484

Nadrowski K, Wirth C, Scherer-Lorenzen M (2010) Is forest diversity driving ecosystem function and service? Curr Opin Env Sust 02:75–79

Nielsen CCN (1995) Recommendations for stabilisation of Norway spruce stands based on ecological surveys. In: Coutts MP, Grace J (eds) Wind and trees, vol 1279. Cambridge University Press, Cambridge, pp 424–435

Nocentini S, Coll L (2013) Mediterranean forests: human use and complex adaptive systems. In: Messier C, Puettmann KJ, Coates KD (eds) Managing forests as complex adaptive systems. Building resilience to the challenge of global change, The Earthscan Forest Library (series). Routledge, New York

Nyland RD (2003) Even- to uneven-aged: the challenges of conversion. For Ecol Manag 171:291–300

O'Hara KL (2014) Multiaged silviculture: managing for complex forest stands. Oxford University Press, Oxford

O'Hara LK (2016) What is close-to-nature silviculture in a changing world? Forestry 89:1–6

O'Hara KL, Ramage BS (2013) Silviculture in an uncertain world: utilizing multi-aged management systems to integrate disturbance. Forestry 86:401–410

O'Hara KL, Leonard LP, Keyes CR (2012) Variable-density thinning and a marking paradox: comparing prescription protocols to attain stand variability in coast redwood. West J Appl For 27(3):143–149

Oliveira AM (1987) The H/D ratio in maritime pine (*Pinus pinaster*) stands. In: Ek AR, Shifley SR, Burk TE (eds) Proceedings of the IUFRO conference Vol. 2 Forest growth modelling and prediction, 23–27 August 1987, Minneapolis. International Union of Forest Research Organizations, Vienna, pp 881–888

Pasztor F, Matulla C, Zuvela-Aloise M, Rammer W, Lexer MJ (2015) Developing predictive models of wind damage in Austrian forests. Ann For Sci 72:289–301

Pausas JG (2004) Changes in fire and climate in the eastern Iberian Peninsula (Mediterranean basin). Clim Chang 63:337–350

Pausas JG, Blade C, Valdecantos A, Seva JP, Fuentes D, Alloza JA, Vilagrosa A, Bautist S, Cortina J, Vallejo VR (2004) Pines and oaks in the restoration of Mediterranean landscapes of Spain: new perspectives for an old practice – a review. Plant Ecol 209:209–220

Peltola H, Nykäinen M-L, Kellomäki S (1997) Model computations on the critical combination of snow loading and windspeed for snow damage of Scots pine, Norway spruce and Birch sp. at stand edge. For Ecol Manag 95:229–241

Persiani AM, Lombardi F, Lunghini D et al (2015) Stand structure and deadwood amount influences saproxylic fungal biodiversity in Mediterranean mountain unmanaged forests. iForest 9:115–124

Petit RJ, Hampe A (2006) Some evolutionary consequences of being a tree. Ann Rev Ecol Evol Sci 37:187–214

Petty J, Swain C (1985) Factors influencing stem breakage of conifers in high winds. Forestry 58(1):75–84

Petty JA, Worrell R (1981) Stability of coniferous tree stems in relation to damage by snow. Forestry 54(2):115–128

Poff RJ (1996) Effects of silvicultural practices and wildfire on productivity of forest soils. Sierra Nevada Ecosystem Project: Final report to Congress, vol. II, Assessments and scientific basis for management options. University of California, Centers for Water and Wildland Resources, Davis

Pons P, Rost J, Tobella C et al (2020) Towards better practices of salvage logging for reducing the ecosystem impacts in Mediterranean burned forests. iForest 13:360–368

Poyatos R, Latron J, Llorens P (2003) Land use and land cover change after agricultural abandonment – the case of a Mediterranean mountain area (Catalan pre-Pyrenees). Mt Res Dev 23:362–368

Preikša Z, Brazaitis G, Marozas V, Jaroszewicz B (2015) Dead wood quality influences species diversity of rare cryptogams in temperate broadleaved forests. iForest 9(2):276–285

Pretzsch H (2019) Transitioning monocultures to complex forest stands in Central Europe: principles and practice. Burleigh Dodds Science Publishing Limited, pp 355–396

Pretzsch H, Biber P, Ďurský J (2002) The single tree-based stand simulator SILVA: construction, application and evaluation. For Ecol Manag 162(1):3–21

Pretzsch H, Grote R, Reineking B, Rötzer TH, Seifert ST (2008) Models for forest ecosystem management: a European perspective. Ann Bot 101(8):1065–1087

Pretzsch H, Block J, Dieler J, Dong PH, Kohnle U, Nagel J, Spellmann H, Zingg A (2010) Comparison between the productivity of pure and mixed stands of Norway spruce and European beech along an ecological gradient. Ann For Sci 67(712)

Pretzsch H, Uhl E, Steinacker L, Moshammer R (2014) Struktur und Dynamik von Bergmischwäldern am Forstbetrieb Schliersee. Exkursionsführer MWW-EF 154:30

Pretzsch H, del Río M, Ammer C et al (2015) Growth and yield of mixed versus pure stands of Scots pine (Pinus sylvestris L.) and European beech (Fagus sylvatica L.) analysed along a productivity gradient through Europe. Eur J For Res 134:927–947

Pretzsch H, Grams T, Häberle KH, Pritsch K, Bauerle T, Rötzer T (2020a) Growth and mortality of Norway spruce and European beech in monospecific and mixed-species stands under natural episodic and experimentally extended drought. Results of the KROOF throughfall exclusion experiment. Trees 34:957–970

Pretzsch H, Hilmers T, Biber P, Avdagic A, Binder F, Bončina A et al (2020b) Evidence of elevation-specific growth changes of spruce, fir, and beech in European mixed-mountain forests during the last three centuries. Can J For Res 50(7):689–703

Pretzsch H, Hilmers T, Uhl E, Bielak K, Bosela M, del Rio M, Dobor L et al (2020c) European beech stem diameter grows better in mixed than in mono-specific stands at the edge of its distribution in mountain forests. Eur J For Res. https://doi.org/10.1007/s10342-020-01319-y

Puettmann KJ, D'Amato AW, Kohnle U, Bauhus J (2009) Individual-tree growth dynamics of mature Abies alba during repeated irregular group shelterwood (Femelschlag) cuttings. Can J For Res 39:2437–2449

Pukkala T (2017) Does management improve the carbon balance of forestry? Forestry 90:125–135

Pukkala T, von Gadov K (2012) Continuous cover forestry. Managing Forest Ecosystems 23. Springer

Pukkala T, Laiho O, Lähde E (2016) Continuous cover management reduces wind damage. For Ecol Manag 372:120–127

Raison RJ, Khanna PK (2011) Chapter 12: Possible impacts of climate change on forest soil health. In: Singh BP et al (eds) Soil health and climate change, Soil biology 29. Springer, Berlin/Heidelberg

Raymond P, Bédard S (2017) The irregular shelterwood system as an alternative to clearcutting to achieve compositional and structural objectives in temperate mixedwood stands. For Ecol Manag 398:91–100

Raymond P, Bédard S, Roy V, Larouche C, Tremblay S (2009) The irregular shelterwood system: review, classification, and potential application to forests affected by partial disturbances. J Forest 107:405–413

Reyer C, Bathgate S, Blennow K, Borges J, Bugmann H, Delzon S et al (2017) Are forest disturbances amplifying or cancelling out climate change induced productivity changes in European forests? Environ Res Lett 12:034027

Rivers M, Beech E, Bazos I, Bogunić F, Buira A, Caković D, Carapeto A, Carta A et al (2019) European red list of trees. IUCN, Cambridge/Brussels. viii + 60pp

Rodríguez-Calcerrada J, Pérez-Ramos IM, Ourcival J-M, Limousin J-M, Joffre R, Rambal S (2011) Is selective thinning an adequate practice for adapting *Quercus ilex* coppices to climate change? Ann For Sci 68(3):575–585

Röhring E, Bartsch N, von Lüpke B (2006) Waldbau auf ökologischer Grundlage. Verlag Eugen Ulmer Stuttgart

Rottmann M (1986) Wind- und Sturmschaden im Wald. Beitrage zur Beurteilung der Bruchgefahrdung, zur Schadensvorbeugung und zur Behandlung sturmgeschadigter Nadelholzbestande. J.D. Sauerläinder's Verlag, Frankfurt am Main, 128 pp

Rötzer T, Seifert T, Gayler S, Priesack E, Pretzsch H (2012) Effects of stress and defence allocation on tree growth: simulation results at the individual and stand level. In: Growth and defence in plants. Springer, Berlin/Heidelberg, pp 401–432

Sáenz-Romero C, Lindig-Cisneros RA, Joyce DG, Beaulieu J, Bradley JC, Jaquish BC (2016) Assisted migration of forest populations for adapting trees to climate change. Revista Chapingo Serie Ciencias Forestales y del Ambiente 22(3):303–323

Santopuoli G, Temperli C, Alberdi I et al (2021) Pan-European sustainable forest management indicators for assessing climate-smart forestry in Europe. Can J For Res 51:1–10

Savill PS (1991) The silviculture of trees used in British forestry. CAB International, 143 pp

Scharlemann JPW, Tanner EVJ, Hiederer R et al (2014) Global soil carbon: understanding and managing the largest terrestrial carbon pool. Carbon Manag 5:81–91

Schelhaas MJ, Kramer K, Peltola H, van der Werf DC, Wijdeven SMJ (2007) Introducing tree interactions in wind damage simulation. Ecol Model 207(2–4):197–209

Schrumpf M, Kaiser K, Schulze E (2014) Soil organic carbon and total nitrogen gains in an old growth deciduous forest in Germany. PLoS One 9

Schulze ED, Aas G, Grimm WG et al (2016) A review on plant diversity and forest management of European beech forests. Eur J For Res 135:51–67

Schütz JP (1987) Auswahl der Auslesebäume in der schweizerischen Auslesedurchforstung. Schweiz Z Forstwes 183:1037–1053

Schütz JP (1989) Der Plenterbetrieb. ETH, Zürich

Schütz JP (1999) Close-to-nature silviculture: is this concept compatible with species diversity? Forestry 72:359–366

Schütz JP (2001a) Der Plenterwald. Parey Buchverlag, Berlin

Schütz JP (2001b) Opportunities and strategies of transforming regular forests to irregular forests. For Ecol Manag 151(1–3):87–94

Schütz JP (2002) Silvicultural tools to develop irregular and diverse forest structures. Forestry 75:329–337

Schütz JP (2003) Der Plenterwald - und weitere Formen strukturierter und gemischter Wälder. Verlag Eugen Ulmer

Schütz JP, Gotz M, Schmid W, Mandallaz D (2006) Vulnerability of spruce (*Picea abies*) and beech (*Fagus sylvatica*) forest stands to storms and consequences for silviculture. Eur J For Res 125(3):291–302

Schütz JP, Saniga M, Diaci J, Vrška T (2016) Comparing close-to-nature silviculture with processes in pristine forests: lessons from Central Europe. Ann For Sci 73:911–921

Schwaiger F, Poschenrieder W, Rötzer T, Biber P, Pretzsch H (2018a) Groundwater recharge algorithm for forest management models. Ecol Model 385:154–164

Schwaiger F, Poschenrieder W, Biber P, Pretzsch H (2018b) Species mixing regulation with respect to forest ecosystem service provision. Forests 9(10):632

Seidl R, Rammer W (2017) Climate change amplifies the interactions between wind and bark beetle disturbances in forest landscapes. Landsc Ecol 32:1485–1498

Seidl R, Rammer W, Lasch P et al (2008) Does conversion of even-aged, secondary coniferous forests affect carbon sequestration? A simulation study under changing environmental conditions. Silva Fenn 42:369–386

Seidl R, Schelhaas M-J, Rammer W, Verkerk PJ (2014) Increasing forest disturbances in Europe and their impact on carbon storage. Nat Clim Chang 4:806–810

Seidl R, Thom D, Kautz M et al (2017) Forest disturbances under climate change. Nat Clim Chang 7:395–402

Sievänen R, Nikinmaa E, Nygren P, Ozier-Lafontaine H, Perttunen J, Hakula H (2000) Components of functional-structural tree models. Ann For Sci 57(5):399–412

Silva C, Holmberg G, Turok J, Stover D, Horstet A (2018) EIP-AGRI focus group forest practices & climate change. MINIPAPER4: climate smart silviculture & genetic resources

Simončič T, Bončina A, Jarni K, Klopčič M (2019) Assessment of the long-term impact of deer on understory vegetation in mixed temperate forests. J Veg Sci 30(1):108–120

Skrzyszewski J (1993) Kształtowanie się zależności pomiędzy żywotnością, cechami morfologicznymi korony i masa systemu korzeniowego a przyrostem promienia na pierśnicy świerka i modrzewia. PhD thesis, Department of Silviculture, University of Agriculture in Krakow (in Polish)

Smith DM, Larson BC, Kelty MJ, Ashton MS (1997) The practice of silviculture: applied forest ecology, 9th edn. Wiley, New York, 537 p

Sohn J, Gebhardt T, Ammer C (2013) Mitigation of drought by thinning. Short-term and long-term effects on growth and physiological performance of Norway spruce (*Picea abies*). For Ecol Manag 308:188–197

Spathelf P, Bolte A (2020) Naturgemäße Waldwirtschaft und Klimawandelanpassung – Kohärenz oder Widerspruch? Jahrbuch Band III 2020 der Nationalparkstiftung Unteres Odertal, Jahresband, 17–27

Spathelf P, Stanturf J, Kleine M et al (2018) Adaptive measures: integrating adaptive forest management and forest landscape restoration. Ann For Sci 75:55

Spellmann H, Albert M, Schmidt M, Sutmöller J, Overbeck M (2011) Waldbauliche Anpassungsstrategien für veränderte Klimaverhältnisse. AFZDerWald 11:19–23

Spiecker H (2004) Norway Spruce conversion- options and consequences. European Forest Institute Research Report, pp 18–269

Stmelf (2018) Jahresbericht 2017 Bayerische Forstverwaltung. Bayerisches Staatsministerium für Ernährung, Landwirtschaft und Forsten, p 59

Stokland JN, Larsson K-H (2011) Legacies from natural forest dynamics: different effects of forest management on wood-inhabiting fungi in pine and spruce forests. For Ecol Manag 261:1707–1721

Sturrock R, Frankel S, Brown A, Hennon P, Kliejunas J, Lewis K, Worrall J, Woods A (2011) Climate change and forest diseases. Plant Pathol 60:133–149

Szwagrzyk J (2020) A healthy forest needs diseased trees (in Polish, English summary). Fragm Florist Geobot Polon 27(1):5–15

Temperli C, Bugmann H, Elkin C (2013) Cross-scale interactions among bark beetles, climate change, and wind disturbances: a landscape modeling approach. Ecol Monogr 83:383–402

Thorn S, Bässler C, Brandl R et al (2018) Impacts of salvage logging on biodiversity – a meta-analysis. J Appl Ecol 55:279–289

Thurm EA, Hernandez L, Baltensweiler A et al (2018) Alternative tree species under climate warming in managed European forests. For Ecol Manag 430:485–497

Tonon G, Dezi S, Ventura M et al (2011) The effect of forest management on soil organic carbon. In: Anonymous sustaining soil productivity in response to global climate change: science, policy, and ethics, pp 225–238

Torresan C, del Río M, Hilmers T et al (2020) Importance of tree species size dominance and heterogeneity on the productivity of spruce-fir-beech mountain forest stands in Europe. For Ecol Manag 457:117716

Vacchiano G, Justin Derose R, Shaw JD, Svoboda M, Motta R (2013) A density management diagram for Norway spruce in the temperate European montane region. Eur J For Res 132:535–549

Valinger E, Lundqvist L, Brandel G (1994) Wind and snow damage in a thinning and fertilisation experiment in *Pinus sylvestris*. Scan J For Res 9:129–134

van Dam BC (2002) EUROPOP: Genetic diversity in river populations of European black poplar for evaluation of biodiversity, conservation strategies, nature development and genetic improvement. In: Dam BC, Bordács S (eds) Genetic diversity in river populations of European black poplar. Proceedings of International Symposium Szekszárd, 16–20 May 2001. Csiszár nyomda, Budapest, pp 15–31

Vedrova EF, Mukhortova LV, Trefilova OV (2018) Contribution of old growth forests to the carbon budget of the Boreal Zone in Central Siberia. Biol Bull 45:288–297

Verkerk PJ, Costanza R, Hetemäki L, Kubiszewski I, Leskinen P, Nabuurs GJ, Potočnik J, Palahi M (2020) Climate-smart forestry: the missing link. For Policy Econ 115:102164

Vilà-Cabrera A, Rodrigo A, Martinez-Vilalta J, Retana J (2012) Lack of regeneration and climatic vulnerability to fire of Scots pine may induce vegetation shifts at the southern edge of its distribution. J Biogeogr 39:488–496

Vitali V, Bauhus J, Büntgen U (2017) Silver fir and Douglas fir are more tolerant to extreme droughts than Norway spruce in south-western Germany. Wiley Online Library. https://doi.org/10.1111/gcb.1377

von Lüpke B (2009) Überlegungen zu Baumartenwahl und Verjüngungsverfahren bei fortschreitender Klimaänderung in Deutschland. Forstarchiv 80:67–75

Walker B, Kinzig A, Langridge J (1999) Plant attribute diversity, resilience, and ecosystem function: the nature and significance of dominant and minor species. Ecosystems 2:95–113

Wang Y, Titus SJ, LeMay VM (1998) Relationships between tree slenderness coefficients and tree or stand characteristics for major species in boreal mixedwood forests. Can J For Res 28:1171–1183

Weed AS, Ayres MP, Hicke JA (2013) Consequences of climate change for biotic disturbances in North American forests. Ecol Monogr 83:441–470

Wermelinger B (2004) Ecoloy and management of the spruce bark beetle Ips typographus – a review of recent research. For Ecol Manag 202:67–82

Wilson JS, Oliver CD (2000) Stability and density management in Douglas-fi r plantations. Can J For Res 30:910–920

Winter S (2012) Forest naturalness assessment as a component of biodiversity monitoring and conservation management. Forestry 85:293–304

Wischmeier WH, Mannering JV (1969) Relation of soil properties to its erodibility. Division S-6 – soil and water management and conservation. Soil Sci Soc Am J 33(1):131–137

Wolf J, Brocard G, Willenbring J, Porder S, Urarte M (2016) Abrupt changes in forest height along a tropical elevation gradient detected using airborne Lidar. Remote Sens 8:864

Yachi S, Loreau M (1999) Biodiversity and ecosystem productivity in a fluctuating environment, the insurance hypothesis. Proc Natl Acad Sci 96:57–64

Yousefpour R, Temperli C, Jacobsen JB et al (2017) Framework for modelling adaptive management and decision-making in forestry under climate change. Ecol Soc 22:40

Zeller L, Pretzsch H (2019) Effect of forest structure on stand productivity in Central European forests depends on developmental stage and tree species diversity. For Ecol Manag 34:193–204

Zwierzyński J, Bednorz L (2012) Regional Programme of Conservation and Restitution of Sorbus torminalis in the Territory of the Regional Directorate of the State Forests in Piła in 2010-2013. Nauka Przyroda Technologie, 6,3,#42

Chapter 9
Smart Harvest Operations and Timber Processing for Improved Forest Management

G. Picchi, J. Sandak, S. Grigolato, P. Panzacchi, and R. Tognetti

Abstract Climate-smart forestry can be regarded as the evolution of traditional silviculture. As such, it must rely on smart harvesting equipment and techniques for a reliable and effective application. The introduction of sensors and digital information technologies in forest inventories, operation planning, and work execution enables the achievement of the desired results and provides a range of additional opportunities and data. The latter may help to better understand the results of management options on forest health, timber quality, and many other applications. The introduction of intelligent forest machines may multiply the beneficial effect of digital data gathered for forest monitoring and management, resulting in forest harvesting operations being more sustainable in terms of costs and environment. The interaction can be pushed even further by including the timber processing industry, which assesses physical and chemical characteristics of wood with sensors to optimize the transformation process. With the support of an item-

---

G. Picchi (✉)
Institute of Bioeconomy of the National Research Council (CNR-IBE),
Sesto Fiorentino, Italy
e-mail: gianni.picchi@cnr.it

J. Sandak
InnoRenew CoE, Izola, Slovenia
e-mail: jakub.sandak@innorenew.eu

S. Grigolato
Department of Land, Environment, Agriculture and Forestry, Università degli Studi di
Padova, Legnaro, PD, Italy
e-mail: stefano.grigolato@unipd.it

P. Panzacchi
Department of Biosciences and Territory, Università degli Studi del Molise, Pesche, Italy

Faculty of Science and Technology, Free University of Bolzano-Bozen, Bolzano, Italy

R. Tognetti
Dipartimento di Agricoltura, Ambiente e Alimenti, Università degli Studi del Molise,
Campobasso, Italy
e-mail: tognetti@unimol.it

317
R. Tognetti et al. (eds.), *Climate-Smart Forestry in Mountain Regions*, Managing
Forest Ecosystems 40, https://doi.org/10.1007/978-3-030-80767-2_9

level traceability system, the same data could provide a formidable contribution to CSF. The "memory" of wood could support scientists to understand the response of trees to climate-induced stresses and to design accordingly an adaptive silviculture, contributing to forest resilience in the face of future changes due to human-induced climate alteration.

Acronyms

AHP	Analytic hierarchy process
CHM	Canopy height model
CSF	Climate-smart forestry
CT	Computed tomography
DBH	Diameter at breast height
DSM	Digital surface model
DTM	Digital terrain model
FI	Forest inventory
GHG	Greenhouse gases
GNSS	Global navigation satellite system
HF	High frequency
LF	Low frequency
LiDAR	Laser imaging detection and ranging
MOE	Modulus of elasticity
MOEd	Dynamic modulus of elasticity
NIR	Near-infrared
OGC	Open Geospatial Consortium
PF	Precision forestry
QR	Quick response (code)
RFID	Radio frequency identification
RGB	Red-green-blue, visible light
SFO	Sustainable forest operations
StandForD	Standard for Forest Machine Data and Communication
SWE	Sensor web enablement (initiative)
TLS	Terrestrial laser scanning
UAV	Unmanned aerial vehicle
UHF	Ultrahigh frequency
UPC	Universal product code
UV	Ultraviolet
VF	Virtual forest

9.1   Climate-Smart Forestry and Forest Operations

Climate change is altering the ecological equilibrium of our planet. Forests are among the most affected ecosystems due to the increasing damages caused directly or indirectly by rising temperatures, such as droughts, wildfires, and pest attacks. On the other hand, in the Land Use, Land Use Change and Forestry sector (LULUCF), forests are considered a key tool in tackling greenhouse gases (GHG) concentration in the atmosphere. In this regard, there is an open debate concerning the most appropriate use and management of forests. Some researchers consider a priority the carbon sequestration capacity of forest ecosystems, suggesting a focus on the accumulation of carbon in forest biomass and soils. This would be achieved with a specific management approach and reduction of timber harvests (Holtsmark 2012). The concern over timber extraction is not only related to illegal logging but also regarding the production of forest biomass for energy (Searchinger et al. 2018). Nevertheless, while illegal logging and deforestation are unanimously regarded as a threat, several authors consider that under appropriate management and regulations, timber and biomass extraction may contribute positively to an overall net carbon control (Kauppi et al. 2018; Favero et al. 2020), e.g., through substitution of fossil fuel-based products.

In this frame, the productive functions of forests are essential to guarantee an effective contribution to GHG mitigation. In fact, they provide renewable materials and fuels (substituting fossil-derived alternatives), food, as well as other non-wood forest products, whose importance may sometimes offset the role of woody materials (Sheppard et al. 2020). The production of goods with direct economic value is essential to support local communities and rural development. This in turn ensures forest management and preservation is economically and technically feasible. For instance, by allowing the presence of local inhabitants, which represent both the workforce and the main users/monitors of forests.

A possible answer to this complex scenario is provided by climate-smart forestry (CSF) (Yousefpour et al. 2018). This is a comprehensive approach, addressing the goals of Sustainable Forest Management (SFM) and the threat of climate change by enabling both forests and society to transform, adapt to, and mitigate climate-induced changes (Bowditch et al. 2020).

The previous and following chapters of this book describe in detail the concept of CSF as well as the solutions envisaged for its application and the tools used for forest monitoring. The present chapter focuses on the contribution that forest operations and, in a broader vision, the whole timber supply chain can provide to CSF application. This contribution can be summarized by two main roles:

– Active forest management
– Production of data at plot and tree level

Even if they stem from the same activities, these two roles can be easily differentiated when considered as part of CSF. In fact, the former is the application in practice of the management strategies, acting "passively" while implementing the guidelines and instructions elaborated by CSF models. The latter, instead, is an active contribution to the development of such models. In fact, the data produced and stored all along the timber supply chain can be used to better understand forest dynamics and improve the very models previously mentioned, closing the loop.

Active forest management is based on the use of specific technologies and techniques. These can be combined to achieve both processes efficiency, social benefits, and minimal environmental impact, implementing the concept of sustainable forest operations (SFO) as introduced by Marchi et al. (2018). SFO deployed in the frame of an adaptive silviculture is a crucial tool to enhance the resilience of forests and to achieve the following goals:

- Increase the sustainability of timber and forest biomass production
- Directly or indirectly enhance or maintain the production of non-wood forest products and indirect services, such as water supply and carbon sequestration
- Implementation of sustainable and smart silviculture practices, such as tree species change, age mixing, density reduction, or salvaging operations

In order to secure a fast and effective implementation of SFO guidelines, sensing and digital technologies are necessary to plan, manage, and execute forest works, creating a link between SFO and precision forestry (PF). The latter can be defined as the application in forestry of geospatial information and remotely sensed data for planning silvicultural operations at the level of stand, sub-stand, or even individual trees (Holopainen et al. 2014). For instance, planning may identify broad restricted areas, as may be the case of riparian vegetation surrounding water bodies enclosed in the forest (e.g., ponds or creeks), but with individual-tree detail level, it can address the protection of individual trees (e.g., trees hosting nests of protected birds' species) as well as meet the precise requirement of nature conservation instances.

Production of data at plot and tree level is one of the distinctive requirements of CSF, which relies on the availability and elaboration of data (including the territorial scale). As described in the chapters of this book, this is essential for understanding the dynamics of forest growth and health and its responses to climate change and to draw models that support management hypotheses. Data is provided by a number of sources, ranging from satellite images to sessile sensors fixed on trees, varying greatly in territorial and time scale. Forest equipment is already an additional source of data, but the potential of machines as carriers of sensors for forest monitoring is still largely unexpressed. Furthermore, the whole timber supply chain, including the highly detailed sensors deployed by the timber transformation industry (e.g., sawmills), provides extremely valuable and accurate data regarding wood development and characteristics. This data could narrate the "memory of wood," relating the stresses and growing conditions experienced in the last decades (or centuries) by an enormous dataset of trees. For this purpose, an effective tracing system from stump to mill is necessary as well as the capacity to store and manage properly a high volume of data.

9.2   Timber Supply Chains and CSF: The Stat of the Art

This section details the practical application of PF in planning and management of forest harvesting. It describes the present sources of data and their commercial application for fleet control and for estimate of timber and biomass yield. Additionally, it discusses the developments expected in the near future and the main research lines currently known. All the information or data infrastructure is also considered from the point of view of CSF, and the potential synergies between commercial operations and forest monitoring activities are highlighted.

9.2.1   From Forest Inventories to the Virtual Forest

At present, forest inventories (FI) are the most common application of precision forestry. For this purpose, visible (red/green/blue, RGB) or hyperspectral images are the main data source, collected from airborne sensors installed on different carriers, such as satellites (long distance), airplanes (medium distance), and unmanned aerial vehicles (UAVs) for short distances (White et al. 2016). Forest inventories are generally integrated with thematic maps, such as soil layers, forest treatments, tree species composition, and any relevant information reported. A detailed forest inventory plays a key role in the management of the forest as well as in its economic valorization (which may aim at producing an income for the owner or simply to cover the costs of the necessary management operations, depending on the function assigned to the forest stand). For this purpose, it includes not only quantitative values, such as timber volume, age distribution, and structure of the forest stand, but also qualitative figures, such as health and fertility of the stands composing the forest.

A further technology with great potential and consolidated application in forest sector is the laser imaging detection and ranging (LiDAR), which has proven particularly effective in large-scale inventories, integrating data elaborated from RGB sensors (Matasci et al. 2018) and multispectral sensors (Puliti et al. 2020). Compared to the latter source of data, LiDAR provides a more reliable 3D information (Noordermeer et al. 2019) and allows a better estimate of the growing stock and the relative value of the harvestable roundwood (Peuhkurinen 2011). The assessment of the quantity and market value of growing stock can be further enhanced thanks to the application of terrestrial laser scanner (TLS). This can integrate both traditional inventories and aerial surveys, returning a detailed inventory including number of trees, shape and taper of stems, crow insertion, and other parameters not detectable with aerial, above-crown sensing (Mengesha et al. 2015). For instance, by integrating RGB images taken from UAVs with TLS scanning, Pichler et al. (2017) increased the effective volume estimation of the trees in the stand from below 80% for the aerial data to over 94% of the combined dataset. This proves the potential of merging data originated from different sensors and integrated with other digital information sources such as topographic or thematic maps (e.g., pedo-logical layers) and traditional forest inventories. Such multisensor data fusion leads to the generation of a digital model of the studied object (Mitchell 2012), in this case the forest. This can be defined as the "virtual forest" (Rossmann et al. 2009), which gathers accurate, consistent, and readily available information regarding the real forest. This information is used to better plan and manage commercial forest operations. Some examples may be given by forest road and skid trail planning and georeferencing (Đuka et al. 2017). The virtual forest and the related database are both generated for and paid by forest management and harvest operations. Yet, from the point of view of CSF, they can be regarded as an excellent and massive data collection infrastructure. In fact, the availability of such an integrated network of sensors can provide stand-specific information extremely useful to monitor the for-est physiology and health parameters (see also Chap. 10 of this book: Tognetti et al. 2021).

9.2.2 Multipurpose Forest Operation Planning

Cost optimization remains the foremost aim (and driver) of forest operation plan-ning. Yet, forest operation plans are also the main tool used to balance forest growth and extraction of woody products, to minimize impact on soil and water resources and to guarantee the conservation of biodiversity.

Over time, the common timber-oriented planning approach has introduced an increasing number of targets, such as minimizing the damages to the remaining standing trees (still a timber-oriented goal). The development of a multi-target for-est operation plan is a complex task, which requires more data regarding the forest and a higher capacity to elaborate and interconnect this information. For this pur-pose, the digital database of the virtual forest (VF) is the perfect source of data, which is not only related to standing trees but also to other descriptors of the forest. For instance, LiDAR data is also used to detail the terrain features through the digi-tal surface model (DSM) and the digital terrain model (DTM). These are used to derive the canopy high model (CHM) but provide also relevant information regard-ing terrain and soil characteristics such as roughness, slope and, with the integration of sampling plots, bearing capacity (Pirotti et al. 2012).

The high-resolution 3D data of the VF combined with multi-spectrum satellite data collected in frequent time series is a key element to optimize planning of forest operations. It allows to minimize the transit for forest machines and to set pre-defined skidding trails (Sterenczak and Moskalik 2015) and cable layout (Bont and Heinimann 2012) that increase efficiency of the operations, reducing costs per cubic meter of timber extracted. As an example, Fig. 9.1 shows a GIS analysis to evaluate alternative forwarder paths in an Alpine environment by considering variables I) slope and roughness of the terrain and II) the density of the forest stand. The vari-ables were determined from LiDAR data: in particular, the digital terrain model (DTM) (a), the canopy height model (CHM) (b), the terrain roughness (c), and the cost surface normalizing and summarizing the previous data (d).

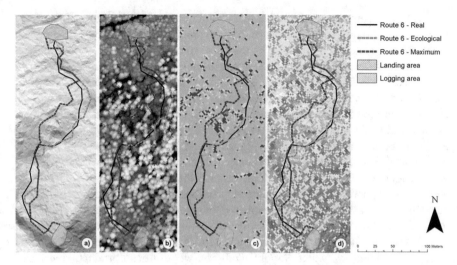

Fig. 9.1 Example of timber hauling operation planning based on digital data. (Credits: Marco Pellegrin, Project NewFor)

Based on the mobility parameters of a medium-size forwarder, two different paths were generated: the ecological path (maximum uphill slope 30% and maximum downhill slope 43%) and the maximum path (maximum uphill slope 35% and maximum downhill slope 67% and a maximum lateral inclination of 4%), a path considering the maximum mobility limits of the machine. In both cases, the maximum lateral inclination was supposed to be 4%. The generated paths were then compared with the real paths of the machine driven by an expert operator.

Thanks to the high detail of the data, plans can get to a tree-level resolution and evaluate the interaction among trees at stand level and the trees' group dynamics (Pirotti et al. 2012; Lindberg and Holmgren 2017). This creates a link with the climate-smart silvicultural prescriptions described in Chap. 8 of this book (Pach et al. 2021) and strengthens the connection that may (and should) exist between the timber supply chains and CSF aims. Examples of this interaction can be provided by forest operation plans tailored to include the priority of habitats conservation, to tackle the spread of forest pests, and to consider the hydrological effect of logging extraction (Blagojević et al. 2019; Görgens et al. 2020). The reduction of soil damage due to forest operation can take advantage from the use of high-resolution data as reported by (Niemi et al. 2017; Salmivaara et al. 2018). Minimizing soil damage is one of the most important targets for the sustainability of forest operation to preserve soil fertility and forest regeneration and to reduce the rate of soil erosion (Cambi et al. 2015; Venanzi et al. 2019), for instance, by developing detailed wet-index and terrain mobility maps (Murphy et al. 2008; Mattila and Tokola 2019). Additionally, in the last two decades, water quality preservation has been introduced as a relevant factor in the planning of forest operations (Keleş and Baskent 2011). This is particularly relevant in mountain areas where the need to maintain wood production while preserving water quality coexists, maximizing the ecosystem services provided by the forest (Ovando and Speich 2020).

The higher complexity of multi-target planning calls for more powerful systems of data interpretation and modeling of forest dynamics. In fact, the more factors and goals are included, the higher will be the effort to develop decision tools. Analytic hierarchy process (AHP) techniques, as well as machine learning fuzzy algorithms, are increasingly deployed for such elaborations. The evaluation of the effects of forest operations in the forest environment can be even more complex when climate change effect is considered (Keenan 2015) with the related nonlinear alterations. On a world scale, climate change is influencing in different ways the operating methods connected to forest harvesting and wood transportation, as well as in the choice of silvicultural addresses and landscape conservation. Consequently, the forest management planning, which includes also the planning of forest operations, needs to be supported by more complex approaches with the introduction of nonlinear approaches and heuristic techniques (Bettinger and Boston 2017). The challenge is to accomplish complex planning, addressing all the risks faced by the forest and all the benefits provided by the same forests such as soil protection, carbon storage, biodiversity, water supply, and non-wood forest products (Matthies et al. 2019).

9.2.3 Sensors on Forest Machines

The appropriate implementation of the forest operation plans based on the FI (or the VF) requires logging machinery or forest teams equipped with geopositioning sensors (at least) and an appropriate interface with the FI (Holopainen et al. 2014). This is even more important when there is an upgrade to the virtual forest and its larger content of digital information. Most modern machines, such as harvesters and forwarders, can already benefit from a digital FI, for instance, with a better organization of harvesting of marked trees or forwarding piled logs along predefined paths. This is possible thanks to the presence onboard of global navigation satellite system (GNSS) receivers, which provide sufficient accuracy on flat terrain, particularly on areas with limited vegetation cover, such as clear-cuts (Valbuena et al. 2012). Digital FIs at present are generally deployed to guide machines through optimal paths to follow. These are elaborated according to predefined criteria, such as minimum cost, minimum impact on soil or standing trees, or a compromise among these (Piragnolo et al. 2019). In optimal cases, where geopositioning is highly reliable and accurate, this application can be stretched to the level of a single tree to cut, implementing a virtual tree marking performed directly on the digital inventory.

This information can be further refined with the information of the exact position of the processing head or the logs grapple (boom's tip) with respect to the machine, leading in optimal conditions to a sub-metric precision in recording the position of the standing tree or the pile of logs (Lindroos et al. 2015). These machines can also feed the virtual forest with additional data collected by the onboard sensors. At present, the parameters measured are the diameters and length of the logs produced by the harvester: values needed to calculate the felled and processed volume (Eriksson and Lindroos 2014). These are generally used for invoicing timber produced and delivered but can also provide a detailed insight of the quantity of roundwood produced in the harvested plot (Rossit et al. 2019) and draw a balance between net annual increment and harvesting of a given forested area. Furthermore, by segregating the assortments, at least in main classes (e.g., pulp-wood vs. industrial timber), it is possible to estimate several characteristics of the harvested trees (Lu et al. 2018) and, possibly in the near future, even to develop forest yield maps (Olivera and Visser 2016).

An additional sensor, which is increasingly installed on forest machines, is the timber crane weighing load cell. It quantifies the exact masses of timber and biomass extracted by forest operations (Laurila and Lauhanen 2012). Currently, this information is used to better plan and link the different operations of the biofuel supply chain (extraction, drying, comminution and transport). It can also contribute to drawing a more precise balance between forest growth and woody products extraction (Fig. 9.2).

Finally, among the last sensors installed in forest machinery, we can report the deployment of cameras used for the automatic detection of stem damage on the trees left in the stand during harvesting. The elaboration of digital pictures, by means of logistic regression models in image processing, allows the operators to identify the damaged (debarked) trees, counting their number and ranking the level of damage (Palander et al. 2019). The system was designed to estimate timber value losses over time and the forest areas potentially susceptible to bark diseases due to damage intensity. A different elaboration of the same dataset of images could return

Fig. 9.2 Modern forest machines feature as a standard georeferencing sensor and wireless data transmission system (1) and encoders (in case of processor heads – 2). Additionally, accurate crane positioning systems (3) and grapple load cells (4) are increasingly used

additional services, for instance, for monitoring phenology, composition, tree cover, and biodiversity of the forest plots, where the machines transit, integrating with remote-sensing data products (see Chap. 11 of this book: Torresan et al. 2021) (Table 9.1).

Clearly, as already discussed, the maximum potential of the data provided by forest machines relies on the possibility to integrate it with other data sources, returning highly detailed estimates of stand characteristics (Barth and Holmgren 2013). Under this perspective, the potential of forest machines, as a carrier of sensors, and its capacity to interact with the virtual forest are still largely unexploited and can provide important benefits both to the efficiency of the supply chain and forest monitoring.

9.3 Marking and Tracking Systems in Forestry

Following the example of precision agriculture and precision farming (Mavridou et al. 2019), the evolution of onboard sensors and communication systems in forest machines has gradually made it possible to measure valuable parameters for each

Table 9.1 Main sensors currently installed on forest machines relevant for precision forestry (current) and CSF (potential) applications

Sensor	Data provided	Current use	CSF contribution	Diffusion on machines
GNSS	Position of machine	Guiding through paths predefined for minimizing costs and/or environmental impact	Georeferencing other machine-sourced data to relate it to the appropriate plot or forest section	Standard in most modern machines
Sensors of crane position	Position of machine	Sub-metric (tree level) positioning	Metric precision relating timber properties provided by onboard sensors (e.g., felled volume)	Installed in most recent machines but not yet designed to store position data
Encoders on processor head	Diameters and length of logs (− > volume)	Volume of timber produced Assortments identification Invoicing	Balance between forest growth and woody products extracted Main physical characteristics of processed stems and proportion of assortments produced	Standard in most modern machines (requires routinely calibration)
Grapple scale	Handled mass	Estimate of green weight of timber and biomass extracted	More precise estimate of balance between forest growth and woody products extracted	Available as optional, present in limited number of machines

produced item, such as logs and wood chips. This has promoted the development of technologies capable of automatically linking, such information to individual timber products, relating each item to its dimensions (returning volume and assortment) and to the position of the machine during the operations. Forest companies are increasingly interested in these systems, since an effective marking (tagging) system of timber products provides both track and trace services, which may significantly reduce the costs of timber production and management:

- The tracking service allows companies involved in the timber supply chain to locate their products at any moment through predefined checkpoints. It not only facilitates invoicing upon delivery of roundwood to end users but also simplifies the management of inventories.
- The tracing service allows companies, and potentially any stakeholder, to know where and when a product had been through the supply chain. It is worth pointing out that the tracing service (traceability) has the capacity to retrieve historical tracking data and appropriately save and store.

In addition to the direct economic savings, the availability of technologies for tree marking and timber product traceability would be a powerful tool for sustainable forest management disclosing or enhancing the following services:

- Certification of timber products and contrast of illegal logging in support to established certification schemes (e.g., PEFC and FSC).
- Allow to plan and execute highly precise forest works according to the guidelines of smart silviculture (described in Chap. 8 of this book: Pach et al. 2021).
- Relate any data produced for productive purposes along the timber supply chain (from stump to mill) to the stand and single tree health and physiology data gathered in the frame of CSF studies.

Tracking systems are common in industrial applications, but to automatically identify single logs throughout the forest supply chain is a challenging task. In timber supply chains, several handling operations may be involved (such as skidding, loading, piling, etc.), causing mechanical impacts and frictions. The capacity of tags or marks to remain attached to the logs is further hindered by the exposure to weather factors, such as rain, frost, and sunlight exposure. These factors, together with the presence of mud, dust, or resins and with the very variable conditions of natural illumination, may challenge readability of ID codes in the case of visual systems. Additionally, the tracking service should be based on unique codes (physical ID number) linked with an ID code stored in a database, allowing it to retrieve the attributes associated with each timber item. A further requirement is that the system must be reliable and relatively inexpensive, allowing to mark the ID of trees and logs without incurring in significant productivity losses for the forest operations. Finally, the stored information must be retrieved automatically at predefined steps of the chain supply with remote systems featuring reading ranges above 1 m (Korten and Kaul 2008).

9.3.1 Current Marking Systems

At present, several simple marking and tracing systems are used in forest management and timber production activities. This is the case in most of the Alpine region, where trees selected for felling are manually marked with log brand hummers in the frame of close-to-nature silviculture (Fig.9.3). In the simplest case, the tool is used to debark a spot of the trunk, creating a mark easily detectable by the chain saw operators. The forester uses the brand, featuring a personal and unique shape, to mark the stump of the tree and certify the selection made. This makes it possible to identify any tree felled other than the selected ones and detect illegal or unauthorized logging.

Due to their low cost and relative robustness in identifying trees intended for timber production, log brands as well as paper or plastic tags are used by a large number of forest companies worldwide to mark trees in the forest but also logs at sorting yards (Murphy et al. 2012; Kaakkurivaara 2019).

Color marking is another very common system, thanks to its fast and practical application. Along with manual marking, color spray enables automated marking. This is done by processor heads equipped with spray nozzles, which quickly mark the processed logs with a combination of up to three colors (Fig. 9.4). The system is generally used to segregate assortments differentiated by length and diameter classes (e.g., pulpwood vs industrial timber), as measured by the onboard sensors

Fig. 9.3 Brand hammer used to mark trees. Note the area debarked with the sharp part of the tool and in the back of it the shape of the brand. Painting and yellow tags were part of a test and would not be used in common tree marking together with brand hammer

Fig. 9.4 On the left, a timber processor crosscutting a tree. Just after this operation, it is possible to mark the cut section with colors. On the right, a processor head equipped with color spray: the blue color tank is visible

simplifying the operations of hauling and piling per assortment. The solution is effective, but it requires visual identification and interpretation by an operator at each step of the process and carries no single-item properties.

In timber supply chains featuring assortments with high unitary value, it is quite common to individually mark each log. Traditionally, this operation is performed manually, and the characteristics to be associated with the log are measured and reported manually in a database. Log brands cannot be used for this purpose, as they cannot define a unique ID for each log. Thus, volume, quality class (based on visual evaluation), species, and forest lot of provenance are associated with a unique alphanumeric code printed on a plastic tag fixed on the log. This information is used to track the log along the logistic process and as a reference for purchasing and invoicing of the timber products. Nevertheless, in spite of their low investment cost, manual-based tracking systems are not sufficient to cope with the recent developments in forest equipment technology.

9.3.2   Optical Marking Systems

When using paper or plastic tags, all the process is manual, as well as the recording of codes at the different steps of the supply chain (e.g., delivery at the yard of the end user). This leads to possible errors while reading or noting the codes, but most of all, it is a time-consuming operation, which may significantly increase the cost of the final assortments (Kaakkurivaara 2019). In the attempt to facilitate its use, some producers print on the plastic tags both the alphanumeric code and a barcode, or just the latter which can be acquired with manual optical readers.

Several types of barcodes are in use worldwide, but the most common types are:

– Universal product code (UPC) with straight vertical lines. It is the most common, consumer-grade standard. The information is featured in a 12 numerical digit code.
– Quick response code (QR) which carries a much larger amount of information compared to the former.

Barcodes are relatively inexpensive as can be printed on any support, usually paper or plastic tags, and applied to the timber products. This technology requires visual connection for code acquisition with smartphones or portable devices, and effective reading can be hindered by inappropriate light conditions or presence of mud and dirt on the barcode. Furthermore, tags can be removed or damaged during handling and transport operations. This last inconvenience can be overcome printing the barcode directly on the wood surface. Spraying ID codes on wood boards is a common tracking system in industrial sawmills and has been successfully tested in forest operations by installing spray nozzles on the chain saw bar of a processor head (Möller et al. 2011). This solution allows to mark the crosscut section of each log with a special barcode while processing and without productivity losses (Figs. 9.5 and 9.6). Clearly, the system has several challenges, among which is the readability of a code printed on the relatively rough surface of a log crosscut with chain saw. Readability can be improved with the addition of microtaggants. These are microscopic particles to be mixed with the paint used for spraying. Being composed by different layers artificially designed, microtaggants provide a unique numeric code (visible with microscopic analysis), which prevents falsification and that may include materials that enhance readability when exposed to UV or laser light sources.

Fig. 9.5 On the left, logs recently hauled marked with plastic tags (orange). On the right, two models of plastic tags with numerical and UPC barcode

Fig. 9.6 Example of UPC barcode on plastic tags (left) and a QR code (right)

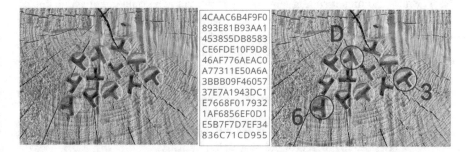

Fig. 9.7 The coded brands punched in the solid wood on the left carry an information carried as a unique ID for the log as simplified on the center and right. (Source www.otmetka.com)

A further development of the barcoding concept in timber industry is provided by the Swedish company Otmetka, which developed a dedicated solution for visually marking single logs. The system, called Woodpecker, is constituted by 12 hammer brands, which are punched into the wood by the harvester head. The hammers can independently rotate, and each position represents an alphanumeric value, generating a unique 12-digit code that identifies the timber item, assigning it an ID that allows retrieving its attributes in the associated database. The marks are impressed in the butt end of the log and can be read with optical technology and software elaboration at the sawmilling facility (Fig. 9.7). Also, this technology has been tested successfully on processor heads, leading to a fast and effective tagging of logs.

Another promising technology for timber tracking is biometrics fingerprinting (Murphy et al. 2012). This technology is based on the identification of the unique characteristics of each log end section (or crosscut section), identified through the analysis of digital images. The system is better suited for industrial, indoor applications due to the more controlled reading conditions (stable light, fixed angle, and distance of reading), but the implementation on processor heads has been conceived in order to secure a tracking service of single logs from the stump to the mill (Schraml et al. 2015). Biometric fingerprinting requires high-quality images of the log ends to discriminate logs. Common cameras providing images in the visible spectra can be deployed, but sensors operating in the hyperspectral and NIRs proved more effective (Schraml et al. 2020). Clearly, the file size of images would be too large to transfer it from the forest to a server and store it in a large database (e.g., storing data for all the logs produced annually by a company). The first elaboration can identify the key identity characteristics of the log end and store the information in much lighter vector files, which can be effectively transferred in a reference database (Fig. 9.8). Deployment of biometrics from stump to mill would provide useful data for the industry, for instance, sorting early in the supply chain the timber assortments, but with slight adjustments of the image analysis software, it would also provide relevant CSF information regarding health and development of the felled trees. Biometric fingerprinting technology is not yet at a commercial stage, but its fast development is expected to soon open a wide range of applications and services to the timber supply chain, forest management, and CSF.

In general terms, all optical systems share the same advantages and drawbacks:

Pros

- Excluding the capital investments, are low-cost solutions (printing material is relatively inexpensive).
- In the case of direct printing, punching, and biometrics, it leaves on the logs no undesired materials for the processing industry.

Cons

- Require visual connection and appropriate illumination conditions for a reliable reading
- Are sensitive to mud and dirt that can strongly reduce readability

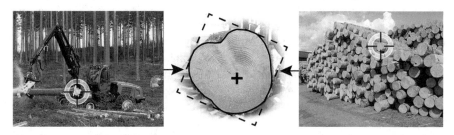

Fig. 9.8 Biometrics fingerprinting is based on the elaboration of visible and measurable characteristics of the log end, compressed in vector-format information. (Credit: Rudolf Schraml)

9.3.3 Radio Frequency Identification (RFID)

Another promising technology for tracking timber products is provided by the radio frequency identification (RFID), based on the capacity to transmit information via electromagnetic waves. A typical RFID system is based on a reader connected to one or more antennas emitting an interrogating signal. The signal is received by tags attached to the item. Tags consist of a transponder and a chip and can be active (with an inner power source) or passive. The latter requires the energy provided by the interrogating wave to return a signal. By EPC standard, the signal returns the identity of the tag, stored in the 12-byte memory of the chip. Disposable, passive RFID tags are preferred in timber supply chains due to their low unitary cost, even if they can store less memory and feature a lower reading distance compared to active tags (Björk et al. 2011). RFID can operate at different frequencies, ranging from low (LF), high (HF), to ultrahigh (UHF). The latter allows effective reading ranges of several meters in operative conditions with passive tags (Tzoulis et al. 2014; Kaakkurivaara and Kaakkurivaara 2019). RFID tags had been successfully used to transfer throughout the sequence of forest operations the valuable information provided by inventory data elaboration, such as bucking instructions for maximum value recovery (Picchi et al. 2015).

Among the main advantages of RFID tags in forestry applications, there is the possibility to perform fast and effective bulk reading (e.g., a whole truckload passing through a mill's gate). Furthermore, they can be used both to mark standing trees (Picchi 2020) and processed logs. Therefore, RFID technology can provide a complete service of timber traceability, which may be a powerful tool in the frame of a chain of custody service, guaranteeing the sustainable production of timber products from the forest to the sawmill and up to the end user (Appelhanz et al. 2015).

Regarding the drawbacks, currently (year 2020) the unitary cost of disposable UHF-RFID tags is relatively high. For custom-made tags, it ranges around 0.6–0.27 €, respectively, for batches of 5000 and 100,000 units (Picchi 2020). Furthermore, mills could be unenthusiastic to deal with tags made of plastic and metal, even if their overall mass is about 1 g. A possible solution could be the use of pulp-neutral material for the shell and radio-reflecting ink for the transponder, even if this would surely increase the cost of tags and probably reduce their performance (Fig. 9.9, Table 9.2).

Pros

- Automatic reading at high distances (3–5 m) of single tags or groups
- Reading not limited by presence of mud, dirt, or obscurity (Picchi et al. 2015)
- Possibility to manually or automatically tag standing trees and logs

Cons

- Unitary cost of tags may not be justified in case of low-value assortments.
- Disposable tags may be not accepted by end users of timber.

Fig. 9.9 On the left, an UHF RFID tag used to mark a standing tree, on the right an UHF RFID applied by the processor head after crosscutting

Table 9.2 Suitability of tracking technologies to applications in forestry and timber supply chains

Technology	Item marked		Application method		Automatic info retrieval		Reading range	Readable with mud/dirt	Resistant to hauling	Quantity of stored data
	Standing trees	logs	Manual	Auto	Forest	Yard/ mill				
Log brands (hammer)	+++	+++	+++	+	−	−	+	−	+++	+
Colors	++	++	+++	++	−	−	++	−	++	+
Barcodes-QR tags	+++	+++	+++	+++	+	+++	+	−	++	+++
Auto-barcoding	−	+++	−	+++	+	+++	+	−	+++	++++
Punching	−	+++	+	+++	+	++	+	−	+++	++++
Biometrics	−	+++	++	+++	+	+++	+	+	+++	+
RFID	+++	+++	++	+++	+++	+++	+++	+++	++	+++

9.4 Timber Industry as a Source of Data for CSF

As mentioned in the previous section, the availability of a reliable tracking service discloses several new opportunities in timber supply chains. Among the most promising applications can be found the synergy between tracking technology and timber grading systems based on fast and nondestructive sensors. At present, these sensors find little application in forest operations: timber assortments can be sorted with machine support (e.g., using the StanForD data and dedicated software in the harvest machine), but quality grading of the logs is performed only visually. Nevertheless, this is a costly and time-demanding operation, which is potentially biased by the subjective judgment of the expert evaluator.

Once in the industry, a large set of sensors is used to precisely quantify the volume of the logs delivered and its quality. Additionally, an increasing number of sensors or analytical techniques are used to optimize the following steps of product transformation in the sawmill or pulp mill. According to Gergel et al. (2019), the sawmilling optimization made possible by a detailed knowledge of the properties and defects of single logs leads to an increased value recovery of 15% and 23%, respectively, for conifer and broadleaf timber. The same authors also highlighted how the status and quality of the standing trees prior to harvest plays a role of the utmost importance in the efficiency and economic balance of the timber supply chain. Nevertheless, in real practice, just empirical sorting models are used for inventorying trees according to their economic value. These return the share of predefined timber assortments with an intrinsic inaccuracy due to the impossibility of detecting internal wood defects from external features. A wide range of sensors could contribute to provide this service early in the supply chain. Their potential is maximum when different sensors work in synergy, contributing to generate a wide and comprehensive picture of the timber quality, for instance, integrating stem shape characteristics as returned by a TLS system (e.g., taper, straightness, crown insertion, etc.) with internal properties as detected by optical sensors on the felled trees (e.g., rot, decay, eccentricity, etc.). In the short term, the fusion of such data provides a more detailed definition of quality of single logs. In the long term, optically sensed characteristics may be used to tune the models interpreting 3D data provided by laser technology, gradually but constantly decreasing the gap between estimated properties and actual characteristics of the assortments. The very same data and approach can also disclose the "memory" of timber, being this the result of the growing conditions that the tree endured along its life. Widening the scope of the data analysis and the inputs of the models, it will be possible to support CSF studies and forecasts by relating the captured wood characteristics with the physiology of the trees and their response to the past stresses.

A recent trend in timber characterization is to deploy sensors as early as possible in the timber supply chain. Although application of timber quality gauges in forest machinery is mostly at the experimental stage, it arises great interest from the industry, since quality determination early in the supply chain would allow a great reduction of procurement costs by delivering just the desired quality to each end user, optimizing the logistics and further increasing the profitability of the industrial transformation. At the same time, the application of sensors for timber quality along the timber production process would bring additional benefits also to CSF applications, providing more data and with higher level of detail.

In addition to UAVs, forest technologies for monitoring tree processes include sensor networks deployed on the tree stems or embedded in the soil layers. The proliferation of these technologies generates a flow of data, which needs to be appropriately investigated through machine learning for automating or responding to disturbance events. The digitalization of forest stands allows forests to operate as technological platforms so that trees function as technical instruments informed by data that are meant to enable precision forestry and practices oriented toward high timber quality.

9.4.1 Wood Properties Relevant for CSF

The industrial conversion of timber highly relies on the intrinsic characteristics and technical properties of this resource, which may be advantageous (or not) depending on the downstream process applied. The set of properties defining the quality of wood must be determined against the industrial requirements, local policy, and environmental constraints to assess its market value.

In a theoretical approach, a healthy tree results in production of the "perfect wood" defined here as a bioresource without wood defects. Perfect wood corresponds to the cylinder with a pith positioned in its geometrical center and containing regular concentric structures corresponding to yearly rings. In practice, perfect wood does not exist and always includes some defects. Here, the "defect" is considered as an undesired imperfection of the regular wood tissues that is a result of diverse stresses and factors affecting tree growth and the morphological constitution of the plant. Such natural features may be highly undesirable, downgrading the industrial value of timber. An example could be a knot that is an imprint of the branch positioned in the trunk. Branches cause the presence of knots that deviate the fiber direction, having a tremendous effect on the mechanical properties of timber, as well as tree stability.

9.4.1.1 Tree-Ring Structure

In temperate forests, the life of trees covers a sequence of several seasons that are recorded in the tree rings. These are structural wood tissues of different properties in spring/summer (early wood) or autumn/winter (late wood). The ring width, its chemical composition, microfibril structure, as well as the ratio of late to early woods may vary, depending on the age of tree, meteorological history, or presence of diverse factors stimulating or inhibiting the tree growth. The natural yearly sequence of air temperature, solar light photoperiod, as well as water stress levels are expressed as the dynamical changes of the phenological events determining specific xylogenetic sequences. Therefore, the tree-ring structure may be considered as a "fingerprint" unique for each plant. This can be used for dating of wood (dendrochronology), determination of the wood origin (dendro-provenance), analysis of the local climate changes (dendro-climatology), or identification of catastrophic events, occurring during the life span of the tree, among others. As such, tree-ring analysis can be considered as one of the most important inputs for the tree growth and health models developed in CSF (see Chap. 7 of this book: Bosela et al. 2021).

Traditionally, tree-ring analysis was performed by visual (microscopic) assessment of tree core samples extracted from the living tree or on cross sections of the log after felling. For that reason, the amount of available information was relatively limited. However, several modern technologies, especially based on the tomography approach (Van den Bulcke et al. 2014), allow mapping of the tree-ring structure without necessity for its cutting or extracting samples. Mobile X-ray computed tomography technology was applied in wood density measurements and moisture content monitoring on standing trees (Raschi et al. 1995; Tognetti et al. 1996). Computer 3D tomography allows locating internal log features that include pith, sapwood, heartwood, knots, and other defects. With the appropriate techniques, it can also return a detailed analysis and densitometry of annual growth rings (Van den Bulcke et al. 2014). Additionally, spectroscopic methods as well as its evolution by means of hyperspectral imaging (Sandak et al. 2020; Schraml et al. 2020) provide a possibility for extraction of till-now not accessible information regarding chemical and physical properties of wood resolved spatially to the tree-ring level.

Nevertheless, from a technical perspective, it is crucial that both the quality of the acquired images and the capacity of the interpretation software provide sufficient information regarding tree rings of the logs (Subah et al. 2017; Cruz-García et al. 2019). The sensors installed on the processor head and operating in the forest on unprepared surfaces (the chain saw cutting surface is relatively rough for this use) may be incapable of returning the required quality for a deep and reliable analysis. On the other hand, they would potentially provide a large dataset, based on all the logs produced in the forest, including those with defects. This contraposes to the higher potential of industrial timber analysis, where more powerful sensors, deployed in a controlled environment, can provide extremely detailed data. Once set up, an automated tree-ring analysis, linked with an effective traceability system, would provide a very large volume of data. Such information may prove extremely valuable for understanding the past dynamics of relatively large areas of forests or to refine and better elaborate the information provided by dendrometers installed in monitoring plots (Cruz-García et al. 2019). For instance, the availability of a large dataset reporting growth pattern of trees of the same species growing in different areas allows to better understand the seasonal and site-specific growth response to drought (Mina et al. 2016) or human-driven factors, such as pollution (Innes and Cook 1989) or wildfires (Walker et al. 2017). This information can be further elaborated, helping to understand the response to stress of forests with different characteristics, such as density (Sun et al. 2020) and elaborate guidelines for CSF implementation. Although measurements of stem growth characteristics in temporal detail (e.g., radial increment, slow vs. fast growing trees) are important to understand wood properties, these properties change during wood formation in response to changing environmental conditions. Indeed, vessel conduit dimensions, cell wall thickness, and the relative proportions of different xylem cell types vary during the growing season. Changes in stem size detected through dendrometers can be associated with tree water status (Zweifel et al. 2007), with daily fluctuations being related to physiological parameters (e.g., leaf water potential, whole plant transpiration) and environmental conditions (e.g., evaporative demand and air temperature) (Giovannelli et al. 2007; Tognetti et al. 2009). These stem radius changes provide a sensitive indicator of the combined effects of actual radial growth and stem water storage and release (Drew and Downes 2009). Stem size variation provides indications of water stress thresholds in tree species and is potentially useful in threshold analysis for binary classification and determining the influence of thermal or moisture cycles on elastic shrinkage (Cocozza et al. 2009, 2012). Coding of the dendrometer signal helps quantify stem (or log) sensitivity to environmental fluctuations (moisture, temperature), as well as synchronize time series related to wood properties and climatic events, and to identify time lags of environmental effects on wood traits (Cocozza et al. 2018).

Ideally, the study of tree reactions to stresses should be based on the analysis of tree-ring development of trunk sections located at different heights. In fact, while the common dendrochronology focuses on basal sections, where all the rings are represented (since the early development of the tree), in physiology studies, the rings grown higher in the tree's crown may better describe the stresses suffered by the plant. For instance, in the case of *Picea abies* growing in the Alps, the phloem's growth at the base of the trunk is completed already in July, while at higher levels of the tree, it keeps growing for the whole summer. Thus, the rings developed higher on the stem provide more effective evidence of the growing conditions of the mid-late growing season. If the tracking system records the order of production of each log, their original position in the trunk can be located and the sawmill's data (e.g., tomography) can be related to sections corresponding to different heights of the tree.

9.4.1.2   Timber Density

Density of wood (or specific gravity) is a metaparameter determining several technical characteristics of natural resources, considered as the most relevant wood quality descriptor (Zobel and Jett 1995). It expresses the amount of wood substance contained in a given volume. Even if density is not directly affecting wood properties (it is a property quantifier), it is highly correlated to the majority of wood assets (chemical, physical and mechanical), the yield of production, and the overall "wood quality" in general. The density distribution differs within a single plant, both along the tree height and the trunk diameter. Despite being a property of remarkably high native variance and heritability, it can be related at stand level to growth conditions or silvicultural interventions (Briggs et al. 2008), thus providing valuable information if appropriate reference values can be defined. Density is often measured in sawmill, but the interest to discriminate the timber products with the desired properties as early as possible within the supply chain stimulated the development of portable instruments. These can be deployed both for assessing the characteristics of standing trees (Paradis et al. 2013) and for automatically measuring each log with gauges installed directly in the processor head (Walsh et al. 2014a).

The effect of tree-ring width on tree-ring wood density depends on the species (conifers vs. broadleaves, fast- vs. slow-growing species), the timing of climatic events influencing growth throughout the growing season and the general fertility of the site. In particular, intra- and interspecific interactions may affect the radial growth and wood density of individual species growing in mixture when compared to its monoculture (Zeller et al. 2017). The acoustic sensing technology, such as the modulus of elasticity of wood (MOE) and the dynamic modulus of elasticity (MOEd), allows the estimation of intrinsic wood properties for standing trees, stems, and logs. These parameters depend on wood density and are fundamental for the evaluation of wood quality, providing information related to wood anatomy and tree physiology (Russo et al. 2020).

9.4.1.3   Chemical Composition of Wood

From the chemical viewpoint, wood is a natural composite of three biopolymers, including cellulose, lignin and hemicellulose. These are major constitutive chemical components, with their specific ratio varying between wood species, forest types, and within individual trees. In temperate areas, the chemical composition differs substantially along the tree height and its radius following the natural lifetime sequence of the periods when the tree grows fast (spring) or forms more mechanically resistant morphological structures (autumn). The variation of chemical composition can also be noticed at the level of the yearly ring that reflects the combined effects of the season and plant development stage as well as any other stresses for the tree due to biotic or abiotic factors. In addition to cellulose, lignin, and hemicellulose, small amounts of minerals and extractive components are present in natural wood. The latter are particularly relevant despite their low concentrations. In fact, extractives may significantly affect the suitability of wood resources for a given conversion process or affect the durability of wood-derived products.

All the chemical components form larger macromolecules, such as microfibrils, that are combined at different scales as fibrils, cells, and yearly rings constituting a hierarchical structure. The specific physical properties of wood are, therefore, highly dependent on the scale of observation (nano, micro, macro). It implies the necessity for adjusting measurement procedures and instrumentation for determination of desired chemical/physical characteristics and material properties. Such information has not yet been used for CSF applications. This is probably due to the cost and time delay of wet chemistry analysis. The availability of fast and nondestructive sensors, such as hyperspectral cameras, may disclose a new source of information to understand the health and growing conditions of trees.

9.4.2   Wood Defects

In contrast to the perfect wood, the real trunk contains diverse imperfections recording all the lifelong-related growth conditions, perturbations, or stresses. In the timber industry, these are defined as wood defects and may include numerous features differentiating defected from the perfect wood (Kimbar 2011). The European standard EN 1927-1:2008 provides a systematic methodology for identification and quantification of the log/wood defects that are later used for determination of the quality class. The following sections report the most relevant defects from the perspective of CSF.

9.4.2.1 Resin Pockets

Resin pockets are small gaps within the structure of the xylem filled with resin. Wood development due to tree growth usually occludes them within a few years after their formation. Resin pockets are a significant technological defect for the timber industry, particularly for joinery and furniture applications, due to the release of resin over time from the finished products. Resin pockets are common in conifers with resin canals (such as *Picea*, *Pinus*, *Larix*, and *Pseudotsuga* genera) but may also be the consequence of stress. In the latter case, they are commonly related to animal or insect attacks, sites exposed to strong winds and storm damages. Research demonstrated that water stress is to be regarded as the most relevant factor leading to the development of this wood defect (Seifert et al. 2010; Jones et al. 2013), although other factors, such as excessive growth rate or share of defect core over total diameters at breast height (DBH), may contribute to pocket formation (Woollons et al. 2008). Availability of data on resin pockets may be a useful tool for identifying historical occurrence of water stress, areas with frequent wind gusts, and, possibly, the impact of forest management on certain growing conditions. Clearly, the interpretation of resin pocket presence is much more significant if associated with other parameters detected by the sensors on processed and sawn timber, namely, tree-ring development. Furthermore, tree rings are present throughout the whole trunk, but their size increases from the base upward and from the core outward (Gjerdrum and Bernabei 2007). Thus, for a correct interpretation of their occurrence, it is important to know the position in the stem of the timber sample considered. This is possible only with an accurate traceability system capable of relating each log sourced from a tree with each other according to their sequential order.

Like resin pockets, resinous wood is a zone within trunk volume with exceptionally high content of extractive components and resins. It is present only in conifer species with resin channels. The usual causes for resinous wood formation are responses to the microorganism activity (especially parasite fungus) or to the damage of wood induced by mechanical actions. The analogy for self-protection by the resin release in some broadleaves is a gum production that can be triggered by the frost, wound, or microorganism attack.

9.4.2.2   Cracks

Checks, splits, and shakes (commonly defined as cracks) are separations or ruptures of the wood tissue in the longitudinal plane that normally occur along fibers in the radial or tangential sections of the trunk. These may appear on the cross section of the tree or on the log side circumference. Checks are separations of the fibers that do not extend through the timber from one face to another. Splits, however, extend the material discontinuity from the one log face to another. Shakes are separations or weaknesses of fiber bond, between or through the annual rings. There are diverse sources of stresses occurring to wood during life cycle. The most relevant are growth and drying stresses, beside thermal, frost, wind-, solar-, or lightning-induced tensions. Shakes may originate from causes other than drying stresses, e.g., from careless felling, where internal stresses existing in the living tree are released when the tree is felled. Checks, shakes, and splits are present although they may not be visible (closed checks and closed splits). Cracks may close up if the dry timber is subsequently exposed to damp conditions, but once the fibers have separated, they cannot join together again. A great threat for the tree after a crack occurring is elevated risk of decay fungi spore access to the unprotected wood tissue. The presence of cracks is an important limitation for the downstream conversion, especially if combined with other defects, such as spiral grain. Due to specific properties, cracks are relatively easily detectable even if not visible on the surface of log. This is due to the discontinuity of the material and related change of the material stiffness (natural frequency) or elasticity (stress wave propagation velocity).

9.4.2.3   Reaction Wood

Reaction wood is a type of defect that tends to form in trees growing in a leaning posture. This may be caused by exposure to strong winds or because the tree grows on a slope. Reaction wood in coniferous trees is formed on the lower side of the lean and is called "compression wood." It is often characterized by a dense hard brittle grain that contains very high content of lignin that increases wood resistance for compression. On the contrary, the broadleaves create reaction wood referred to as "tension wood" that is positioned on the upper side of the lean. It contains higher content of cellulose that increases wood resistance to tension stresses. Properties of compression wood are considerably different from those of normal mature wood. Compression wood tracheids, for example, are about 30% shorter than normal. In addition, compression wood contains about 10% less cellulose and 8–9% more lignin and hemicelluloses than in normal wood. These factors reduce the desirability of compression wood for pulp and paper manufacture. It is also less suitable as sawn timber since it shows a lower strength, stiffness, and dimensional stability, resulting in a decrease in yield of high-quality end products.

Data related to reaction wood can be a particularly useful tool in the frame of CSF in mountain areas, as it records specific reaction of trees to environmental conditions. According to Łszczyńska et al. (2019), data regarding the frequency and characteristics of reaction wood in forested plots may be used to assess the landslide hazard risk. In fact, in case of slight land movements, trees are tilted causing leaning and thus the formation of reaction wood. The same phenomena also lead to the formation of eccentric piths, which can be detected by tree-ring analysis. The vertical stability of a tree can be assessed using an automatic accelerometer (gyroscopic sensor), which measures the position and oscillation in three axes (Matasov et al. 2020), therefore, providing useful information on the effect of wind exposure on the tree aerial architecture and species-specific biomechanics.

9.4.2.4   Rot

Fungi decay is considered as the most problematic biological threat degrading wood at all stages of its life cycle, including postharvesting. Due to its chemical composition, wood is an optimal source of nutrition for fungi and, therefore, it will be a subject of extensive degradation whenever favorable conditions for growth of fungi occur. These include temperature ranges from 20 to 30 °C with a wood moisture content ranging from 20 to 50%. Diverse species of decaying fungi are specialized in degradation of specific wood polymers resulting in different degradation results. This led to the classification of fungi into three major groups: white rot, brown rot, and grey rot. The fungi spores may access wood by several ways, including root, broken branch, damaged leader, or scar on the stem. Cracks, wounds, or any other exposed surfaces of the wood. The presence of birds nesting in hollows in the tree is a certain sign of the progressing decay deterioration. Rot is the wood defect for which the quality grade reduction is obligatory. Logs cut out from older trees are more likely to contain developed rot. The final stage of rot is a complete or extensive material loss forming internal cavities. Plants have developed several mechanisms to defend themselves against decaying fungi. For instance, diverse chemical substances synthetized by trees are natural biocides, such as tannins, resins, or gum.

9.4.2.5   Knots

A knot in the tree is the portion of a branch or limb that has been surrounded by subsequent xylem growth during the tree life. Knots form morphological structures starting from the pith and by following the radial direction reaching the log surface. There are more than 50 types of knots that are classified according to the size, decay presence, and location and distribution within the stem.

The size, type, and distribution of knots have the most important impact on log quality and are the main consideration when applying grading rules. The severity of the grain deflection caused by the knot is correlated with its size. In any case, the presence of knots changes the anatomical structure of surrounding wood (reaction wood presence and grain deviation), as well as its chemical composition (extremely high content of extractive components). The distribution of knots in logs depends on the species and characteristics of the growing site. It is also determined by the growth characteristics of the tree and the tree age. The detailed knowledge regarding knots in trees is highly relevant for CSF. Fortunately, there are several scanning techniques for automatic detection and classification of knots that are implemented during forest operations, log sorting/grading, as well as downstream conversion.

9.4.2.6 Shape Imperfections

Any deviation of the tree form from the perfect cylinder is considered as a shape imperfection, reducing the yield of product that can be obtained in the sawmill. Sweep, excessive taper, bulges, swell, flanges, and out-of-roundness are some of the most important shape imperfections. All these defects can be measured both on the standing tree and on the processed log by means of laser triangulation scanners, photogrammetry, and TLS. In some cases, imperfections can be reported also by the standard measurement equipment installed on most modern timber processors.

Sweep is a bow-like bend in the trunk of a tree diverging the trunk from the straight and vertical theoretical axis of the tree. The presence of sweep is a result of diverse factors, including slope of the terrain, temporary loading due to snow or wind, mechanical damages, or insect activity.

Tree taper is defined as the gradual reduction of the log diameter along its length. It is a natural feature of each living tree, even if logs with a high degree of taper are considered as having a poor form. Although taper cannot be eliminated, it is possible to minimize it by means of appropriate silvicultural activities. In fact, the extent of taper depends not only on the tree species, local climate, age of the tree, soil fertility, and terrain irregularity but also on the density of trees within the surrounding forest stand.

Bulge is an enlargement of the tree diameter forming a barreling shape. It is a natural feature when occurring at the bottom of the tree, assuming reasonable progress of the diameter changes. On the contrary, when occurring in the higher part of the trunk it is frequently associated with fungi or bacteria attack.

Canker is a defected wood in a form of gnarls or volume losses, both attributed to the phytopathological changes triggered by an activity of microorganisms (fungi or bacteria). In contrast to swollen wood, the tissue of the canker is abnormal and sick.

Out-of-roundness is a shape imperfection appearing as an elliptical cross section of the log. It is frequently associated with the eccentric or double pith. The usual consequence of the pith shift is the presence of reaction (tension or compression) wood. Another reason for the out-of-roundness is the partial damage of the cambium due to mechanical or phytopathological injury.

Such stem defects can be related to the influence of various environmental factors, supporting CSF studies to better understand tree growth dynamics. According to Schneider (2018), the hydraulic and biomechanic theories are the most widely used. Both theories underline the importance of crown dimensions in determining tree form, as confirmed by Kidombo and Dean (2018). Yet, climatic variables such as total summer precipitation and mean winter temperature may have a higher influence on tree taper than the average wind speed (Schneider 2018). This underlines that the complex tree growth dynamics require a holistic analysis of all climatic factors and physiological processes to understand stem formation.

9.4.3   Sensors for Timber Quality Assessment

Once the timber reaches the mill, a large array of possible sensors can be deployed to analyze the quality and volume of the logs to be processed. There are several technical solutions available for wood defects detection or quality grading of logs. Some of the most compatible with CSF requirements are presented in Fig. 9.10.

The length of log combined with its diameter is a basic merchantable property. As mentioned in Sect. 9.4.2, all the processors used for trees harvesting and delimbing are equipped with a measurement *wheel and optical encoder* (Mederski et al. 2018) (Fig. 9.10a). Sensors used for log diameter measurements are usually *absolute encoders* configured as protractors, which are integrated with both delimbing knives and feed rollers and measure these rotation angles (Fig. 9.10b). An advantage of such a measurement system is a possibility for a continuous determination of the diameter change trends along the three when passing the processor head. Combined length and diameter measurements provide not only highly precise information regarding a single log volume but can also quantify taper. Such a measurement approach is capable of fast data acquisition and straightforward integration with a database when converted to StanForD file format.

The *light curtain* is a simple optical measurement system, where the dimensional information regarding the object is determined by illumination (or shadowing) of electronic photodetectors (Fig. 9.10c). It was a very popular solution for the size sorting of logs supplied to the sawmill. However, nowadays it is replaced by the *3D laser triangulation systems* relying on the image analysis of the structured light profiles (usually laser) that appear as deformed when observed from an angle (Siekański et al. 2019). The advanced analysis of the surface texture allows identification of some wood defects appearing as particular textures of the bark.

Implementation of *X-ray scanners* for monitoring of logs is the industrial solution for detection of the majority of wood defects not visible on the log surface (Fig. 9.10d). It allows for identification of deviations of the wood properties along the log, without the possibility to localize the depth position of each feature. For that reason, this setup is frequently duplicated to provide a possibility for better recognition of the defect location within the log section. The ultimate solution for the X-ray imaging of logs is *X-ray computed tomography (CT) scanner* (Fig. 9.10k). This technology allows straightforward detection and mapping of internal defects, such as butt decay, voids, cracks, or inclusions. The resolution of images allows dendrometric analysis of tree rings and other refined measurements (Van den Bulcke et al. 2014; Stängle et al. 2015; Rais et al. 2017; Gergel et al. 2019). Appropriate data analysis may allow for the information collected on the logs to be related back and integrated with the information on the original tree. An example of this application is provided by Stängle et al. (2014), which compared data from TLS stem and branch scar analysis with X-ray computed tomography (CT), and Uner et al. (2009) using X-ray CT to highlight the effect of thinning on timber density. Thanks to the presence of several X-ray CT operatives in sawmills around the globe, and to their high-speed analysis (up to 180 m/min), this technology is a highly promising data source for CSF applications.

As in the case of drilling resistance, it is possible to indirectly assess mechanical strength of wood by measuring cutting forces occurring when crosscutting logs with a chain saw (Fig. 9.10f). This can be implemented by integrating *load cells* with the cutting unit or by measuring other effects, such as electrical power consumption or *oil pressure changes in the hydraulic circuit* of the saw motor (Sandak et al. 2019). Another possibility for adopting cutting resistance analysis for log characterization is to measure the forces required for delimbing. It is clear that a healthy and big branch results in much higher cutting resistance than delimbing of the small and dry branch. In any case, value obtained for a chain saw or delimbing knife can be considered only as an estimate and indirect quantification of the wood suitability; however, it is useful to identify some critical wood defects, such as decay or butt rot.

Stress wave propagation velocity is a highly useful tool for identification of the stiffness and modulus of elasticity (Walsh et al. 2014b) of logs derived from harvested trees (Fig. 9.10g). In that case, the sensor can be installed directly on the processor head and perform analysis before a crosscutting operation. A similar approach is used for determination of the mechanical properties of logs by measuring these *natural frequencies*. In that case, the log is induced to vibrate by an impact, and the vibrational or acoustic response is measured with specific detectors (Fig. 9.10h). These systems can be implemented both, as a part of the processor head configuration or component of the log sorting line in the sawmill.

The advanced algorithms used in image analysis enable mimicking of human vision and thus more effective automatic detection of wood imperfections or presence of defects. Other uses of the images is to implement a fingerprinting approach for log traceability and authentications (Schraml et al. 2020). The *scanning with cameras* can be performed on the log crosscut end (Raatevaara et al. 2020) (Fig. 9.10i) or on the side of the log (Shenga et al. 2015) along its circumference (Fig. 9.10j). The spectral information collected may include *monochromatic, RGB color, multispectral, or hyperspectral images*.

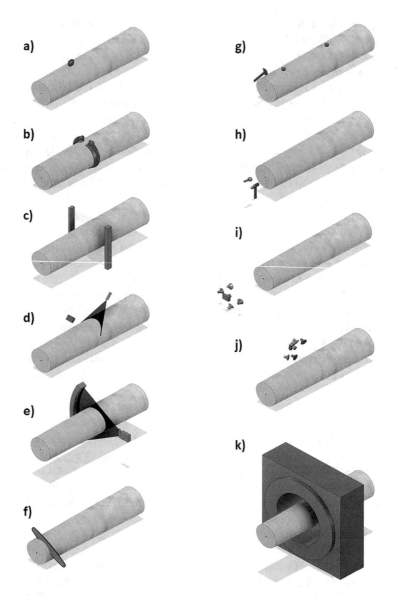

Fig. 9.10 Selected sensing techniques used for assessment of log characterization during forest operations and log sorting in sawmill: (**a**) measurement of lengths with rotary encoder, (**b**) measurement of the log diameters profile by two encoders integrated with debranching knives, (**c**) measurement of the log diameter with light curtain, (**d**) 3D log surface scanning with laser triangulation system, (**e**) determination of the X-ray attenuation for internal defects detection of density profiling, (**f**) measurement of mechanical properties by assessment of cutting forces when cross cutting logs with chain saw, (**g**) modulus of elasticity assessment by stress wave velocity, (**h**) free vibrations for determination of modulus of elasticity, (**i**) camera vision system for cross section scanning, analysis of tree rings and fingerprinting of logs for traceability, (**j**) camera vision system for identification of wood defects present on the log side surface, (**k**) computed X-ray tomography for 3D visualization and detection of internal defects

9.4.3.1 Emerging Scanning Technologies

The scanning technologies presented in the previous section represent the industrial state-of-the-art solutions and are commercially available on the market. However, there are several promising techniques with great potential for migration from the laboratory testing into practical applications both in forest and in timber transformation industries. Some examples of the most suited for the CSF are briefly described below.

Spectroscopy, including visible, near-infrared and mid-infrared ranges is an analytical technique quantifying interaction of the electromagnetic radiation (light) with the matter. The light can be scattered, absorbed, transmitted, reflected, or trans-flected from the measured surface. The specific pattern of that interaction is recorded as a spectrum. Light absorbance, especially in the infrared range, is highly related to the chemical composition (functional groups with dipole momentum) of the sample. Therefore, by assessing light spectra reflected from the wood sample, it is possible to determine its chemical composition and some physical properties. An extensive reference dataset and advanced multivariate data analysis are indispensable to assure the high reliability of prediction models (Sandak et al. 2016b). However, if properly calibrated, spectroscopic evaluation of wood can provide a rapid and very low-cost assessment of a broad range of the properties, including species, provenance, chemical composition, physical properties, or suitability for diverse conversion paths (Sandak et al. 2017). Portable NIR spectrometers even allow scanning of standing trees or fallen logs directly in forest (Sandak et al. 2016a).

An evolution of the spectroscopy toward a space resolved map of spectra is implemented as *multispectral* or *hyperspectral imaging*. The difference between both is in the number of spectral bands constituting the spectrum, that is, <10 in the case of multispectral imaging. This technology allows rapid mapping of the chemical properties of the material over the object surface (Thumm et al. 2010). There are several properties of wood that can be assessed with the help of this technique, such as moisture content, chemical anisotropy of the constitutive polymers at diverse heights of trees (Meder and Meglen 2012), and fiber angle direction (Ma et al. 2017), among others. The limitation of hyperspectral imaging is relatively high investment cost, fragility of the optical instruments when integrating with forest operations, and very high amount of generated data that requires refined IT systems and algorithms.

Only a few wood scanning techniques allow scanning of the bulk interior. In contrast to harmful ionizing radiation of X- and gamma rays, microwaves are considered as safe and easy to apply in the scanning systems. A great advantage is that, assuming sufficient power of emitter, microwaves penetrate wood bulk, and these interactions with the matter can be interpreted as attenuation, phase shift, or polarity change. These wave properties are directly correlated to the wood density, wood moisture content, and grain angle direction (Schajer and Orhan 2005). For that reason, it is possible to simultaneously measure all the above wood properties. *Microwave scanners* can be implemented as an array that, in consequence, allows spatially resolved maps of wood assets (Table 9.3).

Table 9.3 Summary of sensor technologies to determine tree/timber characteristics

	Tree dimensions	Tree-ring structure	Timber density	Chemical composition	Wood defects					Shape imperfections	Mechanical damages
					Resin pockets	Cracks	Reaction wood	Rot	Knots		
Dimension measurement	+++									+++	
Satellite and drone imaging	+								+	+	+
Terrestrial LiDAR	++								++	++	+
Vision systems and photogrammetry	++	++	+	+		++	++	++	++	+++	++
Light curtain	+++									++	
3D laser triangulation systems	+++					+			++	+++	+++
X-ray scanner	++	++	++		+	+	+	++		++	+
CT scanner	+++	+++	+++		+++	+++	++	+++	+++	+++	+
Stress wave propagation velocity	+++		+		+	++	+	+++	++		
Natural vibrational frequencies			+			+		+++	++	+	
Acoustic tomography	+	+	++		+	++	+	+++	++		
Electrical resistivity tomography	+	+	+		+	+		+++	+		
Stiffness of the standing tree	+		++		+	+		+		+	
Cutting forces			++		+	+		++	+		
Drilling resistance	+	+++	++		+	++	+	+++	+		
Spectroscopy		+	+++	+++	+	+	++	+++	++		+
Hyperspectral imaging	++	+++	+++	+++	+++	+++	+++	+++	+++		+
Microwaves		+	++		+	+		+	+		

9.4.4   *Intelligent Forest Machines, the Way Ahead*

As described in the previous sections of this chapter, the availability of digital forest data makes it possible to optimize forest works from the operational cost and environmental impact point of view. Traceability tools further increase the precision of the operations and guarantee a full control of the woody products, whose value is effectively assessed and maximized by an array of sensors deployed in the mills. From the management (and CSF) point of view, the most effective solution is to link the databases of the forest inventories and the derived tracking systems with the log scanners operating at the sawmill. Nevertheless, a further significant improvement could be achieved with the determination of timber characteristics early in the supply chain, enhancing the sorting of logs and increasing the overall value of the derived timber products (Taube et al. 2020). The fast development of sensors makes this challenge possible, as it had been demonstrated by the EU project SLOPE (https://cordis.europa.eu/project/id/604129), funded under the seventh Framework Program. Within the project, the potential of the Virtual Forest, the tracking systems, and the timber sensors had been combined in a technological showcase, proving the feasibility of this concept.

The virtual forest was generated integrating information from satellite, UAVs, and TLS surveying systems, combining macro and local analysis in the characterization of the forest resources in mountain areas. Data was stored in a dedicated database provided with a 3D interface, which allowed users to navigate into the virtual forest, estimate the volume, and value of timber in a selected area as well as to plan accordingly the harvest operations by cable yarding (Fig. 9.11).

The cloud point generated by TLS surveys was used to characterize the trunks in high detail. A dedicated software matched the shape characteristics of the trees with the timber assortments and locally accepted their value, returning the maximum value recovery conditions. Those were appended in the database as bucking instructions to be transferred from the forest to the processor head. For this purpose, standing trees were georeferenced and marked with RFID UHF tags whose ID was linked to the data and bucking instructions of each tree (Fig. 9.12).

All forest operations were performed by prototypes of intelligent machines, namely, a cable yarder and a processor head operating at landing. The former detected the RFID and weight of the load, while the latter by reading the ID of the tree could acquire the crosscutting positions for optimal bucking. Additionally, the processor head installed several sensors for timber quality assessment as described in Sandak et al. (2019).

In detail, the processor deployed the following systems (Fig. 9.13):

- Load cells and hydraulic pressure sensors (1) for estimate of a branch index and the approximate position of knots on the trunk
- Stress wave and free vibration measurement systems (2) for timber density assessment
- Near-infrared (NIR) and hyperspectral imaging systems (3) for characterization of crosscut section and detection of defects (e.g., rot, resin pockets, etc.)
- RFID-UHF reader (4) for acquisition of tree ID and retrieval in the VF database of the cutting instructions elaborated for maximum value recovery

Fig. 9.11 3D visualization of forest (image combined with aerial picture). Marked trees are visualized in two colors: green for the trunk section detected by TLS and white for the estimate of the higher part of the trunk as estimated by UAV image analysis. (Source Project SLOPE)

Fig. 9.12 Trees felled with motor-manual chain saws are marked with RFID for tracking along the supply chain and for transfer of cutting instructions to the processor head at landing

Additionally, the processor could mark each new log with an RFID tag, providing a complete tracking system capable of linking the original standing tree (through the first tag attached on it) to each log delivered to the sawmill.

The whole system proved effective in assessing the quality of timber at roadside. The optical sensors (not installed in the prototype) could identify decay spots and would warn the operator to recalculate the cutting instructions, which were based on the shape of the standing tree. From the point of view of a CSF application of this data, the system provides a further source of information which, when fully integrated with other sensor data, adds to the tool kit available for CSF planning and operations. Currently, only the logs with sufficient market value reach the sawmill and can be further analyzed with industrial sensors. However, the presence of

Fig. 9.13 Prototype of intelligent processor head equipped with sensors for timber quality assessment and tracking of logs

sensors on the processor head would provide data of lower quality (due to the work conditions), but as this is data from all the trees and logs produced in the stand, it is thus more representative of the general health of the trees. Once a track and trace system are fully integrated within the timber supply chain, it is possible to relate the information regarding each standing tree (and the logs produced from it) with the analysis performed in the sawmill. If the same trees were included in a network of sensors, the historical data collected by microcontrollers up to the tree felling could be related and integrated with all the information provided by all the sensors installed on the forest machines or the sawmilling facilities (Fig. 9.14).

The availability of an infrastructure for sensing, wireless transfer, and cloud-elaboration of data developed for forest operations (and maintained by its revenues) is a clear opportunity for CSF. In fact, the network of sensors deployed for forest monitoring and management purposes could rely on this infrastructure for data transfer, storage, and elaboration. Furthermore, by accessing the databases of the virtual forest, it will be possible to integrate, validate, and broaden the data provided by the climate-smart sensors. The data used for operations planning and generated by machine sensors can be integrated with a unique, flexible system serving different long-, medium-, and short-term purposes, such as in situ forest monitoring (see Chap. 10 of this book: Tognetti et al. 2021), large-scale surveying of CSF indicators (see Chap. 11 of this book: Torresan et al. 2021), establishing a CSF network (see

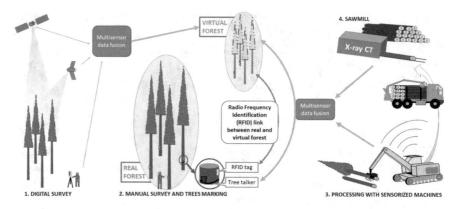

Fig. 9.14 Suitability of sensing technologies to timber characterization for industrial processing and for CSF applications

Chap. 5 of this book: Pretzsch et al. 2021), or implementing climate-smart silvicultural operations (see Chap. 8 of this book: Pach et al. 2021).

In turn, the data gathered for forest monitoring can be used for planning close-to-nature forestry operations. This improves the capacity to meet multiple goals, such as economic value, biodiversity, and forest resilience. Martín-Fernández and García-Abril Martín (2005) proved that if extensive and high-quality data is available, such planning can be performed at the level of individual trees by means of appropriate algorithms (iterative conditional mode), which otherwise could be applied only on small, intensively managed forest properties.

9.5 Data Management in Timber Production and CSF

As a final consideration, it is important to highlight how an essential aspect of the synergy between forest operations and CSF is the transfer, storage, elaboration, and harmonization of data collected by different sensors for different purposes (planning, production, invoicing, monitoring, forecast of health and growth dynamic, etc.). A further level of complexity is given by the perspective and capability of the different players of the supply chain, which produce and use data with different timing, purposes, technical tools, and frequency. This complexity is further increased by the additional target posed by CSF in the hypothesis of creating a unique system of data generation, elaboration, and exchange with a multipurpose vision.

Mortality and weakening control is accomplished by gathering and analyzing data with long-term field plots, where fixed low-cost networks of sensors are installed (see Chap. 10; Tognetti et al. 2021). Data can be effectively integrated with nondestructive sensors for timber quality, which identify timber quality parameters, such as defects, that can provide useful information regarding the health state of forests. In this sense, sensors installed on forest machines may bring a more valuable contribution than sawmill equipment in spite of the unavoidable lower precision. In fact, only quality timber is delivered to industrial users and analyzers. On the contrary, machine-based sensors are used to discriminate the logs among the supply chains of industrial timber, pulp, and biomass. Even if more detailed control will be dedicated to high-quality material, data must be generated for the selection of different classes, thus providing also valuable information regarding health parameters (rot, rings development, resistance wood, wood density, etc.).

A prerequisite for allowing applications and systems to communicate with each other in an agile and flexible way is the interoperability between systems and interfaces used. For this purpose, the Open Geospatial Consortium (OGC) and ISO created a web service interface standard for publishing, accessing, and visualizing spatiotemporal information (de la Beaujardiere 2006). In particular, the Sensor Web Enablement Initiative (SWE) promoted by the OGC proposed standards designed to collect data collected by sensors in a standardized way and augment the sensor data with the spatiotemporal dimension (Bröring et al. 2011). Thus, any machine control data or timber log data, mostly in the format of the Standard for Forest machine Data and Communication (StanForD), can be coupled with a spatial and temporal reference (Purfürst and Lindroos 2011). StandForD constitutes a de facto standard that covers all types of data communication present in forest machines and would probably require an upgrade, enabling it to exchange data provided by the new sensors that will soon be deployed on forest machines. By adopting the standards of SWE, it would also guarantee a standardized transmission, storage, and dissemination of the sensor data. This would be a fundamental step toward a wider use of the information produced, paving the way for a real integration of timber production and CSF.

References

Appelhanz S, Osburg V-S, Toporowski W, Schumann M (2015) Traceability system for capturing, processing and providing consumer-relevant information about wood products: system solution and its economic feasibility. J Clean Prod (forthcoming). https://doi.org/10.1016/j.jclepro.2015.02.034

Barth A, Holmgren J (2013) Stem taper estimates based on airborne laser scanning and cut-to-length harvester measurements for pre-harvest planning. Int J For Eng 24:161–169. https://doi.org/10.1080/14942119.2013.858911

Bettinger P, Boston K (2017) Forest planning heuristics-current recommendations and research opportunities for s-metaheuristics. Forests 8:476. https://doi.org/10.3390/f8120476

Björk A, Erlandsson M, Häkli J et al (2011) Monitoring environmental performance of the forestry supply chain using RFID. Comput Ind 62:830–841. https://doi.org/10.1016/j.compind.2011.08.001

Blagojević B, Jonsson R, Björheden R et al (2019) Multi-criteria decision analysis (MCDA) in Forest operations-an Introductional review. Šumarski fakultet Sveučilišta u Zagrebu

Bont L, Heinimann HR (2012) Optimum geometric layout of a single cable road. Eur J For Res 131:1439–1448. https://doi.org/10.1007/s10342-012-0612-y

Bosela M, Merganičová K, Torresan C, et al (2021) Modelling future growth of mountain forests under changing environments. In: Managing Forest Ecosystems, Vol. 40, Tognetti R,

Smith M, Panzacchi P (Eds) Climate-Smart Forestry in Mountain Regions. Springer Nature, Switzerland, AG

Bowditch E, Santopuoli G, Binder F et al (2020) What is climate-smart forestry? A definition from a multinational collaborative process focused on mountain regions of Europe. Ecosyst Serv 43:101113. https://doi.org/10.1016/j.ecoser.2020.101113

Briggs DG, Thienel G, Turnblom EC, et al (2008) Influence of thinning on acoustic velocity of douglas-fir trees in Western Washington and Western Oregon. Symp A Q J Mod Foreign Lit

Bröring A, Echterhoff J, Jirka S et al (2011) New generation sensor web enablement. Sensors 11:2652–2699. https://doi.org/10.3390/s110302652

Cambi M, Certini G, Neri F, Marchi E (2015) The impact of heavy traffic on forest soils: a review. For Ecol Manag 338:124–138. https://doi.org/10.1016/j.foreco.2014.11.022

Cocozza C, Lasserre B, Giovannelli A et al (2009) Low temperature induces different cold sensitivity in two poplar clones (Populus×canadensis Mönch "I-214" and P. deltoides Marsh. 'Dvina'). J Exp Bot 60:3655–3664. https://doi.org/10.1093/jxb/erp212

Cocozza C, Giovannelli A, Lasserre B et al (2012) A novel mathematical procedure to interpret the stem radius variation in olive trees. Agric For Meteorol 161:80–93. https://doi.org/10.1016/j.agrformet.2012.03.016

Cocozza C, Tognetti R, Giovannelli A (2018) High-resolution analytical approach to describe the sensitivity of tree–environment dependences through stem radial variation. Forests 9:134. https://doi.org/10.3390/f9030134

Cruz-García R, Balzano A, Čufar K et al (2019) Combining dendrometer series and Xylogenesis imagery – DevX, a simple visualization tool to explore plant secondary growth phenology. Front For Glob Chang 2:1–13. https://doi.org/10.3389/ffgc.2019.00060

de la Beaujardiere J (2006) OpenGIS® web map server implementation specification

Drew DM, Downes GM (2009) The use of precision dendrometers in research on daily stem size and wood property variation: a review. Dendrochronologia 27:159–172. https://doi.org/10.1016/j.dendro.2009.06.008

Đuka A, Grigolato S, Papa I et al (2017) Assessment of timber extraction distance and skid road network in steep karst terrain. IForest 10:886–894. https://doi.org/10.3832/ifor2471-010

Eriksson M, Lindroos O (2014) Productivity of harvesters and forwarders in CTL operations in northern Sweden based on large follow-up datasets. Int J For Eng 25:179–200. https://doi.org/10.1080/14942119.2014.974309

Favero A, Daigneault A, Sohngen B (2020) Forests: carbon sequestration, biomass energy, or both? Sci Adv 6:1–14. https://doi.org/10.1126/sciadv.aay6792

Gergel T, Bucha T, Gejdoš M, Vyhnáliková Z (2019) Computed tomography log scanning – high technology for forestry and forest based industry. Cent Eur For J 65:51–59. https://doi.org/10.2478/forj-2019-0003

Giovannelli A, Deslauriers A, Fragnelli G et al (2007) Evaluation of drought response of two poplar clones (Populus x canadensis Mönch "I-214" and P. deltoides Marsh. 'Dvina') through high resolution analysis of stem growth. J Exp Bot 58:2673–2683. https://doi.org/10.1093/jxb/erm117

Gjerdrum P, Bernabei M (2007) Three-dimensional model for size and location of resin pockets in stems of Norway spruce. Holz als Roh und Werkst 65:201–208. https://doi.org/10.1007/s00107-006-0158-0

Görgens EB, Mund JP, Cremer T et al (2020) Automated operational logging plan considering multi-criteria optimization. Comput Electron Agric 170:105253. https://doi.org/10.1016/j.compag.2020.105253

Holopainen M, Vastaranta M, Hyyppä J (2014) Outlook for the next generation's precision forestry in Finland. Forests 5:1682–1694. https://doi.org/10.3390/f5071682

Holtsmark B (2012) Harvesting in boreal forests and the biofuel carbon debt. Clim Chang 112:415–428. https://doi.org/10.1007/s10584-011-0222-6

Innes JL, Cook ER (1989) Tree-ring analysis as an aid to evaluating the effects of pollution on tree growth. Can J For Res 19:1174–1189. https://doi.org/10.1139/x89-177

Jones TG, Downes GM, Watt MS et al (2013) Effect of stem bending and soil moisture on the incidence of resin pockets in radiata pine. NZ J For Sci 43:1–14. https://doi.org/10.1186/1179-5395-43-10

Kaakkurivaara N (2019) Possibilities of using barcode and RFID technology in Thai timber industry. Maejo Int J Sci Technol 13:29–41

Kaakkurivaara T, Kaakkurivaara N (2019) Comparison of radio frequency identification tag housings in a tropical forestry work environment. Aust For 00:1–8. https://doi.org/10.1080/00049158.2019.1678797

Kauppi P, Hanewinkerl M, Lundmark T, et al (2018) Climate smart forestry in Europe. European Forest Institute

Keenan RJ (2015) Climate change impacts and adaptation in forest management: a review. Ann For Sci 72:145–167. https://doi.org/10.1007/s13595-014-0446-5

Keleş S, Baskent EZ (2011) Joint production of timber and water: a case study. Water Policy 13:535–546. https://doi.org/10.2166/wp.2011.125

Kidombo SD, Dean TJ (2018) Growth of tree diameter and stem taper as affected by reduced leaf area on selected branch whorls. Can J For Res 48:317–323. https://doi.org/10.1139/cjfr-2017-0279

Kimbar R (2011) Wady drewna (in Polish). Osie

Korten S, Kaul C (2008) Application of RFID (radio frequency identification) in the timber supply chain. Croat J For Eng 29:85–94

Laurila J, Lauhanen R (2012) Weight and volume of small-sized whole trees at different phases of the supply chain. Scand J For Res 27:46–55. https://doi.org/10.1080/02827581.2011.629621

Lindberg E, Holmgren J (2017) Individual tree crown methods for 3D data from remote sensing. Curr For Rep 3:19–31. https://doi.org/10.1007/s40725-017-0051-6

Lindroos O, Ringdahl O, Pedro LH et al (2015) Estimating the position of the harvester head – a key step towards the precision forestry of the future? Croat J For Eng 36:147–164

Łszczyńska K, Malik I, Wistuba M, Krąpiec M (2019) Assessment of landslide hazard from tree-ring eccentricity and from compression wood – a comparison. Geol Q 63:296–301. https://doi.org/10.7306/gq.1472

Lu K, Bi H, Watt D et al (2018) Reconstructing the size of individual trees using log data from cut-to-length harvesters in Pinus radiata plantations: a case study in NSW, Australia. J For Res 29:13–33. https://doi.org/10.1007/s11676-017-0517-1

Ma T, Inagaki T, Tsuchikawa S (2017) High spatial resolution and non-destructive evaluation of wood density and microfibril angle by NIR hyperspectral imaging. NIR News 28:7–12. https://doi.org/10.1177/0960336017703259

Marchi E, Chung W, Visser R et al (2018) Sustainable Forest operations (SFO): a new paradigm in a changing world and climate. Sci Total Environ 634:1385–1397. https://doi.org/10.1016/j.scitotenv.2018.04.084

Martín-Fernández S, García-Abril A (2005) Optimisation of spatial allocation of forestry activities within a forest stand. Comput Electron Agric 49:159–174. https://doi.org/10.1016/j.compag.2005.02.012

Matasci G, Hermosilla T, Wulder MA et al (2018) Large-area mapping of Canadian boreal forest cover, height, biomass and other structural attributes using Landsat composites and lidar plots. Remote Sens Environ 209:90–106. https://doi.org/10.1016/j.rse.2017.12.020

Matasov V, Marchesini LB, Yaroslavtsev A et al (2020) IoT monitoring of urban tree ecosystem services: possibilities and challenges. Forests 11:775. https://doi.org/10.3390/F11070775

Matthies BD, Jacobsen JB, Knoke T et al (2019) Utilising portfolio theory in environmental research – new perspectives and considerations. J Environ Manag 231:926–939

Mattila U, Tokola T (2019) Terrain mobility estimation using TWI and airborne gamma-ray data. J Environ Manag 232:531–536. https://doi.org/10.1016/J.JENVMAN.2018.11.081

Mavridou E, Vrochidou E, Papakostas GA et al (2019) Machine vision systems in precision agriculture for crop farming. J Imaging 5. https://doi.org/10.3390/jimaging5120089

Meder R, Meglen R (2012) Near infrared spectroscopic and hyperspectral imaging of compression wood in Pinus radiata D. Don J Near Infrared Spectrosc 20:583. https://doi.org/10.1255/jnirs.1001

Mederski PS, Bembenek M, Karaszewski Z et al (2018) Investigation of log length accuracy and harvester efficiency in processing of oak trees. Croat J For Eng 39:173–181

Mengesha T, Hawkins M, Nieuwenhuis M (2015) Validation of terrestrial laser scanning data using conventional forest inventory methods. Eur J For Res 134:211–222

Mina M, Martin-Benito D, Bugmann H, Cailleret M (2016) Forward modeling of tree-ring width improves simulation of forest growth responses to drought. Agric For Meteorol 221:13–33. https://doi.org/10.1016/j.agrformet.2016.02.005

Mitchell HB (2012) Data fusion: concept and ideas. pp. 347, Springer, ISBN 978-3-642-27222-6

Möller B, Wikander J, Hellgren M (2011) A field-tested log traceability system. For Prod J 61:466–472. https://doi.org/10.13073/0015-7473-61.6.466

Murphy PNC, Ogilvie J, Castonguay M et al (2008) Improving forest operations planning through high-resolution flow-channel and wet-areas mapping. For Chron 84:568–574. https://doi.org/10.5558/tfc84568-4

Murphy G, Clark JA, Pilkerton S (2012) Current and potential tagging and tracking Systems for Logs Harvested from Pacific Northwest Forests. West J Appl For 27:84–91. https://doi.org/10.5849/wjaf.11-027

Niemi MT, Vastaranta M, Vauhkonen J et al (2017) Airborne LiDAR-derived elevation data in terrain trafficability mapping. Scand J For Res 32:762–773. https://doi.org/10.1080/02827581.2017.1296181

Noordermeer L, Bollandsås OM, Ørka HO et al (2019) Comparing the accuracies of forest attributes predicted from airborne laser scanning and digital aerial photogrammetry in operational forest inventories. Remote Sens Environ 226:26–37. https://doi.org/10.1016/j.rse.2019.03.027

Olivera A, Visser R (2016) Development of forest-yield maps generated from Global Navigation Satellite System (GNSS)-enabled harvester StanForD files: preliminary concepts. NZ J For Sci 46:1–10. https://doi.org/10.1186/s40490-016-0059-x

Ovando P, Speich M (2020) Optimal harvesting decision paths when timber and water have an economic value in uneven forests. Forests 11:1–26. https://doi.org/10.3390/F11090903

Pach M, Bielak K, Bončina A, et al (2021) Climate-smart silviculture in mountain regions. In: Managing Forest Ecosystems, Vol. 40, Tognetti R, Smith M, Panzacchi P (Eds) Climate-Smart Forestry in Mountain Regions. Springer Nature, Switzerland, AG

Palander TS, Eronen JP, Peltoniemi NP et al (2019) Improving a stem-damage monitoring system for a single-grip harvester using a logistic regression model in image processing. Biosyst Eng 180:36–49. https://doi.org/10.1016/j.biosystemseng.2019.01.011

Paradis N, Auty D, Carter P, Achim A (2013) Using a standing-tree acoustic tool to identify forest stands for the production of mechanically-graded lumber. Sensors (Switzerland) 13:3394–3408. https://doi.org/10.3390/s130303394

Peuhkurinen J (2011) Estimating tree size distributions and timber assortment recoveries for wood procurement planning using airborne laser scanning. University of Eastern Finland

Picchi G (2020) Marking standing trees with RFID tags. Forests 11:1–13. https://doi.org/10.3390/f11020150

Picchi G, Kühmaier M, Marques JDD (2015) Survival test of RFID UHF tags in timber harvesting operations. Croat J For Eng 36:165–174

Pichler G, Poveda Lopez JAA, Picchi G et al (2017) Comparison of remote sensing based RFID and standard tree marking for timber harvesting. Comput Electron Agric 140:214–226. https://doi.org/10.1016/j.compag.2017.05.030

Piragnolo M, Grigolato S, Pirotti F (2019) Planning harvesting operations in forest environment: remote sensing for decision support. ISPRS Ann Photogramm Remote Sens Spat Inf Sci IV-3(W1):33–40. https://doi.org/10.5194/isprs-annals-IV-3-W1-33-2019

Pirotti F, Grigolato S, Lingua E et al (2012) Laser scanner applications in Forest and environmental sciences. Ital J Remote Sens 44:109–123. https://doi.org/10.5721/ItJRS20124419

Puliti S, Hauglin M, Breidenbach J et al (2020) Modelling above-ground biomass stock over Norway using national forest inventory data with ArcticDEM and Sentinel-2 data. Remote Sens Environ 236:111501. https://doi.org/10.1016/j.rse.2019.111501

Pretzsch H, del Río M, Giammarchi F, Uhl E, Tognetti R (2021) Changes of tree and stand growth. Review and implications. In: Managing Forest Ecosystems, Vol. 40, Tognetti R, Smith M, Panzacchi P (Eds) Climate-Smart Forestry in Mountain Regions. Springer Nature, Switzerland, AG

Purfürst T, Lindroos O (2011) The correlation between long-term productivity and short-term performance ratings of harvester operators. Croat J For Eng 32:509–519

Raatevaara A, Korpunen H, Mäkinen H, Uusitalo J (2020) Log end face image and stem tapering indicate maximum bow height on Norway spruce bottom logs. Eur J For Res 139:1079–1090. https://doi.org/10.1007/s10342-020-01309-0

Rais A, Ursella E, Vicario E, Giudiceandrea F (2017) The use of the first industrial X-ray CT scanner increases the lumber recovery value: case study on visually strength-graded Douglas-fir timber. Ann For Sci 74:1–9. https://doi.org/10.1007/s13595-017-0630-5

Raschi A, Tognetti R, Ridder H-W, Beres C (1995) Water in the stems of sessile oak (Quercus petraea) assessed by computer tomography with concurrent measurements of sap velocity and ultrasound emission. Plant Cell Environ 18:545–554. https://doi.org/10.1111/j.1365-3040.1995.tb00554.x

Rossit DA, Olivera A, Viana Céspedes V, Broz D (2019) A big data approach to forestry harvesting productivity. Comput Electron Agric 161:29–52. https://doi.org/10.1016/j.compag.2019.02.029

Rossmann J, Schluse M, Schlette C (2009) The virtual forest: robotics and simulation technology as the basis for new approaches to the biological and the technical production in the forest. In: WMSCI 2009 – the 13th world multi-conference on systemics, cybernetics and informatics, jointly with the 15th international conference on information systems analysis and synthesis, ISAS 2009 – Proceedings, pp 33–38

Russo D, Marziliano PA, Macrì G et al (2020) Tree growth and wood quality in pure vs. mixed-species stands of european beech and Calabrian pine in Mediterranean mountain forests. Forests 11:6. https://doi.org/10.3390/F11010006

Salmivaara A, Miettinen M, Finér L et al (2018) Wheel rut measurements by forest machine-mounted LiDAR sensors – accuracy and potential for operational applications? Int J For Eng 00:1–12. https://doi.org/10.1080/14942119.2018.1419677

Sandak A, Sandak J, Böhm K, Hinterstoisser B (2016a) Near infrared spectroscopy as a tool for in – field determination of log/biomass quality index in mountain forests. J Near Infrared Spectrosc 24:587–594. https://doi.org/10.1255/jnirs.1231

Sandak J, Sandak A, Meder R (2016b) Assessing trees, wood and derived products with near infrared spectroscopy: hints and tips. J Near Infrared Spectrosc 24:485–505. https://doi.org/10.1255/jnirs.1255

Sandak A, Sandak J, Waliszewska B et al (2017) Selection of optimal conversion path for willow biomass assisted by near infrared spectroscopy. IForest 10:506–514. https://doi.org/10.3832/ifor1987-010

Sandak J, Sandak A, Marrazza S, Picchi G (2019) Development of a sensorized timber processor head prototype – part 1: sensors description and hardware integration. Croat J For Eng 40:25–37

Sandak J, Sandak A, Zitek A et al (2020) Development of low-cost portable spectrometers for detection of wood defects. Sensors (Switzerland) 20. https://doi.org/10.3390/s20020545

Schajer GS, Orhan FB (2005) Microwave non-destructive testing of wood and similar orthotropic materials. Subsurf Sens Technol Appl 6:293–313. https://doi.org/10.1007/s11220-005-0014-z

Schneider R (2018) Understanding the factors influencing stem form with modelling tools. Springer, Cham, pp 295–316

Schraml R, Charwat-Pessler J, Petutschnigg A, Uhl A (2015) Towards the applicability of biometric wood log traceability using digital log end images. Comput Electron Agric 119:112–122. https://doi.org/10.1016/j.compag.2015.10.003

Schraml R, Entacher K, Petutschnigg A et al (2020) Matching score models for hyperspectral range analysis to improve wood log traceability by fingerprint methods. Mathematics 8. https://doi.org/10.3390/MATH8071071

Searchinger TD, Beringer T, Holtsmark B et al (2018) Europe's renewable energy directive poised to harm global forests. Nat Commun 9:10–13. https://doi.org/10.1038/s41467-018-06175-4

Seifert T, Breibeck J, Seifert S, Biber P (2010) Resin pocket occurrence in Norway spruce depending on tree and climate variables. For Ecol Manag 260:302–312. https://doi.org/10.1016/j.foreco.2010.03.024

Shenga PA, Bomark P, Broman O (2015) External log scanning for optimizing primary breakdown of tropical hardwood species. In: 22nd international wood machining seminar. Quebec, Canada, pp 65–72

Sheppard JP, Chamberlain J, Agúndez D et al (2020) Sustainable Forest management beyond the timber-oriented status quo: transitioning to co-production of timber and non-wood forest products – a global perspective. Curr For Rep 6:26–40. https://doi.org/10.1007/s40725-019-00107-1

Siekański P, Magda K, Malowany K et al (2019) On-line laser triangulation scanner for wood logs surface geometry measurement. Sensors 19:1074. https://doi.org/10.3390/s19051074

Stängle SM, Brüchert F, Kretschmer U et al (2014) Clear wood content in standing trees predicted from branch scar measurements with terrestrial LiDAR and verified with X-ray computed tomography. Can J For Res 44:145–153

Stängle SM, Brüchert F, Heikkila A et al (2015) Potentially increased sawmill yield from hardwoods using X-ray computed tomography for knot detection. Ann For Sci 72:57–65. https://doi.org/10.1007/s13595-014-0385-1

Sterenczak K, Moskalik T (2015) Use of LIDAR-based digital terrain model and single tree segmentation data for optimal forest skid trail network. iForest Biogeosci For 8:661–667. https://doi.org/10.3832/ifor1355-007

Subah S, Dermninder S, Sanjeev C (2017) An interactive computer vision system for tree ring analysis. Curr Sci 112:1262–1265

Sun SJ, Lei S, Jia HS et al (2020) Tree-ring analysis reveals density-dependent vulnerability to drought in planted Mongolian pines. Forests 11:1–17. https://doi.org/10.3390/f11010098

Taube P, Orłowski KA, Chuchała D, Sandak J (2020) The effect of log sorting strategy on the forecasted lumber value after sawing pine wood. Acta Fac Xylologiae Zvolen 62:89–102. https://doi.org/10.17423/afx.2020.62.1.08

Thumm A, Riddell M, Nanayakkara B et al (2010) Near infrared hyperspectral imaging applied to mapping chemical composition in wood samples. J Near Infrared Spectrosc 18:507–515. https://doi.org/10.1255/jnirs.909

Tognetti R, Raschi A, Beres C et al (1996) Comparison of sap flow, cavitation and water status of Quercus petraea and Quercus cerris trees with special reference to computer tomography. Plant Cell Environ 19:928–938. https://doi.org/10.1111/j.1365-3040.1996.tb00457.x

Tognetti R, Giovannelli A, Lavini A et al (2009) Assessing environmental controls over conductances through the soil-plant-atmosphere continuum in an experimental olive tree plantation of southern Italy. Agric For Meteorol 149:1229–1243. https://doi.org/10.1016/j.agrformet.2009.02.008

Tognetti R, Valentini R, Belelli Marchesini L, Gianelle D, Panzacchi P, Marshall JD (2021) Continuous monitoring of tree responses to climate change for smart forestry – a cybernetic web of trees. In: Managing Forest Ecosystems, Vol. 40, Tognetti R, Smith M, Panzacchi P (Eds) Climate-Smart Forestry in Mountain Regions. Springer Nature, Switzerland, AG

Torresan C, Luyssaert S, Filippa G, Imangholiloo M, Gaulton R (2021) Remote sensing technologies for assessing climate-smart criteria in mountain forests. In: Managing Forest Ecosystems, Vol. 40, Tognetti R, Smith M, Panzacchi P (Eds) Climate-Smart Forestry in Mountain Regions. Springer Nature, Switzerland, AG

Tzoulis IK, Andreopoulou ZS, Voulgaridis E (2014) Wood tracking information systems to confront illegal logging. J Agric Inform 5:9–17

Uner B, Oyar O, Var AA, Altnta OL (2009) Effect of thinning on density of Pinus nigra tree using X-ray computed tomography. J Environ Biol 30:359–362

Valbuena R, Mauro F, Rodríguez-Solano R, Manzanera JA (2012) Partial least squares for discriminating variance components in global navigation satellite systems accuracy obtained under scots pine canopies. For Sci 58:139–153. https://doi.org/10.5849/forsci.10-025

Van den Bulcke J, Wernersson ELG, Dierick M et al (2014) 3D tree-ring analysis using helical X-ray tomography. Dendrochronologia 32:39–46. https://doi.org/10.1016/j.dendro.2013.07.001

Venanzi R, Picchio R, Grigolato S, Latterini F (2019) Soil and forest regeneration after different extraction methods in coppice forests. For Ecol Manag 454. https://doi.org/10.1016/j.foreco.2019.117666

Walker XJ, Mack MC, Johnstone JF (2017) Predicting ecosystem resilience to fire from tree ring analysis in black spruce forests. Ecosystems 20:1137–1150. https://doi.org/10.1007/s10021-016-0097-5

Walsh D, Strandgard M, Carter P (2014a) Evaluation of the Hitman PH330 acoustic assessment system for harvesters. Scand J For Res 29:593–602. https://doi.org/10.1080/02827581.2014.953198

Walsh D, Strandgard M, Carter P (2014b) Evaluation of the Hitman PH330 acoustic assessment system for harvesters. Scand J For Res 29:593–602. https://doi.org/10.1080/02827581.2014.953198

White JC, Coops NC, Wulder MA et al (2016) Remote sensing technologies for enhancing forest inventories: a review. Can J Remote Sens 42:619–641. https://doi.org/10.1080/07038992.2016.1207484

Woollons R, Manley B, Park J (2008) Factors influencing the formation of resin pockets in Pruned radiata pine butt logs from New Zealand. NZ J For Sci 38:323–334

Yousefpour R, Augustynczik ALD, Reyer CPO et al (2018) Realizing mitigation efficiency of European commercial forests by climate smart forestry. Sci Rep 8:1–11. https://doi.org/10.1038/s41598-017-18778-w

Zeller L, Ammer C, Annighöfer P et al (2017) Tree ring wood density of Scots pine and European beech lower in mixed-species stands compared with monocultures. For Ecol Manag 400:363–374. https://doi.org/10.1016/j.foreco.2017.06.018

Zobel BJ, Jett JB (1995) The importance of wood density (specific gravity) and its component parts. In: Genetics of wood production. Springer, Berlin/Heidelberg

Zweifel R, Steppe K, Sterck FJ (2007) Stomatal regulation by microclimate and tree water relations: interpreting ecophysiological field data with a hydraulic plant model. J Exp Bot 58:2113–2131. https://doi.org/10.1093/jxb/erm050

Chapter 10
Continuous Monitoring of Tree Responses to Climate Change for Smart Forestry: A Cybernetic Web of Trees

Roberto Tognetti, Riccardo Valentini, Luca Belelli Marchesini, Damiano Gianelle, Pietro Panzacchi, and John D. Marshall

Abstract Trees are long-lived organisms that contribute to forest development over centuries and beyond. However, trees are vulnerable to increasing natural and anthropic disturbances. Spatially distributed, continuous data are required to predict mortality risk and impact on the fate of forest ecosystems. In order to enable monitoring over sensitive and often remote forest areas that cannot be patrolled regularly, early warning tools/platforms of mortality risk need to be established across regions. Although remote sensing tools are good at detecting change once it has occurred, early warning tools require ecophysiological information that is more easily collected from single trees on the ground.

Here, we discuss the requirements for developing and implementing such a tree-based platform to collect and transmit ecophysiological forest observations and

R. Tognetti (✉)
Dipartimento di Agricoltura, Ambiente e Alimenti, Università degli Studi del Molise, Campobasso, Italy
e-mail: tognetti@unimol.it

R. Valentini
DIBAF – Department for Innovation in Biological, Agri-Food and Forest Systems, University of Tuscia, Viterbo, Italy

Smart Urban Nature Laboratory, RUDN University, Moscow, Russia
e-mail: rik@unitus.it

L. B. Marchesini · D. Gianelle
Department of Sustainable Agro-Ecosystems and Bioresources, Fondazione Edmund Mach, San Michele all'Adige, Italy
e-mail: luca.belellimarchesini@fmach.it; damiano.gianelle@fmach.it

P. Panzacchi
Department of Biosciences and Territory, Università degli Studi del Molise, Pesche, Italy

Faculty of Science and Technology, Free University of Bolzano-Bozen, Bolzano, Italy

J. D. Marshall
Department of Forest Ecology and Management, SLU, Umea, Sweden
e-mail: john.marshall@slu.se

environmental measurements from representative forest sites, where the goals are to identify and to monitor ecological tipping points for rapid forest decline. Long-term monitoring of forest research plots will contribute to better understanding of disturbance and the conditions that precede it. International networks of these sites will provide a regional view of susceptibility and impacts and would play an important role in ground-truthing remotely sensed data.

10.1 Ground-Based Measures of Forest Ecophysiological Indicators for Climate Smartness

A set of criteria and indicators have been proposed, by which the "climate smartness" of a forest can be assessed (Bowditch et al. 2020; Santopuoli et al. 2020). Likewise, Bussotti and Pollastrini (2017) proposed a mix of traditional and novel indicators of forest health, at tree and stand levels, to support visual tree assessment, as well as to improve the prediction of stand dynamics and forest productivity under climate change in European forests.

The indicators are quantitative or qualitative variables that are evaluated periodically to reveal the direction of change with respect to these criteria (Bowditch et al. 2020; see also Chaps. 3 and 2 of this book: del Río et al. 2021; Weatherall et al. 2021, respectively). Within a particular management framework, one begins by choosing which forest processes are relevant for the criteria of "climate smartness." Forestry has traditionally privileged tree growth and wood production as main management goals, assuming that productivity is the ultimate indicator of tree responses to environmental conditions. However, climate change has challenged this view due to uncertainties in disturbance-growth relationships related to climatic variability and extreme weather events. In addition, management now addresses trade-offs between different forest functions and services (Thom and Seidl 2016; Albrich et al. 2018). The widened horizon of modern forest management is well recognized and interpreted by the Sustainable Development Goals 13 and 15 (United Nations 2015) as sustainably managed forests are instrumental to combat climate change and its impacts; to protect, restore, and promote sustainable use of terrestrial ecosystems; to strive against desertification; to halt and reverse land degradation; and to halt biodiversity loss.

Research in forest ecosystems has mostly focused on ring-width time series, forest stand yield measured on plots, or long-term successional dynamics (Harmon and Pabst 2015). These do not require frequent sampling. However, mechanistic analysis of climate-driven and disturbance-related events (e.g., droughts, fires, windthrows, outbreaks) requires direct and frequent repeated observations of processes related to forest demography and resilience (i.e., mortality and recruitment) to identify the causes. Therefore, parameters that reveal ecophysiological status become more valued than in traditional forest monitoring. Within a certain range of climatic conditions (short- to midterm), ecophysiological traits and growth patterns follow climate variability, with species plasticity allowing trees to recover from climate perturbations. However, extreme events may trigger anomalous physiological responses beyond the safe operation mode, leading to irreversible changes and eventually causing the death of trees (Fig. 10.1). The exit of tree responses from the safe operation mode is often difficult to detect without a long-term and high-frequency record of tree functions. Autonomous sensor networks may produce information valuable to monitor tree status, allowing foresters to make informed management decisions.

Numerous experiments and observational studies have been established to address global change-related questions across multiple temporal and spatial scales (Halbritter et al. 2020). In particular, studies on forest decline aim to establish the causal relations, to unravel the climate drivers, and to understand the ecological processes related to trees' mortality. Nevertheless, some ecological processes are more sensitive to changes in extremes than in mean values (Allen et al. 2010; Hansen et al. 2012), including important effects of microclimate. For example, extreme temperatures combined with prolonged drought have been implicated as drivers of forest die-off (Adams et al. 2017).

There is growing scientific interest in forest reactions to drought across different biomes to discern which growth features or functional traits best characterize different species-specific responses to these climate extremes (Lindner et al. 2010;

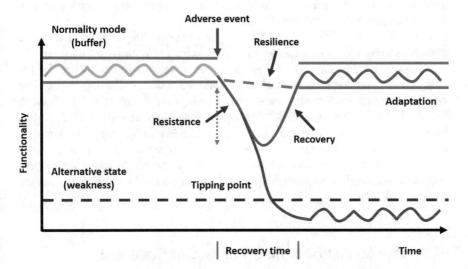

Fig. 10.1 The normal or safe operation mode (normality mode) of the single tree can be perturbed by climatic events or environmental disturbances, leading to anomalous physiological responses beyond the buffer space. Eventually, extreme events (single or series) may provoke persistent changes in the short- to midterm; recovery to an alternative stable state may occur within the resistance limits of the species. Indeed, as climate changes over longer time frames and tree populations display some degree of adaptation, the normality mode may adjust accordingly. The "tipping point," which prevents the tree from recovering physiological functions and triggers tree decline, varies with species and environment, and is not easy to predict

Anderegg et al. 2015). For example, in Mediterranean forests, water availability is the major tree growth constraint, and drought conditions are predicted to increase (Giorgi and Lionello 2008). Nevertheless, the responses of these forests to such extreme climatic events are poorly understood, because controlled field experiments able to mimic drought conditions are costly and difficult to operate on a large scale without introducing environmental modifications. Adaptive forest management strategies to combat climate change need a clear framework of indicators useful for predicting the different components of tree resilience and the ability of trees to recover after disturbance (or their mortality). Therefore, we may pose the following question: how can the observational approach be linked with datasets gathered from in situ experiments, products of hypothesis testing, to detect critical changes in ecological conditions and to determine the ways in which those changes impact ecosystem functions?

Adaptive management practices aimed at combating forest decline need to implement real-time control of the environment and a quick response to changing growth conditions. In this context, a range of sensors is needed to provide a picture of interactions occurring between data and to enable key forest indicators to be identified. A selection of indicators, enabling the assessment of climate smartness of forests at stand level, is provided in Table 10.1; measurable parameters, data solutions, and monitoring tools are also reported. Data from monitoring networks and model forecasts are essential instruments, both to understand forest ecosystem responses to rapid environmental variation and to support forest decision-makers under a climate change scenario (Lindner et al. 2014).

The "smartness" of climate-smart forestry (CSF) comes in part from its ability to predict and respond to changes in stand dynamics using early warning signals, which precede the occurrence of unwanted events, such as forest decline. Large-scale and long-term forest monitoring networks have been collecting information for characterizing forest responses to global change (structure, function, damage, diversity), e.g., CTFS-ForestGEO (Anderson-Teixeira et al. 2015), ICP Forests (http://icp-forests.net/), and eLTER (https://www.lter-europe.net/). However, a mechanistic understanding of forest adjustment to global change is still missing. In this context, a new observational and experimental paradigm based on biogeographic scale, single-tree, high-frequency, and long-term monitoring is required (Steppe et al. 2016).

10.2 Tree Mortality, Tipping Points, and Resilience

Climate scenarios for the next decades predict warmer temperatures, greater vapor pressure deficits, and more frequent and severe drought spells and heat waves than experienced in the recent (Sillmann et al. 2013a, b). These changes are expected to result in increased frequency, intensity, and duration of drought (Polade et al. 2014). Intensifying impacts of drought events on tree functionality have been recently observed across biomes (e.g., Shestakova et al. 2019). Drought episodes interact

Table 10.1 Selected climate-smart forestry (CSF) indicators, stand-level measurable parameters, cybertechnologies for data collection and transmission, and related techniques and tools

Climate-smart forestry indicators	Description based on MCPFE and Forest Europe	Source (CLIMO or Forest Europe, FE)	Criteria classification	Parameters	Cybertechnology for data collection	Cybertechnology for data transmission	Measurement techniques	Tools
1.3 Age structure and/ or diameter distribution	Age structure and/or diameter distribution of forest and other wooded land, classified by availability for wood supply	FE	Management	Tree diameter, TRI	Cable	LoRa, wireless Internet	Dendrochronology, forest mensuration	Diameter tape, tree ring analysis
1.4 Carbon stock	Carbon stock and carbon stock changes in forest biomass, forest soils, and harvested wood products	FE	Ecological	GPP=WUEi*gs, NPP, carbon stock partitioning	Cable	LoRa, wireless internet	Stable isotopes, sap flow = WUE/gs	Dendrometers, tree ring analysis, tree talker, stable isotope (13C) gas analyzer
2.1 Deposition of air pollutants	Deposition and concentration of air pollutants on forest and other wooded land	FE	Ecological	Photochemical reflectance index	Wireless	LoRa, wireless Internet	Spectral reflectance	Spectral reflectance sensor

(continued)

Table 10.1 (continued)

Climate-smart forestry indicators	Description based on MCPFE and Forest Europe	Source (CLIMO or Forest Europe, FE)	Criteria classification	Parameters	Cybertechnology for data collection	Cybertechnology for data transmission	Measurement techniques	Tools
2.2 Soil condition	Chemical soil properties (pH, CEC, C/N, organic C, base saturation) on forest and other wooded land related to soil acidity and eutrophication, classified by main soil types	FE	Ecological	Soil temperature, soil moisture	Radio, cell phone, Wi-Fi, satellite, Ethernet/cable, meteor burst, wireless Internet, telemetry	LoRa, wireless Internet	Electrical capacitance, gamma attenuation, soil heat flux, time-domain reflectometry (TDR), ground-penetrating radar (GPR), infrared thermal imaging (TIR)	Soil temperature-moisture probe, soil temperature-moisture profile probe
2.3 Defoliation	Defoliation of one or more main tree species on forest and other wooded land in each of the defoliation classes	FE	Ecological	Photochemical reflectance index, leaf temperature	Wireless	LoRa, wireless Internet	Spectral reflectance, thermal resistance, thermocouple, infrared, infrared thermal imaging (TIR)	Spectral reflectance sensor, infrared thermometer, UAV
2.4 Forest damage	Forest and other wooded land with damage, classified by primary damaging agent (abiotic, biotic, and human induced)	FE	Ecological	Photochemical reflectance index	Wireless	LoRa, wireless Internet	Spectral reflectance	Spectral reflectance sensor
4.1 Tree species composition	Area of forest and other wooded land, classified by number of tree species occurring	FE	Management	Number	Wireless	LoRa, wireless Internet	Laser	Fieldmap

Climate-smart forestry indicators	Description based on MCPFE and Forest Europe	Source (CLIMO or Forest Europe, FE)	Criteria classification	Parameters	Cybertechnology for data collection	Cybertechnology for data transmission	Measurement techniques	Tools
4.2 Regeneration	Total forest area by stand origin and area of annual forest regeneration and expansion	FE	Management	Area	Wireless	LoRa, wireless Internet	Laser	LiDAR-UAV
4.4 Introduced tree species	Area of forest and other wooded land dominated by introduced tree species	FE	Management	Area, crown	Wireless	LoRa, wireless Internet	Laser	LiDAR-UAV
4.5 Deadwood	Volume of standing deadwood and of lying deadwood on forest and other wooded land	FE	Ecological	Volume, 3D point cloud	Wireless	LoRa, wireless Internet	Laser	TLS
5.1 Protective forests – soil, water, and other ecosystem functions, and infrastructures	Area of forest and other wooded land designated to prevent soil erosion, preserve water resources, maintain other protective functions, protect infrastructure, and manage natural resources against natural hazards	FE	Ecological	Canopy cover	Wireless	LoRa, wireless Internet	Laser	LiDAR-UAV
7.2 Slenderness coefficient	The ratio of tree total height to diameter outside bark at 1.3 m above ground level	CLIMO	Management	DBH, height	Wireless	LoRa, wireless Internet	Laser, modeling	TLS, Fieldmap, DTM-CHM

(continued)

Table 10.1 (continued)

Climate-smart forestry indicators	Description based on MCPFE and Forest Europe	Source (CLIMO or Forest Europe, FE)	Criteria classification	Parameters	Cybertech nology for data collection	Cybertech nology for data transmission	Measurement techniques	Tools
7.3 Vertical distribution of tree crowns	Distribution of tree crown in the vertical space. It can be measured in terms of layers (one, two, multiple), or in terms of ratio between tree height and crown length	CLIMO	Management	Stand vertical structure, 3D point clod	Wireless	LoRa, wireless Internet	Laser	LiDAR, TLS
4.92 Horizontal distribution of tree crowns	Canopy space filling and can be expressed in measure of density of tree crowns, such as crown area, tree crown diameter. It can be also expressed in measure of density of trees, such as trees per hectare, basal area per hectare (in this case the horizontal distribution refers to the tree)	CLIMO	Management	Tree density, 3D point cloud	Wireless	LoRa, wireless Internet	Laser	LiDAR, TLS

with heat waves, possibly inducing die-off events (Allen et al. 2010; Anderegg et al. 2016). Direct effects on tree physiological functions (runaway embolism and/or carbon starvation and their interactions) may kill trees (McDowell et al. 2013) (Fig. 10.2). Globally, tree mortality is expected to increase because of biogeochemical and biophysical climatic feedback following shifts in land carbon and energy balance (Bonan 2008). Reorienting forestry systems to support sustainable forest management under the new realities of climate change needs an integrated understanding of tree adaptation to climate change under field conditions and explicit testing of plasticity-growth relationships for sustaining productivity under more extreme climatic conditions (Millar and Stephenson 2015).

Examples of climate-smart measures include, among others, managing forest disturbances and extreme events; selecting resilient trees and implementing forest reserves; combining carbon storage, sequestration, and substitution; using forest bioenergy and wood in the construction sector; and valuing ecosystems and their services that help halt land degradation. In order to withstand the changing climate, forest ecosystems need to be healthy and strong. Forest health, described by the functional envelope for disease-free trees at the individual level (Hartmann et al. 2018), can be monitored by determining mortality rates that deviate from normal background mortality rates (excess deaths). Recording the normal space of operation and the detection of functional anomalies requires long-term and high-frequency monitoring of trees in forest ecosystems (Trumbore et al. 2015). Abiotic and biotic factors make the tree mortality process complicated. The failure of hydraulic

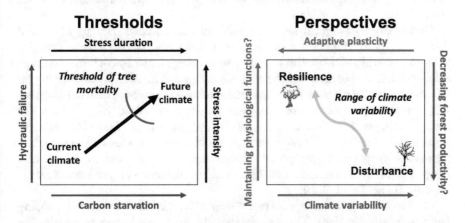

Fig. 10.2 Despite rapid directional environmental changes, forest managers struggle against environmental changes to maintain forests within historical ranges of conditions. However, forests are inherently unstable under climate change, and, beyond a certain threshold, substantial mortality occurs, with an abrupt loss of forest functions and services. Drought may cause tree mortality directly or indirectly through increased vulnerability to insects or pathogens. Although drought-induced mortality is expected to occur more frequently at the southern range limits of tree species, tree death may increase regardless of location. Should forest managers anticipate and assist forest transition by reducing the probability of sudden die-back (e.g., thinning to reduce competition for resources, establishing species adapted to future conditions), the transition will be gradual rather than abrupt, and ecosystem services will be maintained at a higher (although reduced) level

systems and the depletion of carbon reserves determine the physiological response of trees to drought and the pathway of drought-induced mortality (Choat et al. 2018). Vulnerability to pests and pathogens adds to abiotic stress, causing physiological decline and physical damage.

In order to implement CSF practices and assess forest ecosystem resilience, drivers of forest dynamics, indicators of environmental disturbances, and the occurrence of tree mortality need to be selected and monitored (Bowditch et al. 2020). To this end, trends and their directions in tree traits in response to disturbance events can be considered to assess changes in temporal (and spatial) synchrony associated with time series of ecophysiological and growth data and used as early warning signals of mortality risk (Cocozza et al. 2009, 2012; Fierravanti et al. 2015; Cailleret et al. 2016, 2017). Phase synchronization of time series relevant for signal analysis may help understand the relationships between fluctuations in functional traits and impacts of environmental drivers (Perone et al. 2016; Cocozza et al. 2018). Early warning signals of forest systems that are approaching a critical transition are caused by the gradual decrease in the recovery rate after a disturbance event (Wissel 1984; Drake and Griffen 2010; Veraart et al. 2012; Dai et al. 2012; Jarvis et al. 2016). Under increasing levels of stress (e.g., drought), damaged trees are no longer able to use natural resources. Interacting stressors, hence, may lead to system failure (Anderegg et al. 2012). The accumulated physiological damage may cross the tipping point and trigger tree mortality. This critical transition is caused by the combined changes in the intensity, frequency, and duration of stress factors (Dakos et al. 2015) and high sensitivity of the tree to these specific stresses (Brandt et al. 2017).

Models of physiological processes may provide an understanding of mechanisms underlying responses to climate in forest trees that experience drought-induced mortality (McDowell et al. 2013). When physiological models fail, however, empirical data are useful to determine mechanisms and thresholds that may trigger tree mortality. Estimates of the relationships among evaporative demand (dry season), water supply (wet season), and tree growth may help develop indices that capture mortality (Park Williams et al. 2013). Integration of mechanistic approaches with empirical observations can be achieved with specific studies of tree growth in permanent sample plots (prospective studies, e.g., (Cocozza et al. 2016); see also Chap. 5 of this book: Pretzsch et al. 2021) and tree ring analyses (retrospective studies, e.g., Tognetti et al. 2019).

Prior to tree mortality, an ecosystem may cross a critical transition (a tipping point) in forest functions. These might include, for example, runaway embolism caused by drought stress (Tyree and Sperry 1988) or crown damage caused by wind or snow loading (Peltola and Kellomäki 1993; Nadrowski et al. 2014). The exact location of the tipping point depends on species- and stress-specific sensitivity. Although dynamic phenomena are intrinsically difficult to observe, efficient monitoring of spatially and temporally dynamic phenomena is possible through multiscale sampling schemes based on a coarse-to-fine hierarchy system (Rundel et al. 2009). With this approach, the region of interest can be identified and surveyed through low-spatial/time resolution sensors (e.g., airborne surveys and sparse ground-based devices), and selected areas that need high-resolution and real-time observation can be monitored. Mobile nodes may supplement fixed sensors to adapt sampling protocols and instrument modalities (Jordan et al. 2007). Similarly, ecosystem flux studies, which started in the late 1980s (Baldocchi et al. 1987; Jarvis et al. 1989), utilized fixed experimental infrastructures to measure the net ecosystem gas exchange by eddy covariance methods across a global network of terrestrial ecosystems. Because of the difficulty of moving the complex flux infrastructures from site to site for short campaigns, low-cost eddy covariance setups were developed and deployed as roving towers for characterizing the spatial variability at the landscape level (Cavaleri et al. 2008; Markwitz and Siebicke 2019).

Tree mortality and forest dieback may themselves be considered tipping points (Cailleret et al. 2019), which, once passed, may induce further major changes in the system's dynamics. Tipping points are difficult to predict due to species interactions and driver stochasticity. Understanding these thresholds would help predict the circumstances under which trade-offs between different forest functions are minimal and, therefore, when their simultaneous provisioning, that is, ecosystem multifunctionality, is amplified (Gamfeldt et al. 2013; Baeten et al. 2019). Long-term and high-resolution data, in combination with modelling exercises and short-term experiments, may help explain the mechanisms behind tipping points and runaway perturbations. As an example, time-series data may integrate flux tower measurements, which cover only the last three decades, and forest inventories, which have multiannual gaps between successive samplings.

The tipping point can be reached after a series of extreme events, which may vary in duration among tree species and environmental conditions or can be induced by sequential exposure to extreme events (memory or legacy effect) (Fig. 10.3). The extent by which climate extremes impact functional processes and resistance/recovery in tree patterns is also dependent on forest structure (age, height, and diameter classes), genotypic and phenotypic profiles, soil characteristics, and degree and type of disturbances (windstorms, fires, droughts, outbreaks) (Kannenberg et al. 2019). The comprehension of these dynamics, as well as the identification of potential early warning signals in trees, preceding the occurrence of irreversible tree decline (tipping point), requires a new monitoring paradigm based on large-scale, single-tree, high-frequency, and long-term monitoring. This will allow us to follow tree dynamics under climate change in real time at a resolution and accuracy that cannot always be provided through forest inventories or remote sensing.

Measuring forest ecosystem performance in response to changing environmental conditions and detecting threshold responses may improve predictions of tree resilience to disturbance and provide early warning signs of forest transitions (Munson et al. 2018). Critical environmental conditions, such as warming-induced drought stress (e.g., Allen et al. 2010), may shift trees and forests into a different state. Since the returning of the environmental condition to the pre-stress level does not necessarily result in the previous tree or forest state, forcing management to maintain stands within their historic ranges of variability may result in substantial tree mortality and forest dieback once a threshold is exceeded, with a consequent loss of

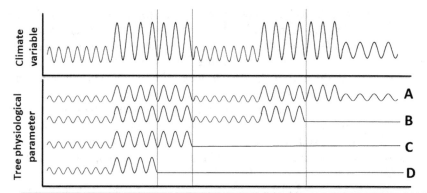

A Full tree functionality recovery after perturbations, tree still alive
B Full tree functionality recovery after one perturbation, failure after two perturbations
C Failure of tree after one perturbation
D Failure of the tree after one perturbation over a shorter time period
Red lines are the tipping points for each case scenario

Fig. 10.3 Physiological responses to climatic perturbations could be defined as the normal or safe operation mode of the single tree (**a**). However, extreme events might lead to anomalous physiological responses beyond the safe operation mode, leading to persistent irreversible changes, tree decline, and tree death. The tipping point, which triggers the exit of tree responses from the safe operation mode, is often not easy to detect without a long-term and high-frequency record of tree functions. The tipping could be reached after a series of extreme days, which might vary among species and conditions (**c** and **d**) or be induced by sequential exposure to extreme events in time (memory effect) (**b**)

forest ecosystem services (Millar and Stephenson 2015). Climate-smart forestry anticipates tree and stand instability in the new environmental condition, facilitating forest adaptation by promoting species mixtures and silvicultural practices aimed at reducing the competition for water and nutrients, thereby ensuring the provision of ecosystem services.

Resource availability strongly influences biogeochemical cycles shaping ecosystem resilience to environmental changes and hence the avoidance of tipping points. Changes in climate and other large-scale environmental alterations (e.g., nitrogen deposition) affect forest ecosystems worldwide (Lindner et al. 2010). At the local scale, these changes magnify the effects of disturbance events and changes in land-use practices, inducing land cover changes and vegetation shifts (Millar and Stephenson 2015).

Although protecting intact forests, restoring degraded forests and managing sustainably productive forests are essential issues to ensure carbon storage, and many other ecosystem services (Pan et al. 2011), forests, and forestry also provide forcing and feedbacks to climate, affecting the exchange of energy and water between land surfaces and the atmosphere (Naudts et al. 2016). In fact, forests influence climate in different and contrasting ways by storing large amounts of carbon (assimilating CO_2), masking the high albedo of snow (warming climate), and sustaining the hydrologic cycle through evapotranspiration (cooling climate) (Bonan 2008). Indeed, the effect of competing processes (carbon emission vs. albedo increase from land-use changes) is large in temperate and boreal latitudes of Europe, where forests have been cleared for agriculture (with an increase in surface albedo), offsetting the warming due to deforestation (Luyssaert et al. 2018). However, in the tropics, forest loss leads to additional warming. Forest resilience to drought and the interaction of disturbances with climate (e.g., fires, pollutants), as well as the effect of deforestation on cloud formation, affect carbon sequestration potential and evaporative cooling of tropical forests.

Inferring the direction of causal dependence between drivers and processes within complex mosaics of forest stands is challenging. Across regions and species, trees that died during drought events were found to be less resilient to stress conditions occurring previously relative to co-occurring resilient trees of the same species (DeSoto et al. 2020). Therefore, widespread (in space) and continuous (in time) monitoring of individual functionality should be planned for describing the causal relationships between climatic patterns or environmental disturbances and tree resilience/vulnerability.

Droughts are linked to a wide range of climatic conditions, such as increased mean and maximum air temperatures, which increase evapotranspiration rate and vapor pressure deficit, with variable impacts on tree functioning across different forest types (Choat et al. 2012; Rita et al. 2020). When coping with drought stress, trees must finely tune the loss of water (transpiration) and the uptake of carbon (growth). Although trees may adjust to extreme conditions, it is not clear whether rapid physiological adjustments in stress tolerance occur in response to heat waves and/or drought spells or whether this is an effective protectant during the extreme events that are predicted to occur in the future (O'Sullivan et al. 2017). Yet, it is unknown whether acclimation to long-term warming modifies the physiological performance of trees during an extreme event (Teskey et al. 2015).

Tree water and carbon management strategies vary with species (e.g., regulation of water potential, vulnerability to xylem embolism, pattern of carbon allocation, etc.), but a clear framework of indicators useful for predicting the different components of tree resilience and the capacity of trees to recover after disturbance (or their mortality) is still missing. Similarly, the relative influence of specific climate parameters on forest decline is poorly understood (Park Williams et al. 2013). Specific functional traits for adapting to climate change and coping with environmental disturbances include tree height, wood density, seed size, specific leaf area, resprout ability, bark thickness, and rooting depth (Aubin et al. 2018). However, a combination of ecophysiological indicators, measured continuously and representing the coupling of tree productivity and water relations, would best explain the tipping point of tree resilience/mortality, predicting the probability of departure from the safe operational space.

10.3 From Tree Observation to Functional Understanding

Single-tree characteristics provide information about the response of stands to disturbance events and the growing stock of stands (see Chap. 4 of this book: Temperli et al. 2021). Similar information can be estimated from remote sensing, but the quality, sensitivity, and resolution of the information are not as high. In addition, ecophysiological traits of trees are increasingly recognized as a useful tool to predict vulnerability to disturbance (namely, drought and the drought-induced xylem dysfunction) and to forecast composition, structure, and function of future forests under climate change scenarios (O'Brien et al. 2017). The increased frequency of extreme events and climate anomalies (e.g., late frosts, heavy storms) may produce immediate damage to stands or alter local phenology of trees, leading to increased risks of pest exposure or carbon starvation. However, widespread climate-driven forest die-off from drought and heat stress is expected to have consequences distinct from those of other forest disturbances (Allen et al. 2010).

Luyssaert et al. (2018) argued that Europe should not rely on forest management to mitigate climate change, whereas adaptation to future climate should be favored. Whether this adaptation can be obtained by changes in species composition and/or revision of silvicultural systems over major biogeographic regions needs standardized data collection across field experiments. In particular, ecophysiological responses of fine-scale processes may help to understand regional-scale trajectories of adaptation patterns and long-term consequences. While acknowledging the importance of biophysical effects on climate, Grassi et al. (2019) claimed that the net annual biophysical climate impact of forest management in Europe remains more uncertain than the net atmospheric CO_2 uptake impact.

The primary reason for forest monitoring to move forward and integrate tree-level and landscape-level data is to operate tools in a manner that consistently generate information in a dynamic environment. A number of traits are good indicators of tree responses to resource availability, or biotic disturbance, and data processed by software platforms can be readily converted into descriptions of these traits. Integrating image processing (e.g., scientific digital webcams; Bothmann et al. 2017) with functional monitoring (e.g., sap flow gauges; Flo et al. 2019) provides an example of how different sensors can be linked to address rapid dynamics in plant response to environmental changes. The fast development of advanced equipment and the vast amount of generated data may allow innovative data-driven approaches to replace traditional hypothesis-driven analyses, providing new insights on forest ecophysiology by means of artificial intelligence, e.g., machine learning approaches (Torresan et al. 2021).

A network of sensors and imagers deployed in the forest can be also used to monitor the simultaneous response of interacting variables, partitioning aboveground and belowground dynamics in the soil-plant-atmosphere continuum. Ground-penetrating radar (GPR) (e.g., Lambot et al. 2006) and wireless soil moisture sensor networks (Rosenbaum et al. 2012) allow the assessment of spatial patterns of soil moisture and soil hydraulic properties, which may integrate measurements of hydraulic redistribution by deep roots, following reversal in sap flow (Oliveira et al. 2005). Cosmic ray sensors provide soil moisture measurements for a footprint with a radius of approximately 300 m and a vertical depth of up to 70 cm (Zreda et al. 2012; Baatz et al. 2014). In drought-stress physiology, in particular, questions about the proportion of water sources accessed by plants during the season can be answered by tracing stable isotopes of hydrogen and oxygen ($^2H/^1H$, $^{16}O/^{18}O$) in the water molecule (Dawson et al. 2002). Relatively cheap and transportable instruments, made available by recent technical development, allow measurement of the stable isotope composition of different waters, including transpired and leaf water, directly in the field (Cernusak et al. 2016; Marshall et al. 2020).

Stable isotopes can be used to trace the uptake and movement of water through the tree, interpreting temporal and spatial variation between neighboring plants. For example, walnut trees were reported to extract water from deeper soils compared to the Italian alder in a mixed plantation in central Italy (Lauteri et al. 2005). In contrast, black walnut was found to extract water from shallow soils compared to a hybrid poplar (*Populus deltoides* x *Populus nigra* clone) in an agroforestry system in Ontario, Canada (Link et al. 2015). Switches between different soil water sources may also occur as a function of seasonal patterns (dry vs. wet periods) or weather events (high vs. low soil moisture) (Sun et al. 2011). Given that transpiration is strongly controlled by water supply and demand, stable isotopes of hydrogen and oxygen in plant organic matter (e.g., leaf tissues, tree rings) reflect the environmental conditions (particularly the evaporative demand) in which the tree grew and the biophysical response to those conditions. Schwendenmann et al. (2010) observed that a higher proportion of deep-water uptake associated with more foliage cover in the dry season (phenological stage), as well as higher sap flux densities and water use rates (transpiration rate). Age and size of trees also have an impact on soil water-use depths and dynamics. The development of technologies for quantifying stable isotope ratios of transpired water and water extracted from plant tissues provides a means to understand the environmental and physiological controls over leaf hydraulics. Labelling experiments, in which labelled water (with D_2O) is added to the soil surface, may further illuminate patterns of water uptake (Koeniger et al. 2010).

Digital sensors open new opportunities for low-cost measurements of vertical soil moisture storage and temperature, including vertical and horizontal patterns of root water uptake (Blonquist et al. 2005; Nadezhdina et al. 2006). Full-range tensiometers (filled with a polymer solution) can be used to measure the soil water matric potential directly in forest, in the range of 0–2 MPa with enough accuracy and low maintenance (Bakker et al. 2007). Estimates of soil hydraulic properties are, however, critical for understanding drought-induced changes in soil hydrological processes, including water infiltration, surface runoff, water retention, moisture content, and solute transport (Robinson et al. 2019), as well as plant transpiration, the principal component of the hydrologic cycle.

The energy associated with water transpired by plants and evaporated directly from wet surfaces (the latent heat flux) is a fundamental component of the Earth's surface energy balance. Soil moisture and evaporative demand affect transpiration, which is the dominant component of the latent heat flux in areas covered by forest. Eddy covariance technique can be used to measure the latent heat flux above the forest canopy, but it does not distinguish between transpiration and evaporation. In this sense, sap flow (sap flux when referred to an area, e.g., conducting sapwood or transpiring foliage) measurements may help disentangle transpiration and evaporation, as well as determine species-specific contributions (for a comparison of sap flow methods, see Steppe et al. 2010; Cermák et al. 2015; Poyatos et al. 2016; Halbritter et al. 2020). Soil properties (e.g., water holding capacity, water content) and plant traits (e.g., sap flow rate, water potential) can be used to derive relative extractable water and water stress indices.

Tree growth dynamics and biomass increment are of high importance as indicators of forest condition in long-term forest monitoring (Dobbertin et al. 2013; see also Chaps. 6 and 7: Pretzsch et al. 2021; Bosela et al. 2021) and of potential uptake of CO_2 by forest ecosystems (Law et al. 2018). Stem radial growth and seasonal cambial rhythm are strongly dependent on environmental factors and, as such, good indicators of tree vitality and of tree responses to stress factors, such as drought (Zweifel 2016; Prislan et al. 2019). Furthermore, strong relationships between annual tree biomass increment and yearly net ecosystem productivity measurement have been observed (Teets et al. 2018). Living trees have similar utility as living laboratories in enabling forest researchers and operators to document and assess the response of trees to climate change in real-life contexts (Farrell et al. 2015). Diel patterns in stem diameter variations (radial growth, water content) and plant water dynamics (sap flow, gas exchange) can be related to mechanisms controlling water and carbon balance and their seasonal variation (Fig. 10.4). In connecting different devices, computer-assisted continuous monitoring of individual trees is essential for the major facets of detection, prediction, and adaptation associated with climate change.

Environmental changes regulate ecosystem processes. Periodic, stochastic, and catastrophic variations in environmental conditions produce, respectively, stress, noise, and disturbance (Sabo and Post 2008). In response to environmental fluctuations, trees generate periodic signals that delineate the boundaries of normal operation. Outside the envelope of normal operation, functional processes in trees (e.g., water and sugar transport between plant organs) may collapse, leading to tree mortality. Sap flow gauges and dendrometers are tools that can be used to monitor the synchronicity of tree signals and environmental fluctuations (Cocozza et al. 2009), providing continuous information on hydraulic safety and carbon status.

Sap flow dynamics can be related to stem diameter variations, considering radial flow of water between xylem and phloem (Steppe et al. 2016). Radial water flow causes changes in stem water capacitance, highlighting functional links between phloem and xylem (Pfautsch et al. 2015), facilitated by wood anatomical traits (parenchyma cells). Complementary measurements of stem tissue moisture can be used to derive the relative water content (i.e., the difference between fresh weight and dry mass, divided by the difference between turgid weight and dry mass of the tissues), an indicator of water stress, which trees try to maintain as constant as possible or above species-specific irreversible thresholds of

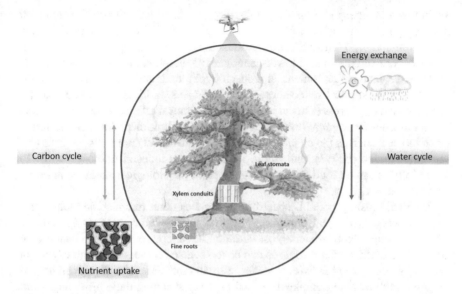

Fig. 10.4 The tree biogeophysical-chemical unit. Ecophysiological processes influence, over time and from tissue to tree level, biogeophysical processes (surface energy fluxes, the hydrologic cycle) and biogeochemical processes (the carbon cycle, the nutrient cycle), as well as biogeographical processes (land use, vegetation dynamics). Single-tree observation provides data for process integration at fine scales, while remote-sensing monitoring is important for scaling indicators to landscape levels. Unmanned aerial vehicles (UAVs, e.g., drones) equipped with miniaturized sensors may map landscape features at high spatial and temporal resolution. Imagery from UAVs may help derive tree growth and monitor forest health (e.g., healthy, dead, or stressed/infested trees)

dehydration (Martinez-Vilalta et al. 2019). Further, transportable computed tomography may represent a powerful tool for measuring density distributions and water contents in the xylem with high spatial resolution in the field (Raschi et al. 1995; Tognetti et al. 1996).

Spectral properties of leaves, based on reflectance-absorbance of light by pigments, may add information on the health status of the forest canopy (Rautiainen et al. 2018). Field spectroscopy provides a cost-effective and practical means to monitor forest functioning with a capacity to upscale to airborne and satellite imagery. Comparing measurements taken with below-canopy sensors, used to measure inside the forest, with reference sensors, located above the forest canopy, may help disentangle the seasonal contribution of understory vegetation to forest reflectance. While multispectral cameras can be used to derive plot-level spectral vegetation indices (SVIs) from discrete spectral wavelengths, hyperspectral analysis of leaf-level photosynthetic parameters has technical challenges (e.g., data storage, sensor availability).

Extensive within-canopy light gradients importantly affect the photosynthetic productivity of leaves in different canopy positions and lead to light-dependent increases in foliage photosynthetic capacity per area (Niinemets et al. 2015).

Within-canopy changes in leaf dry mass per unit area, leaf nitrogen content, and nitrogen partitioning among proteins of the photosynthetic machinery determine the within-canopy photosynthetic modifications. The sun-exposed upper-canopy leaves differ from the shaded lower-canopy leaves in their chlorophyll and nitrogen contents, relative water content, and specific leaf area, and these variations influence the foliar spectral reflectance. Since leaf traits and leaf reflectance co-vary across the canopy layers (Gara et al. 2018), leaf spectral reflectance can be valuable for monitoring the canopy level variation due to environmental stress and reflectance indices, such as the enhanced vegetation index (EVI), normalized difference vegetation index (NDVI), and, more recently, solar-induced fluorescence (SIF). These indices can be used for assessing the plant physiological status by proximal or remote sensing.

Proximal sensing (portable spectrometers and cameras mounted on mobile platforms, towers, or drones) provides validation for the large-scale air-/spaceborne remote sensing, taking advantage of variation in canopy reflectance (Gamon et al. 2019), though the spatial resolution can be too coarse for measuring photosynthetic capacity at the scale of individual leaves in small plots. Fractal analysis can be used to assess architectural complexity based on laser scanning data, providing a link between single-tree canopy attributes and plot-level structural complexity. Combination of structural data (e.g., proximal spectrometry) and ecophysiological measurements (e.g., sap flow) is a valid tool for scaling purposes. The positive relationships between the structural heterogeneity and complexity of forest stands and their functions and services provide a link between proximal spectrometry and forest management (Seidel et al. 2019).

10.4 Experimental Field Trials

It remains difficult to use discrete sampling strategies to address long-term response to multiple stress conditions, relationship between stress response and tree growth, and early detection of plant stress conditions. Understanding rapid changes in functional signals requires quantitative continuous monitoring of both plant physiology and environment conditions. Remote sensing techniques are low in spatial or temporal resolutions, or do not provide timely response to events that influence plant physiology. Therefore, sensors continuously monitoring physiological and environmental parameters (e.g., plant water status, soil moisture content, stem diameter variation, spectral reflectance properties), which are either fixed on plant organs with fixtures or placed in their close proximity, may allow communication with trees.

At heavily instrumented sites, field-portable instruments for analyzing stable isotope compositions may become useful for determining spatial patterns of root water extraction at varying soil depths with succeeding phenological stages (Liu et al. 2019), thus complementing plant transpiration measurements (Nadezhdina et al. 2010; Rothfuss and Javaux 2017). Canopy transpiration flux can be combined with water-use efficiency, as inferred from carbon isotope analysis, to infer gross primary productivity (GPP) of forest canopies (Klein et al. 2016; Vernay et al. 2020).

Continuous measurement of soil respiration can be coupled with chamber CO_2 measurement systems (Tang and Baldocchi 2005), as well as tree- and canopy-scale rates of CO_2 uptake derived by sap flow time series in combination with ^{13}C data, to determine temporal (and spatial) dynamics in autotrophic vs. heterotrophic respiration. Multispectral and/or hyperspectral imaging systems may provide for automated detection of living root dynamics (Bodner et al. 2018), though establishing a sensor network belowground requires considering trade-offs between expensive vs. low-cost multimodal minirhizotrons (Rahman et al. 2020). In this sense, preliminary work with GPR would gather initial imaging analysis of coarse root turnover (Stover et al. 2007), in order to integrate soil texture and soil microclimate (temperature, moisture) and contribute to determine the positions of soil sensor nodes in patchy forest stands (Rundel et al. 2009).

Although stands are the logical operational units for forestry, within-stand variability often hinders identification of the causal relation between mortality episodes and stochastic events (i.e., disturbances). Indeed, a comprehensive assessment of how natural disturbances determine the decline and death of individual trees across sites is still missing. We argue that high frequency and real-time sensor-based measurements of ecophysiological parameters in combination with long-term ecological and silvicultural field-scale studies would enhance our capacity to identify early warning signals in trees, preceding the occurrence of irreversible tree decline, and, thus, monitor forest dieback at sites that are distributed strategically across biogeographic regions. These networks should be able to characterize the spatial and temporal scales of disturbance events.

Observational studies and in situ experiments identify cause-effect relationships, which can be conveniently implemented in ecological syntheses and model exercises to understand interactions between global drivers and change processes. Yet, understanding how functional traits vary among genotypes (tree species or populations) and to what extent this variation has adaptive value is central to CSF. Long-term provenance field trials established in the twentieth century have been conducted to assess genetic diversity in forest tree species. Their coordination may become important in providing data to address climate- and disturbance-related questions for forest productivity and determine species or provenance adaptation to changing environmental conditions.

10.5   Networked Sensors and Wireless Communication at a Site

Low-power communication networks may support data transfer over large distances (kilometers) (Talla et al. 2017). Electro-biochemical devices may run on starch in plants, the most widely used energy storage compound in nature (Zhu et al. 2014).

Potentially, they contain an energy storage density of one order of magnitude higher than that of lithium-ion batteries. Microclimatic sensors can, therefore, be deployed in remote areas and receive continuous electricity supply from trees within dense canopies to run electronics for long-term sampling and monitoring, where solar power is not sufficient and other communication methods are not feasible (Allan et al. 2018). With Internet of things (IoT) technology, many of these networked devices can be connected wirelessly (e.g., temperature sensors, camera traps, and acoustic monitors) and, therefore, able to communicate with each other and transmit data to central nodes.

To reach the ambitious goal of introducing massive data observation and analysis, it is necessary to deploy a great number of specifically designed sensors, connect them in clouds in real time, and analyze the collected data by using big data analytics and machine learning algorithms. Deploying a standardized cybernetic web of specifically designed low-cost sensors will provide real-time access to environmental data from established forest research sites and help identify tree nonlinear responses beyond the safe operation mode (Fig. 10.1), as well as triggering thresholds. A critical feature of a network of sites that are digitally connected is the visualization of records and data storytelling to engage researchers, stakeholders, educators, and the public with climate-smart forests. However, wired systems are costly and energy-demanding, and their use in remote sites is limited (Torresan et al. 2021). Advancements in wireless communication and sensor technologies provide researchers with flexible and scalable tools to monitor smart forestry systems. Agrometeorological data by wireless technology has been implemented in climate-smart agriculture and integrated pest management (Asseng et al. 2016; Marchi et al. 2016), allowing for the control of farming operations based on spatial data (Kaivosoja et al. 2014).

Modern forestry needs to address questions on continuous monitoring and assessing of climate smartness in forests and the impact of disturbance, using the most recent tree-based tools and proximal sensing techniques, combined with field surveys. The complex terrain of mountain regions complicates the study of climate-related disturbances that challenge tree physiology and forest productivity. These forests show large variation in tree density, species composition, and carbon stocks that can hardly be derived from coarse-scale forest inventory and remote sensing (Pan et al. 2011). Rather, fine-scale measurements of ecophysiological traits on individual trees add to leaf- and landscape-level studies, integrating the texture for a comprehensive understanding of forest dynamics (Beer et al. 2010; Brown et al. 2016).

Effects of slope, aspect, and topographic complexity on shaping species-specific physiological responses of mountain forests to seasonal variation in air temperature and soil moisture can be better characterized through instrumented experimental plots. Indeed, mountain forests are subject to landscape-scale differences in soil structure, moisture availability, and energy input that do not apply to plant communities in flat terrain (Zapata-Rios et al. 2015; Wei et al. 2018). Recent development in flexible electronics, sensor designs, and wireless communications is leading to the development of a new generation of sensing devices (e.g., Zhao et al. 2019), which may further advance low-cost and low-power monitoring of microclimate and ecophysiological changes across diverse environmental conditions.

10.6   Measurement Harmonization, Data Integration, and Interoperability Across Sites

The tree-scale measurements emphasized here would be most valuable as part of a larger integrated network. Ground data can be conveniently coupled with standardized observations from highly instrumented research infrastructures. Research infrastructures of multisite networks may provide data on biogeochemical monitoring and allow us to envisage future trajectories of forest-climate relations (Vicca et al. 2018). For example, research infrastructures and networks, such as NEON (https://www.neonscience.org/), collect empirical data of carbon and water fluxes from forest stands and their response to environmental changes in different biogeographic regions (Hinckley et al. 2016; Richter et al. 2018).

Representative forest ecosystem sites can be part of a global Earth observatory, consisting of many well-equipped and similarly equipped ground stations around the world that track key ecosystems fully and continuously (Kulmala 2018). Observational data from these stations can be linked to remote sensing imagery, knowledge from laboratory experiments, and computer modelling simulations to create a coherent dataset, which can be explored in different directions and for specific purposes. Data or product users may include researchers, benefiting from a comprehensive dataset to explore new avenues in the analysis of forest ecosystem functionality and its feedback loops with climate. Other users might include the public and private sector interested in providing diagnostic products, such as early warning alerts for forest managers (e.g., forest fire risk, pest outbreak risk, tree mortality risk, etc.) or ecosystem service assessment for decision-making (payments for ecosystem services).

Such an observational system cannot operate effectively and efficiently without considering data quality standards along the whole pipeline, starting from instrumental measurements up to the processed outputs or products available for different user needs. First, instruments need to be calibrated and harmonized and measurement protocols standardized. Professional staff is needed to install and maintain the instrumentation at the sites, with less assistance required the higher the level of power autonomy, signal stability, and automation of the data collection and transmission. Data processing workflows need to be harmonized across the site network and require the implementation of a raw data quality control (QC) that arises from data quality assessment (QA) procedures agreed and adopted by research scientists operating in the same community. Quality control steps include, for instance, data timestamp verification, elimination of duplicated records, and signal despiking. Obtained raw data time series should, when necessary, be converted to standard physical variables, or further post-processed to produce standard variables, parameters, and indicators of interest. This last set of operations is fundamental to guarantee consistency in the scientific data output across the monitoring network; they underpin data interoperability, defined as the possibility of readily connecting different databases on separate hardware/software systems, and perform data retrieval, analysis, and other applications without regard to the boundaries between the systems (National Research Council 1995).

In an extended forest ecosystem monitoring framework, reducing semantic differences between data from disparate sources (naming conventions, fundamental differences in temporal and spatial scale) means approaching the full interoperability among ground-based monitoring datasets and between these and gridded products (remote and proximal sensing, model simulations). However, differences in technical details at software or hardware level, such as communication protocols and ways of structuring and indexing databases, may hamper the way forward. If, on the one hand, spatial and temporal aggregation of tree-level data into larger scales would allow the comparison with variables typical of forest plot- or catchment-scale observations, information at the original and finest level of detail should be archived and available.

Accessing site information at the single tree scale, including accurate georeferencing of observations, can be fundamental to support climate-smart precision forestry. Yet, the importance of archiving data, as retrieved from the source, lies in the possibility of reprocessing datasets whenever methodological updates are required or a different output standard is chosen to improve data interoperability. Accessing primary data would also give the possibility to scientists to analyze data and develop new products that flow along the virtual line connecting the monitored ecosystem sites to the archives and data users, thus generating more trust about the reliability and utility of the data. It is worth noting that these issues have previously been dealt with by the remote sensing and eddy flux communities.

Comparing functional traits among sites remains challenging due to the large variability in environmental conditions (soil, microclimate, topography, etc.) that modify resource availability (e.g., soil pH, species mixture, terrain slope) and due to species-specific strategies of resource acquisition (e.g., root depths, leaf traits). Integrating field measurements and model representations is not a straightforward exercise (Vicca et al. 2018), though important for understanding processes that occur at various spatial and temporal scales. Nonetheless, the simultaneous measurement of key physiological traits with resource availability indicators may help reduce the caveats associated with any single measurement. Improved capability to record slow and subtle physiological changes and plant-environment interactions is particularly important when comparing stress resilience within and among sites toward an integrated impact assessment of stress events.

A cybernetic web of trees monitors the response of forests to environmental change in near real time. This requires that the data collected by environmental sensors from core sites should be transmitted through wireless technologies (Wi-Fi, LoRa) to a single data concentration point from which collected records are, in turn, transmitted to a data archive (server) through the Internet. These sites should be distributed strategically across major biogeographic regions and forest types. Such a technological platform may combine high-frequency (seconds to days) sensor-based monitoring (e.g., physiological processes) with middle-frequency (weeks to seasons) stand-scale observations and more traditional low-frequency (annual to decadal) forest-level mensuration, in order to respond rapidly to environmental changes and monitor long-term ecological processes.

Studying ecophysiological responses of forest trees enables the prediction of thresholds and, therefore, when changes can be expected in the functioning of individual trees and forest stands. For example, scaling up to stand-level transpiration from measurements on individual trees can be difficult due to errors related to intrinsic wood properties and method characteristics (Vandegehuchte and Steppe 2013; Poyatos et al. 2016; Flo et al. 2019). Scale-up steps from tree to plot level include selecting representative trees for stem diameter classes (depending on the general research objective and species mixture), measuring sapwood area and sap flow radial profiles, quantifying transpiration for all trees in the plot expressed per unit leaf area, and gap-filling data (Ford et al. 2007). Transpiration of the whole stand can then be derived by estimating sapwood area from the diameter distribution of the stand.

Advances in information technology and electronic engineering have prompted the development of smart sensor networks to address complex ecological questions. The proliferation of digital devices allows the creation of cybernetic infrastructures of highly instrumented sites, with advanced storage capacity, data handling, and processing tools, even in mountain environments. Computerized monitoring units can capture and remotely transmit continuous data from a forest site to a remote server over long periods (Sethi et al. 2018). In CSF, a wireless monitoring system is envisaged to obtain field ecological parameters and provide disturbance-related early warning signals in real time. However, autonomous systems for acquiring data should not have high unit costs (Aide et al. 2013) or require complex communication systems (Saito et al. 2015). A new generation of sensors is now accessible for collecting and transmitting physiological data to control units in real time, from an integrated research and monitoring climate-smart forest network, in order to assess tree and forest functionality. A cybernetic web of instrumented trees may provide data on environmental change and alerts at a critical value. In this context, each monitoring unit uploads data from a mobile network of capturing sensors and conveys information for processing and displaying (Fig. 10.5).

Modular multifunctional devices can be developed for the real-time monitoring of tree physiology. An example is represented by the TreeTalker device (Valentini et al. 2019), which measures plant water transport, stem radial fluctuations, leaf spectral characteristics, stem moisture content, tree stem tilting, and environmental microclimatic parameters. It is intended to be deployed on tree clusters and transmit data using IoT technologies, providing cost-effective data. The low-power requirements of the devices are met by high-efficiency batteries and embedded solar panels, which confer power autonomy to the system and allow its deployment in remote and off-grid areas, reducing the need for frequent system maintenance and maintaining the operativity of all the sensors. A large-scale, single-tree, high-frequency, and long-term monitoring network of ecophysiological parameters is represented

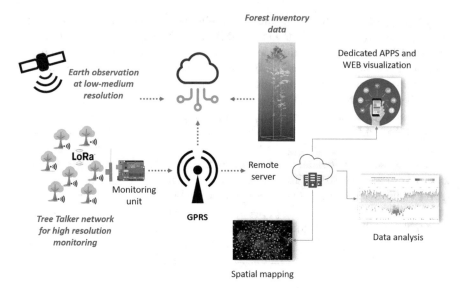

Fig. 10.5 The cybernetic web of modular multifunctional devices (the biogeochemical unit) includes nodes: (**a**) a common suite of low-cost sensors for biological, physical, and chemical measurements, (**b**) real-time data delivery to a single web access point, and (**c**) interactive data visualization and content for scientists, educators, and the public. This networked device allows for data acquisition, processing, and management. Data collected by the device platform and transferred to the cloud can be combined with earth observation datasets and/or forest inventory data. With cyberinfrastructures, near real-time access to all data streams from sensor networks is possible. Therefore, instrument failures, power interruptions, and calibration errors can be quickly identified and corrected, minimizing major data gaps

by forest monitoring research projects in several countries, including China, Italy, Russia, and Spain (Valentini et al. 2019). Based on this example of integrated device technology, a set of variables for identifying drivers of physiological disturbance and a list of measurements and tools for collecting data from experimental forest stands can be outlined (Table 10.2) (other variables can be included to merge diverse approaches). This integrated framework of structural and functional components at monitoring sites is intended to describe the health status of a forest and may feed into climate-smart forest indicators.

Major limitations to continuous monitoring of tree physiological functions are generated by the elevated costs of multi-sensor devices, which are usually energy- and labor-demanding. Current tree monitoring refers to limited sets of devices and trees and/or campaigns in space and time. The TreeTalker network represents a large-scale monitoring system of individual trees in forest plots distributed across a latitudinal gradient. This approach takes advantage of the IoT cyber ecosystem of interconnected sensors and the radio LoRa protocols for data transmission and access to cloud services. The duration of the measurement periods of variables, the acquisition intervals of data, and the frequency of data transmission are customizable, allowing flexible instrument configuration, depending on specific monitoring

Table 10.2 Measurement variables, sensors, and methods for in situ monitoring of climate-smart forest status, considering the stand-level structures and processes and their components

Climate-smart forest	Static and dynamic components	Measurement variables	In situ sensors and methods
Forest structures	Stand heterogeneity	Canopy height	Forest inventory, TLS
		Crown depth	Forest inventory, TLS
		Tree height	Forest inventory, TLS
		DBH and basal area	Forest inventory, TLS
		Species composition	LiDAR and spectral data
		Tree density	Survey, TLS
		Canopy gaps, crown transparency	Survey, TLS, spectral reflectance
	Biotic diversity	Microhabitats	Survey, TLS
		Land cover	ULS
		Species diversity	TLS, survey
		Saproxylic insects	Traps and analysis
		Saproxylic fungi	Survey and analysis
		Lichens	Survey and analysis
		Vertebrates	Counts, camera traps, GPS telemetry
Forest processes	Energy budget	Solar radiation	Pyranometer, light meters
		Albedo	Pyranometer, light meters
		Soil heat flux	Heat flux plate, distributed temperature sensors
		CO_2 and H_2O atmospheric concentrations	Portable GHG gas analyzer
		Latent and sensible heat fluxes	Modeling and land surface temperatures
		LAI	Plant canopy analyzer
		Leaf temperature	Thermal resistance, thermocouple, infrared thermal imaging (TIR)
	Water budget	Precipitation, wind, evaporation, temperature, humidity, snow depth	Pluviometer, anemometer, thermometer, hygrometer, optical sensor
		Transpiration	Sap flow meter
		Throughfall and stemflow	Collectors and samplers
		VPD	Multiparameter probes
		Soil moisture	TDR, electrical capacitance, gamma attenuation
		Soil texture and depth	Shortwave infrared reflectance, ground-penetrating radar (GPR)
		Leaf spectral properties	Spectroradiometer

(continued)

Table 10.2 (continued)

Climate-smart forest	Static and dynamic components	Measurement variables	In situ sensors and methods
	Carbon cycle	*GPP*	Sap flow and stable isotopes (GPP=WUEi*gs)
		Respiration	CO2 flux system
		NPP, aboveground and belowground	Dendrometers, dendrochronology, minirhizotrons, GPR
		SOC	Spectroscopy
		Soil CO_2 flux	CO_2 flux system
		Photosynthesis	CO_2 and H2O flux system, stable isotopes
		Deadwood	Survey, TLS
	Nutrient cycling	*Atmospheric deposition*	Deposition samplers and analysis
		Nutrient uptake	Hyperspectral vegetation indexes, stable isotope labeling
		Soil organic matter	Spectroscopy
		Decomposition and mineralization	Litter bags
		Nitrates and phosphorus	Nitrate and phosphorus sensors
		Litter production	Litter traps
		Soil solution chemistry	Soil solution samplers and analysis

requirements and expected power autonomy. Each single device includes a set of low-cost sensors capable of monitoring tree functions continuously: (1) tree radial growth, as indicator of photosynthetic carbon allocation in biomass; (2) sap flow, as indicator of tree transpiration and functionality of xylem transport; (3) stem wood temperature and xylem water content, as indicators of heat storage and water status of the plant; (4) light penetration through the canopy, as indicator of absorbed radiation fraction; (5) light spectral components, as related to foliage dieback, phenology, and physiology; (6) plant stability (angular deviation of the trunk from the normal along three coordinate axes), related to tree stem tilting, as a result of the momentum exerted by wind on tree canopies and estimated using an automatic accelerometer (gyroscopic sensor); and (7) air temperature and relative humidity in the proximity of the tree trunk, at device installation height (typically 1.3 m), as indicator of tree surrounding microclimate. Each tree can transmit high-frequency data on the web cloud with a unique IoT identifier. This networked device deploys a range of digital sensors, featuring continuous operability and automatic transmission of real-time monitoring data, which provides the basis for translating functional variables into decision support indicators and new research questions (Bayne et al. 2017; Subashini et al. 2018; Valentini et al. 2019).

10.7   Strengths and Limitations

Deploying a standardized cybernetic web of specifically designed low-cost sensors may provide real-time access to environmental data from established forest research sites and help detect nonlinear responses beyond the safe operation mode. Multifunctional devices, based on IoT systems, for the real-time observation of physical and biological parameters of trees can be considered a solution to provide efficient monitoring of forest health. In addition, with the increasing amount of data captured during forest surveys, monitoring systems are becoming important factors in decision-making for management. Modular multifunctional devices allow for long-term (months to years) data collection and observation of a single stand or multiple stands. The distributed nature of a wireless sensor network combined with the spatial resolution of remote sensing data will let a large forest area of study to be monitored in sufficient detail to offer new insights into functional traits and ecosystem services. Spatial links between the data at different scales, stand to landscape, will support researchers in increasing the spatial extent of datasets and performing spatially explicit analyses and predictions. New opportunities emerge to scale up ecological information about the tree-environment interactions at a fine scale, promoting knowledge of forest responses to climate change over coarse scale. Obviously, new technologies come with trade-offs, and integration with traditional inventory data collection is advised when planning forest surveys and monitoring campaigns. Proliferation of digital tools and technologized forest also have political and social impacts that need to be considered (Gabrys 2020). Indeed, forests provide key products and services and are crucial to mitigate global change, contributing to biogeochemical cycles and species diversity. However, though halting deforestation and contributing to reforestation are key to meet international goals (Griscom et al. 2017), climate benefits from carbon sequestration can be offset by environmental disturbances, which are also increasing.

Recent technological advances in instrumentation for measuring physiological ecology variables at experimental sites allow merging information into monitoring data collected in other research infrastructures (Haase et al. 2018). Though sites may differ in the temporal and spatial resolution of instrumentation and in the research questions addressed, modular research platforms may form a multilevel system of distributed monitoring sites, integrating site-specific data source and environmental stratification. Examples of initiatives that have been developed to watch trees grow and function in real time include TreeWatch.net (https://treewatch.net/) and TreeNet (https://treenet.info/) monitoring and modeling networks (Steppe et al. 2016; Zweifel et al. 2016). A global compilation of whole-plant transpiration data from sap flow measurements has been presented by Poyatos et al. (2020), with the aim of harmonizing individual datasets supplied by contributors worldwide (SAPFLUXNET), including subdaily time series of sap flow and ancillary data (https://sapfluxnet.creaf.cat/). Distributed research infrastructures, such as ICOS (https://www.icos-ri.eu/) and FLUXNET (https://fluxnet.fluxdata.org/), generate data and integrate knowledge on biogeochemical cycles and of their perturbations with high operating costs and complex instrumentations (Franz et al. 2018; Rebmann et al. 2018). The TRY database of plant traits (https://www.try-db.org) aims to improve the availability and accessibility of plant trait data for ecology and earth system sciences (Kattge et al. 2020). In this context, selection of key variables documenting early warning signals for critical forest status in highly instrumented sites (tree mortality, biodiversity change) would provide useful directions. Research integration will allow us to better understand the factors driving changes in species diversity, the effects of extreme events on tree productivity, the impacts of disturbances on forest function, and the interactions between short- and long-term trends. Data integration will also facilitate upscaling measurements from local conditions to addressing challenges from global objectives (Fig. 10.6).

The close link between physical properties of the forest canopy (e.g., leaf surface temperature, leaf pigment absorption, chlorophyll fluorescence emission, latent heat flux, etc.) with plant functioning opens a wide range of applications and methods to monitor forest health remotely. However, remote sensing methods may lack adequate resolution for application at the range edge of species distribution. Similarly, the eddy covariance method measures the net effects of a forest upwind of the sensor, ignoring individual trees or species within the stand. These methods are, therefore, unsuited to detect early signs of ecophysiological stress when the functional response of trees differs among ages or species, leading to a compensatory effect at the stand level. Since CSF has the ambition to tailor adaptive silviculture to ensure the resilience of individual trees and species, a more highly resolved

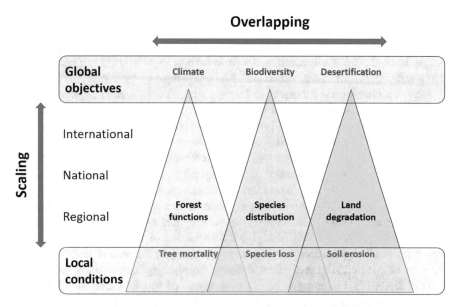

Fig. 10.6 Translation from local conditions (stand-based measurements) to global objectives (global convention requirements) should account for trade-offs and synergies between forest capacity to store carbon, adapt to climate change, and provide products and services

diagnosis of tree decline/mortality is needed. Indeed, risk assessments (diagnosis) and optimal treatments (therapy) require individualized analysis for individual trees exposed to multiple stresses. Effective monitoring of tree responses to environmental disturbance in marginal regions (e.g., mountain areas, range edges) is of critical importance in order to predict and manage threats to tree populations. Therefore, combining remote sensing observations with ground-based methods can be the most effective means of monitoring resilience and vulnerability of forest trees and ecosystems. Stand-based networks committed to long-term monitoring may provide representative datasets (e.g., tree biomass, tree mortality), which become useful for validation of forest modeling exercises and remote sensing missions (Chave et al. 2019). In this context, forest inventories and terrestrial laser scanning (TLS) surveys may contribute with accurate measurements of individual tree traits (e.g., volume, height, allometry, etc.) and forest stand structure, for modelling purposes (Calders et al. 2018). Detailed datasets of 3D vegetation structure from below (Kunz et al. 2019), provided by TLS, can be used for assessing canopy space filling, detecting leaf flush, monitoring tree growth, and deriving microclimate at plot scale. Therefore, information provided by TLS at the stand level may link ground-based measurements and integrate forest structural changes mapped by airborne laser scanning (ALS) from above (Marvin et al. 2016).

Wireless sensor network approach has only recently become cost-effective because of the availability of simple, inexpensive devices. But it also depends on a common, convenient platform for data processing and visualization. Such a platform would ease the use of data for storytelling aimed to engage researchers, stakeholders, educators, and the public with climate-smart forests. We propose using tree-based tools, proximal sensing techniques, and networking tools and coupling them to traditional field surveys and remote sensing in order to address the data needs of continuous monitoring and assessment of climate smartness and the impact of disturbance.

Acknowledgments   J.D.M. and R.T. were supported by *Progetto bilaterale di Grande Rilevanza* (MAECI) Italy-Sweden "Natural hazards in future forests: how to inform climate change adaptation." P.P. was supported by COST (European Cooperation in Science and Technology) Action CLIMO (Climate-Smart Forestry in Mountain Regions – CA15226). R.V. was supported by RSF project #19-77-30012.

References

Adams HD, Zeppel MJB, Anderegg WRL et al (2017) A multi-species synthesis of physiological mechanisms in drought-induced tree mortality. Nat Ecol Evol 1:1285–1291. https://doi.org/10.1038/s41559-017-0248-x

Aide TM, Corrada-Bravo C, Campos-Cerqueira M et al (2013) Real-time bioacoustics monitoring and automated species identification. PeerJ 1:e103. https://doi.org/10.7717/peerj.103

Albrich K, Rammer W, Thom D, Seidl R (2018) Trade-offs between temporal stability and level of forest ecosystem services provisioning under climate change. Ecol Appl 28:1884–1896. https://doi.org/10.1002/eap.1785

Allan BM, Nimmo DG, Ierodiaconou D et al (2018) Futurecasting ecological research: the rise of technoecology. Ecosphere 9:e02163. https://doi.org/10.1002/ecs2.2163

Allen CD, Macalady AK, Chenchouni H et al (2010) A global overview of drought and heat-induced tree mortality reveals emerging climate change risks for forests. For Ecol Manag 259:660–684. https://doi.org/10.1016/j.foreco.2009.09.001

Anderegg WRL, Berry JA, Field CB (2012) Linking definitions, mechanisms, and modeling of drought-induced tree death. Trends Plant Sci 17:693–700. https://doi.org/10.1016/j.tplants.2012.09.006

Anderegg WRL, Hicke JA, Fisher RA et al (2015) Tree mortality from drought, insects, and their interactions in a changing climate. New Phytol 208:674–683. https://doi.org/10.1111/nph.13477

Anderegg WRL, Klein T, Bartlett M et al (2016) Meta-analysis reveals that hydraulic traits explain cross-species patterns of drought-induced tree mortality across the globe. Proc Natl Acad Sci 113:5024–5029. https://doi.org/10.1073/pnas.1525678113

Anderson-Teixeira KJ, Davies SJ, Bennett AC et al (2015) CTFS-ForestGEO: a worldwide network monitoring forests in an era of global change. Glob Change Biol 21:528–549. https://doi.org/10.1111/gcb.12712

Asseng S, McIntosh PC, Thomas G et al (2016) Is a 10-day rainfall forecast of value in dry-land wheat cropping? Agric For Meteorol 216:170–176. https://doi.org/10.1016/j.agrformet.2015.10.012

Aubin I, Boisvert-Marsh L, Kebli H, McKenney D, Pedlar J, Lawrence K, Hogg EH, Boulanger Y, Gauthier S, Ste-Marie C (2018) Tree vulnerability to climate change: improving exposure-based assessments using traits as indicators of sensitivity. Ecosphere 9:e02108. https://doi.org/10.1002/ecs2.2108

Baatz R, Bogena HR, Hendricks Franssen H-J et al (2014) Calibration of a catchment scale cosmic-ray probe network: a comparison of three parameterization methods. J Hydrol 516:231–244. https://doi.org/10.1016/j.jhydrol.2014.02.026

Baeten L, Bruelheide H, van der Plas F et al (2019) Identifying the tree species compositions that maximize ecosystem functioning in European forests. J Appl Ecol 56:733–744. https://doi.org/10.1111/1365-2664.13308

Bakker G, van der Ploeg MJ, de Rooij GH et al (2007) New polymer tensiometers: measuring matric pressures down to the wilting point. Vadose Zone J 6:196–202. https://doi.org/10.2136/vzj2006.0110

Baldocchi DD, Verma SB, Anderson DE (1987) Canopy photosynthesis and water-use efficiency in a deciduous forest. J Appl Ecol 24:251–260. https://doi.org/10.2307/2403802

Bayne K, Damesin S, Evans M (2017) The internet of things – wireless sensor networks and their application to forestry. N Z J For 61(5):37–41

Beer C, Reichstein M, Tomelleri E et al (2010) Terrestrial gross carbon dioxide uptake: global distribution and covariation with climate. Science 329:834–838. https://doi.org/10.1126/science.1184984

Blonquist JM, Jones SB, Robinson DA (2005) Standardizing characterization of electromagnetic water content sensors part 2. Evaluation of seven sensing systems. Vadose Zone J 4:1059–1069. https://doi.org/10.2136/vzj2004.0141

Bodner G, Nakhforoosh A, Arnold T, Leitner D (2018) Hyperspectral imaging: a novel approach for plant root phenotyping. Plant Methods 14:84. https://doi.org/10.1186/s13007-018-0352-1

Bosela M, Merganičová K, Torresan C, et al (2021) Modelling future growth of mountain forests under changing environments. In: Managing Forest Ecosystems, Vol. 40, Tognetti R, Smith M, Panzacchi P (eds) Climate-Smart Forestry in Mountain Regions. Springer Nature, Switzerland, AG

Bonan GB (2008) Forests and climate change: forcings, feedbacks, and the climate benefits of forests. Science 320:1444–1449

Bothmann L, Menzel A, Menze BH et al (2017) Automated processing of webcam images for phenological classification. PLoS One 12:e0171918. https://doi.org/10.1371/journal.pone.0171918

Bowditch E, Santopuoli G, Binder F et al (2020) What is Climate-Smart Forestry? A definition from a multinational collaborative process focused on mountain regions of Europe. Ecosyst Serv 43:101113. https://doi.org/10.1016/j.ecoser.2020.101113

Brandt LA, Butler PR, Handler SD et al (2017) Integrating science and management to assess forest ecosystem vulnerability to climate change. J For 115:212–221. https://doi.org/10.5849/jof.15-147

Brown TB, Hultine KR, Steltzer H et al (2016) Using phenocams to monitor our changing earth: toward a global phenocam network. Front Ecol Environ 14:84–93. https://doi.org/10.1002/fee.1222

Bussotti F, Pollastrini M (2017) Traditional and novel indicators of climate change impacts on European forest trees. Forests 8:137. https://doi.org/10.3390/f8040137

Cailleret M, Bigler C, Bugmann H et al (2016) Towards a common methodology for developing logistic tree mortality models based on ring-width data. Ecol Appl 26:1827–1841. https://doi.org/10.1890/15-1402.1

Cailleret M, Jansen S, Robert EMR et al (2017) A synthesis of radial growth patterns preceding tree mortality. Glob Change Biol 23:1675–1690. https://doi.org/10.1111/gcb.13535

Cailleret M, Dakos V, Jansen S et al (2019) Early-warning signals of individual tree mortality based on annual radial growth. Front Plant Sci 9. https://doi.org/10.3389/fpls.2018.01964

Calders K, Origo N, Burt A et al (2018) Realistic forest stand reconstruction from terrestrial LiDAR for radiative transfer modelling. Remote Sens 10:933. https://doi.org/10.3390/rs10060933

Cavaleri MA, Oberbauer SF, Ryan MG (2008) Foliar and ecosystem respiration in an old-growth tropical rain forest. Plant Cell Environ 31:473–483. https://doi.org/10.1111/j.1365-3040.2008.01775.x

Cermák J, Nadezhdina N, Trcala M, Simon J (2015) Open field-applicable instrumental methods for structural and functional assessment of whole trees and stands. IForest Biogeosci For 8:226. https://doi.org/10.3832/ifor1116-008

Cernusak LA, Barbour MM, Arndt SK et al (2016) Stable isotopes in leaf water of terrestrial plants. Plant Cell Environ 39:1087–1102. https://doi.org/10.1111/pce.12703

Chave J, Davies SJ, Phillips OL et al (2019) Ground data are essential for biomass remote sensing missions. Surv Geophys 40:863–880. https://doi.org/10.1007/s10712-019-09528-w

Choat B, Jansen S, Brodribb TJ et al (2012) Global convergence in the vulnerability of forests to drought. Nature 491:752–755. https://doi.org/10.1038/nature11688

Choat B, Brodribb TJ, Brodersen CR et al (2018) Triggers of tree mortality under drought. Nature 558:531–539. https://doi.org/10.1038/s41586-018-0240-x

Cocozza C, Lasserre B, Giovannelli A et al (2009) Low temperature induces different cold sensitivity in two poplar clones (Populusxcanadensis Mönch 'I-214' and P. deltoides Marsh. 'Dvina'). J Exp Bot 60:3655–3664. https://doi.org/10.1093/jxb/erp212

Cocozza C, Giovannelli A, Lasserre B et al (2012) A novel mathematical procedure to interpret the stem radius variation in olive trees. Agric For Meteorol 161:80–93. https://doi.org/10.1016/j.agrformet.2012.03.016

Cocozza C, Palombo C, Tognetti R et al (2016) Monitoring intra-annual dynamics of wood formation with microcores and dendrometers in Picea abies at two different altitudes. Tree Physiol 36:832–846. https://doi.org/10.1093/treephys/tpw009

Cocozza C, Tognetti R, Giovannelli A (2018) High-resolution analytical approach to describe the sensitivity of tree–environment dependences through stem radial variation. Forests 9:134. https://doi.org/10.3390/f9030134

Dai L, Vorselen D, Korolev KS, Gore J (2012) Generic indicators for loss of resilience before a tipping point leading to population collapse. Science 336:1175–1177. https://doi.org/10.1126/science.1219805

Dakos V, Carpenter SR, van Nes EH, Scheffer M (2015) Resilience indicators: prospects and limitations for early warnings of regime shifts. Philos Trans R Soc B Biol Sci 370:20130263. https://doi.org/10.1098/rstb.2013.0263

Dawson TE, Mambelli S, Plamboeck AH et al (2002) Stable isotopes in plant ecology. Annu Rev Ecol Syst 33:507–559. https://doi.org/10.1146/annurev.ecolsys.33.020602.095451

del Río M, Pretzsch H, Bončina A, et al (2021) Assessment of indicators for climate smart management in mountain forests. In: Managing Forest Ecosystems, Vol. 40, Tognetti R, Smith M, Panzacchi P (eds) Climate-Smart Forestry in Mountain Regions. Springer Nature, Switzerland, AG

DeSoto L, Cailleret M, Sterck F et al (2020) Low growth resilience to drought is related to future mortality risk in trees. Nat Commun 11:545. https://doi.org/10.1038/s41467-020-14300-5

Dobbertin M, Neumann M, Schroeck H-W (2013) Chapter 10 – Tree growth measurements in long-term Forest monitoring in Europe. In: Ferretti M, Fischer R (eds) Developments in environmental science. Elsevier, Amsterdam, pp 183–204

Drake JM, Griffen BD (2010) Early warning signals of extinction in deteriorating environments. Nature 467:456–459. https://doi.org/10.1038/nature09389

Farrell C, Szota C, Arndt SK (2015) Urban plantings: 'living laboratories' for climate change response. Trends Plant Sci 20:597–599. https://doi.org/10.1016/j.tplants.2015.08.006

Fierravanti A, Cocozza C, Palombo C et al (2015) Environmental-mediated relationships between tree growth of black spruce and abundance of spruce budworm along a latitudinal transect in Quebec, Canada. Agric For Meteorol 213:53–63. https://doi.org/10.1016/j.agrformet.2015.06.014

Flo V, Martinez-Vilalta J, Steppe K et al (2019) A synthesis of bias and uncertainty in sap flow methods. Agric For Meteorol 271:362–374. https://doi.org/10.1016/j.agrformet.2019.03.012

Ford CR, Hubbard RM, Kloeppel BD, Vose JM (2007) A comparison of sap flux-based evapotranspiration estimates with catchment-scale water balance. Agric For Meteorol 145:176–185. https://doi.org/10.1016/j.agrformet.2007.04.010

Franz D, Acosta M, Altimir N et al (2018) Towards long-term standardised carbon and greenhouse gas observations for monitoring Europe's terrestrial ecosystems: a review. Int Agrophys 32:439–455. https://doi.org/10.1515/intag-2017-0039

Gabrys J (2020) Smart forests and data practices: from the internet of trees to planetary governance. Big Data Soc 7:2053951720904871. https://doi.org/10.1177/2053951720904871

Gamfeldt L, Snäll T, Bagchi R et al (2013) Higher levels of multiple ecosystem services are found in forests with more tree species. Nat Commun 4:1340. https://doi.org/10.1038/ncomms2328

Gamon JA, Somers B, Malenovský Z et al (2019) Assessing vegetation function with imaging spectroscopy. Surv Geophys 40:489–513. https://doi.org/10.1007/s10712-019-09511-5

Gara TW, Darvishzadeh R, Skidmore AK, Wang T (2018) Impact of vertical canopy position on leaf spectral properties and traits across multiple species. Remote Sens 10:346. https://doi.org/10.3390/rs10020346

Giorgi F, Lionello P (2008) Climate change projections for the Mediterranean region. Glob Planet Change 63:90–104. https://doi.org/10.1016/j.gloplacha.2007.09.005

Grassi G, Cescatti A, Matthews R et al (2019) On the realistic contribution of European forests to reach climate objectives. Carbon Balance Manag 14:8. https://doi.org/10.1186/s13021-019-0123-y

Griscom BW, Adams J, Ellis PW et al (2017) Natural climate solutions. Proc Natl Acad Sci 114:11645–11650. https://doi.org/10.1073/pnas.1710465114

Haase P, Tonkin JD, Stoll S et al (2018) The next generation of site-based long-term ecological monitoring: linking essential biodiversity variables and ecosystem integrity. Sci Total Environ 613–614:1376–1384. https://doi.org/10.1016/j.scitotenv.2017.08.111

Halbritter AH, Boeck HJD, Eycott AE et al (2020) The handbook for standardized field and laboratory measurements in terrestrial climate change experiments and observational studies (ClimEx). Methods Ecol Evol 11:22–37. https://doi.org/10.1111/2041-210X.13331

Hansen J, Sato M, Ruedy R (2012) Perception of climate change. Proc Natl Acad Sci 109:E2415–E2423. https://doi.org/10.1073/pnas.1205276109

Harmon ME, Pabst RJ (2015) Testing predictions of forest succession using long-term measurements: 100 yrs of observations in the Oregon Cascades. J Veg Sci 26:722–732. https://doi.org/10.1111/jvs.12273

Hartmann H, Moura CF, Anderegg WRL et al (2018) Research frontiers for improving our understanding of drought-induced tree and forest mortality. New Phytol 218:15–28. https://doi.org/10.1111/nph.15048

Hinckley E-LS, Anderson SP, Baron JS et al (2016) Optimizing available network resources to address questions in environmental biogeochemistry. Bioscience 66:317–326. https://doi.org/10.1093/biosci/biw005

Jarvis PG, Morison JIL, Chaloner WG et al (1989) Atmospheric carbon dioxide and forests. Philos Trans R Soc Lond Ser B Biol Sci 324:369–392. https://doi.org/10.1098/rstb.1989.0053

Jarvis L, McCann K, Tunney T et al (2016) Early warning signals detect critical impacts of experimental warming. Ecol Evol 6:6097–6106. https://doi.org/10.1002/ece3.2339

Jordan BL, Batalin MA, Kaiser WJ (2007) NIMS RD: a rapidly deployable cable based robot. In: Proceedings 2007 IEEE international conference on robotics and automation, pp 144–150

Kaivosoja J, Jackenkroll M, Linkolehto R et al (2014) Automatic control of farming operations based on spatial web services. Comput Electron Agric 100:110–115. https://doi.org/10.1016/j.compag.2013.11.003

Kannenberg SA, Novick KA, Alexander MR et al (2019) Linking drought legacy effects across scales: from leaves to tree rings to ecosystems. Glob Change Biol 25:2978–2992. https://doi.org/10.1111/gcb.14710

Kattge J, Bönisch G, Díaz S et al (2020) TRY plant trait database – enhanced coverage and open access. Glob Change Biol 26:119–188. https://doi.org/10.1111/gcb.14904

Klein T, Rotenberg E, Tatarinov F, Yakir D (2016) Association between sap flow-derived and eddy covariance-derived measurements of forest canopy CO2 uptake. New Phytol 209:436–446. https://doi.org/10.1111/nph.13597

Koeniger P, Leibundgut C, Link T, Marshall JD (2010) Stable isotopes applied as water tracers in column and field studies. Org Geochem 41:31–40. https://doi.org/10.1016/j.orggeochem.2009.07.006

Kulmala M (2018) Build a global earth observatory. Nature 553:21–23. https://doi.org/10.1038/d41586-017-08967-y

Kunz M, Fichtner A, Härdtle W et al (2019) Neighbour species richness and local structural variability modulate aboveground allocation patterns and crown morphology of individual trees. Ecol Lett 22:2130–2140. https://doi.org/10.1111/ele.13400

Lambot S, Slob EC, Vanclooster M, Vereecken H (2006) Closed loop GPR data inversion for soil hydraulic and electric property determination. Geophys Res Lett 33. https://doi.org/10.1029/2006GL027906

Lauteri M, Alessio GA, Paris P (2005) Using oxygen stable isotopes Tto investigate the soil-plant-atmosphere hydraulic continuum in complex stands of walnut. Acta Hortic 223–230. https://doi.org/10.17660/ActaHortic.2005.705.27

Law BE, Hudiburg TW, Berner LT et al (2018) Land use strategies to mitigate climate change in carbon dense temperate forests. Proc Natl Acad Sci 115:3663–3668. https://doi.org/10.1073/pnas.1720064115

Lindner M, Maroschek M, Netherer S et al (2010) Climate change impacts, adaptive capacity, and vulnerability of European forest ecosystems. For Ecol Manag 259:698–709. https://doi.org/10.1016/j.foreco.2009.09.023

Lindner M, Fitzgerald JB, Zimmermann NE et al (2014) Climate change and European forests: what do we know, what are the uncertainties, and what are the implications for forest management? J Environ Manag 146:69–83. https://doi.org/10.1016/j.jenvman.2014.07.030

Link CM, Thevathasan NV, Gordon AM, Isaac ME (2015) Determining tree water acquisition zones with stable isotopes in a temperate tree-based intercropping system. Agrofor Syst 89:611–620. https://doi.org/10.1007/s10457-015-9795-9

Liu Y, Zhang X, Zhao S et al (2019) The depth of water taken up by walnut trees during different phenological stages in an irrigated arid hilly area in the Taihang Mountains. Forests 10:121. https://doi.org/10.3390/f10020121

Luyssaert S, Marie G, Valade A et al (2018) Trade-offs in using European forests to meet climate objectives. Nature 562:259–262. https://doi.org/10.1038/s41586-018-0577-1

Marchi S, Guidotti D, Ricciolini M, Petacchi R (2016) Towards understanding temporal and spatial dynamics of Bactrocera oleae (Rossi) infestations using decade-long agrometeorological time series. Int J Biometeorol 60:1681–1694. https://doi.org/10.1007/s00484-016-1159-2

Markwitz C, Siebicke L (2019) Low-cost eddy covariance: a case study of evapotranspiration over agroforestry in Germany. Atmos Meas Tech 12:4677–4696. https://doi.org/10.5194/amt-12-4677-2019

Marshall JD, Cuntz M, Beyer M et al (2020) Borehole equilibration: testing a new method to monitor the isotopic composition of tree xylem water in situ. Front Plant Sci 11. https://doi.org/10.3389/fpls.2020.00358

Martinez-Vilalta J, Anderegg WRL, Sapes G, Sala A (2019) Greater focus on water pools may improve our ability to understand and anticipate drought-induced mortality in plants. New Phytol 223:22–32. https://doi.org/10.1111/nph.15644

Marvin DC, Koh LP, Lynam AJ et al (2016) Integrating technologies for scalable ecology and conservation. Glob Ecol Conserv 7:262–275. https://doi.org/10.1016/j.gecco.2016.07.002

McDowell NG, Fisher RA, Xu C et al (2013) Evaluating theories of drought-induced vegetation mortality using a multimodel–experiment framework. New Phytol 200:304–321. https://doi.org/10.1111/nph.12465

Millar CI, Stephenson NL (2015) Temperate forest health in an era of emerging megadisturbance. Science 349:823–826. https://doi.org/10.1126/science.aaa9933

Munson SM, Reed SC, Peñuelas J et al (2018) Ecosystem thresholds, tipping points, and critical transitions. New Phytol 218:1315–1317. https://doi.org/10.1111/nph.15145

Nadezhdina N, Čermák J, Gašpárek J et al (2006) Vertical and horizontal water redistribution in Norway spruce (Picea abies) roots in the Moravian Upland. Tree Physiol 26:1277–1288. https://doi.org/10.1093/treephys/26.10.1277

Nadezhdina N, David TS, David JS et al (2010) Trees never rest: the multiple facets of hydraulic redistribution. Ecohydrology 3:431–444. https://doi.org/10.1002/eco.148

Nadrowski K, Pietsch K, Baruffol M et al (2014) Tree species traits but not diversity mitigate stem breakage in a subtropical Forest following a rare and extreme ice storm. PLoS One 9:e96022. https://doi.org/10.1371/journal.pone.0096022

National Research Council (1995) Finding the forest in the trees: the challenge of combining diverse environmental data. National Academies Press, Washington, DC

Naudts K, Chen Y, McGrath MJ et al (2016) Europe's forest management did not mitigate climate warming. Science 351:597–600. https://doi.org/10.1126/science.aad7270

Niinemets Ü, Keenan TF, Hallik L (2015) A worldwide analysis of within-canopy variations in leaf structural, chemical and physiological traits across plant functional types. New Phytol 205:973–993. https://doi.org/10.1111/nph.13096

O'Brien MJ, Engelbrecht BMJ, Joswig J et al (2017) A synthesis of tree functional traits related to drought-induced mortality in forests across climatic zones. J Appl Ecol 54:1669–1686. https://doi.org/10.1111/1365-2664.12874

O'Sullivan OS, Heskel MA, Reich PB et al (2017) Thermal limits of leaf metabolism across biomes. Glob Change Biol 23:209–223. https://doi.org/10.1111/gcb.13477

Oliveira RS, Dawson TE, Burgess SSO, Nepstad DC (2005) Hydraulic redistribution in three Amazonian trees. Oecologia 145:354–363. https://doi.org/10.1007/s00442-005-0108-2

Pan Y, Birdsey RA, Fang J et al (2011) A large and persistent carbon sink in the world's forests. Science 333:988–993. https://doi.org/10.1126/science.1201609

Park Williams A, Allen CD, Macalady AK et al (2013) Temperature as a potent driver of regional forest drought stress and tree mortality. Nat Clim Chang 3:292–297. https://doi.org/10.1038/nclimate1693

Peltola H, Kellomäki S (1993) A mechanistic model for calculating windthrow and stem breakage of Scots pines at stand age. Silva Fenn. https://doi.org/10.14214/sf.a15665

Perone A, Lombardi F, Marchetti M et al (2016) Evidence of solar activity and El Niño signals in tree rings of Araucaria araucana and A. angustifolia in South America. Glob Planet Change 145:1–10. https://doi.org/10.1016/j.gloplacha.2016.08.004

Pfautsch S, Hölttä T, Mencuccini M (2015) Hydraulic functioning of tree stems—fusing ray anatomy, radial transfer and capacitance. Tree Physiol 35:706–722. https://doi.org/10.1093/treephys/tpv058

Polade SD, Pierce DW, Cayan DR et al (2014) The key role of dry days in changing regional climate and precipitation regimes. Sci Rep 4:4364. https://doi.org/10.1038/srep04364

Poyatos R, Granda V, Molowny-Horas R et al (2016) SAPFLUXNET: towards a global database of sap flow measurements. Tree Physiol 36:1449–1455. https://doi.org/10.1093/treephys/tpw110

Poyatos R, Granda V, Flo V et al (2020) Global transpiration data from sap flow measurements: the SAPFLUXNET database. Earth Syst Sci Data Discuss:1–57. https://doi.org/10.5194/essd-2020-227

Pretzsch H, del Río M, Giammarchi F, Uhl E, Tognetti R (2021a) Changes of tree and stand growth. Review and implications. In: Managing Forest Ecosystems, Vol. 40, Tognetti R, Smith M, Panzacchi P (eds) Climate-Smart Forestry in Mountain Regions. Springer Nature, Switzerland, AG

Pretzsch H, Hilmers T, Uhl E, et al (2021b) Efficacy of trans-geographic observational network design for revelation of growth pattern in mountain forests across Europe. In: Managing Forest Ecosystems, Vol. 40, Tognetti R, Smith M, Panzacchi P (eds) Climate-Smart Forestry in Mountain Regions. Springer Nature, Switzerland, AG

Prislan P, Gričar J, Čufar K et al (2019) Growing season and radial growth predicted for Fagus sylvatica under climate change. Clim Chang 153:181–197. https://doi.org/10.1007/s10584-019-02374-0

Rahman G, Sohag H, Chowdhury R et al (2020) SoilCam: a fully automated minirhizotron using multispectral imaging for root activity monitoring. Sensors 20:787. https://doi.org/10.3390/s20030787

Raschi A, Tognetti R, Ridder H-W, Berés C (1995) The use of computer tomography in the study of pollution effects on oak trees. Agric Mediterr Special Volume:298–306

Rautiainen M, Lukeš P, Homolová L et al (2018) Spectral properties of coniferous forests: a review of in situ and laboratory measurements. Remote Sens 10:207. https://doi.org/10.3390/rs10020207

Rebmann C, Aubinet M, Schmid H et al (2018) ICOS eddy covariance flux-station site setup: a review. Int Agrophys 32:471–494. https://doi.org/10.1515/intag-2017-0044

Richter DD, Billings SA, Groffman PM et al (2018) Ideas and perspectives: strengthening the biogeosciences in environmental research networks. Biogeosciences 15:4815–4832. https://doi.org/10.5194/bg-15-4815-2018

Rita A, Camarero JJ, Nolè A et al (2020) The impact of drought spells on forests depends on site conditions: the case of 2017 summer heat wave in southern Europe. Glob Change Biol 26:851–863. https://doi.org/10.1111/gcb.14825

Robinson DA, Hopmans JW, Filipovic V et al (2019) Global environmental changes impact soil hydraulic functions through biophysical feedbacks. Glob Change Biol 25:1895–1904. https://doi.org/10.1111/gcb.14626

Rosenbaum U, Bogena HR, Herbst M et al (2012) Seasonal and event dynamics of spatial soil moisture patterns at the small catchment scale. Water Resour Res 48. https://doi.org/10.1029/2011WR011518

Rothfuss Y, Javaux M (2017) Reviews and syntheses: isotopic approaches to quantify root water uptake: a review and comparison of methods. Biogeosciences 14:2199–2224. https://doi.org/10.5194/bg-14-2199-2017

Rundel PW, Graham EA, Allen MF et al (2009) Environmental sensor networks in ecological research. New Phytol 182:589–607. https://doi.org/10.1111/j.1469-8137.2009.02811.x

Sabo JL, Post DM (2008) Quantifying periodic, stochastic, and catastrophic environmental variation. Ecol Monogr 78:19–40. https://doi.org/10.1890/06-1340.1

Saito K, Nakamura K, Ueta M et al (2015) Utilizing the cyberforest live sound system with social media to remotely conduct woodland bird censuses in Central Japan. Ambio 44:572–583. https://doi.org/10.1007/s13280-015-0708-y

Santopuoli G, Temperli C, Alberdi I et al (2020) Pan-European sustainable forest management indicators for assessing Climate-Smart Forestry in Europe1. Can J For Res. https://doi.org/10.1139/cjfr-2020-0166

Schwendenmann L, Veldkamp E, Moser G et al (2010) Effects of an experimental drought on the functioning of a cacao agroforestry system, Sulawesi, Indonesia. Glob Change Biol 16:1515–1530. https://doi.org/10.1111/j.1365-2486.2009.02034.x

Seidel D, Annighöfer P, Stiers M et al (2019) How a measure of tree structural complexity relates to architectural benefit-to-cost ratio, light availability, and growth of trees. Ecol Evol 9:7134–7142. https://doi.org/10.1002/ece3.5281

Sethi SS, Ewers RM, Jones NS et al (2018) Robust, real-time and autonomous monitoring of ecosystems with an open, low-cost, networked device. Methods Ecol Evol 9:2383–2387. https://doi.org/10.1111/2041-210X.13089

Shestakova TA, Voltas J, Saurer M et al (2019) Spatio-temporal patterns of tree growth as related to carbon isotope fractionation in European forests under changing climate. Glob Ecol Biogeogr 28:1295–1309. https://doi.org/10.1111/geb.12933

Sillmann J, Kharin VV, Zhang X et al (2013a) Climate extremes indices in the CMIP5 multimodel ensemble: part 1. Model evaluation in the present climate. J Geophys Res Atmo 118:1716–1733. https://doi.org/10.1002/jgrd.50203

Sillmann J, Kharin VV, Zwiers FW et al (2013b) Climate extremes indices in the CMIP5 multimodel ensemble: part 2. Future climate projections. J Geophys Res Atmos 118:2473–2493. https://doi.org/10.1002/jgrd.50188

Steppe K, De Pauw DJW, Doody TM, Teskey RO (2010) A comparison of sap flux density using thermal dissipation, heat pulse velocity and heat field deformation methods. Agric For Meteorol 150:1046–1056. https://doi.org/10.1016/j.agrformet.2010.04.004

Steppe K, von der Crone JS, De Pauw DJW (2016) TreeWatch.net: a water and carbon monitoring and Modeling network to assess instant tree hydraulics and carbon status. Front Plant Sci 7. https://doi.org/10.3389/fpls.2016.00993

Stover DB, Day FP, Butnor JR, Drake BG (2007) Effect of elevated Co2 on coarse-root biomass in Florida Scrub detected by ground-penetrating radar. Ecology 88:1328–1334. https://doi.org/10.1890/06-0989

Subashini MM, Das S, Heble S et al (2018) Internet of things based wireless plant sensor for smart farming. Indones J Electr Eng Comput Sci 10:456–468. https://doi.org/10.11591/ijeecs.v10.i2.pp456-468

Sun S-J, Meng P, Zhang J-S, Wan X (2011) Variation in soil water uptake and its effect on plant water status in Juglans regia L. during dry and wet seasons. Tree Physiol 31:1378–1389. https://doi.org/10.1093/treephys/tpr116

Talla V, Hessar M, Kellogg B et al (2017) LoRa backscatter: enabling the vision of ubiquitous connectivity. Proc ACM Interact Mob Wearable Ubiquitous Technol 1:105:1–105:24. https://doi.org/10.1145/3130970

Tang J, Baldocchi DD (2005) Spatial–temporal variation in soil respiration in an oak–grass savanna ecosystem in California and its partitioning into autotrophic and heterotrophic components. Biogeochemistry 73:183–207. https://doi.org/10.1007/s10533-004-5889-6

Teets A, Fraver S, Hollinger DY et al (2018) Linking annual tree growth with eddy-flux measures of net ecosystem productivity across twenty years of observation in a mixed conifer forest. Agric For Meteorol 249:479–487. https://doi.org/10.1016/j.agrformet.2017.08.007

Temperli C, Santopuoli G, Bottero A, et al (2021) National Forest Inventory data to evaluate Climate-Smart Forestry. In: Managing Forest Ecosystems, Vol. 40, Tognetti R, Smith M, Panzacchi P (eds) Climate-Smart Forestry in Mountain Regions. Springer Nature, Switzerland, AG

Teskey R, Wertin T, Bauweraerts I et al (2015) Responses of tree species to heat waves and extreme heat events. Plant Cell Environ 38:1699–1712. https://doi.org/10.1111/pce.12417

Thom D, Seidl R (2016) Natural disturbance impacts on ecosystem services and biodiversity in temperate and boreal forests. Biol Rev 91:760–781. https://doi.org/10.1111/brv.12193

Tognetti R, Raschi A, Béres C et al (1996) Comparison of sap flow, cavitation and water status of Quercus petraea and Quercus cerris trees with special reference to computer tomography. Plant Cell Environ 19:928–938. https://doi.org/10.1111/j.1365-3040.1996.tb00457.x

Tognetti R, Lasserre B, Di Febbraro M, Marchetti M (2019) Modeling regional drought-stress indices for beech forests in Mediterranean mountains based on tree-ring data. Agric For Meteorol 265:110–120. https://doi.org/10.1016/j.agrformet.2018.11.015

Torresan C, Benito Garzon M, O'Grady M et al (2021) A new generation of sensors and monitoring tools to support climate-smart forestry practices. Can J For Res. https://doi.org/10.1139/cjfr-2020-0295

Trumbore S, Brando P, Hartmann H (2015) Forest health and global change. Science 349:814–818. https://doi.org/10.1126/science.aac6759

Tyree MT, Sperry JS (1988) Do woody plants operate near the point of catastrophic xylem dysfunction caused by dynamic water stress? Plant Physiol 88:574–580

United Nations (ed) (2015) Transforming our world: the 2030 agenda for sustainable development. United Nations, New York

Valentini R, Marchesini LB, Gianelle D et al (2019) New tree monitoring systems: from Industry 4.0 to Nature 4.0. Ann Silvic Res 43:84–88. https://doi.org/10.12899/asr-1847

Vandegehuchte MW, Steppe K (2013) Sap-flux density measurement methods: working principles and applicability. Funct Plant Biol 40:213–223. https://doi.org/10.1071/FP12233

Veraart AJ, Faassen EJ, Dakos V et al (2012) Recovery rates reflect distance to a tipping point in a living system. Nature 481:357–359. https://doi.org/10.1038/nature10723

Vernay A, Tian X, Chi J et al (2020) Estimating canopy gross primary production by combining phloem stable isotopes with canopy and mesophyll conductances. Plant Cell Environ 43:2124–2142. https://doi.org/10.1111/pce.13835

Vicca S, Stocker BD, Reed S et al (2018) Using research networks to create the comprehensive datasets needed to assess nutrient availability as a key determinant of terrestrial carbon cycling. Environ Res Lett 13:125006. https://doi.org/10.1088/1748-9326/aaeae7

Weatherall A, Nabuurs G-J, Velikova V, et al (2021) Defining Climate-Smart Forestry. In: Managing Forest Ecosystems, Vol. 40, Tognetti R, Smith M, Panzacchi P (eds) Climate-Smart Forestry in Mountain Regions. Springer Nature, Switzerland, AG

Wei L, Zhou H, Link TE et al (2018) Forest productivity varies with soil moisture more than temperature in a small montane watershed. Agric For Meteorol 259:211–221. https://doi.org/10.1016/j.agrformet.2018.05.012

Wissel C (1984) A universal law of the characteristic return time near thresholds. Oecologia 65:101–107. https://doi.org/10.1007/BF00384470

Zapata-Rios X, McIntosh J, Rademacher L et al (2015) Climatic and landscape controls on water transit times and silicate mineral weathering in the critical zone. Water Resour Res 51:6036–6051. https://doi.org/10.1002/2015WR017018

Zhao Y, Gao S, Zhu J et al (2019) Multifunctional stretchable sensors for continuous monitoring of long-term leaf physiology and microclimate. ACS Omega 4:9522–9530. https://doi.org/10.1021/acsomega.9b01035

Zhu Z, Kin Tam T, Sun F et al (2014) A high-energy-density sugar biobattery based on a synthetic enzymatic pathway. Nat Commun 5:3026. https://doi.org/10.1038/ncomms4026

Zreda M, Shuttleworth WJ, Zeng X et al (2012) COSMOS: the COsmic-ray soil moisture observing system. Hydrol Earth Syst Sci 16:4079–4099. https://doi.org/10.5194/hess-16-4079-2012

Zweifel R (2016) Radial stem variations – a source of tree physiological information not fully exploited yet. Plant Cell Environ 39:231–232. https://doi.org/10.1111/pce.12613

Zweifel R, Haeni M, Buchmann N, Eugster W (2016) Are trees able to grow in periods of stem shrinkage? New Phytol 211:839–849. https://doi.org/10.1111/nph.13995

Chapter 11
Remote Sensing Technologies for Assessing Climate-Smart Criteria in Mountain Forests

Chiara Torresan, Sebastiaan Luyssaert, Gianluca Filippa, Mohammad Imangholiloo, and Rachel Gaulton

Abstract Monitoring forest responses to climate-smart forestry (CSF) is necessary to determine whether forest management is on track to contribute to the reduction and/or removal of greenhouse gas emissions and the development of resilient mountain forests. A set of indicators to assess "the smartness" of forests has been previously identified by combining indicators for sustainable forest management with the ecosystem services. Here, we discuss the remote sensing technologies suitable to assess those indicators grouped in forest resources, health and vitality, productivity, biological diversity, and protective functions criteria. Forest cover, growing stock, abiotic, biotic, and human-induced forest damage, and tree composition indicators can be readily assessed by using established remote sensing techniques. The emerging areas of phenotyping will help track genetic resource indicators. No single existing sensor or platform is sufficient on its own to assess all the individual CSF

C. Torresan (✉)
Institute of BioEconomy (IBE) – National Research Council of Italy,
San Michele all'Adige (TN), Italy
e-mail: chiara.torresan@cnr.it

S. Luyssaert
Department of Ecological Sciences, Faculty of Sciences, University of Amsterdam,
Amsterdam, The Netherlands
e-mail: s.luyssaert@vu.nl

G. Filippa
ARPA Valle d'Aosta, Climate Change Unit, Saint-Christophe (AO), Italy
e-mail: g.filippa@arpa.vda.it

M. Imangholiloo
Department of Forest Sciences, University of Helsinki, Helsinki, Finland
e-mail: mohammad.imangholiloo@helsinki.fi

R. Gaulton
School of Natural and Environmental Sciences, Newcastle University,
Newcastle upon Tyne, UK
e-mail: rachel.gaulton@newcastle.ac.uk

© The Author(s) 2022   399
R. Tognetti et al. (eds.), *Climate-Smart Forestry in Mountain Regions*, Managing
Forest Ecosystems 40, https://doi.org/10.1007/978-3-030-80767-2_11

indicators, due to the need to balance fine-scale monitoring and satisfactory coverage at broad scales. The challenge of being successful in assessing the largest number and type of indicators (e.g., soil conditions) is likely to be best tackled through multimode and multifunctional sensors, increasingly coupled with new computational and analytical approaches, such as cloud computing, machine learning, and deep learning.

11.1 Introduction

Climate-smart forestry (CSF), as defined by Bowditch et al. (2020), consists of forest management practices that should enable both forests and society to transform, adapt to, and mitigate climate-induced changes. This definition is not far from the European Forest Institute (EFI) interpretation. Indeed, in EFI's vision, CSF is an approach built on practices and active forest management targeted at reducing and/or removing greenhouse gas emissions to mitigate climate change, building resilient forests, and sustainably increasing forest productivity and incomes (Nabuurs et al. 2017; Kauppi et al. 2018). The economic dimension in the EFI's point of view substitutes the social dimension of CSF on which Bowditch et al. (2020) focused. These two dimensions do not exclude each other: practices to stimulate forest productivity should not conflict with forestry practices aimed at growing forests able to contribute to the well-being of the people.

To determine whether forest management is on track to meet the goals of forest adaptation and mitigation to climate change, monitoring the forest response to practices applied during years of climate-smart forest management is necessary. Bowditch et al. (2020) selected a set of indicators to assess "the smartness" of forests, induced by forest management activities carried out in response to climate changes, by combining the pan-European indicators for sustainable forest management (SFM) (FOREST EUROPE 2015) with the ecosystem services defined by the European Environment Agency in the Common International Classification of Ecosystem Services (CICES V5.1 2018, Haines-Young and Potschin 2018). The full list of indicators is reported in Chap. 2 of the book (Weatherall et al. 2021) together with their classification in core and peripheral groups according to their importance to assess the provision of forest ecosystem services.

Remote sensing, "as the practice of deriving information about the Earth's land and water surfaces using images acquired from an overhead perspective, using electromagnetic radiation in one or more regions of the electromagnetic spectrum, reflected or emitted from the Earth's surfaces" (Campbell and Wynne 2011), can contribute to quantifying CSF indicators. As a general consideration, the benefits of remote sensing to monitor the forests as a result of the application of CSF practices are related to full coverage of forested areas in a relatively short time, repeatability of measurements, and availability of data for remote or inaccessible terrestrial areas (Koch 2015). Remote sensing plays an important role in mountain forest monitoring, i.e., forests at an elevation of 2500 m a.s.l. or higher, irrespective of the slope, or on land with an elevation of 300–2500 m and a slope with sharp changes in elevation within a short distance (Kapos et al. 2000). Because of their steep slopes and often-extreme climates and weather events, mountain forests are fragile ecosystems. Under a global change scenario, remote sensing technologies allow more complete spatial and temporal monitoring of climate-smart forests and forestry (e.g., to prevent and contrast illegal logging), including those in inaccessible mountain environments. Mountains are often data-scarce regions due to their remoteness and the harsh environment: in these contexts, remote sensing may provide one of the few methods for assessing the state of dynamic changes occurring in mountain forests (Weiss and Walsh 2009). Indeed, remote sensing overcomes the challenges of collecting field data in rugged terrain and the constraints imposed by the seasonality of access to many mountain environments. Generally, remote sensing in mountain areas is very similar to remote sensing elsewhere, but the complex topography common to mountainous regions, i.e., slope with sharp changes in elevation within a short distance, introduces several challenges unique to these environments (Weiss and Walsh 2009). Remote sensing products over mountain regions come with a larger measurement error than remote sensing products over flat terrain due to topographic effects (Li et al. 2014). In the case of satellite microwave radiometric data, for example, the error is particularly correlated to the mean values of the height and slope within the radiometric pixel, as well as to the standard deviation of the aspect and local incidence angle (Li et al. 2014). In optical images, corrections in preprocessing are in general required to reduce the spectral biases due to the topographic features that led to aspect-dependent illumination and reflectance differences, shadowing, and geometric distortion (Weiss and Walsh 2009). In other remote sensing data, such as the radio detection and ranging (RADAR), topography can result in distortions, such as foreshortening and layover on slopes and in areas of shadow that are not measured.

When assessing the CSF indicators in mountain forests by remote sensing, we have to consider that the temporal scale of monitoring needs to be adjusted for different indicators to ensure early detection of change is possible. Specific focus should be put on those indicators sensitive to climate change. Forest-based climate change indicators should complement SFM indicators by capturing the effects of climate change on the forest environment and the forest sector (Lorente et al. 2018).

In this chapter, we briefly describe the key aspects of remote sensing techniques for monitoring the climate smartness of forests. Next, we consider the techniques suitable to quantify indicators of forest resources, health and vitality, productivity, biological diversity, and protection considering specific challenges in mountain regions. Finally, considerations on future developments to assess climate smartness criteria in mountain forests are provided.

11.2   Remote Sensing of CSF Criteria in Mountain Forests: An Overview

Pan-European Criteria and Indicators (PECI) have proved to be a very helpful tool in providing solid information as the basis for the sustainable management of the forests in the pan-European region between policymakers, the private sector, and civil society over the years (FOREST EUROPE 2015). The role of the CSF indicators selected by Bowditch et al. (2020) is in line with the role of PECI.

The relevance of remote sensing in quantifying the CSF indicators is linked to the possibility to extract relevant variables from remotely sensed data. In some cases, it may be possible to make relevant direct measurements, but often remote sensing proxies can be used to represent indicator values (Ghaffarian et al. 2018). For example, from tree crown delineation processes applied to light detection and ranging (LiDAR) data, tree crowns can be segmented as well as tree height quantified, and, through allometric equations, the volume can be successively estimated. For these reasons, when using remote sensing data, it is important to identify the information to be derived from the data and the kind of product and information to be delivered as an expression of CSF indicator. Besides, the coverage of remote sensing data has to be investigated. While for satellite images, the coverage should be not a problem, in the case of LiDAR data, availability could be sparse in the area of interest, and the timing and frequency of data acquisition could differ among different areas. Despite their importance, terrestrial remote sensing techniques, such as terrestrial photogrammetry and terrestrial laser scanning (TLS), are not included in this chapter. The description of the development of TLS as a plot-scale measurement tool can be found in Newnham et al. (2015), and the current state of the art in the utilization of close-range sensing in forest monitoring is summarized in Vastaranta et al. (2020). For the sake of clarity, in close-range sensing are included technologies, such as terrestrial and mobile laser scanning as well as unmanned aerial vehicles (UAV), which are mainly used for collecting detailed information from single trees, forest patches or small forested landscapes (Vastaranta et al. 2020). It is worth underlining here that, based on the current published scientific literature, the capacity to characterize changes in forest ecosystems using close-range sensing has been recognized (Vastaranta et al. 2020) and, among close-range sensing techniques, terrestrial laser scanning should be viewed as a disruptive technology that requires a rethink of vegetation surveys and their application across a wide range of disciplines (Newnham et al. 2015). These technologies are potentially game-changing but outside the scope of this chapter. Here, we focus on the systems carried on spaceborne and airborne (both manned and unmanned) platforms.

11.3 Remote Sensing of Climate Smartness According to the Forest Resources

11.3.1 Defining Forest Resources in the Context of Climate Smartness

The area covered by forests is likely to change as the climate changes. There are also likely to be shifts in forest types due to changing temperatures and precipitation regimes. Forest area is expected to contract in the mountain and boreal regions and to expand in the temperate zone (Lucier et al. 2009; Wang et al. 2019). Natural changes in climate that occurred in past geological eras have determined analogous changes in forest cover, but for the present era, it will be difficult to isolate climate change from the other factors that are affecting the range of forest area (Lucier et al. 2009). Boreal forests are expected to move north due to climate change. Temperate forests are also expected to increase their area to the north but to a greater extent than boreal forests, which will reduce the total area of boreal forests (Burton et al. 2010).

Interactions among the impacts of climate change, land-use conversion, and unsustainable land-use practices are expected. Changes in water availability will be a key factor in the survival and growth of many forest species, although the response to prolonged droughts will vary among species and also among varieties of the same species (Lucier et al. 2009). Climate change will increase the risk of frequent and more intense fires, especially in areas where it leads to lower precipitation or longer dry periods, as in boreal forests (Burton et al. 2010), and forests in the Mediterranean and subtropical regions (Fischlin et al. 2009).

CSF is needed to increase the total forest area and avoid deforestation and to facilitate the use of wood products that store carbon and substitute emission-intensive fossil and nonrenewable products and materials (Verkerk et al. 2020). Deforestation and forest degradation account for about 12% of global anthropogenic carbon emissions, which is second only to fossil fuel combustion (Calders et al. 2020). Those emissions are partially compensated by forest growth, forestation, and the rebuilding of soil carbon pools following afforestation.

As forest resources are important for climate change mitigation, timely and accurate information about their status is needed. Indeed, assessing forest resources means assessing their extent in terms of area and their distribution, the volume of standing trees, and the carbon stock in woody biomass and soil. As a consequence, the maintenance and the appropriate enhancement of forest resources and their contribution to global carbon cycles are assessed by indicators that quantify the forest area, growing stock, carbon stock, and age structure and/or diameter distribution (FOREST EUROPE 2015).

11.3.2 Appropriate Remote Sensing Methods for the Monitoring of Forest Resource Indicators

Advances in remote sensing technologies drive innovations in forest resource assessments and monitoring at varying scales. Data acquired with spaceborne and airborne platforms provide us with higher spatial resolution, more frequent coverage, and increased spectral information than was available previously (Calders et al. 2020), allowing for frequent updates of forest information layers. Optical spaceborne sensors represent a consolidated opportunity to augment traditional data sources for large-area and sample-based forest inventories, especially for inventory updates (Falkowski et al. 2009). For example, Kempeneers et al. (2012) derived two pan-European forest maps and forest-type maps for the years 2000 and 2006 from MODIS medium-resolution, optical satellite imagery using an automatic processing technique. Knorn et al. (2009) produced a map of forest/non-forest cover of large areas in the Carpathian Mountains using chain classification of neighboring Landsat satellite images. High-resolution layers of tree cover density, dominant leaf type, and forest type are derived from semiautomatic classification algorithms applied on Sentinel-1 and Sentinel-2 images every 3 years (the first products were delivered in 2006, the last in 2018). These products, representing the status and evolution of the forest surface, are used for the assessment of pan-European forest resources (Copernicus Emergency Management Service 2020). Tree cover mapping based on Sentinel-2 images demonstrated high thematic overall accuracy in Europe, i.e., up to 90% (Ottosen et al. 2020). Among those based on aerial platforms, LiDAR or airborne laser scanning (ALS), typically multiphoton LiDAR, has become an operational technology in mapping, and it is used for inventorying forests. The feasibility of using single-photon LiDAR (SPL) for land cover classification has been recently studied in North Europe (Matikainen et al. 2020). The application of algorithms to LiDAR data, most of them based on geometric characteristics of point clouds, including mathematical morphology, and adaptive and robust filtering, allows separate vegetation points from ground points in a mountainous environment. The filtering process is an essential step for the generation of the digital terrain model (DTM), and makes possible the estimation of canopy height and the production of the canopy height model (CHM). LiDAR data can also enhance the capability to discriminate forest areas in satellite images, for example, QuickBird imagery (Hilker et al. 2008) and Sentinel-2 images (Fragoso-Campón et al. 2020), by fusion in the satellite data of metrics concerning the height.

Growing stock, i.e., the stem volume of living trees, is a basic variable to assess forest resources, and it is used as a basis for estimating the amount of carbon accumulated in living trees, thereby allowing for the assessment of harvesting possibilities and risks of disturbance (FOREST EUROPE 2015). Using satellite images, Päivinen et al. (2009) produced broadleaf, coniferous, and total growing stock maps for the pan-European forest area by combining the NOAA-AVHRR imagery and statistics derived from national forest inventories of European countries. Gallaun et al. (2010) did the same using MODIS imagery. At a smaller scale, Mura et al. (2018) used Sentinel-2 imagery to estimate growing stock volume in two forest areas in Italy, and for comparison, they used Landsat 8 OLI and RapidEye images. Since the application of ALS in forestry, models trained with local inventory data have been widely applied for growing stock estimation (Næsset 1997; Maltamo et al. 2006; Dalponte et al. 2009; Corona et al. 2014). The inference of growing stock volume is carried out by regression models built to correlate values of LiDAR metrics to the values of the ground-truth volume. The metrics can be extracted from the raw (point or waveform) LiDAR data or from the CHM, at tree or plot level, following an individual tree crown (ITC) or an area-based approach (AB). Examples of metrics computable from the ALS point cloud data are aboveground elevation of highest return, height percentiles, coefficient of variation of return height, skewness and kurtosis of returns height, non-ground percentage of total returns, etc. In the CHM, metrics, such as height per pixel, coefficient of variation of the height per pixel, or the sum of the heights of all the pixels in the plot, are calculated and used in the model building. The Geoscience Laser Altimeter System (GLAS) was a laser-ranging instrument able to provide large footprint waveform LiDAR datasets for global observations of Earth, which was aboard ICESat from 2003 to 2009. GLAS data have been used to extract canopy height and map the growing stock at 1 km spatial resolution for Spanish forest areas in combination with ground forest inventory data (Sánchez-Ruiz et al. 2016).

Carbon stock, i.e., the quantity of carbon in forest biomass, dead organic matter and soil, and harvested wood products, is linked to society's efforts to mitigate climate change by reducing the net emissions of greenhouse gases to the atmosphere (FOREST EUROPE 2015). In the past, the use of satellite imagery to assess forest biomass, and consequently to estimate the carbon stock, was mainly based on the normalized difference vegetation index (NDVI, the index that quantifies vegetation by measuring the difference between near-infrared, which vegetation strongly reflects, and red light, which vegetation absorbs) datasets. For example, at regional scale, NDVI and enhanced vegetation index (EVI) extracted from MODIS images in combination with field data were used to model carbon stock in aboveground biomass of European beech forest in central Italy (Taghavi-Bayat et al. 2012). Attempts to explore the spatiotemporal changes in carbon stock have been conducted overlaying the vegetation maps of a region and NDVI datasets (Shi and Liu 2017). It is worth underlining that NDVI is largely determined by canopy dynamics, which from an ecological point of view have very little to do with dead wood, litter, and soil carbon. In addition, Hasenauer et al. (2017) highlighted that local daily climate data should be used, and stand density effects should be addressed to obtain realistic forest productivity estimates when using satellite imagery. Hence, these kinds of products should no longer be considered data products, but they have become model products. These shortcomings contributed to the BIOMASS Earth Explorer satellite of the European Space Agency (ESA) being selected to perform a global survey of Earth's forests and see how they change, thanks to the data that will be used in carbon cycle calculations, over the course of BIOMASS's 5-year mission set to start on 2022. With this launch, a fully polarimetric P-band synthetic aperture radar (SAR) will be available for the first time in space. Mutual gains will be made by combining BIOMASS data with data from other missions that will measure forest biomass, structure, height, and change, including the NASA Global Ecosystem Dynamics Investigation (GEDI) LiDAR after its launch in December 2018, and the NASA-ISRO NISAR L- and S-band SAR, due for launch in 2022 (Quegan et al. 2019). Limitations of these missions have to be taken into consideration, for example, in the case of GEDI, the fact that it samples about 4% of the Earth's land surface between 51.6° N and S latitude (Dubayah et al. 2020). Airborne S-Band SAR data have been used to estimate forest aboveground biomass in temperate mixed forests of the UK (Ningthoujam et al. 2016), while integrated spaceborne SAR data from COSMO-SkyMed (X band) and ALOS PALSAR (L band) with field inventory have been used to estimate the forest aboveground carbon (AGC) stock by Sinha et al.

(2019). However, mapping biomass in mountain regions can be challenging as many regions include steep topography, making the use of RADAR data complex; for this, Mitchard et al. (2012) proposed to use a combination of terrain-corrected L-band RADAR data (ALOS PALSAR), spaceborne LiDAR data (ICESat GLAS), and ground-based data as a solution to this problem. Referring to ALS data, the most used approach for estimating carbon stocks is similar to that described for the estimation of the growing stock: it involves computing statistics from ALS point clouds for a specific pixel of forested land and relating these to carbon estimates obtained from field plots in a regression framework (Jucker et al. 2017). Currently, efforts are moving from this AB approach toward a tree-centric approach for integrating tree-level ALS data into biomass monitoring programs. Two solutions are most commonly applied: the first is to use tree height and crown dimensions, computed for a single tree after its segmentation, to predict diameters, allowing the biomass to be estimated using existing allometric equations (e.g., Dalponte and Coomes 2016); the second is to develop equations that estimate biomass directly from tree height and crown size, thus bypassing diameter altogether (Jucker et al. 2017).

Regarding the carbon in soils, Rasel et al. (2017) explored the possibility of developing a model based on variables (i.e., elevation, forest type, and aboveground biomass) extracted from LiDAR data and WorldView-2 imagery to estimate soil carbon stocks. It is evident from this kind of approach that soil carbon content cannot be measured directly by LiDAR data; hence, the problem is to understand how much modelling is acceptable to still consider its result an observational product.

With reference to the age structure indicator, information concerning the age-class structure of forests, and for uneven-aged forests, their diameter distributions, is important for understanding the history of forests and their likely future development, for assessing the harvesting potential, and for providing insights into biodiversity and recreation, which are generally more favorable in uneven-aged and old even-aged forests than in young even-aged forests (FOREST EUROPE 2015). It is known that the diameter of a tree can generally be modelled as a function of tree height or tree crown or measures related to stand structure (Filipescu et al. 2012) and derived from LiDAR data (Thomas et al. 2008; Salas et al. 2010; Bergseng et al. 2015; Spriggs et al. 2017; Arias-Rodil et al. 2018). Recently, harvester-mounted and ALS data (Maltamo et al. 2019) as well as SPOT-5 satellite imagery and field sample data (Peuhkurinen et al. 2018) have been used to estimate stand-level stem diameter distribution. Forest types (i.e., pole-stage, young, adult, mature, and old-growth forests) have been predicted using classification trees from LiDAR data (Torresan et al. 2016). Global forest canopy height products have been derived from GLAS, revealing a global latitudinal gradient in canopy height, increasing toward the equator, as well as coarse forest disturbance patterns (Simard et al. 2011), and also from MODIS and GLAS data using image segmentation (Lefsky 2010). Global canopy cover distributions were analyzed using observations from GLAS (Tang et al. 2019), and it was discovered that the estimates were sensitive to canopy cover dynamics even over dense forests with cover exceeding 80% and were able to better characterize biome-level gradients and canopy cover distributions than the existing products derived from conventional optical remote sensing.

ALS is a powerful source of data to compute the new indicators of climate smartness defined by Bowditch et al. (2020) related to vertical and horizontal forest structure. The first experiences in the application of ALS data for the assessment of vertical and horizontal forest structure are attributable to Friedlaender and Koch (2000). Successively, vertical distribution of tree crowns in terms of layers has been analyzed and characterized by Zimble et al. (2003) using ALS-derived tree heights, which allowed detecting differences in the continuous nature of vertical structure forest and specifically allowed two classes of vertical forest structure to be distinguished. Vertical distribution has been derived in multi-story stands, stratifying the ALS point cloud to canopy layers and segmenting individual tree crowns within each layer using a DSM-based tree segmentation method (Hamraz et al. 2017). Horizontal and vertical distribution of forest canopy has been derived using two point clouds from a UAV: one obtained by applying the Structure from Motion technique to digital photographs and the other one obtained from a LiDAR system (Wallace et al. 2016). Results indicate that both techniques are capable of providing information that can be used to describe canopy properties in areas of relatively low canopy closure. A comparison between waveform ALS data and discrete return ALS data, using TLS data as an independent validation, to describe the 3D structure of vegetation canopies (Anderson et al. 2015) highlighted that discrete return ALS data provide more biased and less consistent measurements of woodland canopy height than waveform ALS data. Besides, discrete return ALS data performed poorly in describing the canopy understory, compared to waveform data, but waveform ALS carried a higher data processing cost.

The slenderness coefficient, i.e., the ratio of tree total height to DBH, is a fundamental attribute for determining tree and stand stability (Vincent et al. 2012), but despite using very high-density LiDAR point cloud from aerial platforms, the stem is not sufficiently visible for accurate DBH extraction. This is only really feasible from ground-based LiDAR or photogrammetry data.

11.4 Remote Sensing of Climate Smartness According to the Forest Health and Vitality

11.4.1 Defining Health and Vitality in the Context of Climate Smartness

The state of health and vitality of a forest is determined by considering various factors, such as age, structure, composition, function, vigor, the presence of unusual levels of insects or disease, and resilience to disturbance. Climate change may have profound impacts on the health and vitality of the forests. In some cases, vitality may increase due to a combination of carbon dioxide fertilization and a more favorable climate. However, in many cases, benefits of carbon dioxide increase on tree growth may be outweighed by increasing drought- and heat-induced tree mortality (Allen et al. 2010), and increasing temperatures can favor the growth of insect populations that are particularly detrimental to the health of forests composed of few tree species (Lucier et al. 2009), as in the case of alpine forests affected by the spruce bark beetle (*Ips typographus* L.). Longer harvesting periods, increased storm damage, and longer spore-production season seem to be the causes of the increase in infestations of root and bud rot by the fungus *Heterobasidion parviporum* Niemelä & Korhonen in coniferous forests in North Europe (Burton et al. 2010).

The criterion of forest health and vitality addresses one of the main concerns of the European countries at the start of the pan-European process. This criterion includes indicators of soil conditions; forest damage by abiotic, biotic, and human-induced agents; defoliation; deposition; and concentration of air pollutants on forests (FOREST EUROPE 2015).

11.4.2 Appropriate Remote Sensing Methods for the Monitoring of Health and Vitality Indicators

Indicators of soil conditions, defined in terms of carbon, water, and nutrient concentrations, are generally not directly measurable by remote sensing techniques in forest environments, as exposed soil is rarely visible. However, SAR and microwave radiometer systems using long wavelengths (e.g., L-Band) have been used to measure surface (0–5 cm depth) soil moisture under forest canopies, but modelling of the scattering and absorption effects of the canopy is required, and the effect of the litter layer and surface roughness must be considered. Root-zone soil moisture can be estimated by assimilation of such data into land-surface or hydrological models, but such approaches may be less successful in mountainous and densely forested areas (Pablos et al. 2018). A review of modelling the passive microwave signature from land surfaces can be found in Wigneron et al. (2017), and some assessment of accuracy over forested areas is provided in Vittucci et al. (2016). The coarse resolution of passive microwave radiometer satellite missions, such as Soil Moisture Ocean Salinity (SMOS) and Soil Moisture Active Passive (SMAP), largely limits their application to regional scales and above. The potential for the use of active RADAR for soil moisture retrieval has also been explored, with good agreement with ground measurements, using L- and P-band polarimetric airborne SAR in jack pine forest stands (Moghaddam et al. 2000), and for soil moisture variations retrieval from ERS SAR satellite data in a recently burned black spruce forest in Alaska (Kasischke et al. 2007).

Remote sensing can significantly contribute to measuring and monitoring forest damage from abiotic (e.g., drought, winter injury, wind storms, avalanche, landslide, fires, air pollution) and biotic (insect pests, diseases) stresses, which influence key biophysical and biochemical parameters of the tree canopy and structure.

Large-scale, stand-replacing, disturbance events leading to significant loss of tree cover can be readily monitored with satellite or aerial imagery using change detection approaches. Over large areas, extensive satellite image time series, such as the Landsat archive, have been used to map tree cover loss, using methods such as the vegetation change tracker algorithm (Masek et al. 2013), and to capture both slowly evolving and abrupt changes in forest cover using LandTrendr – temporal segmentation algorithm (Kennedy et al. 2010). IKONOS satellite imagery with Tasseled Cap transformation and edge enhancements have been tested for mapping of snow avalanche paths (Walsh et al. 2004) and RapidEye imagery for detecting windthrow damage based on pre- and post-storm object-based change detection (Einzmann et al. 2017). Remote sensing has also been widely applied in monitoring both active fires (using thermal sensors) and fire severity and in detecting fire scar (Szpakowski and Jensen 2019). Although both stand-replacing disturbances and finer-scale abiotic or biotic forest damage and decline can be measured, attribution of the specific cause of observed stress or disturbance can frequently be more challenging. But it can potentially be achieved through consideration of the spectral properties (e.g., dead and burnt materials resulting from fires; McDowell et al. 2015), temporal signatures and specific symptom progression (e.g., for pest or disease; Stone and Mohammed, 2017), and spatial patterns of disturbance. Regarding forest damage by human-induced agents, Kennedy et al. (2007) used distinctive temporal signatures in the progression of spectral properties before and after an event to identify the timing of disturbance events, such as clear-cuts and thinnings from a dense stack of Landsat TM images, and to attribute the type using a series of rules. Hilker et al. (2011) utilized spatial characteristics of disturbed patches (patch size, core area, and contiguity), along with the date of disturbance, to attribute disturbance types in Alberta, Canada, using a regression tree classification method, while Hermosilla et al. (2015) utilized spectral, temporal, and geometric metrics from Landsat time series to attribute disturbance as fire, harvesting, road, and non-stand-replacing changes with a 91.6% accuracy level. Baumann et al. (2014) developed a method to separate windfall disturbance from clearcut forest harvesting activity using Landsat data, after Tasseled Cap transformation, obtaining classification accuracy over 75% for windfall areas and better results for larger disturbance patches. The classification was based on spectral differences between the disturbance types, such as lower brightness (due to shadows from remaining biomass) and higher wetness for the windfall areas. Often ancillary data, such as meteorological observations, information on known disturbance agents or events (e.g., storm paths, species ranges for hosts and pests, indices of fire risk), and additional field-based monitoring observations, are needed to reliably attribute or confirm causes, often through integration with modelling (McDowell et al. 2015).

To monitor changes in forest vitality and stress in individual trees, a range of biophysical and biochemical parameters can be estimated with remote sensing. These include defoliation, alterations in pigment concentrations, reduced photosynthesis and light-use efficiency, changes in water relations and hydraulic transport, including leaf water content and evapotranspiration rates, and changes in leaf cell structure due to senescence or wilting. Detailed reviews of the use of remote sensing systems to detect such changes resulting from pests and disease (Chen and

Meentemeyer 2016) and die-off from abiotic stress, such as drought (Huang et al. 2019), provide an in-depth insight into the capability of different sensor systems. Optical sensors are capable of detecting noticeable changes in foliar color due to changes in pigment concentrations or photosynthetic activity. A wide range of spectral indices have been developed to estimate pigment concentrations (e.g., the carotenoid reflectance index; Gitelson et al. 2002), light-use efficiency (LUE), or photosynthetic activity from multispectral or hyperspectral data, and a detailed summary of such methods in monitoring forest decline and disturbance can be found in Pontius et al. (2020). Widely used LUE indices, such as the photosynthetic reflectance index (PRI; Gamon et al. 1997) and chlorophyll/carotenoid index (CCI; Gamon et al. 2016), have allowed detection of water stress (Hernández-Clemente et al. 2011; Dotzler et al. 2015) and forest pests, e.g., Peña and Altmann (2009) use the PRI to detect aphid-induced stress in the Chilean Andes. Such methods require hyperspectral data, which can be costly and hard to acquire and can be sensitive to illumination conditions and canopy structure. Some early-stage pest and disease symptoms cannot be easily observed in the visible range but can be identified by sensors with the capacity to detect the near- and shortwave- infrared spectrum, where reflectance is strongly influenced by the structure of leaf mesophyll tissue and water content. A common example is the use of the NDVI, which can be sensitive to defoliation and to changing chlorophyll levels but also prone to saturation at high values of leaf area index (LAI). Optical remote sensing has been extensively tested for the survey of damage by mountain pine beetle in the Canadian and American Rocky Mountains, ranging from visual interpretation of aerial photography (e.g., Klein 1982) to the use of Landsat time series and spectral indices related to needle water content for automated mapping of red attack stages (e.g., Skakun et al. 2003). More recently, efforts have been made to detect early "green" attack stages of mountain pine beetle infestation, using hyperspectral data (e.g., Fassnacht et al. 2014) and with Mullen et al. (2018) using high-resolution WorldView-2 satellite imagery to detect differences in spectral properties of individual tree crowns, especially in the near-infrared region. Other remote sensing methods attempt to measure changes in photosynthesis or plant functioning more directly. Thermal measurements are sensitive to changes in leaf temperature caused by reduced evapotranspiration as a result of stomatal closure following water stress. Such measurements have been used in the detection of red-band needle blight in pine from a UAV platform (Smigaj et al. 2019) but are highly sensitive to changes in meteorological conditions and time of acquisition. The ECOSTRESS instrument on the International Space Station aims to provide thermal-derived estimates of water stress over much larger extents. A wide range of optical spectral indices (e.g., shortwave infrared-based normalized difference water index, Gao 1996, or moisture stress index, Hunt and Rock 1989) have been proposed to detect vegetation drought stress, but methods have also been proposed based on passive microwave sensors (e.g., estimation of relative water content based on vegetation optical depth, Rao et al. 2019) and solar-induced chlorophyll fluorescence (SIF) of leaves. SIF results from the emittance of light during photosynthetic activity and, therefore, allows for tracking of photosynthetic activity, phenology, and estimation of GPP, but it is influenced by plant stress when excess light is present and LUE is low. SIF has been used to detect stress due to environmental conditions in forest environments (e.g., the Lägeren forest site in Switzerland, where an abrupt decrease in SIF was shown to relate to a heat wave and an aphid outbreak causing early leaf senescence, Paul-Limoges et al. 2018) and has significant future potential with the expected 2022 launch of the ESA FLEX Earth Explorer mission; however, such measurements are again influenced by the diurnal timing of acquisition and the species being observed.

Defoliation is a key indicator of forest decline and has been extensively monitored using optical, LiDAR, and RADAR data at a range of scales from individual trees to landscape. For example, Olsson et al. (2016) use MODIS NDVI products to monitor insect outbreak-linked defoliation in a subalpine birch forest in Sweden, in near real time. Through the use of an NDVI time series, 74% of defoliation was detected where pixels comprised at least 50% birch forest cover, but with some significant misclassification of undisturbed forest, depending on threshold selection. In contrast, Meng et al. (2018) map defoliation at the individual tree level, based on airborne hyperspectral imaging and LiDAR metrics. They show red-edge and near-infrared wavelength regions to be sensitive to defoliation due to gypsy moth at the crown scale and demonstrate the superior capability of LiDAR structural and intensity metrics to predict leaf area. Meiforth et al. (2020) combine WorldView-2 and LiDAR to detect dieback of New Zealand kauri trees, utilizing spectral indices, such as NDVI and red-green ratio. Including LiDAR structural metrics improved the correlation with graded stress levels but incurs significant additional expense, and steep terrain can cause spatial misalignment with optical data. However, the use of LiDAR to identify and segment individual tree crowns can often be important for the analysis of high-resolution optical data. Recent and next-generation satellite sensors, including Sentinel-2, WorldView-2 and WorldView-3, and GEDI, increase the capacity for monitoring of defoliation across larger spatial scales (Meng et al. 2018).

Pollution can also impact forest health, particularly air pollution from ground-level ozone and excessive nitrogen or sulfur deposition. Remote sensing approaches, including near-infrared aerial photography, airborne hyperspectral systems, and satellite observations, have long been used to monitor forest decline attributed to pollutants including acid rain. For example, Rock et al. (1988) showed a shift in the red-edge location of spruce and fir tree spectra to shorter wavelengths (a so-called blue shift) from airborne hyperspectral data in the presence of forest decline in the American and German mountain sites, believed to be due to pollutants including trace metals. Rees and Williams (1997) monitored the effects of air pollution on terrestrial ecosystems using Landsat-MSS images from 1978 to 1992, to study the impact of sulfur dioxide emissions on boreal forest, while Diem (2002) showed foliar injury related to ozone exposure could be detected from Landsat-MSS vegetation indices. However, the pollution-related decline is often compounded by other stress factors, including insect or disease outbreaks. Remote sensing also has an additional role to play in this area through monitoring of atmospheric conditions and pollutants, including nitrogen dioxide and sulfur dioxide (Martin 2008). For example, ground-level ozone formation can be studied from measurements of precursor compounds (e.g., formaldehyde and nitrogen dioxide) from satellite observations by the ozone monitoring instrument onboard NASA's Aura satellite (Jin et al. 2017).

11.5 Remote Sensing of Climate Smartness According to the Forest Productivity

11.5.1 Defining Forest Productivity in the Context of Climate Smartness

Forest productivity, i.e., the potential of a particular forest stand to produce aboveground wood volume, is affected by climate change to differing extents, according to geographic area, species, stand composition, tree age, soil water retention capacity, and the interactions between these factors (Gao et al. 2019; Ammer 2019; Paquette et al. 2018). Some changes in productivity may be short term, and transitory and previous levels of productivity may be restored once carbon sinks become saturated (Hedin 2015) and water availability scarce. However, in areas where the water is not a limiting factor, there may be an initial increase in growth if there is less waterlogging. Similar reactions have been noted for carbon dioxide (Ollinger et al. 2008) and nitrogen fertilization (LeBauer and Treseder 2008) and increased temperatures (Reich and Oleksyn 2008). While in most temperate areas, forest productivity has been found to increase with higher temperature, which is probably due to carbon dioxide fertilization, in tropical areas, the productivity declines when carbon dioxide saturation is reached (Hubau et al. 2020), probably due to water deficits over extended periods.

The indicators of productive functions of forests aim to assess the quantity and the values of the produced goods and marketed services, as well as to make sure that this productivity is sustainable, using multipurpose management (FOREST EUROPE 2015). The balance between net annual increment and annual felling is the indicator used to understand the forest's potential for wood production and the conditions it provides for biodiversity, health, recreation, and other forest functions. The assessment of the quantity and market values of roundwood is important for wood supply, particularly for marginalized rural areas, where the wood energy chain has been suggested as a means to reactivate forest management and improve the value of forest stands (Vacchiano et al. 2018).

11.5.2 Appropriate Remote Sensing Methods for the Monitoring of the Forest Productivity Indicators

Although remote sensing cannot itself directly measure indicators such as the balance between the net annual increment and annual felling of wood, or the quantity and market values of roundwood available for wood chain supply, or the proportion of forest under a management plan, it is worth underlining that the computation of increment assumes the knowledge of the volume of standing trees, living or dead, which is the growing stock whose estimation using remote sensing was already considered in Sect. 11.3.2.

Net primary production (NPP) can be considered an indicator of productivity, quantifiable using remote sensing data (e.g., Coops 2015): according to this approach, top-of-atmosphere measurements of solar radiance from satellite observation are used to estimate the incoming photosynthetically active radiation (PAR), and, successively, LUE modelling techniques are used to estimate the gross primary production (GPP) and the NPP. It appears clear that PAR can be measured and thus GPP can be estimated, but going from GPP to NPP requires that radiation is modelled or makes assumptions regarding LUE, which depend on field observations. Besides, the application of these models at a higher level of detail requires the availability of high spatial and temporal resolution maps of a series of model drivers (e.g., meteorology, land use) frequently unavailable for mountain areas (Yang et al. 2020). Furthermore, there are few good and complete GPP observations globally, and NPP estimates are typically multi-year averages and cannot be better than field observations (Šímová and Storch 2017). Even the amount of foliage of stands, measured as the LAI, is a key indicator of forest productivity, principally due to its importance for photosynthesis, transpiration, evapotranspiration, and, in turn, GPP. Remote sensing estimation of LAI has been undertaken using several approaches reported in Coops (2015), but LAI measurements based on light absorption have been shown to saturate for values bigger than 4 (Waring et al. 2010); therefore, this aspect limits the value of this product. Remote sensing estimates of LAI and fraction of absorbed PAR can be integrated into process-based forest growth models, such as 3-PG SPATIAL, to model productivity at stand to regional scales and growth variables such as site index (SI), the most common means for quantifying forest stand-level potential productivity, and to estimate impacts of climate change on future productivity (Coops et al. 2011). LAI estimates are routinely available from MODIS satellite data and have become widely used since their release in 2000 (De Kauwe et al. 2011). SI has been compared with GPP estimates obtained from 3-PG SPATIAL using climate variables and MODIS (Weiskittel et al. 2011). Results indicated that a nonparametric model with two climate-related predictor variables explained over 68% and 76% of the variation in SI and GPP, respectively. The relationship between GPP and SI was limited (determination coefficient of 36–56%), while the relationship between GPP and climate (determination coefficient of 76–91%) was stronger than the one between SI and climate (determination coefficient of 68–78%).

In terms of harvesting activity, Smith and Askne (2001) used ERS SAR interferograms over 3 years to detect clearcut areas of a minimum size of 0.4 ha in northern Sweden. Saksa et al. (2003) compared the applicability of Landsat satellite imagery and high-altitude panchromatic aerial orthophotos using digital change detection methods in detecting clearcut areas in a boreal forest. MODIS images were used by Bucha and Stibig (2008) for the detection and monitoring of forest clearcuts in the boreal forest in north-west Russia, while Lambert et al. (2015)

proposed the usage of MODIS NDVI time series in south-western Massif Central Mountains in France to detect clearfelling. Panagiotidis et al. (2019) detected fallen logs from high-resolution UAV images at plot level (i.e., 2500 m² plot size) in flat topographic conditions and open canopy cover.

Climatic changes affect the treeline location, causing shifts, and identification and quantification of treeline dynamics are critical. An approach for the identification of shifts in the treeline altitudes for a period of four decades based on NDVI, land surface temperature (LST) data, air temperature data, and forest stand maps has been developed for the mountain forest of Cehennemdere in Turkey, showing a geographical expansion of the treeline in both the highest altitudes and the lowest altitudes (Arekhi et al. 2018). The integration of in situ dendrometric data with analyses from dendrochronological samples, high-resolution 3D UAV photos, and new satellite images has been found a solution to study the dynamics and underlying causal mechanisms of any treeline movement and growth changes in mountain forests located in Central and East Asia (Cazzolla Gatti et al. 2019).

11.6 Remote Sensing of Climate Smartness According to the Forest Biological Diversity

11.6.1 Defining Forest Biological Diversity in the Context of Climate Smartness

Different species have individual climatic ranges, in which they remain competitive with other plant species, can adapt to environmental change, and respond to increased insect attacks, disease, and adverse environmental conditions and anthropogenic influences. Some species will adapt better than others to changing conditions, which will lead to changes in the composition of forests, instead of geographic shifts in forest types (Breshears et al. 2008). In general, tree species are likely to move to higher latitudes or altitudes due to global warming (Rosenzweig et al. 2007; Breshears et al. 2008).

Climate change drives phenological changes, i.e., changes in seasonal timing of life-history events, in many tree species observed in phenological gardens (Seppälä et al. 2009). A phenological garden contains a selection of species and a selection of clonal strains among each species. In order to minimize non-climatic influences on plant development, a network of phenological gardens with the same species and the same clones should be set up as to constitute a network of plants. The observers at the gardens have detailed, illustrated instructions that describe exactly the phenophases to be reported. Phenological gardens, therefore, assure a maximum of exact observations with controlled internal plant conditions. An adjacent meteorological station is essential for later correlations among the results from a phenological garden and for possible physiological modelling (Schnelle and Volkert 1974). The highest number of changes and the most significant changes were noted in phenological gardens located at higher latitudes (Seppälä et al. 2009). There is evidence that part of this phenological change, such as dates of spring bud break and flowering, which can affect productivity and carbon sequestration potential, and autumnal foliar coloration, may be due to increasing atmospheric carbon dioxide concentrations as well as warming (Seppälä et al. 2009).

Biodiversity remains an important topic for forest policy and management in Europe (FOREST EUROPE 2015). Forest biological diversity criterion encompasses indicators referring to tree species composition, introduced tree species, and threatened forest species. It also includes indicators that quantify the area of forests by the class of naturalness and fragmentation, protected forests, forests for the conservation and utilization of tree genetic resources, seed production, and new forests. The average volume of deadwood, both standing and lying, is also an indicator of biological diversity.

11.6.2   Appropriate Remote Sensing Methods for the Monitoring of Biological Diversity Indicators

The assessment of biological diversity can benefit from the availability of different sources of remotely sensed data and from the opportunities for automated forest interpretation at the tree level, i.e., the possibility to delineate and classify tree species. Whatever approach is used for tree detection, crown area delineation, and species classification, remote sensing offers strong potentiality for the assessment of tree species composition and introduced tree species. Various experiences of tree crown detection, delineation, and species classification had applied digital aerial images (e.g., Brandtberg 2002; Haara and Haarala 2002; Erikson 2004; Korpela 2004), but LiDAR represents the most effective source of data for detecting and delineating trees (e.g., Heinzel and Koch 2012; Maltamo et al. 2009). The integration of ALS data with aerial high-resolution multispectral or hyperspectral images has been tested for tree crown delineation and tree species classification (e.g., Zarea and Mohammadzadeh 2016; Dalponte et al. 2019; Weinstein et al. 2019) as well as the integration of ALS data with high-resolution aerial near-infrared images (Persson and Holmgren 2004). Satellite imagery, e.g., high spatial resolution 8-Band WorldView-2 and 5-band RapidEye, has proven to be valid in species classification (Immitzer et al. 2012).

There are many studies that have modelled patterns in spectral diversity and species richness, paralleling those of biochemical diversity, demonstrating a linkage between the taxonomic and remotely sensed properties of forest canopies (e.g., Asner et al. 2009; Warren et al. 2014; Roth et al. 2015). The approach, ushered in Asner and Martin (2008) and called "spectranomics," resulting from the combination of science and technology, emerged from aspects of established remote sensing research with new ideas to causally link the biochemistry, spectroscopy, taxonomy, and community ecology of canopies. A review of the history of remote sensing approaches for biodiversity estimation, specifically focused on relating spectral diversity to biodiversity at different scales, with the summarization of the pros and cons of different methods in remote sensing of plant biodiversity can be found in Wang and Gamon (2018). The step from individual trees to the area of forest classified by the number of tree species occurring or to the area of forest land dominated by introduced tree species is quite short.

Despite progress in tree crown delineation and species classification, the quantification of indicators, such as the number of threatened forest species, classified according to the International Union for Conservation of Nature (IUCN) Red List categories in relation to the total number of forest species, could be challenging. Indeed, although the application of techniques – such as support vector machines and Gaussian maximum likelihood with leave-one-out-covariance algorithm classifiers – in hyperspectral imaging and LiDAR data allows classifying classes of species (Dalponte et al. 2008), the classification of all single species contained in the IUCN Red List categories is not yet operationally applied as in the case of species categories.

The amount and variability of deadwood in a forest stand are important indicators of forest biodiversity because deadwood provides critical habitat for thousands of species in forests (Bater et al. 2009; Sandström et al. 2019). The current situation and new perspectives of remote sensing application for deadwood identification and characterization have been summarised by Marchi et al. (2018a, b). Infrared aerial photos are suitable for mapping and quantifying single standing dead trees, i.e., snags (Bütler et al. 2004; Bütler and Schlaepfer 2004). The identification of standing deadwood using LiDAR data has started to be addressed around 10 years ago, due to the increase in high-quality and high-density data availability, as well as the availability of segmentation methods required to work directly with the point cloud. The moderate capacity of LiDAR remote sensing to estimate the distribution of standing dead tree classes in forest stands has been demonstrated (correlation coefficient of 0.61; Bater et al. 2009), as well as the good capacity of high-spatial resolution aerial photos taken from a UAV to survey fallen trees in deciduous broadleaved forests (trees with a diameter bigger than 30 cm or longer than 10 m were identified with a rate equal to 80%, but many trees that were narrower or shorter were missed; Inoue et al. 2014). In general, failure in identifying snags, logs, and stumps may be due to the similarity of fallen trees to trunks and branches of standing trees or masking by standing trees. The noise near the soil also affects the detection performance of the trees lying in the ground.

Indicators of regeneration, naturalness, landscape patterns, protected forests, and genetic resources are related to a landscape scale of analysis instead of a single-tree scale, and they can be assessed using approaches of classification applied to remote sensing data, both to produce maps or to obtain statistics. The emitted and/or reflected radiance from the canopy of the regenerated forest, which is related to the biophysical properties of the vegetation, such as leaf and wood biomass, can be used to identify the forest regenerative stage after, for example, a clearance (Lucas et al. 2000) or wildfire (Morresi et al. 2019) using satellite imagery. Multispectral aerial images in the green, red, and near-infrared spectral bandwidths have been used to monitor regenerating forests using an automated tree detection-delineation algorithm by Pouliot et al. (2005). Sentinel-2 imagery with ALS data was integrated for inventorying regeneration stands by the estimation of sapling density (Landry et al. 2020). The application of optical UAV-based imagery for the inventory of natural regeneration in post-disturbed forests has been tested by Röder et al. (2018). The assessment of the status of forest regeneration using aerial photogrammetry and UAV was carried out by Goodbody et al. (2018). Moreover, young and advanced regenerating forests were studied using UAV-based hyperspectral imagery and photogrammetric point clouds (Imangholiloo et al. 2019) and multispectral ALS data (Imangholiloo et al. 2020), both under leaf-off and leaf-on conditions in boreal forests. Similarly, UAV data have been applied for regeneration assessments in seedling stands (Puliti et al. 2019; Castilla et al. 2020; Green and Burkhart, 2020).

Landsat satellite images constitute a major data source for spatial patterns of fragmentation that allow disturbances in protected areas to be identified (Nagendra et al. 2013). In particular, multitemporal Landsat data resulted in valid satellite imagery to analyze patterns of forest fragmentation by the extraction of landscape metrics, which convey significant information on biophysical changes associated with forest fragmentation at broad scales (Fuller 2001). ALS and SAR data, especially when used synergistically with optical data, allow the detection of changes in the three-dimensional structure of forests, facilitating their classification in different classes of naturalness (Hirschmugl et al. 2007).

Protected forests to conserve biodiversity are a land-use and not a land cover category because the protection of an area is given by the humans according to the use they want to do and by the law. For this reason, remote sensing can be used for mapping large intact forest areas or adjusting the boundaries of areas already protected or detecting major changes in land cover within protected areas. This aspect is relevant to underpin the development of a general strategy for nature conservation at the global and regional scales (Willis 2015). For example, Potapov et al. (2008) introduced a new approach for mapping large intact forest landscapes using existing fine-scale maps and global coverage of high-spatial resolution satellite imagery. Gillespie and Willis (2015) reviewed advances and limitations in spaceborne remote sensing that can be applied to all terrestrial protected areas around the world for baseline vegetation mapping, land cover classifications, invasive species, and degradation identification, monitoring forest ecosystems and land cover dynamics using time series data.

The same consideration can be made for the areas managed for conservation and utilization of forest tree genetic resources or managed for seed production, which are land-use rather than land cover categories that require additional data, to which remote sensing can potentially contribute. Indeed, remote sensing cannot do genotyping, but it can contribute to phenotyping, therefore providing indicators of genetic resilience, e.g., variations in stress responses. This is an emerging area for remote sensing. Most of the studies have been carried out in tree nurseries, experimental plots, or crop trees rather than in the field. For example, Ludovisi et al. (2017) used thermal imaging acquired by UAV for high-throughput field phenotyping of black poplar response to drought, enabling highly precise and efficient, nondestructive screening of genotype performance in large plots. Phenotyping of individual trees in situ in planted forests helping to track genetic performance was conducted by Dungey et al. (2018). But there have been "phenotyping platforms," which involve the phenotyping of whole forests eventually down to the individual tree level, proposed for whole-forest phenotyping to better quantify the key drivers of forest productivity to inform and optimize future breeding and deployment programs (Dungey 2016). Improved UAV-based measurements of tree health (fluorescence, canopy temperature, structure traits from LiDAR, etc.) could have much potential for linking genetic and phenotypic traits in terms of stress responses to climate change.

11.7 Remote Sensing of Climate Smartness According to the Forest Protective Function

11.7.1 Defining Forest Protective Function in the Context of Climate Smartness

The contribution of forests to water and soil protection has long been recognized. Forests can play a vital role in preventing soil erosion and protecting water supplies (FOREST EUROPE 2015). In addition, a wide variety of man-made infrastructure relies on the protection provided by forests. Such protective functions are mostly found in mountainous areas or areas subject to extreme climatic conditions. The importance of forests with protective functions has increased in the last decades due to settlement pressure, climate change, and the high vulnerability of society in mountain regions (Bauerhansl et al. 2010). However, foresters and hydrologists still debate the nature of the influence that forests have on water regulation. Climate change may make the role of forests in water regulation and soil protection more important, but the capacity of forests to fulfill this role may also be affected. Reductions in rainy season flows and increases in dry season flows are of little value when total annual rainfall is low and significant quantities of water are lost through evapotranspiration and are consumed by forests. Unmanaged forests supply high levels of climate regulation and erosion regulation, while best practice management slightly improves water regulation (Seidl et al. 2019).

In this context, information on the spatial distribution of protective forests designated to prevent soil erosion, preserve water resources, protect infrastructure, and manage natural resources against natural hazards (i.e., avalanche, snow gliding, rockfall, landslide, flood, water erosion, wind erosion, karstification) becomes essential.

11.7.2 Appropriate Remote Sensing Methods for the Monitoring of Protective Function Indicators

To assess the forest protective function against natural hazards, model simulations are often used. Variables used in the models range from topographic features (e.g., altitude, slope and slope gradient, orientation to wind and orientation to the sun, aspect, plan curvature) to forest characteristics (e.g., crown cover, stem per hectare, gap width) to individual tree features (e.g., tree species, tree diameter, and height) (Bigot et al. 2009). In this context, remote sensing techniques are an indispensable supplement of data, which can be integrated with inventory field data for the identification of protective effects on large areas (Bauerhansl et al. 2010), having in mind that the protective function strongly depends on small-scale local conditions, such as terrain and stand characteristics (Teich and Bebi 2009).

Since the 2000s, Bebi et al. (2001) and Teich and Bebi (2009) used aerial photographs for assessing structures in mountain forests (e.g., canopy density, crown closure, trees in clusters) as a basis for investigating the forests' protective function. Dorren et al. (2006) assessed protection forest structure with ALS in mountain terrain, and, according to their results, ALS provides excellent input data for 3D natural hazard simulation models, even in steep terrain. Monnet et al. (2010) assessed the potential of ALS for estimating stand parameters required as input data for rockfall simulation models or more generally for quantifying the rockfall protection function of forests. In the case of falling rocks, the topography determines its occurrence, the direction, as well as the velocity of a falling rock; therefore, DEM is the most relevant dataset for simulation studies (Dorren et al. 2004). Maroschek et al. (2015) combined remote sensing data (i.e., a LiDAR-based normalized crown model and a volume map) and inventories to generate realistic fine-grained forest landscapes with single-tree level information as input to the spatially explicit hybrid model used to assess the protective effect of the vegetation. Brožová et al. (2020) determined forest parameters for avalanche simulation using remote sensing data, specifically photogrammetry-based vegetation height model and LiDAR-based vegetation height model.

Hydrological models are commonly used to study both the flow and the quality of water. Forest cover, tree height, LAI, and sky view factor are the four main structural parameters used by hydrological models of forests (Varhola and Coops 2013). Varhola and Coops (2013) estimated these four watershed-level distributed forest structure metrics using LiDAR and Landsat data. They found a high correlation between forest spectral indices and structural metrics, and they successfully modelled the four metrics by Landsa-LiDAR calibrations. Satellite-derived land cover data, as well as data on snow cover, LAI, evapotranspiration, and surface soil moisture, are widely integrated into catchment and regional- and global-scale hydrological modelling (Xu et al. 2014). Concerning the role of forest in preventing soil erosion, de Asis and Omasa (2007) estimated vegetation parameters for modelling soil erosion using Landsat ETM data in a linear spectral mixture analysis at a catchment scale.

11.8 Remote Sensing of Climate Smartness According to Socioeconomic Function

The impacts of climate change on the forest sector will increase in strength, but they will vary in space and over time, depending on the geographical region and the crop type (Viccaro et al. 2019). According to the expected future scenarios for Europe, in the Mediterranean area, there will be a reduction of the forest capital, caused by a reduced water supply, while in the North of Europe, there will be an expansion of forests, both in terms of surface and species and an extended growing season due to more favorable soil temperature and moisture conditions and the higher supply of carbon dioxide for photosynthesis (Viccaro et al. 2019). The expected global increase in wood production together with provisional and local increased availability of wood in the market due to windstorms could lead to lower prices, which would benefit consumers. However, lower prices and regionally differentiated impacts on productivity will have varying effects on incomes and employment derived from timber.

The socioeconomic function criterion assesses the economic value of forests (FOREST EUROPE 2015); indicators related to wood consumption, trade in wood (i.e., import and export), and the share of wood energy in total primary energy supply are contemplated together with accessibility for recreation (i.e., right of access, provision of facilities, and intensity of use). The first three indicators are not directly measurable with remote sensing techniques as wood consumption cannot be estimated from areas felled due to the international market in timber. But, where trees are grown specifically for a local market, e.g., as a biomass energy crop such as short rotation willow, remote sensing can potentially provide input layers for field trial scale analyses and modelling (Castaño-Díaz et al. 2017). The last indicator refers to the right to access the forests for recreational purposes, which is related to the forest holdings: remote sensing technique cannot support the assessment of this indicator.

11.9 Conclusion

The potentiality of the remote sensing as a source of data for the assessment of climate-smart criteria and indicators in mountain forests has been demonstrated for many of the indicators examined in the above sections. Indicators of forest resources, such as the spatial extent of forest cover that can be used to assess the spatial dynamics of that cover, can be monitored using data from optical sensors, which may provide information on the amount of foliage and its biochemical properties, with optical wavelengths both absorbed and scattered by leaves, needles, and branches that make up a forest canopy. Vegetation indices extracted from optical satellite imagery have been used to model carbon stock in aboveground biomass for many years. More recently, the capacity emerged to assess the 3D forest structure, using data acquired from LiDAR and SAR sensors, due to the ability of the waves to penetrate through the canopy and backscatter from branches and stems and so to provide a geometrical description of the forest. LiDAR allows the estimation of the volume of living trees and of carbon stock by models that assume relationships between growing and carbon stock and LiDAR-derived canopy metrics. The measurements provided by optical and radar sensors are different due to divergences in wavelength, imaging geometry, and technical implementation, but, when combined, they can be complementary when the sensitivity to individual 3D forest structure components is considered. This combination of sensors and observation platforms holds a lot of promise for the future.

Indicators of health and vitality can be monitored using data from sensors able to acquire plant reflectance in the visible to shortwave-infrared regions, which provide information in the biochemical properties of foliage and plant fluorescence, which are indicative of the capacity and functioning of the photosynthetic apparatus. Through multisensor data and consideration of spatial and temporal patterns of changes, abiotic and biotic causes of disturbance or stress can be identified.

The specific indicators of productive function, such as increment and felling and roundwood, are not directly measurable. NPP, as an indicator globally considered an expression of forest productivity, is modelled using optical satellite imagery data. These models have been successfully used on a global and continental scale to estimate daily and annual NPP from satellite data with low ground resolutions, but the availability of maps at high level of detail related to model drivers, such as precipitation, temperature, and land use, is required to produce accurate estimation through these models. Indicators of the productive function of forest can be quantified using data acquired with microwave sensors that can provide information on woody biomass.

Indicators of biological diversity criterion, such as tree composition, deadwood, etc., can be assessed with optical, LiDAR, RADAR, and SAR. Although hyperspectral holds a lot of promise, few applications, using its full potential, have so far emerged. Spectral diversity is also increasingly considered as a measure of biodiversity at a range of scales, but tree species mapping over large area is not still an operational procedure mostly due to costs and limited extent of airborne hyperspectral.

LiDAR provides the most suitable data to be inputted in models that quantify the protective function of forests against, for example, rockfall. Data extracted from optical satellite imagery can also be used to assess the effect of natural disturbances on the functionality of direct protection forests.

In the context of an increasing amount of remote sensing data generated by multiple airborne and spaceborne platforms, integrated multisensor frameworks will be required to move beyond the state of the art in forest monitoring (Lehmann et al. 2015). This kind of approach allows balancing the needs for fine-scale monitoring of distribution patterns and satisfactory coverage at broad scales (He et al. 2019). For example, with specific reference to the assessment of the forest health and vitality, Lausch et al. (2018) underline that it is becoming evident that no existing monitoring approach, technique, model, or platform is sufficient on its own to monitor, model, forecast, or assess forest health and its resilience. Hence, a multisource forest health monitoring network for the twenty-first century, which couples different monitoring approaches, is a viable strategy that has to be supported. In addition to multisensor and multiplatform approaches, recently, a significant amount of work has been carried out in the field of multimode sensors and multifunctional sensors (Majumder et al. 2019). Such approaches are increasingly coupled with new computational and analytical approaches, such as cloud computing, machine learning, and deep learning, able to handle big datasets and diverse data sources. From this perspective, monitoring the effects of climate-smart forest management interventions requires smart deployment and the use of remote sensing technologies, which can open up new opportunities in the assessment of the capabilities of the forests to transform, adapt to, and mitigate climate-induced changes throughout the selected indicators.

References

Allen CD, Macalady AK, Chenchouni H, Bachelet D, McDowell N, Vennetier M, Kitzberger T, Rigling A, Breshears DD, Hogg EH (Ted), Gonzalez P, Fensham R, Zhang Z, Castro J, Demidova N, Lim JH, Allard G, Running SW, Semerci A, Cobb N (2010). A global overview of drought and heat-induced tree mortality reveals emerging climate change risks for forests. For Ecol Manag 259:660–684. https://doi.org/10.1016/j.foreco.2009.09.001

Ammer C (2019) Diversity and forest productivity in a changing climate. New Phytol 221:50–66. https://doi.org/10.1111/nph.15263

Anderson K, Hancock S, Disney M, Gaston KJ (2015) Is waveform worth it? A comparison of LiDAR approaches for vegetation and landscape characterization. Remote Sens Ecol Conserv 2(1):5–15. https://doi.org/10.1002/rse2.8

Arekhi M, Yesil A, Ozkan UY, Balik Sanli F (2018) Detecting treeline dynamics in response to climate warming using forest stand maps and Landsat data in a temperate forest. For Ecosyst 5:23–36. https://doi.org/10.1186/s40663-018-0141-3

Arias-Rodil M, Diéguez-Aranda U, Álvarez-González JG et al (2018) Modeling diameter distributions in radiata pine plantations in Spain with existing countrywide LiDAR data. Ann For Sci 75:36–47. https://doi.org/10.1007/s13595-018-0712-z

Asner GP, Martin RA (2008) Airborne spectranomics: mapping canopy chemical and taxonomic diversity in tropical forests. Front Ecol Environ 7(5):269–276. https://doi.org/10.1890/070152

Asner GP, Martin RE, Ford AJ, Metcalfe DJ, Liddell MJ (2009) Leaf chemical and spectral diversity in Australian tropical forests. Ecol Appl 19(1):236–253. https://doi.org/10.1890/08-0023.1

Bater CW, Coops NC, Gergel SE et al (2009) Estimation of standing dead tree class distributions in northwest coastal forests using lidar remote sensing. Can J For Res 39:1080–1091. https://doi.org/10.1139/X09-030

Bauerhansl C, Berger F, Dorren L et al (2010) Development of harmonized indicators and estimation procedures for forests with protective functions against natural hazards in the alpine space (PROALP). European Commission, Joint Research Centre, Institute for Environment and Sustainability. Office for Official Publications of the European Communities. © European Communities, 2010. https://doi.org/10.2788/51473

Baumann M, Ozdogan M, Wolter PT et al (2014) Landsat remote sensing of forest windfall disturbance. Remote Sens Environ 143:171–179. https://doi.org/10.1016/j.rse.2013.12.020

Bayat AT, van Gils H, Weir M (2012) Carbon stock of European Beech forest; a case at M. Pizzalto, Italy. APCBEE Procedia 1:159–168. https://doi.org/10.1016/j.apcbee.2012.03.026

Bebi P, Kienast F, Schönenberger W (2001) Assessing structures in mountain forests as a basis for investigating the forests' dynamics and protective function. For Ecol Manag 145:3–14. https://doi.org/10.1016/S0378-1127(00)00570-3

Bergseng E, Ørka HO, Næsset E, Gobakken T (2015) Assessing forest inventory information obtained from different inventory approaches and remote sensing data sources. Ann For Sci 72:33–45. https://doi.org/10.1007/s13595-014-0389-x

Bigot C, Dorren LKA, Berger F (2009) Quantifying the protective function of a forest against rockfall for past, present and future scenarios using two modelling approaches. Nat Hazard 49:99–111. https://doi.org/10.1007/s11069-008-9280-0

Bowditch E, Santopuoli G, Binder F et al (2020) What is Climate-Smart Forestry? A definition from a multinational collaborative process focused on mountain regions of Europe. Ecosyst Serv 43:101113. https://doi.org/10.1016/j.ecoser.2020.101113

Brandtberg T (2002) Individual tree-based species classification in high spatial resolution aerial images of forests using fuzzy sets. Fuzzy Sets Syst 132:371–387. https://doi.org/10.1016/S0165-0114(02)00049-0

Breshears DD, Huxman TE, Adams HD, Zou CB, Davison JE (2008) Vegetation synchronously leans upslope as climate warms. Proceedings of the National Academy of Sciences 105(33):11591–11592. https://doi.org/10.1073/pnas.0806579105

Brožová N, Fischer JT, Bühler Y et al (2020) Determining forest parameters for avalanche simulation using remote sensing data. Cold Reg Sci Technol 172:102976. https://doi.org/10.1016/j.coldregions.2019.102976

Bucha T, Stibig HJ (2008) Analysis of MODIS imagery for detection of clear cuts in the boreal forest in north-west Russia. Remote Sens Environ 112:2416–2429. https://doi.org/10.1016/j.rse.2007.11.008

Burton PJ, Bergeron Y, Bogdansky BEC, Juday GP et al (2010) Sustainability of boreal forests and forestry in a changing environment. In: Mery G, Katila P, Galloway G, Alfaro R, Kanninen M, Lobovikov M, Varjo J (eds) Forests and society responding to global drivers of change, vol 25. IUFRO World, Series, pp 249–282

Bütler R, Schlaepfer R (2004) Spruce snag quantification by coupling colour infrared aerial photos and a GIS. For Ecol Manag 195:325–339. https://doi.org/10.1016/j.foreco.2004.02.042

Bütler R, Angelstam P, Ekelund P, Schlaepfer R (2004) Dead wood threshold values for the three-toed woodpecker presence in boreal and sub-Alpine forest. Biol Conserv 119:305–318. https://doi.org/10.1016/j.biocon.2003.11.014

Calders K, Jonckheere I, Nightingale J, Vastaranta M (2020) Remote sensing technology applications in forestry and REDD+. Forests 11:10–13. https://doi.org/10.3390/f11020188

Campbell JB, Wynne RH (2011) Introduction to remote sensing, 5th edn. Guilford Press, New York. ISBN:9781609181765

Castaño-Díaz M, Álvarez-Álvarez P, Tobin B, Nieuwenhuis M, Afif-Khouri E, Cámara-Obregón A (2017) Evaluation of the use of low-density LiDAR data to estimate structural attributes and biomass yield in a short-rotation willow coppice: an example in a field trial. Ann For Sci 74:69. https://doi.org/10.1007/s13595-017-0665-7

Castilla G, Filiatrault M, McDermid GJ, Gartrell M (2020) Estimating individual conifer seedling height using drone-based image point clouds. Forests 11:924. https://doi.org/10.3390/f11090924

Cazzolla Gatti R, Callaghan T, Velichevskaya A et al (2019) Accelerating upward treeline shift in the Altai Mountains under last-century climate change. Sci Rep 9:1–13. https://doi.org/10.1038/s41598-019-44188-1

Chen G, Meentemeyer RK (2016) Remote sensing of forest damage by diseases and insects. 145–162. https://doi.org/10.1201/9781315371931-9

Coops NC (2015) Characterizing forest growth and productivity using remotely sensed data. Curr For Rep 1:195–205. https://doi.org/10.1007/s40725-015-0020-x

Coops NC, Gaulton R, Waring RH (2011) Mapping site indices for five Pacific Northwest conifers using a physiologically based model. Appl Veg Sci 14:268–276. https://doi.org/10.1111/j.1654-109X.2010.01109.x

Copernicus Emergency Management Service. https://emergency.copernicus.eu/. Accessed 06/29/2020

Corona P, Fattorini L, Franceschi S et al (2014) Estimation of standing wood volume in forest compartments by exploiting airborne laser scanning information: model-based, design-based, and hybrid perspectives. Can J For Res 44:1303–1311. https://doi.org/10.1139/cjfr-2014-0203

Dalponte M, Coomes DA (2016) Tree-centric mapping of forest carbon density from airborne laser scanning and hyperspectral data. Methods Ecol Evol 7:1236–1245. https://doi.org/10.1111/2041-210X.12575

Dalponte M, Bruzzone L, Gianelle D (2008) Fusion of hyperspectral and LIDAR remote sensing data for classification of complex forest areas. IEEE Trans Geosci Remote Sens 46:1416–1427. https://doi.org/10.1109/TGRS.2008.916480

Dalponte M, Bruzzone L, Dalponte M et al (2009) Analysis on the use of multiple returns LiDAR data for the estimation of tree stems volume. IEEE J Sel Top Appl Earth Obs Remote Sens 2:310–318. https://doi.org/10.1109/JSTARS.2009.2037523

Dalponte M, Frizzera L, Gianelle D (2019) Individual tree crown delineation and tree species classification with hyperspectral and LiDAR data. PeerJ 6:e6227. https://doi.org/10.7717/peerj.6227

de Asis AM, Omasa K (2007) Estimation of vegetation parameter for modeling soil erosion using linear Spectral Mixture Analysis of Landsat ETM data. ISPRS J Photogramm Remote Sens 62:309–324. https://doi.org/10.1016/j.isprsjprs.2007.05.013

De Kauwe MG, Disney MI, Quaife T, Lewis P, Williams M (2011) An assessment of the MODIS collection 5 leaf area index product for a region of mixed coniferous forest. Remote Sens Environ 115(2):767–780. https://doi.org/10.1016/j.rse.2010.11.004

Diem JE (2002) Remote assessment of forest health in southern Arizona, USA: evidence for ozone-induced foliar injury. Environ Manag 29:373–384. https://doi.org/10.1007/s00267-001-0011-5

Dorren LKA, Maier B, Putters US, Seijmonsbergen AC (2004) Combining field and modelling techniques to assess rockfall dynamics on a protection forest hillslope in the European Alps. Geomorphology 57:151–167. https://doi.org/10.1016/S0169-555X(03)00100-4

Dorren L, Maier B, Berger F (2006) Assessing protection forest structure with airborne laser scanning in steep mountainous terrain. Paper presented at the Workshop on 3D Remote Sensing in Forestry, 14th–15th Feb 2006, Vienna 238–242

Dotzler S, Hill J, Buddenbaum H, Stoffels J (2015) The potential of EnMAP and sentinel-2 data for detecting drought stress phenomena in deciduous forest communities. Remote Sens 7:14227–14258. https://doi.org/10.3390/rs71014227

Dubayah R, Blair JB, Goetz S et al (2020) The global ecosystem dynamics investigation: high-resolution laser ranging of the Earth's forests and topography. Sci Remote Sens 1:100002. https://doi.org/10.1016/j.srs.2020.100002

Dungey H (2016) Forest genetics for productivity – the next generation. New Zeal J For Sci 46:40490. https://doi.org/10.1186/s40490-016-0081-z

Dungey HS, Dash JP, Pont D et al (2018) Phenotyping whole forests will help to track genetic performance. Trends Plant Sci 23:854–864. https://doi.org/10.1016/j.tplants.2018.08.005

Einzmann K, Immitzer M, Böck S et al (2017) Windthrow detection in european forests with very high-resolution optical data. Forests 8:1–26. https://doi.org/10.3390/f8010021

Erikson M (2004) Species classification of individually segmented tree crowns in high-resolution aerial images using radiometric and morphologic image measures. Remote Sens Environ 91:469–477. https://doi.org/10.1016/j.rse.2004.04.006

Falkowski MJ, Wulder MA, White JC, Gillis MD (2009) Supporting large-area, sample-based forest inventories with very high spatial resolution satellite imagery. Prog Phys Geogr 33:403–423. https://doi.org/10.1177/0309133309342643

FAO (2015) Knowledge reference for national forest assessments. http://www.fao.org/3/a-i4822e.pdf

Fassnacht FE, Latifi H, Ghosh A et al (2014) Assessing the potential of hyperspectral imagery to map bark beetle-induced tree mortality. Remote Sens Environ 140:533–548. https://doi.org/10.1016/j.rse.2013.09.014

Filipescu CN, Groot A, MacIsaac DA et al (2012) Prediction of diameter using height and crown attributes: a case study. West J Appl For 27:30–35. https://doi.org/10.1093/wjaf/27.1.30

Fischlin A, Ayres M, Karnosky D, Kellomäki S, Louman B, Ong C, Plattner C, Santoso H, Thompson I, Booth T, Marcar N, Scholes B, Swanston C, Zamolodchikov D (2009) Future environmental impacts and vulnerabilities. In: Seppala R, Buck A, Katila P (eds) Adaptation of forests and people to climate change. IUFRO World Series 22. Geist and Lambin, 2002

FOREST EUROPE (2015) State of Europe's Forests:2015

Fragoso-Campón L, Quirós E, Mora J, Gutiérrez Gallego JA, Durán-Barroso P (2020) Overstory-understory land cover mapping at the watershed scale: accuracy enhancement by multitemporal remote sensing analysis and LiDAR. Environ Sci Pollut Res 27:75–88. https://doi.org/10.1007/s11356-019-04520-8

Friedlaender H, Koch B (2000) First experience in the application of laser scanner data for the assessment of vertical and horizontal forest structures. Int Arch Photogramm Remote Sensing XXXIII(Part B7, ISPRS Congr XXXIII):693–700

Fuller DO (2001) Forest fragmentation in Loudoun County, Virginia, USA evaluated with multitemporal Landsat imagery. Landscape Ecology 16:627–642. https://link.springer.com/article/10.1023/A:1013140101134

Gallaun H, Zanchi G, Nabuurs GJ et al (2010) EU-wide maps of growing stock and above-ground biomass in forests based on remote sensing and field measurements. For Ecol Manag 260:252–261. https://doi.org/10.1016/j.foreco.2009.10.011

Gamon JA, Serrano L, Surfus JS (1997) The photochemical reflectance index: an optical indicator of photosynthetic radiation use efficiency across species, functional types, and nutrient levels. Oecologia 112:492–501. https://doi.org/10.1007/s004420050337

Gamon JA, Huemmrich KF, Wong CYS et al (2016) A remotely sensed pigment index reveals photosynthetic phenology in evergreen conifers. Proc Natl Acad Sci USA 113:13087–13092. https://doi.org/10.1073/pnas.1606162113

Gao BC (1996) NDWI – a normalized difference water index for remote sensing of vegetation liquid water from space. Remote Sens Environ 58:257–266. https://doi.org/10.1016/S0034-4257(96)00067-3

Gao WQ, Lei XD, Fu LY (2019) Impacts of climate change on the potential forest productivity based on a climate-driven biophysical model in northeastern China. J For Res. https://doi.org/10.1007/s11676-019-00999-6

Ghaffarian S, Kerle N, Filatova T (2018) Remote sensing-based proxies for urban disaster risk management and resilience: a review. Remote Sens 10:1760–1789. https://doi.org/10.3390/rs10111760

Gillespie TW, Willis KS, Ostermann-Kelm S (2015) Spaceborne remote sensing of the world's protected areas. Prog Phys Geogr 39:388–404. https://doi.org/10.1177/0309133314561648

Gitelson AA, Zur Y, Chivkunova OB, Merzlyak MN (2002) Assessing carotenoid content in plant leaves with reflectance spectroscopy. Photochem Photobiol 75:272. https://doi.org/10.1562/0031-8655(2002)0750272ACCIPL2.0.CO2

Goodbody TRH, Coops NC, Hermosilla T et al (2018) Assessing the status of forest regeneration using digital aerial photogrammetry and unmanned aerial systems. Int J Remote Sens 39:5246–5264. https://doi.org/10.1080/01431161.2017.1402387

Green PC, Burkhart HE (2020) Plantation Loblolly pine seedling counts with unmanned aerial vehicle imagery: a case study. J For 118(5):487–500. https://doi.org/10.1093/jofore/fvaa020

Haara A, Haarala M (2002) Tree species classification using semi-automatic delineation of trees on aerial images. Scand J For Res 17:556–565. https://doi.org/10.1080/02827580260417215

Haines-Young R, Potschin MB (2018) Common international classification of ecosystem services (CICES) V5.1 and guidance on the application of the revised structure. Fabis Consulting Ltd.

The Paddocks, Chestnut Lane, Barton in Fabis, Nottingham, NG11 0AE, UK. Available from www.cices.eu

Hamraz H, Contreras MA, Zhang J (2017) Vertical stratification of forest canopy for segmentation of understory trees within small-footprint airborne LiDAR point clouds. ISPRS J Photogramm Remote Sens 130:385–392. https://doi.org/10.1016/j.isprsjprs.2017.07.001

Hasenauer H, Neumann M, Moreno A, Running S (2017) Assessing the resources and mitigation potential of European forests. Energy Procedia 125:372–378. https://doi.org/10.1016/j.egypro.2017.08.052

He Y, Chen G, Potter C, Meentemeyer RK (2019) Integrating multi-sensor remote sensing and species distribution modeling to map the spread of emerging forest disease and tree mortality. Remote Sens Environ 231:111238. https://doi.org/10.1016/j.rse.2019.111238

Hedin LO (2015) Signs of saturation in the tropical carbon sink Hot on the trail of temperature processing. Nature 519:295–296. https://doi.org/10.1038/519295a

Heinzel J, Koch B (2012) Investigating multiple data sources for tree species classification in temperate forest and use for single tree delineation. Int J Appl Earth Obs Geoinf 18:101–110. https://doi.org/10.1016/j.jag.2012.01.025

Hermosilla T, Wulder MA, White JC et al (2015) Regional detection, characterization, and attribution of annual forest change from 1984 to 2012 using Landsat-derived time-series metrics. Remote Sens Environ 170:121–132. https://doi.org/10.1016/j.rse.2015.09.004

Hernández-Clemente R, Navarro-Cerrillo RM, Suárez L et al (2011) Assessing structural effects on PRI for stress detection in conifer forests. Remote Sens Environ 115:2360–2375. https://doi.org/10.1016/j.rse.2011.04.036

Hilker T, Wulder MA, Coops NC (2008) Update of forest inventory data with lidar and high spatial resolution satellite imagery. Can J Remote Sens 34:5–12. https://doi.org/10.5589/m08-004

Hilker T, Coops NC, Gaulton R, Wulder MA, Cranston J, Stenhouse GB (2011) Biweekly disturbance capture and attribution: case study in western Alberta grizzly bear habitat. J Appl Remote Sens 5(1). https://doi.org/10.1117/1.3664342

Hirschmugl M, Ofner M, Raggam J, Schardt M (2007) Single tree detection in very high resolution remote sensing data. Remote Sens Environ 110:533–544. https://doi.org/10.1016/j.rse.2007.02.029

Huang C-Y, Anderegg WRL, Asner GP (2019) Remote sensing of forest die-off in the Anthropocene: from plant ecophysiology to canopy structure. Remote Sens Environ 231:111233. https://doi.org/10.1016/j.rse.2019.111233

Hubau W, Lewis SL, Phillips OL et al (2020) Asynchronous carbon sink saturation in African and Amazonian tropical forests. Nature 579:80–87. https://doi.org/10.1038/s41586-020-2035-0

Hunt ER, Rock BN (1989) Detection of changes in leaf water content using Near- and Middle-Infrared reflectances. Remote Sens Environ 30:43–54. https://doi.org/10.1016/0034-4257(89)90046-1

Imangholiloo M, Saarinen N, Markelin L et al (2019) Characterizing seedling stands using leaf-off and leaf-on photogrammetric point clouds and hyperspectral imagery acquired from unmanned aerial vehicle. Forests 10:1–17. https://doi.org/10.3390/f10050415

Imangholiloo M, Saarinen N, Holopainen M, Yu X, Hyyppä J, Vastaranta M (2020) Using leaf-off and leaf-on multispectral airborne laser scanning data to characterize seedling stands. Remote Sens 12:3328. https://doi.org/10.3390/rs12203328

Immitzer M, Atzberger C, Koukal T (2012) Tree species classification with Random forest using very high spatial resolution 8-band worldView-2 satellite data. Remote Sens 4:2661–2693. https://doi.org/10.3390/rs4092661

Inoue T, Nagai S, Yamashita S et al (2014) Unmanned aerial survey of fallen trees in a deciduous broadleaved forest in eastern Japan. PLoS One 9:1–7. https://doi.org/10.1371/journal.pone.0109881

Jin X, Fiore AM, Murray LT et al (2017) Evaluating a space-based indicator of surface ozone-NOx-VOC sensitivity over midlatitude source regions and application to decadal trends. J Geophys Res Atmos 122:10439–10461. https://doi.org/10.1002/2017JD026720

Jucker T, Caspersen J, Chave J et al (2017) Allometric equations for integrating remote sensing imagery into forest monitoring programmes. Glob Chang Biol 23:177–190. https://doi.org/10.1111/gcb.13388

Kapos V, Rhind J, Edwards M et al (2000) Developing a map of the world's mountain forests. In: Price MF, Butt N (eds) Forests in sustainable mountain development: a state-of-knowledge report for 2000. CAB International, Wallingford, pp 4–9

Kasischke ES, Bourgeau-Chavez LL, Johnstone JF (2007) Assessing spatial and temporal variations in surface soil moisture in fire-disturbed black spruce forests in Interior Alaska using spaceborne synthetic aperture radar imagery – implications for post-fire tree recruitment. Remote Sens Environ 108:42–58. https://doi.org/10.1016/j.rse.2006.10.020

Kauppi P, Hanewinkel M, Lundmark T, Nabuurs GJ, Peltola H, Trasobares A, Hetemäki L (2018) Climate smart forestry in Europe. European Forest Institute

Kempeneers P, Sedano F, Pekkarinen A, Seebach L, Strobl P, San Miguel-Ayanz J (2012). Pan-European forest maps derived from optical satellite imagery. IEEE Earthzine 5 (2nd quarter theme). https://earthzine.org/pan-european-forest-maps-derived-from-optical-satellite-imagery/

Kennedy RE, Cohen WB, Schroeder TA (2007) Trajectory-based change detection for automated characterization of forest disturbance dynamics. Remote Sens Environ 110:370–386. https://doi.org/10.1016/j.rse.2007.03.010

Kennedy RE, Yang Z, Cohen WB (2010) Detecting trends in forest disturbance and recovery using yearly Landsat time series: 1. LandTrendr – temporal segmentation algorithms. Remote Sens Environ 114:2897–2910. https://doi.org/10.1016/j.rse.2010.07.008

Klein WH (1982) Estimating bark beetle-killed lodgepole pine with high altitude panoramic photography. Photogramm Eng Remote Sens 48:733–737

Knorn J, Rabe A, Radeloff VC et al (2009) Land cover mapping of large areas using chain classification of neighboring Landsat satellite images. Remote Sens Environ 113:957–964. https://doi.org/10.1016/j.rse.2009.01.010

Koch B. (2015). Remote sensing supporting national forest inventories NFA. In Food and Agriculture Organization of the United Nations, Knowledge reference for national forest assessments. Rome: Food and Agriculture Organization of the United Nations (pp. 77–92)

Korpela I (2004) Individual tree measurements by means of digital aerial photogrammetry. In Silva Fennica Monographs 3:93 p

Lambert J, Denux JP, Verbesselt J et al (2015) Detecting clear-cuts and decreases in forest vitality using MODIS NDVI time series. Remote Sens 7:3588–3612. https://doi.org/10.3390/rs70403588

Landry S, St-Laurent M-H, Pelletier G, Villard M-A (2020) The best of both worlds? Integrating Sentinel-2 images and airborne LiDAR to characterize forest regeneration. Remote Sens 2020(12):2440. https://doi.org/10.3390/rs12152440

Lausch A, Borg E, Bumberger J et al (2018) Understanding forest health with remote sensing, Part III: Requirements for a scalable multi-source forest health monitoring network based on data science approaches. Remote Sens 10:1120–1171. https://doi.org/10.3390/rs10071120

LeBauer DS, Treseder KK (2008) Nitrogen limitation of net primary productivity in terrestrial ecosystems is globally distributed. Ecology 89(2):371–379. https://doi.org/10.1890/06-2057.1

Lefsky MA (2010) A global forest canopy height map from the moderate resolution imaging spectroradiometer and the geoscience laser altimeter system. Geophys Res Lett 37:1–5. https://doi.org/10.1029/2010GL043622

Lehmann EA, Caccetta P, Lowell K et al (2015) SAR and optical remote sensing: Assessment of complementarity and interoperability in the context of a large-scale operational forest monitoring system. Remote Sens Environ 156:335–348. https://doi.org/10.1016/j.rse.2014.09.034

Li X, Zhang L, Weihermuller L et al (2014) Measurement and simulation of topographic effects on passive microwave remote sensing over mountain areas: a case study from the tibetan plateau. IEEE Trans Geosci Remote Sens 52:1489–1501. https://doi.org/10.1109/TGRS.2013.2251887

Lorente M, Gauthier S, Bernier P, Ste-Marie C (2018) Tracking forest changes: Canadian Forest Service indicators of climate change. Clim Change:1–15. https://doi.org/10.1007/s10584-018-2154-x

Lucas RM, Honzák M, Curran PJ et al (2000) Mapping the regional extent of tropical forest regeneration stages in the Brazilian Legal Amazon using NOAA AVHRR data. Int J Remote Sens 21:2855–2881. https://doi.org/10.1080/01431160050121285

Lucier A, Ayres M, Karnosky D, Thompson I, Loehle C, Percy K, Sohngen B (2009) Forest responses and vulnerabilities to recent climate change. In: Seppala R, Buck A, Katila P (eds) Adaptation of forests and people to climate change. IUFRO World Series 22. Geist and Lambin, 2002

Ludovisi R, Tauro F, Salvati R et al (2017) UAV-based thermal imaging for high-throughput field phenotyping of black poplar response to drought. Front Plant Sci 8:1–18. https://doi.org/10.3389/fpls.2017.01681

Majumder BD, Roy JK, Padhee S (2019) Recent advances in multifunctional sensing technology on a perspective of multi-sensor system: a review. IEEE Sens J 19:1204–1214. https://doi.org/10.1109/JSEN.2018.2882239

Maltamo M, Eerikäinen K, Packalén P, Hyyppä J (2006) Estimation of stem volume using laser scanning-based canopy height metrics. Forestry 79:217–229. https://doi.org/10.1093/forestry/cpl007

Maltamo M, Peuhkurinen J, Malinen J et al (2009) Predicting tree attributes and quality characteristics of scots pine using airborne laser scanning data. Silva Fenn 43:507–521. https://doi.org/10.14214/sf.203

Maltamo M, Hauglin M, Næsset E, Gobakken T (2019) Estimating stand level stem diameter distribution utilizing harvester data and airborne laser scanning. Silva Fenn 53:1–19. https://doi.org/10.14214/sf.10075

Marchi N, Pirotti F, Lingua E (2018a) Airborne and terrestrial laser scanning data for the assessment of standing and lying deadwood: current situation and new perspectives. Remote Sens 10:1356–1376. https://doi.org/10.3390/rs10091356

Marchi N, Weisberg P, Greenberg J et al (2018b) Remote sensing application for deadwood identification and characterisation. In: Geophysical Research Abstracts of the 20th EGU General Assembly held 4–13 April, 2018 in Vienna, Austria, vol 20, EGU2018-16440-1, p 16440. https://ui.adsabs.harvard.edu/abs/2018EGUGA..2016440M/abstract

Maroschek M, Rammer W, Lexer MJ (2015) Using a novel assessment framework to evaluate protective functions and timber production in Austrian mountain forests under climate change. Reg Environ Chang 15:1543–1555. https://doi.org/10.1007/s10113-014-0691-z

Martin RV (2008) Satellite remote sensing of surface air quality. Atmos Environ 42:7823–7843. https://doi.org/10.1016/j.atmosenv.2008.07.018

Masek JG, Goward SN, Kennedy RE et al (2013) United states forest disturbance trends observed using landsat time series. Ecosystems 16:1087–1104. https://doi.org/10.1007/s10021-013-9669-9

Matikainen L, Karila K, Litkey P et al (2020) Combining single photon and multispectral airborne laser scanning for land cover classification. ISPRS J Photogramm Remote Sens 164:200–216. https://doi.org/10.1016/j.isprsjprs.2020.04.021

McDowell NG, Coops NC, Beck PSA et al (2015) Global satellite monitoring of climate-induced vegetation disturbances. Trends Plant Sci 20:114–123. https://doi.org/10.1016/j.tplants.2014.10.008

Meiforth JJ, Buddenbaum H, Hill J, Shepherd JD, Dymond JR (2020) Stress Detection in New Zealand Kauri Canopies with WorldView-2 Satellite and LiDAR Data. Remote Sens 12:1906. https://doi.org/10.3390/rs12121906

Meng R, Dennison PE, Zhao F et al (2018) Mapping canopy defoliation by herbivorous insects at the individual tree level using bi-temporal airborne imaging spectroscopy and LiDAR measurements. Remote Sens Environ 215:170–183. https://doi.org/10.1016/j.rse.2018.06.008

Mitchard ETA, Saatchi SS, White LJT et al (2012) Mapping tropical forest biomass with radar and spaceborne LiDAR in Lopé National Park, Gabon: overcoming problems of high biomass and persistent cloud. Biogeosciences 9:179–191. https://doi.org/10.5194/bg-9-179-2012

Moghaddam M, Saatchi S, Cuenca RH (2000) Estimating subcanopy soil moisture with radar. J Geophys Res Atmos 105:14899–14911. https://doi.org/10.1029/2000JD900058

Monnet J, Mermin E, Chanussot J, Berger F (2010) Using airborne laser scanning to assess forest protection function against rockfall. Interpraevent International Symposium in Pacific Rim, April 2010, Taipei, Taiwan, pp 586–594. hal-00504706

Morresi D, Vitali A, Urbinati C, Garbarino M (2019) Forest spectral recovery and regeneration dynamics in stand-replacing wildfires of central Apennines derived from Landsat time series. Remote Sens 11:308–325. https://doi.org/10.3390/rs11030308

Mullen K, Yuan F, Mitchell M (2018) The mountain pine beetle epidemic in the Black Hills, South Dakota: the consequences of long term fire policy, climate change and the use of remote sensing to enhance mitigation. J Geogr Geol 10:69. https://doi.org/10.5539/jgg.v10n1p69

Mura M, Bottalico F, Giannetti F et al (2018) Exploiting the capabilities of the Sentinel-2 multi spectral instrument for predicting growing stock volume in forest ecosystems. Int J Appl Earth Obs Geoinf 66:126–134. https://doi.org/10.1016/j.jag.2017.11.013

Nabuurs GJ, Delacote P, Ellison D et al (2017) By 2050 the mitigation effects of EU forests could nearly double through climate smart forestry. Forests 8:1–14. https://doi.org/10.3390/f8120484

Næsset E (1997) Estimating timber volume of forest stands using airborne laser scanner data. Remote Sens Environ 61:246–253. https://doi.org/10.1016/S0034-4257(97)00041-2

Nagendra H, Lucas R, Honrado JP et al (2013) Remote sensing for conservation monitoring: assessing protected areas, habitat extent, habitat condition, species diversity, and threats. Ecol Indic 33:45–59. https://doi.org/10.1016/j.ecolind.2012.09.014

Newnham GJ, Armston JD, Calders K et al (2015) Terrestrial laser scanning for plot-scale forest measurement. Curr For Rep 1:239–251. https://doi.org/10.1007/s40725-015-0025-5

Ningthoujam RK, Balzter H, Tansey K et al (2016) Airborne S-band SAR for forest biophysical retrieval in temperate mixed forests of the UK. Remote Sens 8:1–22. https://doi.org/10.3390/rs8070609

Ollinger SV, Goodale CL, Hayhoe K, Jenkins JP (2008) Potential effects of climate change and rising CO2 on ecosystem processes in northeastern U.S. forests. Mitig Adapt Strateg Glob Chang 13:467–485. https://doi.org/10.1007/s11027-007-9128-z

Olsson PO, Lindström J, Eklundh L (2016) Near real-time monitoring of insect induced defoliation in subalpine birch forests with MODIS derived NDVI. Remote Sens Environ 181:42–53. https://doi.org/10.1016/j.rse.2016.03.040

Ottosen T-B, Petch G, Hanson M, Skjøth CA (2020) Tree cover mapping based on Sentinel-2 images demonstrate high thematic accuracy in Europe. Int J Appl Earth Obs Geoinf 84:101947. https://doi.org/10.1016/j.jag.2019.101947

Pablos M, González-Zamora Á, Sánchez N, Martínez-Fernández J (2018) Assessment of root zone soil moisture estimations from SMAP, SMOS and MODIS observations. Remote Sens 10:981–1000. https://doi.org/10.3390/rs10070981

Päivinen R, Van Brusselen J, Schuck A (2009) The growing stock of European forests using remote sensing and forest inventory data. Forestry 82:479–490. https://doi.org/10.1093/forestry/cpp017

Panagiotidis D, Abdollahnejad A, Surový P, Kuželka K (2019) Detection of fallen logs from high-resolution UAV images. New Zeal J For Sci 49. https://doi.org/10.33494/nzjfs492019x26x

Paquette A, Vayreda J, Coll L et al (2018) Climate change could negate positive tree diversity effects on forest productivity: a study across five climate types in Spain and Canada. Ecosystems 21:960–970. https://doi.org/10.1007/s10021-017-0196-y

Paul-Limoges E, Damm A, Hueni A et al (2018) Effect of environmental conditions on sun-induced fluorescence in a mixed forest and a cropland. Remote Sens Environ 219:310–323. https://doi.org/10.1016/j.rse.2018.10.018

Peña MA, Altmann SH (2009) Use of satellite-derived hyperspectral indices to identify stress symptoms in an Austrocedrus chilensis forest infested by the aphid Cinara cupressi. Int J Pest Manag 55:197–206. https://doi.org/10.1080/09670870902725809

Persson Å, Holmgren J (2004) Tree species classification of individual trees in Sweden by combining high resolution laser data with high resolution near-infrared digital images. Int Arch Photogramm Remote Sens Spat Inf Sci 36:204–207

Peuhkurinen J, Tokola T, Plevak K, Sirparanta S, Kedrov A, Pyankov S (2018) Predicting Tree Diameter Distributions from Airborne Laser Scanning, SPOT 5 Satellite, and Field Sample Data in the Perm Region, Russia. Forests 9:639. https://doi.org/10.3390/f9100639

Pontius J, Schaberg P, Hanavan R (2020) Remote sensing for early, detailed, and accurate detection of forest disturbance and decline for protection of biodiversity. In: Cavender-Bares J, Gamon JA, Townsend PA (eds) Remote sensing of plant biodiversity. Springer, Cham

Potapov P, Yaroshenko A, Turubanova S et al (2008) Mapping the world's intact forest landscapes by remote sensing. Ecol Soc 13(2):51–66. https://doi.org/10.5751/ES-02670-130251

Pouliot DA, King DJ, Pitt DG (2005) Development and evaluation of an automated tree detection-delineation algorithm for monitoring regenerating coniferous forests. Can J For Res 35:2332–2345. https://doi.org/10.1139/x05-145

Puliti S, Solberg S, Granhus A (2019) Use of UAV photogrammetric data for estimation of biophysical properties in forest stands under regeneration. Remote Sens 11:233. https://doi.org/10.3390/rs11030233

Quegan S, Le Toan T, Chave J et al (2019) The European Space Agency BIOMASS mission: Measuring forest above-ground biomass from space. Remote Sens Environ 227:44–60. https://doi.org/10.1016/j.rse.2019.03.032

Rao K, Anderegg WRL, Sala A et al (2019) Satellite-based vegetation optical depth as an indicator of drought-driven tree mortality. Remote Sens Environ 227:125–136. https://doi.org/10.1016/j.rse.2019.03.026

Rasel SMM, Groen TA, Hussin YA, Diti IJ (2017) Proxies for soil organic carbon derived from remote sensing. Int J Appl Earth Obs Geoinf 59:157–166. https://doi.org/10.1016/j.jag.2017.03.004

Rees WG, Williams M (1997) Monitoring changes in land cover induced by atmospheric pollution in the Kola Peninsula, Russia, using Landsat-MSS data. Int J Remote Sens 18:1703–1723. https://doi.org/10.1080/014311697218061

Reich PB, Oleksyn J (2008) Climate warming will reduce growth and survival of Scots pine except in the far north. Ecol Lett 11(6):588–597

Rock BN, Hoshizaki T, Miller JR (1988) Comparison of in situ and airborne spectral measurements of the blue shift associated with forest decline. Remote Sens Environ 24:109–127. https://doi.org/10.1016/0034-4257(88)90008-9

Röder M, Latifi H, Hill S, Wild J, Svoboda M, Brůna J, Macek M, Nováková MH, Gülch E, Heurich M (2018) Application of optical unmanned aerial vehicle-based imagery for the inventory of natural regeneration and standing deadwood in post-disturbed spruce forests. International Journal of Remote Sensing 39(15-16):5288–5309. https://doi.org/10.1080/01431161.2018.1441568

Rosenzweig C, Casassa G, Karoly DJ et al (2007) Assessment of observed changes and responses in natural and managed systems. In: Parry ML, Canziani OF, Palutikof JP, van der Linden PJ, Hanson CE (eds) Contribution of working group II to the fourth assessment report of the intergovernmental panel on climate change. Cambridge University Press, Cambridge, pp 79–131

Roth KL, Roberts DA, Dennison PE, Alonzo M, Peterson SH, Beland M (2015) Differentiating plant species within and across diverse ecosystems with imaging spectroscopy. Remote Sens Environ 167:135–151. https://doi.org/10.1016/j.rse.2015.05.007

Saksa T, Uuttera J, Kolström T et al (2003) Clear-cut detection in boreal forest aided by remote sensing. Scand J For Res 18:537–546. https://doi.org/10.1080/02827580310016881

Salas C, Ene L, Gregoire TG et al (2010) Modelling tree diameter from airborne laser scanning derived variables: a comparison of spatial statistical models. Remote Sens Environ 114:1277–1285. https://doi.org/10.1016/j.rse.2010.01.020

Sánchez-Ruiz S, Chiesi M, Maselli F, Gilabert MA (2016) Mapping growing stock at 1-km spatial resolution for Spanish forest areas from ground forest inventory data and GLAS canopy height. In: Proceedings of SPIE 10005, Earth Resources and Environmental Remote Sensing/GIS Applications VII, 100051I (18 October 2016). https://doi.org/10.1117/12.2241166

Sandström J, Bernes C, Junninen K et al (2019) Impacts of dead wood manipulation on the biodiversity of temperate and boreal forests. A systematic review. J Appl Ecol 56:1770–1781. https://doi.org/10.1111/1365-2664.13395

Schnelle F, Volkert E (1974) International phenological gardens in Europe. The basic network for international phenological observations. In: Lieth H (ed) Phenology and seasonality modeling. Ecological studies (Analysis and synthesis), vol 8. Springer, Berlin/Heidelberg. https://doi.org/10.1007/978-3-642-51863-8_32

Seidl R, Albrich K, Erb K et al (2019) What drives the future supply of regulating ecosystem services in a mountain forest landscape? For Ecol Manag 445:37–47. https://doi.org/10.1016/j.foreco.2019.03.047

Seppälä R, Buck A, Katila P (eds) (2009) Adaptation of forests and people to climate change. A global assessment report. IUFRO World Series Volume 22. Helsinki, 224 p

Shi L, Liu S (2017) Methods of estimating forest biomass: a review. In: Tumuluru JS (ed) Biomass volume estimation and valorization for energy. IntechOpen. https://doi.org/10.5772/65733

Simard M, Pinto N, Fisher JB, Baccini A (2011) Mapping forest canopy height globally with spaceborne lidar. J Geophys Res Biogeosci 116:1–12. https://doi.org/10.1029/2011JG001708

Šímová I, Storch D (2017) The enigma of terrestrial primary productivity: measurements, models, scales and the diversity–productivity relationship. Ecography 40(239–252):2017. https://doi.org/10.1111/ecog.02482

Sinha S, Santra A, Das AK et al (2019) Regression-based integrated Bi-sensor SAR data model to estimate forest carbon stock. J Indian Soc Remote Sens 47:1599–1608. https://doi.org/10.1007/s12524-019-01004-7

Skakun RS, Wulder MA, Franklin SE (2003) Sensitivity of the Thematic Mapper Enhanced Wetness Difference Index (EWDI) to detect mountain pine needle red-attack damage. Remote Sens Environ 86(4):433–443. https://doi.org/10.1016/S0034-4257(03)00112-3

Smigaj M, Gaulton R, Suárez JC, Barr SL (2019) Canopy temperature from an Unmanned Aerial Vehicle as an indicator of tree stress associated with red band needle blight severity. For Ecol Manag 433:699–708. https://doi.org/10.1016/j.foreco.2018.11.032

Smith G, Askne J (2001) Clear-cut detection using ERS interferometry. Int J Remote Sens 22:3651–3664. https://doi.org/10.1080/01431160110040477

Spriggs RA, Coomes DA, Jones TA et al (2017) An alternative approach to using LiDAR remote sensing data to predict stem diameter distributions across a temperate forest landscape. Remote Sens 9:944. https://doi.org/10.3390/rs9090944

Stone C, Mohammed C (2017) Application of remote sensing technologies for assessing planted forests damaged by insect pests and fungal pathogens: a review. Curr For Rep 3:75–92. https://doi.org/10.1007/s40725-017-0056-1

Szpakowski DM, Jensen JLR (2019) A Review of the Applications of Remote Sensing in Fire Ecology. Remote Sens 11:2638. https://doi.org/10.3390/rs11222638

Tang H, Armston J, Hancock S et al (2019) Characterizing global forest canopy cover distribution using spaceborne lidar. Remote Sens Environ 231:111262. https://doi.org/10.1016/j.rse.2019.111262

Teich M, Bebi P (2009) Evaluating the benefit of avalanche protection forest with GIS-based risk analyses – a case study in Switzerland. For Ecol Manag 257:1910–1919. https://doi.org/10.1016/j.foreco.2009.01.046

Thomas V, Oliver RD, Lim K, Woods M (2008) LiDAR and Weibull modeling of diameter and basal area. For Chron 84:866–875. https://doi.org/10.5558/tfc84866-6

Torresan C, Corona P, Scrinzi G, Marsal JV (2016) Using classification trees to predict forest structure types from LiDAR data. Ann For Res 59(2):281–298. https://doi.org/10.15287/afr.2016.423

Vacchiano G, Berretti R, Motta R, Mondino EB (2018) Assessing the availability of forest biomass for bioenergy by publicly available satellite imagery. IForest 11:459–468. https://doi. org/10.3832/ifor2655-011

Varhola A, Coops NC (2013) Estimation of watershed-level distributed forest structure metrics relevant to hydrologic modeling using LiDAR and Landsat. J Hydrol 487:70–86. https://doi. org/10.1016/j.jhydrol.2013.02.032

Vastaranta M, Saarinen N, Yrttimaa T, Kankare V (2020) Monitoring forests in space and time using close-range sensing. https://doi.org/10.20944/preprints202002.0300.v1

Verkerk PJ, Costanza R, Hetemäki L et al (2020) Climate-smart forestry: the missing link. For Policy Econ 115:102164. https://doi.org/10.1016/j.forpol.2020.102164

Viccaro M, Cozzi M, Fanelli L, Romano S (2019) Spatial modelling approach to evaluate the economic impacts of climate change on forests at a local scale. Ecol Indic 106:105523. https:// doi.org/10.1016/j.ecolind.2019.105523

Vincent G, Caron F, Sabatier D, Blanc L (2012) LiDAR shows that higher forests have more slender trees. Bois Forets des Tropiques 66:51–56

Vittucci C, Ferrazzoli P, Kerr Y et al (2016) SMOS retrieval over forests: exploitation of optical depth and tests of soil moisture estimates. Remote Sens Environ 180:115–127. https://doi. org/10.1016/j.rse.2016.03.004

Wallace L, Lucieer A, Malenovský Z et al (2016) Assessment of forest structure using two UAV techniques: a comparison of airborne laser scanning and structure from motion (SfM) point clouds. Forests 7:1–16. https://doi.org/10.3390/f7030062

Walsh SJ, Weiss DJ, Butler DR, Malanson GP (2004) An assessment of snow avalanche paths and forest dynamics using Ikonos satellite data. Geocarto Int 19:85–93. https://doi. org/10.1080/10106040408542308

Wang R, Gamon JA (2018) Remote sensing of terrestrial plant biodiversity. Remote Sens Environ 231:111218. https://doi.org/10.1016/j.rse.2019.111218

Wang C-J, Zhang Z-X, Wan J-Z (2019) Vulnerability of global forest ecoregions to future climate change. Glob Ecol Conserv 20:1–10. https://doi.org/10.1016/j.gecco.2019.e00760

Waring RH, Coops NC, Landsberg JJ (2010) Improving predictions of forest growth using the 3-PGS model with observations made by remote sensing. For Ecol Manag 259:1722–1729. https://doi.org/10.1016/j.foreco.2009.05.036

Warren SD, Alt M, Olson KD, Irl SDH, Steinbauer MJ, Jentsch A (2014) The relationship between the spectral diversity of satellite imagery, habitat heterogeneity, and plant species richness. Ecol Informatics 24:160–168. https://doi.org/10.1016/j.ecoinf.2014.08.006

Weatherall A, Nabuurs G-J, Velikova V et al (2021) Defining Climate-Smart Forestry. In: Managing Forest Ecosystems, Vol. 40, Tognetti R, Smith M, Panzacchi P (Eds): Climate-Smart Forestry in Mountain Regions. Springer Nature, Switzerland, AG

Weinstein BG, Marconi S, Bohlman S et al (2019) Individual tree-crown detection in rgb imagery using semi-supervised deep learning neural networks. Remote Sens 11:1–13. https://doi. org/10.3390/rs11111309

Weiskittel AR, Crookston NL, Radtke PJ (2011) Linking climate, gross primary productivity, and site index across forests of the western United States. Can J For Res 41(8). https://doi. org/10.1139/x11-086

Weiss DJ, Walsh SJ (2009) Remote sensing of mountain environments. Geogr Compass 3:1–21. https://doi.org/10.1111/j.1749-8198.2008.00200.x

Wigneron JP, Jackson TJ, O'Neill P et al (2017) Modelling the passive microwave signature from land surfaces: a review of recent results and application to the L-band SMOS & SMAP soil moisture retrieval algorithms. Remote Sens Environ 192:238–262. https://doi.org/10.1016/j. rse.2017.01.024

Willis KS (2015) Remote sensing change detection for ecological monitoring in United States protected areas. Biol Conserv 182:233–242. https://doi.org/10.1016/j.biocon.2014.12.006

Xu X, Li J, Tolson BA (2014) Progress in integrating remote sensing data and hydrologic modelling. Prog Phys Geogr 38(4):464–498. https://doi.org/10.1177/0309133314536583

Yang H, Hu D, Xu H, Zhing X (2020) Assessing the spatiotemporal variation of NPP and its response to driving factors in Anhui province, China. Environ Sci Pollut Res 27:14915–14932. https://doi.org/10.1007/s11356-020-08006-w

Zarea A, Mohammadzadeh A (2016) A novel building and tree detection method from LiDAR data and aerial images. IEEE J Sel Top Appl Earth Obs Remote Sens 9:1864–1875. https://doi.org/10.1109/JSTARS.2015.2470547

Zimble DA, Evans DL, Carlson GC, Parker RC, Grado SC, Gerard PD (2003) Characterizing vertical forest structure using small-footprint airborne LiDAR. Remote Sens Environ 87(2–3):171–182. https://doi.org/10.1016/S0034-4257(03)00139-1

Chapter 12
Economic and Social Perspective of Climate-Smart Forestry: Incentives for Behavioral Change to Climate-Smart Practices in the Long Term

Veronika Gežík, Stanislava Brnkaľáková, Viera Baštáková, and Tatiana Kluvánková

Abstract In this volume, the concept of climate-smart forestry (CSF) has been introduced as adaptive forest management and governance to address climate change, fostering resilience and sustainable ecosystem service provision. Adaptive forest management and governance are seen as vital ways to mitigate the present and future impact of climate change on forest. Following this trajectory, we determine the ecosystem services approach as a potential adaptive tool to contribute to CSF. Ecosystem services as public or common goods face the traditional social dilemma of individual versus collective interests, which often generate conflicts, overuse, and resource depletion. This chapter focuses on the ecosystem service governance approach, especially on incentive tools for behavioral change to CSF in the long term, which is a basic precondition for the sustainability of ecosystem integrity and functions, as well as ensuring the continuous delivery of ecosystem goods and services, as per the CSF definition. Payments for ecosystem services (PES) are seen as innovative economic instruments when adding a social dimension by involving local communities and their values to ensure the long-term resilience and adaptation of forest ecosystems to climate change. We argue that tackling climate change adaptation requires the behavioral change of ecosystem service providers to a collaborative and integrated PES approach, as also emphasized by the Sustainable Development Goals (SDGs) of the Agenda 2030.

V. Gežík (✉)
Institute for Forest Ecology, Slovak Academy of Sciences, Bratislava, Slovakia

S. Brnkaľáková · V. Baštáková · T. Kluvánková
SlovakGlobe: Slovak Academy of Sciences and Slovak University of Technology, Bratislava, Slovakia
e-mail: brnkalakova@ife.sk; bastakova@3e-institut.sk; tana@cetip.sk

12.1 Introduction

The concept of climate-smart forestry (CSF) has recently been introduced in order to contribute to climate change adaptation and mitigation from the forestry sector that is aimed at reducing GHG emissions, fostering resilience, and increasing productivity and incomes (Nabuurs et al. 2017; Bowditch et al. 2020). Such adaptive management and governance are seen as vital to mitigate the present and future impact of climate change on forest, ensuring the survival of forest stands. A relatively minor change in climate can have a devastating effect upon forest by increasing vulnerability to drought, insect attack, and fire. In this volume, CSF has been understood as adaptive forest management and governance to address climate change, by fostering resilience and balancing climate change mitigation measures with long-term multiple goods and services provision (e.g., biodiversity, water services, or surface cooling) and sustainably increasing forest productivity and long-term environmental benefits and economic welfare based on forestry. Forest management and planning must consider the expected future impact of climate change on, e.g., tree species' distribution, productivity, risk of hazards (fires, pests, etc.), and drought (Schelhaas et al. 2015), while incorporating uncertainty (Lindner et al. 2014) and adaptation and resilience concepts (Lexer and Bugmann 2017). More active and flexible forest management, and the improved protection of forest areas, can not only reduce CO_2 emissions but also significantly reduce fire risk and thus land-use change. This aspect is also stressed by Yousefpour et al. (2018), who highlight that CSF implementation requires multipurpose and diversified forest management strategies. Diversity in forest management can support various ecosystem services and thus contribute to increasing forest resilience to climate change threats.

To deal with the abovementioned challenges in forest management and increase the provision of multiple ecosystem services and thus reach CSF, it is important to understand the complexity of relationships within ecological systems and the complex nature of social-ecological systems. The concept of ecosystem services (ES) for describing the relationship between human societies and the natural environment is historically quite recent (Gomez-Baggethun et al. 2010). It puts emphasis on the values of natural systems and socioecological dynamics in the planning of economic policies by providing incentive for sustainable use and increasing the convergence of sectoral policies. The concept is expected to induce a paradigm shift in the management of natural resources (Cowx and Portocarrero-Aya 2011) and link natural systems and human well-being (Amsworth et al. 2007; Skroch and Lopez-Hoffman 2009) to propose effective strategies for the management of vulnerable natural resources and their ecosystem services, especially under the risk of climate change (Kluvánková et al. 2019; Primmer et al. 2021). Most ES fall within the types of goods that are considered either "common-pool resources" or "public goods" (though ownership of the resource base might be private, public, or communal) characterized by two particular features: excludability and rivalry. If there is no excludability in supply and no rivalry in demand, the goods and services are public (most supporting, regulating, and cultural ecosystem services), whereas if there is no excludability in supply but rivalry in demand, the goods and services are common, which is the case of most provisioning ecosystem services (Farley and Costanza 2010; Ostrom 2010; Muradian and Rival 2012; Muradian and Gómez-Baggethun 2013).

The combination of governance structures and hybrid governance (positioned between markets and hierarchies) is necessary when the provision of particular ES (such as climate regulation) is characterized by the high complexity of their functioning, high levels of uncertainty, imperfect and asymmetric information between transacting parties, and cognitive barriers in assessing the service itself (Williamson 1991; Muradian and Rival 2012; Otto and Chobotová 2013; Kluvánková et al. 2019). Such regime in vulnerable areas can be seen as effective governance for ES governance to overcome the social dilemma of individual interests, directing sectoral policies toward a more integrated approach of EU regions and at a lower cost than hierarchy or market (Muradian and Rival 2012; Kluvánková et al. 2019). In this way, ES governance can contribute to the sustainable management of socio-ecological systems in the long term (Primmer et al. 2021).

The analysis of the emerging concept of PES in recent decades is considered as one of the most promising tools for enhancing or safeguarding the provision of specific or bundled ES. PES schemes are expected to generate a continuous flow of ES in the long term while also maintaining their quality. Although PES has been extensively analyzed in terms of potential positive and negative impacts on the poor (e.g., Suyanto et al. 2007; Bulte et al. 2008; Randrianarison et al. 2017; Blundo-Canto et al. 2018), not enough attention has been paid to examining the role of PES in the context of adaptation and mitigation to climate change (van de Sand 2012). The common pool or public nature of most ES implies that market mechanisms are not always suitable as governance tools of climate change problems, since markets tend to be more effective in dealing with private goods.

Following this trajectory, this chapter focuses on discussing whether PES is an appropriate and promising approach to promote CSF. It contributes to these debates by focusing on the PES design, especially the aspects (Fig. 12.1) that influence the behavioral change to climate-smart practices in the long term, the capacity of actors dealing with climate change, and the continuous delivery of multiple ecosystem goods and services, as per the CSF definition. In this regard, long-term behavioral change is a basic precondition for the sustainability of ecosystem integrity and functions. Moreover, several authors argue that PES can be seen as more effective innovative economic instruments for adaptation to climate change within CSF when adding a social dimension by involving local communities and their values to ensure forest ecosystems' long-term resilience and adaptation to climate change (Sattler and Matzdorf 2013; Brownson et al. 2019). Additionally, we determine that by targeting multiple ES via the CSF approach, it has the potential to contribute to climate change adaptation and mitigation. To tackle the abovementioned issues, the present chapter aims to reveal in which way the PES schemes currently implemented in Europe contribute to climate change mitigation and adaptation.

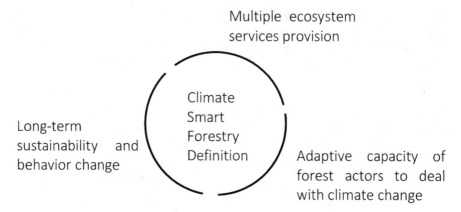

Fig. 12.1 The aspects of PES in the context of adaptation and mitigation to climate change based on CSF definition

The following section reviews what we mean by PES schemes and how this concept has been framed and defined by different scholars. Then, the next subsections focus on the different aspects of CSF definition and especially elaborate on how various PES contribute to enhancements in the provision of multiple ES, how various PES designs and implementations can influence the ability of various actors to deal with climate variability and change, and in what way PES can contribute to providing a long-term incentive mechanism to adopt specific measures for adaptation and mitigation to climate change. The last section summarizes the paper and suggests potential areas for future PES and climate change-related research.

12.2   Payments for Ecosystem Services (PES)

According to Tognetti (2016), forestry measures at the regional scale (i.e., forest mountain areas) should be implemented, along with bridging the gap between local development via FES (forest ecosystem services) provision, ecosystem resilience, and climate change adaptation and mitigation strategies and policies inclusive. The successful development of CSF calls for policymakers to create incentives for investments needed to activate forest management and finance mitigation and adaption measures, which include protecting biodiversity and other ES (Verkerk et al. 2020). A possible solution in that sense is the implementation of PES in CSF (Bartczak and Metelska-Szaniawska 2015; Matthies et al. 2015). According to Wunder (2005), PES are defined as voluntary transactions where a well-defined ES is bought by a buyer (i.e., someone who is willing to pay for it), if and only if the provider secures the provision of such service. However, this prescriptive definition is problematic because it excludes a variety of PES schemes operating under different principles with ill-defined ES or under inefficient provision levels (Muradian et al. 2010). The revised Wunder's definition defines PES as "(1) voluntary transactions (2) between service users (3) and service providers (4) that are conditional on agreed rules of natural resource management (5) for generating offsite services" (Wunder 2015). These transactions are labelled as "Coasean PES" (Coase 1960; Pagiola and Platais 2007) or "private PES" (Wunder 2005). Those PES which satisfy most but not all of Wunder's criteria are generally called "quasi-PES" or "PES-like" (Wunder 2008) and usually come with government intervention that is mostly characterized by subsidies (Vatn 2010; Sattler and Matzdorf 2013).

In reality, few real-world schemes meet all five of Wunder's definition criteria (i.e., voluntariness, clarity in defining ES, conditionality), while the number of PES-like schemes is much larger (Wunder 2008; Kosoy and Corbera 2010; Sattler et al. 2013). The latter can be defined more broadly as a transfer (monetary or nonmonetary) of resources between social actors, which aims to create incentives to align individual and/or collective land-use decisions with social interest in the management of natural resources (Muradian et al. 2010). They include, for example, those programs financially supported by public governments that "buy" ES on behalf of their taxpayers who, strictly speaking, cannot decide whether or not to participate in the program (Russi et al. 2011). This wide range of real-world existing or potential schemes focuses on influencing the ES providers through monetary or in-kind incentives. In most cases, the payment amount is not based on a monetary evaluation of the ES value but rather on lengthy negotiations among providers and users, informed by the opportunity cost associated with the required land-use practices (Russi et al. 2011).

PES and PES-like initiatives are now being promoted around the globe to incentivize the sustainable management of numerous ES (e.g., Kosoy et al. 2007; Pagiola et al. 2007; Muñoz-Piña et al. 2008; Tallis et al. 2008; Wunder and Albán 2008; Wunder 2008; Stanton et al. 2010; Brouwer et al. 2011). However, there are few PES examples in the European Union. Although PES/PES-like schemes have been widely adopted at the local, national, and international levels to reflect FES value in decisionmaking processes (Kemkes et al. 2010), a limited number of studies have so far examined the role of PES in improving the mitigation strategies of forests in Europe, especially considering CSF.

12.2.1   PES and Multiple ES

Designing and implementing policy and market tools to support synergies between ecosystem and ES relationships and reduce trade-offs among them is particularly crucial in forest ecosystems that represent important carbon sinks on the global scale. This makes them key ecosystems relevant for the regulation of ES that contribute to mitigating climate change via carbon uptake from the atmosphere and precipitation reduction of solar heating (Bonan 2008). As also emphasized by the EU Biodiversity Strategy for 2030, EU Forest Strategy and proposed new EU Forest Strategy of the European Green Deal, sustainable forest management and multiple

FES provisions are considered a long-term strategy to mitigate climate change impacts. In CSF definition, the multiplicity of ES provision is an important factor for supporting smart decisionmaking in forestry (Bowditch et al. 2020). Regulatory services, such as water regulation or erosion control, can to some extent buffer the natural and social system against the impacts of climate change (such as floods or droughts). Provisioning services (e.g., food and fiber) can provide an alternative source of food and income in the case of extreme events. Cultural services of forest ecosystems contribute to health and well-being and thus contribute to the social system's adaptive capacity (Millennium Ecosystem Assessment 2005; Locatelli et al. 2008).

Policymakers and forest managers need to make decisions to manage forest ecosystems sustainably, even while a gap remains in our understanding of the ecosystem-ES relationships (Mach et al. 2015; Van Wensem et al. 2016). They should seek to align the economic incentives with regulation to avoid environmentally irresponsible behavior by economic players. The forest environment is heavily exploited for the goods and services it provides and also faces global pressures such as climate change. This adds uncertainty to sustainable management as it is unclear how these pressures affect ecosystem resilience and adaptive capacity to climate change or the services provided by those vulnerable ecosystems (Knights et al. 2013; Mach et al. 2015).

Interactions between ES have been the subject of an increasing number of studies because their understanding is essential to the design and implementation of public policies, management strategies, and PES schemes that can foster the sustainability of ecosystem service provision (Demestihas et al. 2019; Mouchet et al. 2014). However, the commoditization of ES via PES schemes usually entails the identification and commercialization of single services (Muradian and Rival 2012). Such a focus on single services might be problematic due to the existence of trade-offs that may induce changes in the structure and functioning of the resource base, which may in turn jeopardize the supply of other services and even the service whose provision is being promoted (Corbera and Brown 2010; Kosoy and Corbera 2010; Muradian and Rival 2012).

The enhanced provision of a single service can also lead to disadvantages for users at the local scale, as trade-offs are often involved in enhancing different ES for different scale and purposes (Chan et al. 2006). According to Lee and Lautenbach (2016), trade-offs are mostly dominant between regulating and provisioning services. An often-cited example for such trade-offs is the establishment of trees for global benefit (fast-growing tree species, such as eucalyptus for carbon sequestration), which not only might replace more biodiversity-rich areas but could also have implications for the water table and thus increase system sensitivity to drought (van de Sand 2012). Gomez-Baggethun et al. (2011) reported significant cork tree forest destruction and consequent biodiversity loss due to eucalyptus plantation in Doñana National Park, Spain. Such project interventions may affect the flow of provisioning services, as well as the stakeholders whose livelihoods are related to ecosystem production functions. Corbera and Brown (2010) found that in some carbon forestry payment schemes that they reviewed, access to grazing land was restricted, and degradation of soil and vegetation occurred. Thus, although some ES can be increased, others might be eroded, thereby potentially increasing rather than decreasing the vulnerability of those dependent on the services to climate variability and change.

Although markets for multiple services might involve considerable transaction costs, targeting multiple FES rather than single FES (e.g., biodiversity) reduces contradictions among providers and users and may positively affect the transformation from sectoral to ES governance. Targeting multiple ES via payment schemes is particularly relevant as there is a dominance of single provider and multiple users of ES. In many payment schemes, provisioning FES and cultural FES are addressed in combination with other FES rather than separately (ibid.). This can be supported by the research of Brnkaľáková et al. (2019), where they showed that a synergistic relationship was dominant between different regulating services and between different cultural services (positive relationship); yet the review of Lee and Lautenbach (2016) illustrated that the relationship between regulating and provisioning services was trade-off dominated. Comprehensive information is required for well-informed management and policy decisions that take account of ecosystem complexity and relationships among ES.

According to Matthies et al. (2016), the goal of bundling or stacking multiple ES within a single PES scheme is to reduce the risk of adverse intraservice trade-offs, which is done by incentivizing the co-provisioning ES. Moreover, such bundling could decrease marginal service provisioning costs to society per unit of service provided. They highlighted that the stacking of biodiversity conservation and climate change mitigation objectives is possible if appropriate care is taken to determine those management interventions that are complementary to achieving the correct balance between the equitable and aggregate achievement of desired outcomes (ibid.).

Demestihas et al. (2019) in their study analyzed the patterns of ES relationships in agroecosystems to address the challenge of supporting regulating ES while maintaining or enhancing provisioning services. Such an example can be used in promoting PES that focus on the provision of multiple ES. PES that go beyond the food production service may support the provision of multiple non-marketed services (such as soil structure and fertility, water quantity and quality, biological pest control, pollination, and climate regulation through carbon sequestration and greenhouse gas (GHG) mitigation). Despite this recognition, non-marketed ES have been undervalued in policy, which has led to biodiversity loss and ecosystem degradation (TEEB 2009).

When designing PES schemes, it is important to focus on the temporal variation of trade-offs and synergies among multiple ES throughout and after the PES period. Although some PES initially focus on single ES, over time they may also support the provision of other related ES. Even though PES can support several ES, most are designed for each ES separately. However, analyses of the effects of PES schemes should focus on multiple services to allow the capture of trade-offs among them and explore system complexity (Lester et al. 2013; Mach et al. 2015; Cavanagh et al. 2016). Consequently, such analyses will help to quantify and forecast changes to ES under different PES and management measures (Daily et al. 2009; Mach et al. 2015). According to Broszeit et al. (2019), such an approach would ideally help to understand if or why PES or other types of policy interventions that aim to halt biodiversity loss and the decline of ES have failed or succeeded (Carpenter et al. 2009).

In order to achieve successful CSF, implementation of PES requires a balancing act between wood production, biodiversity, and other important ES. According to Verkerk et al. 2020, the optimal balance will vary from country to country and region to region, depending on the socio-ecological and technological framework, climate change impacts, but also cultural aspects. This approach demonstrates a step toward customizing PES within CSF to fit individuals and regional differences and local priorities and capacities to tackle climate change issues.

12.2.2   PES and Adaptive Capacity

European forests currently face changes in ecosystem functionality and resilience due to climate change (e.g., increased severity of disturbances such as hazard risk, storms with consequent insect attacks, fires, drought, etc.) (Schelhaas et al. 2015; Kulakowski et al. 2017). Apart from reducing GHG emissions, Millar et al. (2007) proposed forest management adaptation via behavior change, promoting the resilience of forests and increasing the adaptive capacity of forest users to climate change. Such transformations in turn influence FES availability (Thom and Seidl 2016) and finally affect sustainable development in forest areas (e.g., Beniston 2003).

As stated by CSF definition, the key objective of CSF is to adapt forests and forest management to the gradual changing of climate. PES not only can contribute to increasing forest resilience and adaptive capacity through the provision of multiple ES but could also strengthen the adaptive capacity of users and providers through the way in which PES is implemented and designed. In line with common pool resource theory and numerous empirical evidence (Ostrom 2010), PES can increase the capacity of local governance regimes, strengthen local economies, and improve the social capital that are essential features of adaptive capacity for sustainable long-term FES provision. van de Sand (2012), in her review, has shown that PES can potentially increase the adaptive capacity of involved actors via the establishment of institutional structures, increased access to and generation of financial resources, generation of knowledge between ecosystems and land-use practices, and supporting conflict resolution.

However, the long-term nature of many PES contracts may prevent ES providers from implementing certain adaptation strategies that would involve CSF, changing land-use practices via crop diversification, or leaving the agriculture or forestry sector altogether. According to Chobotova (2013), despite the mixed evidence of PES's role in the long-term behavioral changes of users and providers toward sustainability, significant interest in PES can nevertheless be explained by schemes that encourage greater transparency and more flexibility in allowing actors to reach a certain goal. Moreover, she reported that PES actors have problems with the long-term commitment inherent in PES, due in particular to unclear property rights and rules in use (Schlager and Ostrom 1992), multiple ownership structure, or land rental contracts that are often subject to change and speculation. Most landowners usually rent land to farmers for only shorter periods, with rental periods of less than 5 years, making land users ineligible for PES (ibid.). Short rental periods are generally insufficient for ES provision (e.g., increased biodiversity by restoring species-rich communities), which may demotivate land users from participating in such schemes or implementing certain adaptation strategies. Therefore, PES flexibility is an important factor for implementation.

In some cases, it may be that a tree species is not adapted to changing climatic conditions and has to be replaced by other tree species. In Slovakia, for example, foresters in mountain areas are replacing nonnative coniferous trees by mixed stands of species that are likely to be well adapted to emerging environmental conditions and changing climate (Brnkaláková et al. 2019). CSF in this case suggests selling (in a sustainable manner) timber in order to finance the conversion of disturbance-vulnerable forests to a more resilient new forest type (Yousefpour et al. 2018). If ES providers see the need for and are willing to undertake adaptation measures to climate change yet lack the appropriate means for implementation, there is thus an opportunity to tailor PES compensation in such a way that it provides direct incentives for adaptation measures (van de Sand 2012). PES compensation schemes for the provision of ES and as a direct incentive for adaptation and mitigation measures can often be made in kind, in addition to or instead of using cash payments (e.g., Wunder 2008), which is a strong sign of intrinsic motivation (Muradian et al. 2010). Brnkaľáková et al. (2019) reported that foresters are willing to join PES schemes if they are targeted at the forest machinery important for CSF, as such investment is far beyond their own budget.

Investing in climate-smart practices can result in short-term income losses (Haile et al. 2019), which often inhibits forest actors from investing in adaptation measures, which could generate long-term economic and environmental returns (Neufeldt et al. 2011; Ndah et al. 2014). Also according to Lipper et al. (2011), actors value short-run costs much stronger than longer-term benefits. In both cases, PES can help cover short-run costs and contribute to continuous payoffs, thereby increasing profitability and lowering investment risk (Engel and Muller 2016). Therefore, changing the timing of payments in a PES program may trigger a change in actors' behavior in favor of CSF, which has economic and ecosystem benefits.

If a PES program compensates foresters for the investment costs associated with adopting CSF in initial years when cash outflows characterize the investment, foresters are willing to engage in environmentally conscious practices, and there is a high possibility of large-scale adoption of the innovation across Europe. Within the Iceland PES scheme, each farm's afforestation grant covers 97% of establishment costs, including fencing, trails, site preparation, planting, and precommercial thinning (Brynleifsdóttir 2017; Icelandic Forest Service 2017).

van de Sand (2012) mentioned that payment could take the form of drought-resistant seeds as an adaptation measure against drought or more generally climate variability. Haile et al. (2019), in the case of climate-smart agroforestry, highlighted that farmers are willing to receive a low amount if the mode of payment is food rather than cash; they also reported that the failure of output markets could explain such a preference. In areas characterized by vulnerability to climate shocks and associated severe food shortages, in the absence of well-functioning markets and where cash transfers are vulnerable to price increases of food items, farmers rationally choose the end goods (food) rather than the means (cash) (ibid.).

12.2.3 Long-Term Sustainability

PES have been heralded as an effective strategy to increase tree cover in forest or agricultural landscapes and thus contribute as a climate change mitigation measure, but their efficacy beyond the payment period has rarely been evaluated. The permanence of activities (the extent to which the induced change is permanent after the finalization of funding) in the event of PES reduction is debatable. The temporal limitation of PES schemes has important implications as to whether payments foster an environmental attitude or, in the words of Swart (2003), an attitude of "no pay, no care." Ultimately, the sustainability of environmental outcomes following short-term PES programs needs to be tested rather than assumed (Calle 2020). In recent years, several studies have shown that the permanence of PES interventions is highly context dependent (Prokofieva and Gorriz 2013; Calle 2020).

Whereas the rational choice approach of most PES programs states that permanent outcomes require ongoing payments, forestry PES schemes anticipate that because changing forestry practices would soon become profitable, landowners (or land users) would, therefore, permanently adopt such practices (Calle 2020). According to Prokofieva and Gorriz (2013), permanence beyond an agreement period is not secured especially in the absence of additional financial resources. However, when landowners' private interest is strong or activities are aligned with their personal values, it is expected that high additional costs are not imposed on landowners and any costs complement their activities. This is consistent with the predictions of self-determination theory, according to which behavioral changes outlast the withdrawal of external incentives only when intrinsic motivation is strong enough (e.g., Deci 1971; Green-Demers et al. 1997; Deci and Ryan 2002). On the other hand, according to Frey and Oberholzer-Gee (1997), the use of price incentives needs to be reconsidered in all areas where intrinsic motivation can empirically be shown as important. They suggest that in policy areas where intrinsic motivation does not exist or has already been crowded out, the relative price effect, and thus the use of compensation, is a promising strategy (ibid.), even if payments are short term.

Ezzine-de-Blas et al. (2019) stressed that the willingness to maintain behavior after incentives are discontinued, or "crowding-in," results when satisfaction with the new practices gradually strengthens intrinsic motivations, eventually replacing the external incentive as the main driver of behavioral change. So, the crowding-in effect means that people conform and entrain with a new norm, which may also become a new moral standard (Vatn 2010). Monetary incentives can support long-term sustainability and climate change adaptation, if implemented as a local social norm, and aligned with cultural and interpersonal values. Kolinjivadi et al. (2015) confirm the link between motivation crowding-in and collective PES, if the latter aligns with social norms and the social capital is strong. In a context where social capital is strong (i.e., reciprocity norms exist, people trust each other, and leaders are respected), collective PES can increase intrinsic motivations (Andersson et al. 2018; Bottazzi et al. 2018). Prokofieva and Gorriz (2013) in their review of three European PES forest initiatives claimed that success and durability rely on the strong self-interest of involved forest owners, shared values and priorities, social capital that permits strong local networks to form, positive environmental attitudes, and local networks.

Moreover, crowding-in has been mostly observed for practices that provide long-term financial benefits. On the other hand, motivations for involvement in PES schemes do not relate to purely monetary logic. It has also been seen for practices that are easier to maintain, such as delivery of other services (e.g., water regulation) or in-kind benefits (e.g., soil fertility), or difficult to reverse (e.g., forest reclearing) (Kissinger et al. 2013; Swann 2016; Dayer et al. 2018), such as improved land tenure security and community organization, increased recipient knowledge about the importance of forest conservation (Kosoy et al. 2007; Wunder 2015), and information about (and experience with) the risk and negative effects of climate change.

PES programs can also be considered in the context of the broader effects of climate change. Forest cover or woody vegetation cover can increase slightly as a result of changed climate conditions and consequent rural migration, land abandonment in areas that became too dry for agriculture, and reforestation programs. Where progress in changing forest landscape management is achieved during the short-term payment period and has been retained (i.e., permanence), it suggests that forest owners can understand the benefits of maintaining trees even without payments (crowding-in) (Calle 2020). Whether forester owners opt to maintain reforested areas at the end of contract periods remains to be seen. These longer-term landowner decisions will ultimately determine whether the program is an effective strategy in addressing climate change issues. Assuming that CSF practices are – by definition – more profitable for the farmer in the longer run, temporary payments should be sufficient to induce a permanent change in forestry practices. Also according to Lipper et al. (2011), actors value short-run costs much stronger than longer-term benefits. In both cases, PES can help cover short-run costs and contribute to continuous payoffs, thereby increasing profitability and lowering investment risk (Engel and Muller 2016).

Our literature review confirms that with well-designed PES, some forest landowners will not only *adopt* but also and more importantly *maintain* climate-smart practices. CSF can transform production forests into heterogeneous landscapes that support higher productivity, biodiversity conservation, carbon sequestration, and the flow of ecosystem services and are, therefore, an important component of climate change adaptation and mitigation measures. Since foresters' participation in adaptation and mitigation to climate initiatives remains marginal at best, even short-term PES can be a useful policy tool to facilitate the widespread adoption of CSF.

12.3   Discussion and Conclusions

To date, countries have been more specific in setting ambitious restoration or refor-estation targets than in anticipating or implementing strategies to ensure the longev-ity of their restoration and reforestation outcomes (Calle 2020). The greater the synergies and the fewer the trade-offs between climate policy and other societal, forest-related goals, the more likely climate objectives will be effectively imple-mented and maintained in practice (Nabuurs et al. 2017). Therefore, trade-offs between design features and the quest for particular objectives in PES implementa-tion reflect how specific paradigms or discourses about deforestation, climate change, poverty, or the role of incentives in motivating specific human behaviors need to be understood and thus mainstreamed in such implementation (Moros et al. 2019).

Where payments have been made to individual farmers, a critical issue that emerged from the review is that in many cases, payments were not based on the opportunity costs of ES providers and often tended to diminish over time, thus risk-ing the long-term sustainability of the scheme. In some cases, however, the longev-ity of strategic remnant or restored ecosystems can only be ensured via the strict enforcement of conservation areas or long-term PES schemes.

On the other hand, policies and programs are not commonly funded in perpetuity, and they suffer budgetary and implementation adjustments along the way (Moros et al. 2019). In many of the studies we have reviewed, short-term contracts are found to be essential attributes that positively influence actors' decisions to take up a contractual arrangement. As foresters or other landowners seek more climate-resilient forestry systems, and representatives grapple to meet ambitious national and global climate policy targets, opportunities to align foresters' needs and climate policy goals are emerging. From the policy perspective, short-term PES interventions that can trigger positive and lasting change are especially promising (Calle 2020).

Knowledge continues to be lacking about specific impacts of climate change or the technology and strategies to implement certain adaptation measures (such as CSF) (van de Sand 2012). More needs to be known if more knowledge and informa-tion about climate change can influence the willingness of various actors to adopt climate strategies and measures and the extent to which climate change consider-ation can influence the willingness of ES users to pay for ES provision. Moreover, future research is needed about how specific PES design can contribute to the increased adaptive capacity of different forest actors and the options which exist to provide incentives for adaptation strategies through specific forest PES and alterna-tive governance regimes.

It is important to stress that the role of PES alone in reducing CO_2 emissions from land-use activities with the aim of reaching to carbon neutrality by 2050 is limited. This is due to the potential implementation scale of PES, particularly compared with other initiatives (such as REDD+) or carbon pricing schemes which are almost nonexistent in Europe. We acknowledge that the combination of government regimens and several policy instruments to preserve forests and avoid deforestation should be urgently pursued in the coming years in order to reduce the share of land-use change emissions in Europe, which currently account for 24% of Europe total CO_2 emissions.

Acknowledgement This work was supported by COST (European Cooperation in Science and Technology) - Climate-Smart Forestry in Mountain Regions (CLIMO) (CA15226) and the Scientific Grant Agency of the Ministry of Education, Science, Research and Sport of the Slovak Republic and the Slovak Academy of Sciences (Management of global change in vulnerable areas - VEGA n. 2/0013/17) and (The role of ecosystem services in support of landscape conservation under the global change - VEGA n. 2/0170/21).

References

Amsworth P, Chan K, Daily G et al (2007) Ecosystem-service science and the way forward for conservation. Conserv Biol 21(6):1383–1384

Andersson KP, Cook NJ, Grillos T et al (2018) Experimental evidence on payments for forest commons conservation. Nat Sustain 1:128–135

Bartczak A, Metelska-Szaniawska K (2015) Should we pay, and to whom, for biodiversity enhancement in private forests? An empirical study of attitudes towards payments for forest ecosystem services in Poland. Land Use Policy 48:261–269. https://doi.org/10.1016/j.landusepol.2015.05.027

Beniston M (2003) Climatic change in mountain regions: a review of possible impacts. Clim Chang 59. https://doi.org/10.1023/A:1024458411589f

Blundo Canto G, Bax V, Quintero M et al (2018) The different dimensions of livelihood impacts of Payments for Environmentals Services (PES) schemes: a systematic review. Ecol Econ 149:160–183. https://doi.org/10.1016/j.ecolecon.2018.03.011

Bonan GB (2008) Forests and climate change: Forcings, feedbacks, and the climate benefits of forests. Science 80(320):1444–1449. https://doi.org/10.1126/science.1155121

Bottazzi P, Wiik E, Crespo D et al (2018) Payment for environmental "self- service": exploring the links between farmers' motivation and additionality in a conservation incentive programme in the Bolivian Andes. Ecol Econ 150:11–23

Bowditch E, Santopuoli G, Binder F et al (2020) What is climate-smart forestry? A definition from a multinational collaborative process focused on mountain regions of Europe. Ecosyst Serv 43. https://doi.org/10.1016/j.ecoser.2020.101113

Brnkaľáková S, Udovc A, Kluvánková T et al (2019) Carbon smart forestry in forest commons in Slovakia and Slovenia. Report 5.4 Analytical-Informational Case Studies (Type B/C) led by CETIP and IFE SAS. Social Innovation in Marginalised Rural Areas (SIMRA), p 23

Broszeit S, Beaumont N, Hooper T et al (2019) Developing conceptual models that link multiple ecosystem services to ecological research to aid management and policy, the UK marine example. Mar Pollut Bull 141:236–243. https://doi.org/10.1016/j.marpolbul.2019.02.051

Brouwer R, Tesfaye A, Pauw P (2011) Meta-analysis of institutional-economic factors explaining the environmental performance of payments for watershed services. Environ Conserv 38(4):380–392

Brownson K, Guinessey E, Carranza M et al (2019) Community-Based Payments for Ecosystem Services (CB-PES): implications of community involvement for program outcomes. Ecosyst Serv 39. https://doi.org/10.1016/j.ecoser.2019.100974

Brynleifsdóttir JS (2017) Affecting afforestation. Pan Eur Netw Gov 24:1–2

Bulte EH, Lipper L, Stringer R et al (2008) Payments for ecosystem services and poverty reduction: concepts, issues, and empirical perspectives. Environ Dev Econ 13:245–254. https://doi.org/10.1017/S1355770X08004348

Calle A (2020) Can short-term payments for ecosystem services deliver long-term tree cover change? Ecosystem Services 42(C)

Carpenter SR, Mooney HA, Agard J et al (2009) Science for managing ecosystem services: beyond the millennium ecosystem assessment. Proc Natl Acad Sci 106:1305–1312

Cavanagh RD, Broszeit S, Pilling G et al (2016) Valuing biodiversity and ecosystem services – a useful way to manage and conserve marine resources? Proc R Soc B Biol Sci 283

Chan KMA, Shaw MR, Cameron DR et al (2006) Conservation planning for ecosystem services. PLoS Biology 4(11):379. https://doi.org/10.1371/journal.pbio.0040379

Chobotová V (2013) The role of market-based instruments for biodiversity conservation in Central and Eastern Europe. Ecol Econ 95:41–50

Coase RRH (1960) The problem of social cost. J. Law Econ. 3, 1–44. https://doi.org/10.1086/466560

Corbera E, Brown K (2010) Offsetting benefits? Analysing access to forest carbon. Environ Plan A 42(7):1739–1761

Cowx I, Portocarrero-Aya M (2011) Paradigm shifts in fish conservation: moving to the ecosystem services concept. J Fish Biol 79:1663–1680

Daily GC, Polasky S, Goldstein J et al (2009) Ecosystem services in decision making: time to deliver. Front Ecol Environ 7:21–28

Dayer AA, Lutter SH, Sesser K et al (2018) Private landowner conservation behavior following participation in voluntary incentive programs: recommendations to facilitate behavioral persistence. Conserv Lett 11:1–11

Deci EL (1971) Effects of externally mediated rewards on intrinsic motivation. J Pers Soc Psychol 18:105–115

Deci EL, Ryan RM (Eds.) (2002) Handbook of self-determination research. University of Rochester Press

Demestihas C, Plénet D, Génard M et al (2019) A simulation study of synergies and tradeoffs between multiple ecosystem services in apple orchards. J Environ Manag 236:1–16. https://doi.org/10.1016/j.jenvman.2019.01.073

Engel S, Müller A (2016) Payments for environmental services to promote "climate-smart agriculture"? Potential and challenges. Agric Econ 47:173–184. https://doi.org/10.1111/agec.12307

Eysteinsson T (2017) Forestry in treeless land. Icelandic Forest Service

Ezzine de Blas D, Corbera E, Lapeyre R (2019) Payments for environmental services and motivation crowding: towards a conceptual framework. Ecol Econ 156:434–443

Farley J, Costanza R (2010) Payments for ecosystem services: from local to global. Ecol Econ 69:2060–2068

Frey B, Oberholzer-Gee F (1997) The cost of price incentives: an empirical analysis of motivation crowding out. Am Econ Rev 87(4):746–755

Gómez-Baggethun E, de Groot R, Lomas PL et al (2010) The history of ecosystem services in economic theory and practice: from early notions to markets and payment schemes. Ecol Econ 69:1209–1218. https://doi.org/10.1016/j.ecolecon.2009.11.007

Gómez-Baggethun E, Alcorlo P, Montes C (2011) Ecosystem services associated with a mosaic of alternative states in a Mediterranean wetland: case study of the Doñana Marsh (Southwest Spain). Hydrol Sci J 56(22):1374–1387

Green-Demers I, Pelletier LG, Ménard S (1997) The impact of behavioural difficulty on the saliency of the association between self-determined motivation and environmental behaviours. Can J Behav Sci 29(3):157–166

Haile KK, Tirivayi N, Tesfaye W (2019) Farmers' willingness to accept payments for ecosystem services on agricultural land: the case of climate-smart agroforestry in Ethiopia. Ecosyst Serv 39. https://doi.org/10.1016/j.ecoser.2019.100964

Kemkes RJ, Farley J, Koliba CJ (2010) Determining when payments are an effective policy approach to ecosystem service provision. Ecol Econ 69:2069–2074. https://doi.org/10.1016/j.ecolecon.2009.11.032

Kissinger G, Patterson C, Neufeldt H (2013) Payments for ecosystem services schemes: project-level insights on benefits for ecosystems and the rural poor. ICRAF Working Paper No 172. World Agroforestry Centre, Nairobi

Kluvankova T, Brnkalakova S, Gezik V et al (2019) Ecosystem services as commons? In: Hudson B, Rosenbloom J, Cole DH (eds) Routledge handbook of the study of the commons. Rutledge, New York, pp 208–219

Knights AM, Koss RS, Robinson LA (2013) Identifying common pressure pathways from a complex network of human activities to support ecosystem-based management. Ecol Appl 23:755–765

Kolinjivadi V, Gamboa G, Adamowski J et al (2015) Capabilities as justice: analysing the acceptability of payments for ecosystem services (PES) through' social multi-criteria evaluation. Ecol Econ 118:99–113

Kosoy N, Corbera E (2010) Payments for ecosystem services as commodity fetishism. Ecol Econ 69(6):1228–1236. https://doi.org/10.1016/j.ecolecon.2009.11.002

Kosoy N, Martinez-Tuna M, Muradian R et al (2007) Payments for environmental services in watersheds: insights from a comparative study of three cases in Central America. Ecol Econ 61:446–455. https://doi.org/10.1016/j.ecolecon.2006.03.016

Kulakowski D, Seidl R, Holeksa J et al (2017) A walk on the wild side: disturbance dynamics and the conservation and management of European mountain forest ecosystems. For Ecol Manag 388:120–131. https://doi.org/10.1016/j.foreco.2016.07.037

Lee H, Lautenbach S (2016) A quantitative review of relationships between ecosystem services. Ecol Indic 66:340–351

Lester SE, Costello C, Halpern BS, Gaine SD et al (2013) Evaluating tradeoffs among ecosystem services to inform marine spatial planning. Mar Policy 38:80–89

Lexer MJ, Bugmann H (2017) Mountain forest management in a changing world. Eur J For Res 1:981–982. https://doi.org/10.1007/s10342-017-1082-z

Lindner M, Fitzgerald JB, Zimmermann NE et al (2014) Climate change and European forests: what do we know, what are the uncertainties, and what are the implications for forest management? J Environ Manag 146:69–83. https://doi.org/10.1016/j.jenvman.2014.07.030

Lipper S, Neves B, Wilkes A et al (2011) Climate change mitigation finance for smallholder agriculture. A Guide Book to Harvesting Soil Carbon Sequestration Benefits. FAO, Rome

Locatelli B, Kanninen M, Brockhaus M et al (2008) Facing an uncertain future: how forests and people can adapt to climate change. Forest Perspectives No. 5, Center for International Forestry Research (CIFOR), Bogor, Indonesia

MA (2005) Millennium Ecosystem Assessment: ecosystems and human well-being. https://doi.org/10.1196/annals.1439.003

Mach ME, Martone RG, Chan KMA (2015) Human impacts and ecosystem services: insufficient research for trade-off evaluation. Ecosyst Serv 16:112–120

Matthies BD, Kalliokoski T, Ekholm T et al (2015) Risk, reward, and payments for ecosystem services: a portfolio approach to ecosystem services and forestland investment. Ecosyst Serv 16:1–12. https://doi.org/10.1016/j.ecoser.2015.08.006

Matthies BD, Kalliokoski T, Eyvindson K et al (2016) Nudging service providers and assessing service trade-offs to reduce the social inefficiencies of payments for ecosystem services schemes. Environ Sci Policy 55:228–237. https://doi.org/10.1016/j.envsci.2015.10.009

Millar CI, Stephenson NL, Stephens SL (2007) Climate change and forests of the future: managing in the face of uncertainty. Ecol Appl 17:2145–2151. https://doi.org/10.1890/06-1715.1

Moros L, Vélez MA, Corbera E (2019) Payments for ecosystem services and motivational crowding in Colombia's Amazon Piedmont. Ecol Econ 156:468–488. https://doi.org/10.1016/j.ecolecon.2017.11.032

Mouchet MA, Lamarque P, Martín-López B et al (2014) An interdisciplinary methodological guide for quantifying associations between ecosystem services. Glob Environ Change 28:298–308. https://doi.org/10.1016/j.gloenvcha.2014.07.012

Muñoz-Piña C, Guevara A, Torres JM et al (2008) Paying for the hydrological services of Mexico's forests: analysis, negotiations and results. Ecol Econ 65(4):725–736

Muradian RE, Gómez-Baggethun (2013) The institutional dimension of "Market- based Instruments" for governing ecosystem services. Soc Nat Resour 26(22):1113–1121

Muradian R, Rival L (2012) Between markets and hierarchies: the challenge of governing ecosystem services. Ecosyst Serv 1(1):93–100. https://doi.org/10.1016/j.ecoser.2012.07.009

Muradian R, Corbera E, Pascual U et al (2010) Reconciling theory and practice: an alternative conceptual framework for understanding payments for environmental services. Ecol Econ 69(6):1202–1208

Nabuurs GJ, Delacote P, Ellison D et al (2017) By 2050 the mitigation effects of EU forests could nearly double through climate smart forestry. Forests 8(12):484. https://doi.org/10.3390/f8120484

Ndah HT, Schuler J, Uthes S et al (2014) Adoption potential of conservation agriculture practices in sub-Saharan Africa: results from five case studies. Environ Manag 53:620–635. https://doi.org/10.1007/s00267-013-0215-5

Neufeldt H, Kristjanson P, Thorlakson T et al (2011) Making Climate-Smart Agriculture Work for the Poor World Agroforestry Center, Nairobi

Ostrom E (2010) Beyond markets and states: polycentric governance of complex economic systems. Am Econ Rev 100(3):641–672

Otto MI, Chobotova V (2013) Opportunities and constraints of adopting market governance in protected areas in Central and Eastern Europe. Int J Commons 7(1):34–57

Pagiola S, Platais G (2007) Payments for environmental services: from theory to practice. World Bank, Washington

Pagiola S, Ramírez E, Gobb J et al (2007) Paying for the environmental services of silvopastoral practices in Nicaragua. Ecol Econ 64(2):374–385

Primmer E, Varumo L, Krause T et al (2021) Mapping Europe's institutional landscape for forest ecosystem service provision, innovations and governance. Ecosyst Serv 47:11–15. https://doi.org/10.1016/j.ecoser.2020.101225

Prokofieva I, Gorriz E (2013) Institutional analysis of incentives for the provision of forest goods and services: an assessment of incentive schemes in catalonia (North-East Spain). For Policy Econ 37:104–114. https://doi.org/10.1016/j.forpol.2013.09.005

Randrianarison H, Ramiaramanana J, Wätzold F (2017) When to pay? Adjusting the timing of payments in PES design to the needs of poor land-users. Ecol Econ 138:168–177

Russi D, Corbera E, Puig-Ventosa I et al (2011) Payments for ecosystem services in Catalonia, Spain. A review of experience and potential applications. Span J Rural Dev 2:87–100

Sattler C, Matzdorf B (2013) PES in a nutshell: from definitions and origins to PES in practice-approaches, design process and innovative aspects. Ecosyst Serv 6:2–11. https://doi.org/10.1016/j.ecoser.2013.09.009

Sattler C, Trampnau S, Schomers S et al (2013) Multi-classification of payments for ecosystem services: how do classification characteristics relate to overall PES success? Ecosyst Serv 6:31–45. https://doi.org/10.1016/j.ecoser.2013.09.007

Schelhaas MJ, Nabuurs GJ, Hengeveld G et al (2015) Alternative forest management strategies to account for climate change- induced productivity and species suitability changes in Europe. Reg Environ Chang 15:1581–1594. https://doi.org/10.1007/s10113-015-0788-z

Schlager E, Ostrom E (1992) Property-rights regimes and natural resources: a conceptual analysis. Land Econ 68(3):249. https://doi.org/10.2307/3146375

Skroch M, Lopez-Hoffman L (2009) Saving nature under the big tent of ecosystem services: a response to Adams and Redford. Conserv Biol 24(1):325–327

Stanton T, Echavarria E, Hamilton K et al (2010) State of watershed payments: an emerging marketplace. Ecosystem Marketplace, Washington, DC

Suyanto S, Khususiyah N, Leimona B (2007) Poverty and environmental services: case study in Way Besai watershed, Lampung Province, Indonesia. Ecol Soc 12(2):13

Swann E (2016) What factors influence the effectiveness of financial incentives on long- term natural resource management practice change? Evid Base:1–32

Swart J (2003) Will direct payments help biodiversity? Science 299:1981

Tallis H, Kareiva P, Marvier M et al (2008) An ecosystem services frame- work to support both practical conservation and economic development. Proc Natl Acad Sci 105(28):9457–9464

TEEB (2009) The economics of ecosystems and biodiversity

Thom D, Seidl R (2016) Natural disturbance impacts on ecosystem services and biodiversity in temperate and boreal forests. Biol Rev Camb Philos Soc 91:760–781. https://doi.org/10.1111/brv.12193

Tognetti R (2016) Climate-smart forestry in mountain regions. Impact 2017(3):29–31. https://doi.org/10.21820/23987073.2017.3.29

van de Sand I (2012) Payments for Ecosystem Services in the Context of Adaptation to Climate. Ecol. Soc. 17.

Van Wensem J, Calow P, Dollacker A et al (2016) Identifying and assessing the application of ecosystem services approaches in environmental policies and decision making. Integr Environ Assess Manag 9999:1–10

Vatn A (2010) An institutional analysis of payments for environmental services, Ecological Economics, 69, (6), 1245–1252

Verkerk PJ, Costanza R, Hetemäki L et al (2020) Climate-smart forestry: the missing link. Forest Policy Econ 115. https://doi.org/10.1016/j.forpol.2020.102164

Williamson O (1991) Comparative economic organization: the analysis of discrete structural alternatives. Adm Sci Q 36(2):269–296

Wunder S (2005) Payments for environmental services: some nuts and bolts. CIFOR Occas Pap 42(24). https://doi.org/10.1111/j.1523-1739.2006.00559.x

Wunder S (2008) Necessary conditions for ecosystem service payments, in: Economics and Conservation in the Tropics: A Strategic Dialogue. p. 11. https://doi.org/10.1007/BF01190840

Wunder S (2015) Revisiting the concept of payments for environmental services. Ecol Econ 117:234–243. https://doi.org/10.1016/j.ecolecon.2014.08.016

Wunder S, Alban M (2008) Decentralized payments for environmental services: the cases of Pimampiro and PROFAFOR in Ecuador. Ecol Econ 65:685–698

Yousefpour R, Augustynczik ALD, Reyer CPO et al (2018) Realizing mitigation efficiency of European commercial forests by climate smart forestry. Sci Rep 8(1):345. https://doi.org/10.1038/s41598-017-18778-w

Chapter 13
Assessing the Economic Impacts of Climate Change on Mountain Forests: A Literature Review

Giorgia Bottaro, Paola Gatto, and Davide Pettenella

Abstract The effects of climate change are increasingly more visible on natural ecosystems. Being mountain forest ecosystems among the most vulnerable and the most affected, they appear to be, at the same time, the most suitable for the assessment of climate change effects on ecosystem services. Assuming this, we review the literature on the economic assessment of climate change impacts on European mountain forests. Initially, the trends in the provision of mountain forest ecosystem services are discussed. We, then, considered the effects on forest structure and tree physiology, these two being strictly associated with the capability of the ecosystem to provide ecosystem services. The results have been grouped into a table that displays the trend, the quality and the quantity of the information found. Subsequently, the main methods that can be employed to assess the economic value of the different ecosystem services have been described. For each method, some implementation examples have been introduced to better understand its functioning. Concluding, the main gaps still existing in literature concerning the effects of climate change on ecosystem services provided by mountain forests have been highlighted. Finally, some more considerations about the existing methods for the economic valuation of ecosystem services have been done.

Abbreviations

CICES   Common International Classification of Ecosystem Services
CM   Choice Modelling
CSG   Cost of Substitute Goods
CV   Contingent Valuation
DE   Defensive Expenditures
DIC   Damage and Insurance Costs

G. Bottaro (✉) · P. Gatto · D. Pettenella
Land Environment Agriculture and Forestry Department (TeSAF), University of Padua, Padua, Italy
e-mail: giorgia.bottaro.2@phd.unipd.it; paola.gatto@unipd.it; davide.pettenella@unipd.it

© The Author(s) 2022
R. Tognetti et al. (eds.), *Climate-Smart Forestry in Mountain Regions*, Managing Forest Ecosystems 40, https://doi.org/10.1007/978-3-030-80767-2_13

453

ES   Ecosystem Services
HP   Hedonic Pricing
OC   Opportunity Cost
PF   Production Function
RC   Replacement Cost
TC   Travel Cost Method
TEV   Total Economic Value

13.1 Introduction

Climate change is one of the main drivers of changes in mountain ecosystems and in their capability to provide related services. This is due to their higher vulnerability, compared to other terrestrial ecosystems, to the changes in temperature and precipitation (Beniston 2003). The upper shift of species and consequently their adaptation to changes is limited by the long time span of trees that cannot quickly react to the changes and the limitation of their altitudinal range (Lindner et al. 2010). For these reasons, forests located in mountain areas are among the most appropriate ecosystems for climate change detection (Ding et al. 2016). Moreover, ecosystem services (ES) provided by mountain regions support a large number of components essential for human health and well-being (water, quality of food products, biomass, flood prevention, tourism and recreation, etc.) (Briner et al. 2013).

It is also for this reason that forest ecosystems, in Europe, cover an important role in relation to climate change mitigation and connected strategies. To understand the role of forests in climate change mitigation, see Chap. 15 of this book (Vizzarri et al. 2021).

According to the new version of the Common International Classification of Ecosystem Services (CICES, V5.1, http://cices.eu/resources/; Haines-Young and Potschin 2018), ES can be divided in three main categories: provisioning, regulating and cultural. Provisioning ES are those material services related to the goods provided by ecosystems. Regulative ecosystem services are those intangible services that have a regulative function (e.g. erosion control, water purification, climate control). Finally, cultural ES are those intangible services that comprise aesthetic, spiritual, educational, recreational and touristic value. It is important to quantify and to value the provision of ES through numerical and economic indicators to be able to monitor and compare them. In this way, it is consequently possible to address them in political and economic discourses. ES values can then be presented to stakeholders for understanding and defining trade-off and synergies between material goods and intangible services (Grêt-Regamey et al. 2013). Whether provisioning services are easier to assess and evaluate, most of the regulative and cultural services cannot be measured in market terms, since the methods to quantify them and to assess their values have only recently been developed.

In the valuation of ES, using the terminology of the cost-benefit analysis, a basic distinction should always be made between the financial analysis, which assesses the incurred expenditures and gained revenues, and the economic analysis, which is aimed to detect the real value of ES for the society, taking into account both the positive impacts (benefits) and the negative ones, which cannot be described by market prices (externalities). This kind of analysis tries to include the so-called total economic value (TEV) of ES that incorporates not only their market values (wood, non-wood forest products, water, etc.) but also all their intangible benefits and costs (Thorsen et al. 2014). While the literature connected with the financial analysis of mountain forest ES has a long tradition in terms of the role and importance of provisioning services (with a focus on wood products), the economic analysis of the total value of forest ES in mountain regions has been rarely carried out. Moreover, notwithstanding the mentioned socioeconomic characteristics of mountain forest ES (diversity and multiplicity of the ES, high perceived values, relevance and non-market benefits), there is no systemic analysis of the literature on their economic assessment. The aim of this chapter is to contribute to the existing knowledge through a literature review on the existing economic assessment of climate change effects on mountain forest ES with a special focus on European mountain regions.

13.2 European Mountain Forests

Mountains cover 29% of the EU territory, and, in this area, the most diffuse land use is forest covering 41% of the total mountain areas (Zisenis et al. 2010; Hartl et al. 2014). Global warming does not evenly affect Europe; its impact varies depending on a bioclimatic region allocated at different elevations and latitudes (Rogora et al. 2018). Furthermore, the impact of climate change on forest ecosystems also depends on the bioclimatic zone and on the resulting forest types (Lindner et al. 2007, 2010).

The main European bioclimatic zones are polar, boreal, temperate and Mediterranean. Because of the absence of forest in polar areas, the focus of our research has been on the other three regions. Within the temperate region, an important distinction has been made between the oceanic and the continental subareas. The bioclimatic map of the European countries is presented in Fig. 13.1. Besides, the alpine region was considered to better represent the characteristics of the main European mountain ranges: the Alps, the Pyrenees and the Carpathian (Lindner et al. 2010).

13.3 The Methodological Approach

To estimate the economic impact of climate change on the provision of European mountain forest ES, the chapter was organized in two parts. Firstly, the impact of climate change on forest ES was described. In order to meet this first objective, the analysis of the bibliography included also the effects of climate change on tree physiology and forest structure (Kurbanov et al. 2007), due to the fact that changes in forest structure are strictly related to the ecosystem capability in delivering ES

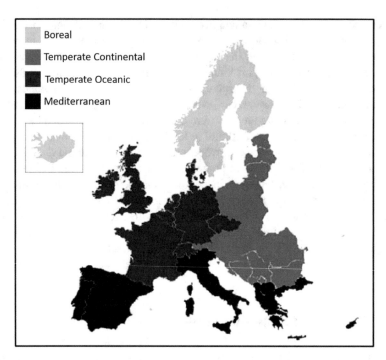

Fig. 13.1 European countries' classification divided by bioclimatic areas. (Modified from Lindner et al. 2010)

(Brockerhoff et al. 2017). This relation has been also analysed in Chap. 6 of this book (Pretzsch et al. 2021). In the analysis of the bibliography, papers predicting the effects of climate change through modelling have not been taken into consideration; this topic is explored in Chap. 7 of this book (Bosela et al. 2021). In the second part, the approaches employed for forest ES quantification and for their economic assessment were analysed. Also in this, case scientific literature was consulted.

The Scopus bibliographic database was used to analyse the literature and as a source of data for our research.

13.4 Climate Change Impacts on the Provision of Mountain Forest ES

In Table 13.1, the data found as a result of the review of the literature are summarized. The trends of the climate change impact on the provision of mountain forest ES are grouped in four categories: "increasing", "decreasing", "stable" and "mixed". Depending on the quantity and quality of the evidence and correlations among them, the data obtained were classified as follows: "well established", "established but incomplete" and "unresolved" using a similar approach used by IPBES (2018).

Table 13.1 Trends on the provision of forest ecosystem services affected by climate change

Forest Ecosystem Services	ES category	forest ES sub-category	Boreal	Temperate Oceanic	Temperate Continental	Mediterranean	Alpine
Provisioning services		Bioenergy production			▼	◆	▼
		Timber production	▲	◆	▼	◆	◆
		Non-wood forest products	▲		▼		▬
Regulating Services	Climate regulation	forests carbon stocks	▼	▼	▼	◆	◆
		Soil carbon stocks	▲			◆	▼
		Albedo	▼				
	Pest control		▼	▼	▼	▼	▼
	Natural hazard regulation	Forest fires/wildfires				▼	▼
		Erosion, avalanche, landslide					▼
		Flooding					▼
	Water quality regulation		▼	▼	▼	▼	▼
	Biodiversity		▲	▲	◆	▼	▼
Cultural Services	Recreation (fishing, nature enjoyment)	Hunting					
		NWFP picking					
	Tourism (skiing)						▼
	Aesthetic / heritage (landscape character, cultural landscapes)						▼

	TREND	CONFIDENCE LEVEL
▲	increasing	well established
▲		established but incomplete
▬	stable	established but incomplete
▼	decreasing	established but incomplete
▼		well established
◆	mixed	unresolved
NA		not enough data

Source: compiled by the authors based on: Kullman (1996), Beniston (2003), Meining et al. (2004),Jolly et al. (2005), De Wit et al. (2006), Friedrichs et al. (2009), Saccone et al. (2009), Tømmervik et al. (2009), Allen et al. (2010), Galiano et al. (2010), Lebourgeois et al. (2010), Scarascia-Mugnozza et al. (2010), Courbaud et al. (2010), Linares and Tiscar (2011), Forsius et al. (2013), Kozlov et al. (2013), Hartl-Meier et al. (2014), Horàk et al. (2014), Prietzel and Christophel (2014), Sarris et al. (2014), Fernàndez- Feurdean et al. (2016), Fernàndez-Martínez and Fleck (2016), Panayotov et al. (2016), Cudlín et al. (2017), Dupire et al. (2017), Fleischer et al. (2017), Vacek et al. (2017), Krupková et al. (2019), Rogora et al. (2018)

Provisioning Services Within this category of ES were considered the ones related to forest biomass (bioenergy and timber production) and non-wood forest products. Water provision is also generally considered dealing with provisioning services. Nevertheless, in this chapter, water issues were considered only in relation to water quality regulation dealing with regulating services.

Changes in net primary production (NPP), which influence timber provision, have different trends in diverse bioclimatic areas. In the Mediterranean region, tree growth increment is negatively affected principally by water scarcity (Linares et al. 2009; Scarascia-Mugnozza et al. 2010; Fyllas et al. 2017; Rogora et al. 2018). The opposite trend has been detected in the boreal region, where temperature tends to be the most limiting factor; in this region, climate change is, therefore, enhancing forest productivity, even if winter frost has a negative impact on it (Kullman 1996). In temperate regions, the trend is more heterogeneous with different impacts according to the local and environmental conditions, especially related to water and temperature trends (Kurbanov and Post 2002; Lindner et al. 2010; Loboda et al. 2017). In the north-western part of the temperate oceanic region, tree growth and increment is slightly higher because of temperature increase, a factor that significantly influences tree response in the area, while in more south-eastern and temperate continental regions, water scarcity negatively affects the radial growth dynamics (Friedrichs et al. 2009; Horak et al. 2014; Panayotov et al. 2016). Finally, considering alpine areas, they are characterized by a general increase in timber production (Rogora et al. 2018), with the presence of an inverse trend where soil moisture is not enough to support a high photosynthetic rate (Meining et al. 2004; Galiano et al. 2010).

Regulating Services Forests play an important role in climate regulation, being able to store CO_2 above the soil level. Moreover, tree canopies can modify the albedo of the land surface. For instance, in the boreal region, the expansion of forests is changing the capacity of forest ecosystems to mitigate climate changes, because forest expansion decreases the albedo of the area (Beniston 2003). Regarding carbon sequestration, the impact of climate change on this ES varies in different regions. In fact, being strictly related to tree growth, stand capacity of stocking carbon follows a pattern similar to stem radial increment. For instance, in temperate continental regions, the carbon uptake is negatively impacted by the higher temperature and lower precipitation, because of the reduction of tree photosynthetic rate (Horak et al. 2014). In alpine and Mediterranean areas, CO_2 absorption can follow different patterns: in some regions, the carbon uptake is enhanced by global warming, due to the longer growing season and the earlier melting of snow or due to the rise up of the timberline (Rogora et al. 2018). In some other regions, a negative impact of climate change on forest carbon stock is recorded due to the lower capacity of forest soils to store organic carbon mainly caused by the accelerated decomposition of soil organic matter (Prietzel and Christophel (2014)). In this second case, Mediterranean mountain forests are generally further limited in their carbon adsorption capability because of water stress (Scarascia-Mugnozza et al. 2010) or insect defoliation (Jacquet et al. 2012).

Another important regulating service provided by mountain forests is pest control. Several studies assessed the expansion of insects' range, winter survival and frequency of pest outbreaks (e.g. Battisti and Larsson 2015; Pureswaran et al. 2018). Generally, pests spread depending on the altitudinal gradient, even if latitudinal expansion seems to be prevalent (Battisti and Larsson 2015). In the Mediterranean region, pest control in mountain forest ecosystems is harder to manage in comparison with the other regions, due to vulnerability of trees caused by water scarcity (Scarascia-Mugnozza et al. 2010).

Biodiversity protection has been considered as a broad ES, more specifically described by lifecycle maintenance, habitat and gene pool protection. Considering these indicators, in the recent years, biodiversity protection has increased in most of Europe. The regions that experienced a decrease in species richness are the Mediterranean region and the alpine region (Pauli et al. 2012).

Climate change also affects the dynamics of disturbances, such as fire, landslide and wind, making forests more vulnerable and affecting their capability of natural hazards regulation. In their paper, Seidl et al. (2007) analysed the correlation between climate change and natural disturbances and argued there was a direct interrelation between them.

13.5   Economic Evaluation of Climate Change Damages in Mountain Forests

The effects of climate change on mountain forest capability to provide ES in the different European bioclimatic areas have been identified in the first part of this chapter. In order to understand which are the consequent economic impacts caused by these changes, it is necessary to estimate the value of the considered forest ES. Several methods have been developed to estimate ES value, and different frameworks have been designed to systematize and to classify them. Hereafter, the framework developed by Masiero et al. (2019) in their manual *Valuing Forest Ecosystem Services: A Training Manual for Planners and Project Developers* has been used as reference (Fig. 13.2).

In the next Sects. (13.5.1 and 13.5.2), the cases founded in literature that use the methods shown in Fig. 13.2 to assess ES value will be presented and described. In

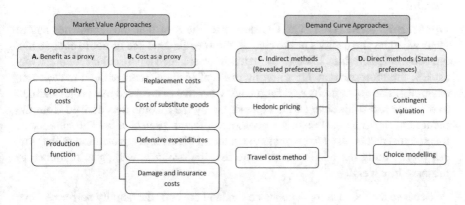

Fig. 13.2 Methods used to evaluate forest ecosystem services. (Masiero et al. 2019)

Sect. 13.5.1, the different ES provided by mountain forests are cross-checked with the methods used for their economic valuation with reference to the literature review.

The methods used for the assessment (financial and economic) of ES values can be divided into two main categories: "market value approaches" and "demand curve approaches". In the first section (13.5.1), those cases where methods included in the first category have been implemented will be described. In the second section (13.5.2), the cases where methods based on the demand curve will be analysed. The examples founded through a literature review have been integrated with the cases described in a database developed within the Gestire project (2015), with the aim to assess the economic values of the Natura 2000 network in Lombardy (Italy).

13.5.1 Market Value Approaches

This category comprises all the methods that are based on the use of values recorded by the market to carry out direct or indirect estimation of the value of mountain forest ES. Market prices can be a good signal of the value of ES being influenced by supply and demand functions, i.e. by the revenue generation capacity, current costs and the preferences of consumers. Compared to the other category of techniques (demand curve approaches), these methods are easier to apply due to the use of already existing values directly assumed from the real market. This is also the reason why the outcomes from the implementation of these methods are considered as "hard results", i.e. connected with real evidence from the market. Nevertheless, in many cases, the results of the market value approaches represent an underestimate of the total economic value (TEV) of mountain forest ES.

(a) *Benefit as a proxy*

The "benefit as a proxy" methods comprise those methods that use the final revenues or economic benefits to estimate the value of goods and services. This category is represented by two main methods: *opportunity cost* and *production function*.

Opportunity Cost (OC) The OC describes the cost that the landowner has to incur when he/she decides not to change the specific land use or not to change his/her economic activities in order to maintain or enhance a particular mountain forest ES, therefore renouncing to an increase of his/her income. Let's make some examples. The *opportunity cost* for a landowner that is involved in a project, aimed to enhance forest biodiversity, is represented by the income loss derived from the reduction on timber harvesting to comply with the project aim. The amount of income lost can be used to estimate the value of biodiversity protection in that forest, representing the additional revenue that the owner is willing to renounce to preserve biodiversity.

Because the OC strictly depends on the land cover or the activity performed in a specific area, its value is related to the local situation (Barton et al. 2013). Some examples of OC application are listed hereafter. Extensive application of this methodology was found in decision-making processes related to forest conservation, biodiversity protection, carbon sequestration or, at the contrary, forest exploitation (Kniivilä and Saastamoinen 2002; Seidl et al. 2007; Schröter et al. 2014; Hily et al. 2015). Similarly, OC approach has been used to consider the effects of land-use changes on the provision of different forest ES in Central and Eastern European countries (Ruijs et al. 2017) or to evaluate the different provision of forest ES, changing from a monoculture to a close-to-nature forest management system in order to estimate its costs and benefits (Schou et al. 2012). Another example in the use of OC is provided by Campos et al. (2020), where it has been used to estimate the environmental income deriving from different activities landowners can implement in a specific ecosystem, cork oak open woodlands, in southern Spain.

Production Function (PF) This method puts in relation a specific forest ES with the production of a specific good associated to the market. The forest ES is viewed as input for the provision of a specific good. The assessment of the economic value of the selected ecosystem service is calculated considering its contribution in the provision of the market good. The value of the forest ES is thus associated with the increase of income generated by the improved production system. In using this approach, it is necessary to know the existing relation between the forest ES and the provided good.

This method has been mainly used in literature for the valuation of regulating forest ES (see Table 13.2). To better understand the operational aspect of PF implementation, some authors that used it are here presented. Gren et al. (2018), for instance, used the *production function* method, in combination with another method called replacement cost that is described within the "cost as a proxy method", in order to assess the impact of pathogens spread in the capability of carbon dioxide sequestration in forest ecosystems. Another good example of the application of this method can be found in Nahuelhual et al. (2007), where PF was selected as a suitable methodology for assessing the economic value of water provision of the Valdivian forests in Chile. A similar application was used by Westling et al. (2020) to estimate the value of water quality regulation of Sweden forests used as input to produce drinkable water.

(b) *Cost as a proxy*

The "cost as a proxy" methods referred to those methods that use the cost incurred in producing a certain good or in substituting it with similar ones, as estimation of the economic value of the considered forest ecosystem service. This category comprises *replacement cost*, *cost of substitute goods*, *defensive expenditures* and *damage and insurance costs* methods.

Replacement Cost[1] (RC) In this approach, the value of the forest ES is associated with the avoided cost to replace the service in case of its loss. In other words, the

---

[1] This method has been described by Forest Europe as "the most realistic method of re-creating non-market benefits" (https://foresteurope.org/overview-valuation-approaches-methods).

Table 13.2 Forest ES and related economic evaluation approaches

ES			Market value analysis						Demand curve approaches			
			Benefit as a proxy		Cost as a proxy				Indirect methods		Direct methods	
Section	ES category	ES sub category	Opportunity costs	Production function	Replacement costs	Cost of substitute goods	Defensive expenditures	Damage and insurance costs	Hedonic pricing	Travel cost method	Contingent valuation	Choice modelling
Provisioning	Bioenergy production		x		x	x					x	x
	Timber production		x		x	x					x	x
	Non-wood forest products		x		x	x					x	x
Regulating	Climate regulation	Forest carbon stocks	x	x	x		x	x			x	x
		Soil carbon stocks	x	x	x		x	x			x	x
		Albedo	x	x	x		x	x			x	x
	Pest control		x	x	x		x	x			x	x
	Natural hazard regulation	Forest fires/ wildfires	x	x	x	x	x	x			x	x
		Erosion, avalanche, landslide	x	x	x	x	x	x	x		x	x
		Flooding	x	x	x	x	x	x			x	x
	Water quality regulation		x	x	x		x	x			x	x
	Biodiversity		x	x	x		x	x			x	x

Cultural	Recreation (hunting, nature enjoyment)	x		x		x		x	x	x
	Tourism	x		x		x		x	x	x
	Aesthetic/heritage (landscape character, cultural landscapes)	x		x		x	x	x	x	x

Source: Compiled by the authors based on the information from Forest Europe (2015) and Masiero et al. (2019)

value of the benefits associated with a certain forest ES is derived from the cost to replace the same benefit with different services or goods.

Several studies have used this methodology to assess forest ES values. Bianchi et al. (2018), for instance, carried out a literature review on the use of different methodologies to measure the value of protection services against rock falls, avalanches and landslides and for investment in flood protection in the Alps, in which RC proved its effectiveness (see also Notaro and Paletto 2012; Häyhä et al. 2015; Getzner et al. 2017; Accastello et al. 2019). Grilli et al. (2015) assessed the values of different forest ES in Italian alpine valleys. *Replacement cost* was also used to evaluate carbon sequestration in different forests, for instance, to assess the value of CO_2 sequestration in Swedish forests (Gren 2015), Italian forests (Notaro et al. 2009) and Iranian forests (Karimzadeh Jafari et al. 2020). The RC method was also used to assess water quality services provided by forests. In their paper, Hunter et al. (2019) estimated the value of water purification by coastal wetland in Louisiana, calculating the costs incurred by the treatment plants to provide drinkable water. A slightly different approach in the implementation of this method can be found in the study of Clinch (2000). In this research, RC was implemented in combination with other methodologies (Contingent Valuation and Damage Cost) to evaluate the Irish national forest plantation programme and assess its negative and positive aspects.

Cost of Substitute Goods (CSG) The rationale behind this methodology is to relate the value of the ecosystem service to the cost that would be necessary to produce a substitute good or service, also called surrogate, fulfilling the same or similar function. This method faces the difficulty of finding appropriate surrogates to forest ES that cover comparable functions. For this reason, the cases in which this method has been used are fewer compared to other methodologies.

An interesting application of this method was done to evaluate the different financial benefits, generated from provisioning ES in a specific region of Nepal, obtained by two different systems of community-based forest management (Acharya et al. 2020).

Defensive Expenditures (DE) This method analyses the value of a specific forest ES, associating it with the cost that would occur to avoid and/or reduce the negative environmental impact caused by the absence of the considered forest ecosystem service or with the hypothetic implementation costs for actions intended for the mitigation or compensation of the consequent damages caused by the absence of the considered ES.

This method was used in Morri et al. (2014) to quantify the monetary value of flood protection by forests of the Apennines in Italy. Moreover, in their paper, Snider et al. (2006) used the DE method to understand if the funds invested by the US federal government in forest fire prevention were effective. The value associated with the actions that had been implemented for forest fire protection by the government was used as a proxy of the value of the forest fire control services (regulating ES).

Damage and Insurance Costs (DIC) Always related to "cost as proxy", this method is used to assess the value of a forest ES, putting it in relation with the expenses incurred because of damages caused, for instance, by natural hazards or the insurance costs paid out as a result of the occurrence of the insured events.

In Pulkrab et al. (2011), this method was used to assess the value of pest control services of forests in the Czech Republic, calculating the damage caused by the outbreak. Similarly, in Gren et al. (2009), DIC was used to assess the damage caused by alien invasive species and their severe effects on biodiversity in Swedish forests. This method was also used to quantify the value of carbon sequestration of German forests (Wüstemann et al. 2014) and Irish forests (Clinch 2000). In Germany, DIC was also implemented to assess the monetary value of flood protection service of riparian forest (Barth and Döll 2016). Through this method, it was possible to estimate the avoided damage costs because of the presence of the riparian forest. The avoided costs were considered as a proxy of the value of the regulating ES itself. Finally, in Pavanelli and Voulvoulis (2019), DIC method was used to assess the value of those ES negatively affected by environmental damages.

13.5.2 Demand Curve Approaches

The second category of methods that allows the assessment of the economic value of forest ES is based on those methods that rely on the demand curve as a proxy. These methods are used whenever the assessment of market values is not applicable and when relevant non-market prices influence the calculation of the total economic value (TEV) of forest ES. This set of approaches works by estimating the value of forest ES through:

– The decisions made by real consumers revealed by their concrete expenditures (so-called indirect methods)
– The declared preferences of the real and potential consumers analysing their willingness to pay for a specific ES (so-called direct methods)

(c) *Indirect methods*

The "indirect methods" are those methods based on the revealed preferences of the end users and consumers. Through consumers' behaviour, in fact, it is possible to indirectly estimate the value they give to ES. In this category, two main methods are present: *hedonic pricing* and *travel cost methods*.

Hedonic Pricing (HP) This technique assumes that the final price of a specific good depends on its internal characteristics but also on some external factors. For instance, if house pricing is taken into consideration, houses with similar characteristics could have different prices. This is also due to the different location of the buildings that can be surrounded by different landscapes. The difference in the price of similar houses, which can be explained by the presence or absence of a specific landscape, is an indicator of the price that people indirectly give to that specific landscape. Dealing with ES, through this method, it is possible to estimate the value of different forest ES. For instance, the aesthetic value of a forest can be estimated as the additional amount people pay to buy a house surrounded by forest instead of a similar, but cheaper, house located in a small city.

A clear example of this is reported in a study implemented in Croatia, where *hedonic pricing* method was applied considering the price of hotel rooms to estimate the touristic value of Mediterranean forest (Marušić et al. 2005). In literature, it is also possible to find studies in which other ES were estimated through this method. For instance, the Austrian Federal Forests commissioned the valuation of the protective functions of forests against landslides, avalanches and rock falls (Getzner et al. 2017). In Switzerland, according to Schläpfer et al. (2015), the value of different landscape amenities (comprising forests) was estimated by analysing the variation in rental prices. According to Sundelin et al. (2015), through the analysis of the values of different forest features (such as fragmentation, density, shape and productivity), it was possible to detect which characteristics of forestland affect the Swedish land value. Finally, in Poland, this method was used to assess the value of the presence of urban forest, considering apartment prices (Łaszkiewicz et al. 2019). Outside Europe (in the USA and Canada), *hedonic pricing* was applied to evaluate, among the others, the impact on cultural ES (touristic and aesthetic services) in forest affected by insect infestation (Price et al. 2010), the value of hunting recreational services (Hussain et al. 2007) and the value of erosion control services provided by the Ohioan forests (Hitzhusen 1999).

Travel Cost Method (TC) In this method, the travel cost incurred by people who want to reach and visit a certain habitat/ecosystem is analysed to derive their willingness to pay for a specific forest ES or a combination of them. Generally, TC is used to estimate the value of cultural ES, specifically those related to tourism and recreation. In the implementation of TC, also the opportunity cost of time is considered in value assessment.

There are numerous applications of this method in assessing mountain forest ES. In Germany, for instance, it was used to estimate the value of cultural ES (recreation) provided by German protected areas (Mayer and Woltering 2018). It was also used to estimate the potential recreational value of forest in the UK (Ezebilo 2016), in the Czech Republic (Melichar 2014; Březina et al. 2019) and in Slovakia (Tutka and Kovalčík 2010). According to Moran et al. (2006), a more detailed TC method was carried out assessing the cultural services provided by Scottish forests, considering the cost of mountain biking as a recreational activity. In other cases implemented in the North America, specifically in the Rocky Mountains in Colorado (USA), this method was applied to assess the impact of forest fire on recreational ES (Loomis et al. 2001) and to assess the effects of tree density, influenced also by insect pests and other hazards, on the demand of recreational services (Walsh et al. 1989).

(d) *Direct methods*

The "direct methods" comprise those methods that collect the willingness to pay off end users in relation to specific ES. The main tools used by this methods' category are questionnaires and surveys that allow directly asking individuals' opinion when different scenarios are presented to them. The two direct methods that are hereafter described are the *contingent valuation* and the *choice modelling*.

Contingent Valuation (CV) This method aims to measure the willingness to accept the loss of a certain ES, if no actions for its provision or enhancement are implemented. Alternatively, it can be used to investigate the end users' willingness to pay for the implementation of actions aimed to support the provision of a specific ES. A representative sample of people who directly or indirectly take advantage of the presence of a defined ES is interviewed. The analysis of their responses allows gathering information about their readiness to accept the loss of the ES or their willingness to pay for its provision through the presentation of different scenarios.

The just described method was used to assess the value of forest recreational services in an Italian alpine valley (Grilli et al. 2014), in Slovakian mountains (Tutka and Kovalčík 2010) and in British woodlands (Christhe et al. 2007). Moreover, in the Appalachian Mountains, it was used to value the health protection function of forest ecosystems (Holmes and Kramer 1996). According to Bastian et al. (2017), CV was one of the methods used to assess the value of forest ES provided by Eastern Ore Mountains (Germany and Czech Republic). In Italy, it was used also to evaluate the aesthetic services of the national forest landscape (Tempesta and Marangon 2004).

Choice Modelling (CM) In *choice modelling* methods, consumers' willingness to pay is detected by asking them to choose the best option among a variety of alternatives. The alternatives are characterized by different attributes of the ES under investigation. One of these attributes is the amount of money people would be willing to pay for the provision of the considered ES (and its attributes). The survey is designed to reveal the value given to the attributes and to their combinations. The assumption under this approach is that forest ES can be subdivided into different attributes. One of the alternatives represents the current situation (baseline), and the other ones correspond to different variations of the selected attributes. Each scenario is associated with a different amount of money that has to be taken into consideration in selecting the most suitable alternative.

Some examples of its application can be mainly found regarding the evaluation of different attributes of a single ES. For instance, it was applied in valuing recreation services in relation to biological impacts (e.g. bark beetle attack) (e.g. in Horne et al. 2005; Christie et al. 2007; De Valck et al. 2014; Arnberger et al. 2018) or in the assessment of biodiversity value (Horne 2006; Czajkowski et al. 2009, 2017; Meyerhoff et al. 2009; Hoyos et al. 2012). It was also applied in the assessment of heritage values (particularly referring to the landscape characters, e.g. Garrod et al. 2009) or to evaluate different forest ES (Gatto et al. 2014; Giergiczny et al. 2015).

13.6 Conclusion

A large variety of studies can be found in the scientific literature about climate change and its impacts on forest ecosystems, but a deep interdisciplinary analysis investigating how global warming is affecting the provision of the different forest ES is still missing. Moreover, the literature highlights a lack of information regarding the impacts of climate change on the provision of certain forest ES, such as cultural ES and some regulating services, for instance, "natural hazard regulation" (see Sect. 13.4.1). Because of the high environmental and climatic variability of mountain regions, it would be necessary to rely on good quality and quantity of primary data to be able to have a comprehensive understanding of the whole phenomenon under discussion. For these reasons, there is the necessity to integrate the existing knowledge with studies aimed to investigate climate change impacts on tree physiology and stand structures and integrating these data with the consequent impact on the provision of forest ES. Changes in their provision significantly influence human livelihood, particularly in mountain areas where the interdependence between human and forest ecosystems is stronger and thus more exposed to changing climate.

Through the literature review, several methods to assess the economic value of goods and services were detected. The most frequently used methods are the *demand curve approaches*. This could be explained by the growing interest in the use of these methods, which makes it possible to assess the non-market value of ES, a value that often is not recorded in the financial analysis. Another explanation could be related to the fact that the *market value approaches*, such as *production function* or *cost of substitute good*, need a profound knowledge about the interrelation between forest functions and the resulting provision of ES, an interrelation that is still not always known or fully understood.

The aim of this chapter was to outline the existing methods used to assess the value of ES and give an overview of the climate change impacts on the mountain forests ES provision capacity. The methods described in this chapter allow the assessment of the TEV of mountain forest ES, which can be used to create a baseline to assess, in the future, how their value will be modified in relation to climatic changes. Implementing similar studies in the long period will allow monitoring the changing value of mountain forest ES, permitting proper estimate of the economic impact of climate change on them.

Moreover, the gathered data could be also used to fill in the knowledge gap existing in the evaluation of specific forest ES in specific areas of interest. In fact, through a different approach, known as *benefit transfer,* it would be possible to use the existing data on ES evaluation to estimate their value in different contexts.

Several databases are already present in the web, gathering valuations that can be used to transfer ES value from a specific geographic area to another. These databases are EUROFOREX (https://www.evri.ca/en), ENVALUE (https://www.environment.nsw.gov.au/envalue), RED Database (http://www.isis-it.net/red/start_search.asp), a database reported by Elsasser et al. (2016) and a database that is the result of a EC-financed research projects (http://ec.europa.eu/environment/enveco/studies.htm).

Concluding, the increasing importance of global warming and the need to estimate its impact on the different forest ES are increasing the attention of the academic world, citizens and policy-makers on it. Because mountain forests are ecosystems suitable for early detection of climate change impacts, they can be an interesting ecosystem to start the analysis about how climate change impacts on the provision of ES. A lot of studies already assess climate change impact on forest structures and tree growth, but it is necessary to improve the analysis of the impacts on ES provision to be able to estimate the economic loss (or gain) due to global warming. A deeper understanding could support the estimation of less evident impacts of climate change on forest ES in different contexts than mountain forest, implementing the *benefit transfer* methods. But it could also be necessary to provide some guidelines to forest managers and public authorities on the suitable forest management needs to minimize the negative impacts of climate change and maximize the positive ones.

References

Accastello C, Bianchi E, Blanc S, Brun F (2019) ASFORESEE: a harmonized model for economic evaluation of forest protection against rockfall. Forests 10. https://doi.org/10.3390/f10070578

Acharya RP, Maraseni T, Cockfield G (2020) Assessing the financial contribution and carbon emission pattern of provisioning ecosystem services in Siwalik forests in Nepal: valuation from the perspectives of disaggregated users. Land Use Policy 95:104647. https://doi.org/10.1016/j.landusepol.2020.104647

Allen CD, Macalady AK, Chenchouni H, Bachelet D, McDowell N, Vennetier M, Kitzberger T, Rigling A, Breshears DD, Hogg EH, Ted, Gonzalez P, Fensham R, Zhang Z, Castro J, Demidova N, Lim JH, Allard G, Running SW, Semerci A, Cobb N (2010) A global overview of drought and heat-induced tree mortality reveals emerging climate change risks for forests. For Ecol Manag 259:660–684. https://doi.org/10.1016/j.foreco.2009.09.001

Arnberger A, Ebenberger M, Schneider IE, Cottrell S, Schlueter AC, von Ruschkowski E, Venette RC, Snyder SA, Gobster PH (2018) Visitor preferences for visual changes in bark beetle-impacted forest recreation settings in the United States and Germany. Environ Manag 61:209–223. https://doi.org/10.1007/s00267-017-0975-4

Barth NC, Döll P (2016) Assessing the ecosystem service flood protection of a riparian forest by applying a cascade approach. Ecosyst Serv 21:39–52. https://doi.org/10.1016/j.ecoser.2016.07.012

Barton DN, Brouwer R, Barton D, Bernasconi P, Blumentrath S, Brouwer R, Oosterhuis F, Pinto R, Tobar DE (2013) Conservation policy instruments

Bastian O, Syrbe RU, Slavik J, Moravec J, Louda J, Kochan B, Kochan N, Stutzriemer S, Berens A (2017) Ecosystem services of characteristic biotope types in the Ore Mountains (Germany/Czech Republic). Int J Biodiversity Sci Ecosyst Services Manag 13:51–71. https://doi.org/10.1080/21513732.2016.1248865

Battisti A, Larsson S (2015) Climate change and insect climate change and insect pests: CABI climate change series. 7(2009):1–15. https://doi.org/10.1079/9781780643786.0000

Beniston M (2003) Climatic change in mountain regions: a review of possible impacts, pp 5–31. https://doi.org/10.1007/978-94-015-1252-7_2

Bianchi E, Accastello C, Trappmann D, Blanc S, Brun F (2018) The economic evaluation of forest protection service against Rockfall: a review of experiences and approaches. Ecol Econ 154:409–418. https://doi.org/10.1016/j.ecolecon.2018.08.021

Bosela M, Merganičová K, Torresan C, et al (2021) Modelling future growth of mountain forests under changing environments. In: Managing Forest Ecosystems, Vol. 40, Tognetti R, Smith M, Panzacchi P (Eds): Climate-Smart Forestry in Mountain Regions. Springer Nature, Switzerland, AG

Březina D, Michal J, Adamec Z, Burdová J (2019) Quantification of the economic value of the recreational function of forests in the territory of Městské lesy Hradec Králové a.s. J For Sci 65:161–170. https://doi.org/10.17221/38/2019-JFS

Briner S, Huber R, Bebi P, Elkin C, Schmatz DR, Grêt-Regamey A (2013) Trade-offs between ecosystem services in a mountain region. Ecol Soc 18. https://doi.org/10.5751/ES-05576-180335

Brockerhoff EG, Barbaro L, Castagneyrol B, Forrester DI, Gardiner B, González-Olabarria JR, Lyver POB, Meurisse N, Oxbrough A, Taki H, Thompson ID, van der Plas F, Jactel H (2017) Forest biodiversity, ecosystem functioning and the provision of ecosystem services. Biodivers Conserv 26:3005–3035. https://doi.org/10.1007/s10531-017-1453-2

Campos P, Álvarez A, Oviedo JL, Mesa B, Caparrós A, Ovando P (2020) Environmental incomes: refined standard and extended accounts applied to cork oak open woodlands in Andalusia. Spain Ecol Indic 117:106551. https://doi.org/10.1016/j.ecolind.2020.106551

Christie M, Hanley N, Hynes S (2007) Valuing enhancements to forest recreation using choice experiment and contingent behaviour methods. J For Econ 13:75–102

Clinch JP (2000) Assessing the social efficiency of temperate-zone commercial forestry programmes: Ireland as a case study. Forest Policy Econ 1:225–241. https://doi.org/10.1016/s1389-9341(00)00029-0

Courbaud B, Kunstler G, Morin X, Cordonnier T (2010) What is the future of the ecosystem services of the Alpine forest against a backdrop of climate change? Revue de Géographie Alpine:0–12. https://doi.org/10.4000/rga.1317

Cudlín P, Klopčič M, Tognetti R, Mališ F, Alados CL, Bebi P, Grunewald K, Zhiyanski M, Andonowski V, La Porta N, Bratanova-Doncheva S, Kachaunova E, Edwards-Jonáová M, Ninot JM, Rigling A, Hofgaard A, Hlasny T, Skalák P, Wielgolaski FE (2017) Drivers of treeline shift in different European mountains. Clim Res 73:135–150. https://doi.org/10.3354/cr01465

Czajkowski M, Buszko-Briggs M, Hanley N (2009) Valuing changes in forest biodiversity. Ecol Econ 68:2910–2917. https://doi.org/10.1016/j.ecolecon.2009.06.016

Czajkowski M, Budziński W, Campbell D, Giergiczny M, Hanley N (2017) Spatial heterogeneity of willingness to pay for forest management. Environ Resour Econ 68:705–727. https://doi.org/10.1007/s10640-016-0044-0

De Valck J, Vlaeminck P, Broekx S, Liekens I, Aertsens J, Chen W, Vranken L (2014) Benefits of clearing forest plantations to restore nature? Evidence from a discrete choice experiment in Flanders. Belgium Landscape and Urban Planning 125:65–75. https://doi.org/10.1016/j.landurbplan.2014.02.006

De Wit HA, Palosuo T, Hylen G, Liski J (2006) A carbon budget of forest biomass and soils in Southeast Norway calculated using a widely applicable method. For Ecol Manag 225:15–26. https://doi.org/10.1016/j.foreco.2005.12.023

Ding H, Chiabai A, Silvestri S, Nunes PALD (2016) Valuing climate change impacts on European forest ecosystems. Ecosyst Serv 18:141–153. https://doi.org/10.1016/j.ecoser.2016.02.039

Dupire S, Curt T, Bigot S (2017) Spatio-temporal trends in fire weather in the French Alps. Sci Total Environ 595:801–817. https://doi.org/10.1016/j.scitotenv.2017.04.027

Elsasser P, Meyerhoff J, Weller P (2016) Ecosystem service valuation studies in Austria, Germany and Switzerland ecosystem service valuation studies in Austria, Germany, and Switzerland

Ezebilo EE (2016) Economic value of a non-market ecosystem service: an application of the travel cost method to nature recreation in Sweden. Int J Biodiversity Sci Ecosyst Serv Manag 12:314–327. https://doi.org/10.1080/21513732.2016.1202322

Fernàndez-Martínez J, Fleck I (2016) Photosynthetic limitation of several representative subalpine species in the Catalan Pyrenees in summer. Plant Biol (Stuttgart, Germany) 18:638–648. https://doi.org/10.1111/plb.12439

Feurdean A, Gałka M, Tanţău I, Geantă A, Hutchinson SM, Hickler T (2016) Tree and timberline shifts in the northern Romanian Carpathians during the Holocene and the responses to environmental changes. Quat Sci Rev 134:100–113. https://doi.org/10.1016/j.quascirev.2015.12.020

Fleischer P, Pichler V, Flaischer P, Holko L, Málic F, Gömöryová E, Cudlín P, Holeksa J, Michalová Z, Homolová Z, Skvarenina J, Střelcová K, Hlaváč P (2017) Forest ecosystem services affected by natural disturbances, climate and land-use changes in the Tatra Mountains. Clim Res 73:57–71. https://doi.org/10.3354/cr01461

Forsius M, Anttila S, Arvola L, Bergström I, Hakola H, Heikkinen HI, Helenius J, Hyvärinen M, Jylhä K, Karjalainen J, Keskinen T, Laine K, Nikinmaa E, Peltonen-Sainio P, Rankinen K, Reinikainen M, Setälä H, Vuorenmaa J (2013) Impacts and adaptation options of climate change on ecosystem services in Finland: a model based study. Curr Opin Environ Sustain 5:26–40. https://doi.org/10.1016/j.cosust.2013.01.001

Friedrichs DA, Trouet V, Büntgen U, Frank DC, Esper J, Neuwirth B, Löffler J (2009) Species-specific climate sensitivity of tree growth in Central-West Germany. Trees Struct Funct 23:729–739. https://doi.org/10.1007/s00468-009-0315-2

Fyllas NM, Christopoulou A, Galanidis A, Michelaki CZ, Dimitrakopoulos PG, Fulé PZ, Arianoutsou M (2017) Tree growth-climate relationships in a forest-plot network on Mediterranean mountains. Sci Total Environ 598:393–403. https://doi.org/10.1016/j.scitotenv.2017.04.145

Galiano L, Martínez-Vilalta J, Lloret F (2010) Drought-induced multifactor decline of scots pine in the Pyrenees and potential vegetation change by the expansion of co-occurring oak species. Ecosystems 13:978–991. https://doi.org/10.1007/s10021-010-9368-8

Garrod G, Ruto E, Snowdon P (2009) Assessing the value of forest landscapes: a choice experiment approach. Arboricultural J 32. https://doi.org/10.1080/03071375.2009.9747573

Gatto P, Vidale E, Secco L, Pettenella D (2014) Exploring the willingness to pay for forest ecosystem services by residents of the Veneto region. Bio-based Appl Econ 3:21–43. https://doi.org/10.13128/BAE-11151

GESTIRE Project (2015) Stima del valore socio-economico della rete Natura 2000 in Lombardia Rapporto finale. Retrieved from https://www.naturachevale.it/gestire/wp-content/uploads/2014/04/Stima-del-valore-socio-economico-della-Rete-Natura-2000-in-Lombardia.pdf

Getzner M, Gutheil-Knopp-Kirchwald G, Kreimer E, Kirchmeir H, Huber M (2017) Gravitational natural hazards: valuing the protective function of alpine forests. Forest Policy Econ 80:150–159. https://doi.org/10.1016/j.forpol.2017.03.015

Giergiczny M, Czajkowski M, Zylicz T, Angelstam P (2015) Choice experiment assessment of public preferences for forest structural attributes. Ecol Econ 119:8–23. https://doi.org/10.1016/j.ecolecon.2015.07.032

Gren IM (2015) Estimating values of carbon sequestration and nutrient recycling in forests: an application to the Stockholm-Malar region in Sweden. Forests 6:3594–3613. https://doi.org/10.3390/f6103594

Gren IM, Isacs L, Carlsson M (2009) Costs of alien invasive species in Sweden. Ambio 38:135–140. https://doi.org/10.1579/0044-7447-38.3.135

Gren IM, Aklilu A, Elofsson K (2018) Forest carbon sequestration, pathogens and the costs of the EU's 2050 climate targets. Forests 9. https://doi.org/10.3390/f9090542

Grêt-Regamey A, Brunner SH, Altwegg J, Bebi P (2013) Facing uncertainty in ecosystem services-based resource management. J Environ Manag 127:S145–S154. https://doi.org/10.1016/j.jenvman.2012.07.028

Grilli G, Paletto A, De Meo I (2014) Economic valuation of forest recreation in an alpine valley. Balt For 20:167–175

Grilli G, Nikodinoska N, Paletto A, De Meo I (2015) Stakeholders' preferences and economic value of forest ecosystem services: an example in the Italian alps. Balt For 21:298–307

Haines-Young R, Potschin M (2018) CICES V5. 1. Guidance on the application of the revised structure. Fabis Consulting, p 53

Hartl-Meier C, Zang C, Büntgen U, Esper J, Rothe A, Göttlein A, Dirnböck T, Treydte K (2014) Uniform climate sensitivity in tree-ring stable isotopes across species and sites in a mid-latitude temperate forest. Tree Physiol 35:4–15. https://doi.org/10.1093/treephys/tpu096

Häyhä T, Franzese PP, Paletto A, Fath BD (2015) Assessing, valuing, and mapping ecosystem services in Alpine forests. Ecosyst Serv 14:12–23. https://doi.org/10.1016/j.ecoser.2015.03.001

Hily E, Garcia S, Stenger A, Tu G (2015) Assessing the cost-effectiveness of a biodiversity conservation policy: a bio-econometric analysis of Natura 2000 contracts in forest. Ecol Econ 119:197–208. https://doi.org/10.1016/j.ecolecon.2015.08.008

Hitzhusen FJ (1999) A resource economic perspective on erosion control for non-economists. Investing in the protection of our environment. In: Proceedings of conference 30, Nashville. International Erosion Control Association

Holmes T, Kramer R (1996) Contingent valuation of ecosystem health. Ecosyst Health 2:58–60

Horák R, Borišev M, Pilipović A, Orlović S, Pajevic S, Nikolić N (2014) Drought impact on forest trees in four nature protected areas in Serbia. Sumarski List 138:301–308

Horne P (2006) Forest owners' acceptance of incentive based policy instruments in forest biodiversity conservation – a choice experiment based approach. Silva Fennica Monographs 40:169–178. https://doi.org/10.14214/sf.359

Horne P, Boxall PC, Adamowicz WL (2005) Multiple-use management of forest recreation sites: a spatially explicit choice experiment. For Ecol Manag 207:189–199. https://doi.org/10.1016/j. foreco.2004.10.026

Hoyos D, Mariel P, Pascual U, Etxano I (2012) Valuing a Natura 2000 network site to inform land use options using a discrete choice experiment: an illustration from the Basque Country. J For Econ 18:329–344. https://doi.org/10.1016/j.jfe.2012.05.002

Hunter RG, Day JW, Wiegman AR, Lane RR (2019) Municipal wastewater treatment costs with an emphasis on assimilation wetlands in the Louisiana coastal zone. Ecol Eng 137:21–25. https:// doi.org/10.1016/j.ecoleng.2018.09.020

Hussain A, Munn IA, Grado SC, West BC, Jones WD, Jones J (2007) Willingness to provide fee-access hunting. For Sci 53:493–506

IPBES (2018) The IPBES regional assessment report on biodiversity and ecosystem services for Europe and Central Asia, p 834

Jacquet JS, Orazio C, Jactel H (2012) Defoliation by processionary moth significantly reduces tree growth: a quantitative review. Ann For Sci 69:857–866. https://doi.org/10.1007/ s13595-012-0209-0

Jolly WM, Dobbertin M, Zimmermann NE, Reichstein M (2005) Divergent vegetation growth responses to the 2003 heat wave in the Swiss Alps. Geophys Res Lett 32: 1–4. https://doi. org/10.1029/2005GL023252

Karimzadeh E, Naghavi H, Adeli K, Latifi H (2020) A nondestructive, remote sensing-based estimation of the economic value of aboveground temperate Forest biomass (case study: Hyrcanian Forests, Nowshahr-Iran). J Sustain For 39

Kniivilä M, Saastamoinen O (2002) The opportunity costs of Forest. Silva Fennica 36:853–865. Retrieved from http://128.214.52.218/silvafennica/full/sf36/sf364853.pdf

Kozlov MV, van Nieukerken EJ, Zverev V, Zvereva EL (2013) Abundance and diversity of birch-feeding leafminers along latitudinal gradients in northern Europe. Ecography 36:1138–1149. https://doi.org/10.1111/j.1600-0587.2013.00272.x

Krupková L, Havránková K, Krejza J, Sedlák P, Marek MV (2019) Impact of water scarcity on spruce and beech forests. J For Res 30:899–909. https://doi.org/10.1007/s11676-018-0642-5

Kullman L (1996) Rise and demise of cold-climate Picea abies forest in Sweden. New Phytol 134:243–256. https://doi.org/10.1111/j.1469-8137.1996.tb04629.x

Kurbanov EA, Post WM (2002) Changes in area and carbon in forests of the Middle Zavolgie: a regional case study of Russian forests. Clim Chang 55:157–171. https://doi.org/10.1023/A: 1020275713889

Kurbanov E, Vorobyov O, Gubayev A, Moshkina L, Lezhnin S (2007) Carbon sequestration after pine afforestation on marginal lands in the Povolgie region of Russia: a case study of the potential for a joint implementation activity. Scand J For Res 22:488–499. https://doi. org/10.1080/02827580701803080

Łaszkiewicz E, Czembrowski P, Kronenberg J (2019) Can proximity to urban green spaces be considered a luxury? Classifying a non-tradable good with the use of hedonic pricing method Ecological Economics 161: 237–247. https://doi.org/10.1016/j.ecolecon.2019.03.025

Lebourgeois F, Rathgeber C, Ulrich E (2010) Sensitivity of French temperate coniferous forests to climate variability and extreme events (Abies alba, Picea abies and Pinus sylvestris). J Veg Sci 21

Linares JC, Tíscar PA (2011) Buffered climate change effects in a Mediterranean pine species: range limit implications from a tree-ring study. Oecologia 167:847–859. https://doi.org/10.1007/s00442-011-2012-2

Linares JC, Camarero JJ, Carreira JA (2009) Interacting effects of changes in climate and forest cover on mortality and growth of the southernmost European fir forests. Glob Ecol Biogeogr 18:485–497. https://doi.org/10.1111/j.1466-8238.2009.00465.x

Lindner M, Seidl R, Lexer MJ, Kremer A (2007) Impacts of climate change on European forests and options for adaptation

Lindner M, Maroschek M, Netherer S, Kremer A, Barbati A, Garcia-Gonzalo J, Seidl R, Delzon S, Corona P, Kolström M, Lexer MJ, Marchetti M (2010) Climate change impacts, adaptive capacity, and vulnerability of European forest ecosystems. For Ecol Manag 259: 698–709. https://doi.org/10.1016/j.foreco.2009.09.023

Loboda T, Krankina O, Savin I, Kurbanov E, Hall J (2017) land management and the impact of the 2010 extreme drought event on the agricultural and ecological systems of European Russia. In: Land-cover and land-use changes in Eastern Europe after the collapse of the Soviet Union in 1991

Loomis J, González-Cabán A, Englin J (2001) Testing for differential effects of forest fires on hiking and mountain biking demand and benefits. J Agric Resour Econ 26: 508–522. https://doi.org/10.22004/ag.econ.31049

Marušić Z, Horak S, Navrud S (2005) The economic value of coastal forests for tourism: a comparative study of three valuation methods. Tourism 53:141–152

Masiero M, Pettenella D, Boscolo M, Barua S., Animon I, Matta JR (2019) Valuing forest ecosystem services: a training manual for planners and project developers. Fao, p 200. Retrieved from http://www.wipo.int/amc/en/mediation/rules

Mayer M, Woltering M (2018) Assessing and valuing the recreational ecosystem services of Germany's national parks using travel cost models. Ecosyst Serv 31:371–386. https://doi.org/10.1016/j.ecoser.2017.12.009

Meining S, Schröter H, von Wilpert K (2004) Waldzustandsbericht 2004 der Forstlichen Versuchs- und Forschungsanstalt Baden-Württemberg, p 54

Melichar J (2014) The economic valuation of the change in forest quality in the Jizerske Hory mountains: a contingent behavior model. In: public recreation and landscape protection – with man hand in hand?

Meyerhoff J, Liebe U, Hartje V (2009) Benefits of biodiversity enhancement of nature-oriented silviculture: evidence from two choice experiments in Germany. J For Econ 15:37–58. https://doi.org/10.1016/j.jfe.2008.03.003

Moran D, Tresidder E, McVittie A (2006) Estimating the recreational value of mountain biking sites in Scotland using count data models. Tour Econ 12:123–135. https://doi.org/10.5367/000000006776387097

Morri E, Pruscini F, Scolozzi R, Santolini R (2014) A forest ecosystem services evaluation at the river basin scale: supply and demand between coastal areas and upstream lands (Italy). Ecol Indic 37:210–219. https://doi.org/10.1016/j.ecolind.2013.08.016

Nahuelhual L, Donoso P, Lara A, Núñez D, Oyarzún C, Neira E (2007) Valuing ecosystem services of Chilean temperate rainforests. Environ Dev Sustain 9:481–499. https://doi.org/10.1007/s10668-006-9033-8

Notaro S, Paletto A (2012) The economic valuation of natural hazards in mountain forests: an approach based on the replacement cost method. J For Econ 18:318–328. https://doi.org/10.1016/j.jfe.2012.06.002

Notaro S, Paletto A, Raffaelli R (2009) Economic impact of forest damage in an alpine environment. Acta Silvatica et Lignaria Hungarica 5:131–143

Panayotov M, Kulakowski D, Tsvetanov N, Krumm F, Berbeito I, Bebi P (2016) Climate extremes during high competition contribute to mortality in unmanaged self-thinning Norway spruce stands in Bulgaria. For Ecol Manag 369:74–88. https://www.dora.lib4ri.ch/wsl/islandora/object/wsl:5547

Pauli H, Gottfried M, Dullinger S, Abdaladze O, Akhalkatsi M, Alonso JLB, Coldea G, Dick J, Erschbamer B, Calzado RF, Ghosn D, Holten JI, Kanka R, Kazakis G, Kollár J, Larsson P, Moiseev P, Moiseev D, Molau U, Mesa JM, Nagy L, Pelino G, Puşcaş M, Rossi G, Stanisci A, Syverhuset AO, Theurillat JP, Tomaselli M, Unterluggauer P, Villar L, Vittoz P, Grabherr G (2012) Recent plant diversity changes on Europe's mountain summits. Science 336:353–355. https://doi.org/10.1126/science.1219033

Pavanelli DD, Voulvoulis N (2019) Habitat equivalency analysis, a framework for forensic cost evaluation of environmental damage. Ecosyst Serv 38:100953. https://doi.org/10.1016/j.ecoser.2019.100953

Pretzsch H, del Río M, Giammarchi F, Uhl E, Tognetti R (2021) Changes of tree and stand growth. Review and implications. In: Managing Forest Ecosystems, Vol. 40, Tognetti R, Smith M, Panzacchi P (Eds): Climate-Smart Forestry in Mountain Regions. Springer Nature, Switzerland, AG

Price JI, McCollum DW, Berrens RP (2010) Insect infestation and residential property values: a hedonic analysis of the mountain pine beetle epidemic. Forest Policy Econ 12:415–422. https://doi.org/10.1016/j.forpol.2010.05.004

Prietzel J, Christophel D (2014) Organic carbon stocks in forest soils of the German alps. Geoderma 221–222:28–39. https://doi.org/10.1016/j.geoderma.2014.01.021

Pulkrab K, Sloup R, Sloup M, Bukáček J (2011) Optimum costs of forest protection according to ecosite classes – optimum nákladů na ochranu lesa podle souborů lesních typů. Zpravy Lesnickeho Vyzkumu 56:65–74

Pureswaran DS, Roques A, Battisti A (2018) Forest insects and climate change. Curr For Rep 4:35–50. https://doi.org/10.1007/s40725-018-0075-6

Rogora M, Frate L, Carranza ML, Freppaz M, Stanisci A, Bertani I, Bottarin R, Brambilla A, Canullo R, Carbognani M, Cerrato C, Chelli S, Cremonese E, Cutini M, Di Musciano M, Erschbamer B, Godone D, Iocchi M, Isabellon M, Magnani A, Mazzola L, Morra di Cella U, Pauli H, Petey M, Petriccione B, Porro F, Psenner R, Rossetti G, Scotti A, Sommaruga R, Tappeiner U, Theurillat JP, Tomaselli M, Viglietti D, Viterbi R, Vittoz P, Winkler M, Matteucci G (2018) Assessment of climate change effects on mountain ecosystems through a cross-site analysis in the Alps and Apennines. Sci Total Environ 624:1429–1442. https://doi.org/10.1016/j.scitotenv.2017.12.155

Ruijs A, Kortelainen M, Wossink A, Schulp CJE, Alkemade R (2017) Opportunity cost estimation of ecosystem services. Environ Resour Econ 66:717–747. https://doi.org/10.1007/s10640-015-9970-5

Saccone P, Delzon S, Jean-Philippe P, Brun JJ, Michalet R (2009) The role of biotic interactions in altering tree seedling responses to an extreme climatic event. J Veg Sci 20:403–414. https://doi.org/10.1111/j.1654-1103.2009.01012.x

Sarris D, Christopoulou A, Angelonidi E, Koutsias N, Fulé PZ, Arianoutsou M (2014) Increasing extremes of heat and drought associated with recent severe wildfires in southern Greece. Reg Environ Chang 14:1257–1268. https://doi.org/10.1007/s10113-013-0568-6

Scarascia-Mugnozza G, Callegari G, Veltri A, Matteucci G (2010) Water balance and forest productivity in mediterranean mountain environments. Ital J Agron 5:217–222. https://doi.org/.https://doi.org/10.4081/ija.2010.217

Schläpfer F, Waltert F, Segura L, Kienast F (2015) Valuation of landscape amenities: a hedonic pricing analysis of housing rents in urban, suburban and periurban Switzerland. Landsc Urban Plan 141:24–40. https://doi.org/10.1016/j.landurbplan.2015.04.007

Schou E, Jacobsen JB, Kristensen KL (2012) An economic evaluation of strategies for transforming even-aged into near-natural forestry in a conifer-dominated forest in Denmark. Forest Policy Econ 20:89–98. https://doi.org/10.1016/j.forpol.2012.02.010

Schröter M, Rusch GM, Barton DN, Blumentrath S, Nordén B (2014) Ecosystem services and opportunity costs shift spatial priorities for conserving forest biodiversity. PLoS One 9. https://doi.org/10.1371/journal.pone.0112557

Seidl R, Rammer W, Jäger D, Currie WS, Lexer MJ (2007) Assessing trade-offs between carbon sequestration and timber production within a framework of multi-purpose forestry in Austria. For Ecol Manag 248:64–79. https://doi.org/10.1016/j.foreco.2007.02.035

Snider G, Daugherty PJ, Wood DB (2006) The irrationality of continued fire suppression: an avoided cost analysis of fire hazard reduction treatments versus no treatment. J For 104:431–437. https://doi.org/10.1093/jof/104.8.431

Sundelin J, Högberg T, Lönnstedt L (2015) Determinants of the market price of forest estates: a statistical analysis. Scand J For Res 30

Tempesta T, Marangon F (2004) The total economic value of Italian forest landscapes * Tiziano Tempesta – Department TESAF University of Padova (Italy) Francesco Marangon – DSE University of Udine (Italy). MIUR – PRIN 2003 Project. "Landscape and Environmental Actions within the Regional Rural Development Policies"

Thorsen BJ, Mavsar R, Tyrväinen L, Prokofieva I, Stenger A (2014) The provision of Forest ecosystem services volume I: quantifying and valuing non-marketed ecosystem services. The provision of Forest ecosystem services volume I : quantifying and valuing non-marketed ecosystem services, Vol. I

Tømmervik H, Johansen B, Riseth JÅ, Karlsen SR, Solberg B, Høgda KA (2009) Above ground biomass changes in the mountain birch forests and mountain heaths of Finnmarksvidda, northern Norway, in the period 1957-2006. For Ecol Manag 257:244–257. https://doi.org/10.1016/j.foreco.2008.08.038

Tutka J, Kovalčík M (2010) Possibilities of valuation of recreation in forests – Mo ž nosti hodnotenia rekrea č nej funkcie lesov Rekreace a ochrana p ř írody. In: Rekreace a Ochrana Prirody – Sborník Prispevku

Vacek Z, Vacek S, Bulušek D, Podrázský V, Remeš J, Král J, Putalová T (2017) Effect of fungal pathogens and climatic factors on production, biodiversity and health status of ash mountain forests. Dendrobiology 77:161–175. https://doi.org/10.12657/denbio.077.013

Vizzarri M, Pilli R, Korosou A, Frate L, Grassi G (2021) The role of forests in climate change mitigation: the EU context. In: Managing Forest Ecosystems, Vol. 40, Tognetti R, Smith M, Panzacchi P (Eds): Climate-Smart Forestry in Mountain Regions. Springer Nature, Switzerland, AG

Walsh R, Ward F, Olienyk J (1989) Recreational demand for trees in national forests. J Environ Manag 28:255–268

Westling N, Stromberg PM, Swain RB (2020) Can upstream ecosystems ensure safe drinking water – insights from Sweden. Ecol Econ 169:106552. https://doi.org/10.1016/j.ecolecon.2019.106552

Wüstemann H, Meyerhoff J, Rühs M, Schäfer A, Hartje V (2014) Financial costs and benefits of a program of measures to implement a National strategy on biological diversity in Germany. Land Use Policy 36:307–318. https://doi.org/10.1016/j.landusepol.2013.08.009

Zisenis M, Dominique R, Price M, Torre-Marin A, Jones-Walters L, Halada L, Gajdos P, Oszlanyi J, Carver S (2010) 10 messages for 2010 Mountain Ecosyst 16. https://doi.org/10.2800/54895

Chapter 14
Review of Policy Instruments for Climate-Smart Mountain Forestry

Lenka Dubova, Lenka Slavikova, João C. Azevedo, Johan Barstad, Paola Gatto, Jerzy Lesinski, Davide Pettenella, and Roar Stokken

Abstract Implementing the Climate-Smart Forestry (CSF) concept into practice requires interaction among key stakeholders, especially forest owners and managers, policymakers (or regulators in general), forest consultants, and forest users. But what could be the most effective policy instruments to achieve climate smartness in mountain forests? Which ones would be the most acceptable for forest owners? And for the local forest communities? Should they be designed and implemented with the use of participatory approaches or rather on a top-down basis? This chapter summarizes key policy instruments structured in three subsequent categories: command-and-control, voluntary market-based instruments, and community cooperation. It provides examples of their functioning in the forestry sector and discusses their suitability for the implementation of climate smart forestry. It appears that there are

L. Dubova (✉) · L. Slavikova
Institute for Economic and Environmental Policy, Faculty of Social and Economic Studies
J. E. Purkyne University, Usti nad Labem, Czech Republic
e-mail: lenka.dubova@ujep.cz; lenka.slavikova@ujep.cz

J. C. Azevedo
Instituto Politécnico de Bragança, Escola Superior Agrária, Campus de Santa Apolónia, Bragança, Portugal
e-mail: jazevedo@ipb.pt

J. Barstad
University College for Green Development, Bryne, Norway
e-mail: johan@hgut.no

P. Gatto · D. Pettenella
Dipartimento Territorio e Sistemi Agroforestali, Università degli Studi di Padova, Agripolis, Padova, Italy
e-mail: paola.gatto@unipd.it; davide.pettenella@unipd.it

J. Lesinski
Department of Forest Biodiversity, Institute of Forest Ecology and Silviculture, University of Agriculture, Krakow, Poland
e-mail: jerzy.lesinski@urk.edu.pl

R. Stokken
Volda University College, Department for Social Sciences and History, Volda, Norway
e-mail: roar.stokken@hivolda.no

© The Author(s) 2022
R. Tognetti et al. (eds.), *Climate-Smart Forestry in Mountain Regions*, Managing Forest Ecosystems 40, https://doi.org/10.1007/978-3-030-80767-2_14

many policy instruments used with varying degrees of success such as forest concessions or voluntary certification schemes. A wide range of instruments are responding to direct regulation; this has been seen as insufficient to deal with natural hazards and calamities.

Abbreviations

CAP	Common Agricultural Policy
CSF	Climate-Smart Forestry
CFT	Forest Territory Charters
CoIN	Collaborative Innovation Networks
CSR	Corporate social responsibility
EC	European Commission
ES	Ecosystem Service
EU	European Union
FNCoFor	Forest-owning Communes represented by their Federation
FOA	Forest Owners' Association
FSC®	the Forest Stewardship Council
MF	Model Forest
NFO	National Forest Office
NGO	Nongovernmental Organization
NPG	New Public Governance
NPM	New Public Management
PAM	Public Administration and Management
PEFC™	the Programme for the Endorsement of Forest Certification
PES	Payment for Ecosystem Services
RESP	Rural Environment Protection Scheme
RRI	Responsible Research and Innovation
SFM	Sustainable Forest Management
TPA	Traditional Public Administration

14.1   Introduction

It is generally assumed that sustainable forest management (SFM) can play a significant role in climate change mitigation (e.g., Makundi 1997; FAO 2018). Moreover, according to the analysis of diverse services schemes (e.g., payments for watershed), different types of novel institutional setups based on the establishment of financial incentives to natural resources owners or managers for ecosystem service (ES) provision provide co-benefits, such as carbon mitigation and others (UNECE and FAO 2018). In accordance with Climate-Smart Forestry (CSF) (see Chap. 2 of this book: Weatherall et al. 2021), this chapter presents and describes tools for achieving SFM and the provision of different types of forest ES, including (but not exclusively) carbon mitigation.

According to existing institutional economics and governance structures (e.g., Thompson 2003; Lemos and Agrawal 2006; Vatn 2010; Muradian and Rival 2012; UNECE and FAO 2018), environmental governance instruments can be divided into 3 main types: 1) hierarchies (governmental command-and-control), 2) market-based (voluntary exchange), and 3) community-based (cooperation networks). Currently and most typically, these can be combined into hybrid systems (Muradian and Rival 2012; Pahl-Wostl 2019).

One possible way of categorizing these systems is through the identification of key characteristics or dimensions. Pahl-Wostl (2019) highlights the *role of government*, divided into three types: 1) setting the rules, 2) being a partner to society, 3) delivering services to society. Such a classification could be viewed as being consistent with the distinguishing between top-down or bottom-up approaches (e.g., Hollberg et al. 2019) or with the distinguishing of *motivation*: external, as based on political targets (e.g., Braune and Wittstock 2011) or internally self-motivated. The same holds for the distinguishing among the subjects: individual providers of ES, local communities, regions (local government), or the state.

While these aspects are important when economists describe how the state governs resources of economic nature, political scientists attend to how the state relates to and interacts with the citizens. Within this tradition (e.g., Bryson et al. 2014; Torfing 2018; Amdam 2020; Sørensen 2020), the public sector is often portrayed on the basis of Osborne (2006, 2010) and thus depicted through three different regimes: Traditional Public Administration (TPA), New Public Management (NPM), and New Public Governance (NPG). Relating to the characteristics used within the economic domain, these three regimes represent "archetypes," which often coexist or overlap, also with economic means and instruments, within Public Administration and Management (PAM).

Within TPA, the role of the state is to be a unitary and hierarchical organization, focusing upon the policy system, emphasizing policy implementation and understanding nonpublic partners as potential elements or disturbances in the policy system. Due to the hierarchical logic of this regime, citizens are framed as someone who *ought to* obey the state. This regime is often understood as bureaucratic in the negative sense of the word (Amdam 2020). Since citizens ought to obey the state, this regime is aligned with the "setting rules" approach described by Pahl-Wostl (2019).

Further, NPM can be understood as a reaction to TPA and, as such, reflects a wish to modernize the public sector. Within NPM, the state is understood as a disaggregated organization with efficacy resting upon competition and market-place mechanisms. The focus is upon intraorganizational management and emphasis is on service inputs and outputs. Nonpublic partners are thus understood as independent contractors within the competitive market. In line with the private sector analogy, citizens are generally framed as customers with the ability to choose. This regime is often criticized for entailing fragmentation of power and responsibility (Amdam

2020). Due to the role of the state as being a service provider to the public, this regime aligns with the delivering services to society approach (Pahl-Wostl 2019).

NPG can, as the newest regime, be viewed as a reaction to experienced NPM shortcomings, and as a strategy to comply with citizens demanding more influence than through the ballot box alone. This regime, therefore, emphasizes governing on the basis of equality, dialogue, and coproduction (Sørensen 2020). Within this framework, the state is seen as a pluralist unit, governed from a neocorporatist point of view, where trust and relational contracts regulate service processes and outcomes. Nonpublic partners are seen as preferred suppliers, often interdependent agents, with active and ongoing relationships in already established networks. Citizens are framed as someone that can exert influence; thus, the regime emphasizes processes where citizens can have influence beyond the limits of the representative democracy, aligning this regime with the "be a partner to society" approach (Pahl-Wostl 2019).

While the key elements of TPA and "setting rules" remain law, rules, guidelines, bureaucracy and the hegemony of professionals in service delivery, NPM and "delivering services to society" pays attention to lessons learned from the private sector; cost management, input and output control, competition, and contracts. In contrast to these, NPG and "be a partner to society" focuses upon the active citizen that can exert his/her influence upon both state and own situation. Thus, this last regime is often viewed as the preferred tool for mending the distrust to politicians presently found throughout western societies, often leading to populism and polarization. Implementing processes and mindsets belonging to this last regime can therefore be of great importance in the field of forestry to avoid such destructive processes. All relevant stakeholders ought thus to be involved in the innovative development processes of instruments meant to stimulate CSF.

Based on literature review and schematic structure of environmental governance from Lemos and Agrawal (2006), Fig. 14.1 shows the classification of key policy instruments that can be used to address the implementation of the CSF concept.

As mentioned above, the three major types of governance structures are *hierarchy* (*command-and-control*), *voluntary* (*market-based*), and *community cooperation*. Hierarchy or *command-and-control* instruments are based on the power of the state to regulate private actions through laws and regulations (e.g., Vatn 2010). An example is direct regulation, for example, in the form of obligatory, minimal percentages of given tree species planted during reforestation.

Hierarchy command-and-control instruments represent direct regulation by local, state, or national governing bodies. However, some types of regulation could stimulate or frame certain market-based tools (e.g., the European Commission (EC) regulations on ecolabel, mark of origin, organic products) and even some community tools (e.g., the EC regulations related to the so-called Leader approach).

Market-based instruments are based on voluntary exchange. These could include regulations or institutions that affect behavior through market signals (Stavins 1998). An example of such instruments may be payment for ecosystem services (PES). Most authors (e.g., UNECE and FAO 2018; Prokofieva and Gorriz 2013)

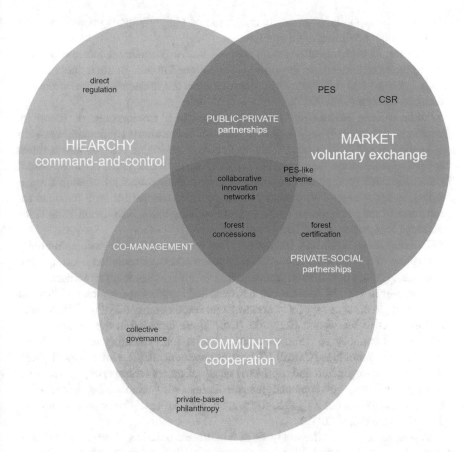

Fig. 14.1 Classification of key policy instruments with potential application within CSF (*PES = Payments for Ecosystem Services; CSR = Corporate Social Responsibility*). (Based on Lemos and Agrawal 2006: Fig. 1; and Vatn 2010)

have adopted the PES definition from Wunder (2005). According to his definition, PES must comply with five characteristics (Wunder 2005, p. 2): "(1) a voluntary transaction, in which (2) a well-defined environmental service (or a land-use likely to secure this) (3) is bought by at least one buyer (4) from at least one provider (5) if and only if its provision can be secured (conditionality)." If a PES scheme does not fulfil all (but at the minimum one) of the conditions, it is called *a PES-like* scheme (Wunder 2005). Some PES-like schemes or mechanisms may fulfil the criteria of the command-and-control category rather than market-based. In practice, PES schemes are mostly hybrid systems based on both market and nonmarket instruments (e.g., Higgins et al. 2014; UNECE and FAO 2018).

Community-based governance is based on cooperation with and between local actors, creating a sense of community ownership. Collective governance is an example of community governance. Private-based philanthropy is another example of an instrument building upon community.

However, instruments often simultaneously hold aspects that reflect and are constituted by more than one of the structures mentioned above. As such, these instruments are to be found in the intersections between the main governance structures described by Vatn (Vatn 2010).

Public-private partnership is created through a combination of hierarchy (command-and-control) and market instruments. It represents cooperation between state agencies and market actors (Lemos and Agrawal 2006). An example is forest concession, a temporary allocation of typically state-owned public forest resources to another actor (companies, communities, or nongovernmental organizations – NGOs), in order to regulate timber production to be compatible with the need for protection of the ecosystems, and the rights of indigenous people and local communities (Tegegne et al. 2018).

Comanagement connects hierarchical governance structure and community cooperation and represents cooperation between state agencies and communities. Such examples are information campaigns or voluntary certifications schemes.

As shown in Fig. 14.1, there can be identified instruments, which combine all three major governance types. Corporate social responsibility (CSR) and in particular Collaborative Innovation Networks (CoINs) are included in this category. The CSR concept connects elements of voluntary behavior, mandatory responsibilities, and societal expectations (Carroll 1999; Brammer et al. 2012). CoIN is mostly perceived as a self-organized and motivated group of actors (enterprises, universities, research institutes, governments, intermediaries, etc.), elaborating a collective vision, collaborating not only face-to-face, but also through Web or other platforms (Gloor 2007; Wu et al. 2019).

As seen in Fig. 14.1, different types of instruments based on different governance structures can be identified. We recognize primarily instruments, such as direct regulation, PES and PES-like schemes, certification, information campaigns, collaborative innovation networks and private-based philanthropy, or corporate social responsibility.

14.2 Direct Regulation

Direct regulation (or direct control or administrative instruments of regulation) is based on the legitimacy and power of a state to regulate private actions (Holzinger and Knill 2004; Vatn 2010). Legal regulations may come under legislation of local, state, regional, or European Union (EU) governments. In every country, the direct regulation in forestry exists in some form – it is mainly based on protection of forest in terms of determination of what is permitted and what is illegal (McManus 2009). The underlying principle is to set standards or targets (e.g., for lodging, maintenance, renewal, etc.) by a government authority that has to be complied with. If the standards or targets are not fulfilled, negative sanctions and penalties may occur

(Baldwin et al. 2011). Direct regulation has been used since the beginning of environmental policy making by industrialized countries in the 1970s (Böcher 2012). Its advantage is speed and clearness. In case of functioning control, the meeting of preset targets is expected. On the other hand, fixed standards and rules are reactive and rigid. If fulfilled, no other motivations for desirable behavior exist. Economists stress the problem of high compliance costs and expensive control that results in generally low economic efficiency of this kind of intervention (Tietenberg and Lewis 2018).

The example of national regulation of forestry and forest management is the Federal Forest Act in Germany. German forest policy defines rules relating to forest, forestry, and timber utilization. The aim is to deal with conflicts of interests between the various economic, ecological, and social demands. The act is accompanied by other acts (e.g., Act on Forest Reproductive Material, the Timber Trade Safeguard, the Federal Hunting Act, and others). Comparable acts can be found in other European countries. Acts can be supplemented by various strategies, since these general acts mainly do not deal with changing social and environmental conditions, such as climate change. An example from Germany in this context is the Forest Strategy 2020. However, since this strategy addresses climate change mitigation and adaptation broadly, it does not state specific goals for mountain forestry (Federal Ministry of Food, Agriculture and Consumer Protection 2011).

National rules for afforestation might be considered relevant direct regulation within climate change policies. In many countries, however, including Czech Republic, the landowner cannot freely decide to change the arable land or pasture into forest. The Czech Act on the protection of the agricultural land fund defines the rules for exemption of land from the fund and levies for exemption of agricultural land. For afforestation of agricultural land consent of the state administration authority is always required and levies for exemption are set. Levies are not set only for land in the lowest levels of protection (fourth and fifth level of protection). It is also required to change the land to land for forest function in case of afforestation of land in the third level of protection. In other words, the owner cannot use the land for plantation or forest monocultures. Agricultural land in the first and second level protection may be withdrawn only in cases where the other public interest significantly outweighs the public interest in the protection of the agricultural land fund.

On the other hand, forest expansion has been one of the most prominent land use changes encouraged also by the introduction of a legal framework for afforestation payments in the reform of EU's Common Agricultural Policy (CAP) (Kotecký 2015). However, while the governments in the EU had built up a legislative framework for afforestation support scheme, this is not a direct regulation. By contrast, there are also additional agri-environmental subsidies paid under Rural Environment Protection Scheme (RESP) adopted with the reform of CAP in 1993. Nevertheless, the legislation and direct regulation, for example, in Ireland, caused decreasing interest of farmers in planting trees, since land under RESP could be withdrawn after 5 years, while the decision of afforestation was irreversible under current legislation (McCarthy et al. 2003).

In the European Union, the formulation of forest policies is the competence of the Member States. Different examples of direct regulations at the EU level affecting forestry are hidden in other directives. One such example is The Fauna-Flora-Habitats Directive (92/43/EEC – FFH or Habitat Directive). Very limited requirements for forest management can be identified in the Directive (Winkel et al. 2009), but Rosenkranz et al. (2014) investigated income losses in forestry due to the implementation of Habitat Directive caused by reduction of the amount of harvested timber. Since approximately 5.2% of the total German forest area had not been subject to a special protection status before (Sippel 2007), annual income losses for the enterprises averaged 31 to 39 €/ha (Rosenkranz et al. 2014). Direct regulation can, therefore, by spillover effects, be an efficient means for the implementation of CSF. The aim of the Habitats Directive is biodiversity protection regardless of implementing CSF.

It can be stated that direct regulation is, in principle, efficient until the calamity. Among the most common disasters in European forests induced by climate changes are fires, pest outbreaks (e.g., bark beetle), windstorms, ice storms, and drought (Seidl et al. 2014). Pests have especially caused problems in mountain forests of East-Central Europe, since increasing temperatures (among other factors) created better conditions for forest pests at higher mountain sites (Battisti and Larsson 2015). Unfortunately, there is no evidence of direct regulation dealing with CSF. Also, there is a lack of direct regulation that deals with prevention of climate change related issues. Mourao and Martinho (2016) examined the effectiveness of Portugal's direct regulation, regarding fires in forests. Their findings are in line with the claim that direct regulation is mostly reactive. Although there have been problems associated with the frequency and incidence of forest fires since 1980, Portuguese legislative documents on forest fires follow the evolution of fire. This direct regulation *tends to be unable to prevent the dimensions of abiotic forest factors* (Mourao and Martinho 2016: 476). They added that both legal instruments instability and conflicts are other problems that do not contribute to achieving desired results (Mourao and Martinho 2019). Conversely Fernandes et al. (2017) claim that crises should be an opportunity, or trigger, for change or reforms. However, they also noted that sudden and reactive changes in (fire-related) legislation can cause confusion (Jensen and McPherson 2008) by reason of the bounded rationality and selective attention to relevant information (Busenberg 2004). It can be stated that direct regulation cannot prevent any problems or calamities, but it should prevent an escalation of the problems.

14.3 Collective Governance

Since the 1970s, there has been a growing awareness of negative social and environmental problems caused by attempts to control very complex natural systems. There has also been increasing criticism of conventional command-and-control-based resource management, and innovative approaches to understanding and managing environmental and social systems have emerged (Secco et al. 2011; Paavola and Hubacek 2013). An adaptive collaborative approach did not take hierarchical management as a starting point but, on the contrary, positions itself by introducing the concept of "bottom-up" management (Fraser et al. 2006). It is expected that actors traditionally involved in forest-related decision-making also will address and apply new concepts and tools, such as public participation, shared responsibility, and networking in governance mechanisms (Secco and Petenella 2006; Secco et al. 2011; Lovrić et al. 2012; Niedziałkowski et al. 2012, 2015; Paloniemi et al. 2015; Primmer et al. 2015). Depending on the sector and context, forest governance may be dominated by private actors, NGOs, public authorities, or others, which are a mixture of theoretically equal actors. The difference in governance modes "lies simply in who is involved in making common choices" and how the one's involvement is managed.

In Northern and North-Western European countries, private ownership of forestland ranges from about 40% to 80%, and in Eastern European countries from 10% to 60%. In Western and Central Europe, small (up to 5 ha) land-holdings represent about 85% of all forest owners (Schmithüsen and Hirsch 2010). In countries with a large share of private forest owners, high fragmentation, and large number of owners, various forms of decentralized partnerships or even private governance are more common (Beland Lindahl et al. 2017).

These private forests are very important as they have significant potential for the production of wood and nonwood products, provision of multiple ecosystem services, including carbon storage and biodiversity conservation (Osman-Elasha et al. 2009). On the other hand, their role in the wood market has been, and still is, small. In order to strengthen their position, the first Forest Owners' Associations (FOAs) have emerged in Europe already at the end of XIX, and beginning of XX century and they have become very common over time (Weiss et al. 2011; Kronholm 2015; Aurenhammer et al. 2017). The organization and functioning of FOAs reflects the way forests are governed, which is based on the cooperation of local actors who create a sense of community ownership. The term "collective governance" was introduced by Ostrom (1990), while Armitage (2005) presented the principles of community resource management.

There are numerous definitions of forest governance, but the definition proposed by Lockwood (2010), that is, "the interaction between structures, processes and traditions that determine how power and accountability are exercised, how decisions are made and how citizens or other stakeholders have their say," seems to be the most appropriate to describe collective governance. This term, however, does not mean either equal sharing of power or a high level of stakeholder involvement.

The community forestry's basics include such objectives as empowering forest users, and improving the condition of the forests, by means of managing forests by local stakeholders, for commercial and noncommercial purposes (Vizzari et al. 2012). FOAs regulate the use of the natural resources and take care of all aspects of community life (Vizzari et al. 2012). They also focus on sustainable collective governance outcomes, such as the quality of the decision-making, efficiency of the implementation process, and sensitivity to different context-specific aspects. Thus, when evaluating community forestry outcomes, it is especially important to state whether this governance mode has delivered what was expected (Maryudi et al. 2012).

If any conflict emerges, FOAs try to overcome it taking into consideration interactions among ecological functioning, social structures, and stakeholders' participation (Paavola and Hubacek 2013). The gained expertise is invaluable in the process of continuous collective learning (Secco and Petenella 2006; Sandström 2009; Saarikoski et al. 2012). Broad argumentation, inclusive stakeholder participation, and genuine knowledge sharing are considered to contribute to positive outcomes (e.g., Reed 2008).

As mentioned before, the community forests are managed by the forest owners themselves. In France, however, the FOAs' forests are used by the forest owners, but they are managed by the National Forest Office (NFO). For the discussions with NFO, the forest-owning communes are represented by their federation (FNCoFor) (Chauvin 2012). Such a share of decision-making between the private and the state actors results in the comanagement governance mode.

The "forest territory charters" (CFTs), that is, participative strategic processes carried out at a local level, were instituted on the basis of the French forest law of 2001. The CFTs were conceived as a policy tool for promotion of the principles of SFM in community forestry. Different stakeholders at all scales have seized this tool to adapt it to their own strategies. The CFTs have influenced these policies in the long run, even changing economic conditions and governments (Chauvin 2012). A mission of coordination of the CFTs, given to the FNCoFor by the national government, resulted in a conversion of the FNCoFor toward a reliable intermediate to promote the CFTs as a bottom-up approach. It induced a firm empowerment of the FNCoFor, evolving from a simple political representation, in a state-steered device, to a strong lobby system, shaped as a support network with expertise in both forestry and rural development, while NFO had to withdraw to technical matters notably due to the costs of participation (Chauvin 2012). It is not clear, however, whether the FOAs governance mode is the same as before the CFT initiative was undertaken or if it has changed into collective governance.

In 1992, the federal government of Canada set up a collaborative governance initiative through the Model Forest Program (LaPierre 2002; Parkins et al. 2016). The government's idea behind the initiative was to allow an initial 10 Model Forests (MFs) across the country to act as demonstration sites for SFM, while the government initially provided core funding and leadership. Such a share of responsibilities between the MFs and government might be identified as the comanagement. The mandated structure called for an industrial partner as well as a commitment to accommodate not only the local residents and representatives of the government, but also First Nations, NGOs, and individual persons interested in the MF goals and their mode of action (Hall and Bonnell 2004; Rametsteiner 2009; Parkins et al. 2016). Yet, the presence of the industrial partner in the MF partnership resulted in a change of the governance mode from comanagement to corporate social responsibility (CSR).

Since the concept has shown itself to be flexible and adaptive to its setting (Besseau et al. 2002; Ho et al. 2014; IMFN 2019), what once began in Canada is now the world's largest network consisting of more than 60 MFs across the World, of which 10 have been established in the EU countries (IMNF, 2020). Each MF is unique but all of them are governed by the same principles of trust, transparency, and collaborative decision-making (IMFN 2019, 2020). MFs are proven examples for good governance effectiveness and efficiency as well as shared stewardship of forests and the larger landscapes that surround them (Rametsteiner 2009). It has to be noted that the Canadian suggestion of incorporating the industrial partner has been adopted just by a few MFs outside Canada (Shingo 2017). On the contrary, the majority of the MFs adopted the collective governance mode, since they are not any longer financed nor led by the respective national states but, by acquiring a legal identity, they were able to function, thanks to grants awarded by foundations, aid agencies, and the local, national, and international institutions and organizations.

14.4   Voluntary Certification Schemes

Forest certification is a voluntary market-based instrument to achieve SFM. Voluntary instruments aim to change the forest owners' and forest managers' behavior without the force of law (van der Heijden 2018). Originally, it was introduced in the early 1990s to address concerns of deforestation and forest degradation, to protect forest biodiversity and to mitigate illegal logging (e.g., Cashore et al. 2006; Stupak et al. 2011), especially in the tropics (Rametsteiner and Simula 2003). The goal of labelling wood products based on forest certification is an identification of products from well-managed forests (Elliot and Schlaepfer 2001). Such initiatives are always addressed by independent third parties (Elliot andet Schlaepfer 2001; Buliga and Nichiforel 2019), mostly nongovernmental organizations.

The aim of the forest certification is to promote SFM with two tools: criteria and indicators (Rametsteiner and Simula 2003). There are currently several international certification schemes defining specific sets of standards for sustainable forest management. Two forest certification schemes are present across Europe: the Forest Stewardship Council (FSC®) and the Programme for the Endorsement of Forest Certification (PEFC™) (Buliga and Nichiforel 2019). PEFC is an international umbrella organization for approving national schemes. Most countries are involved in these two processes. Both schemes include criteria for best practices in forest management, covering environmental, and social and economic aspects. The area of the world's forests covered by these two schemes increased from 14 million hectares in 2000 to almost 420 million hectares in 2014 (FAO 2016: 26). Even though the original purpose for the establishment of such a scheme was to protect mainly tropical forests, most certified forests are in the temperate and boreal zones, with Europe as the most important region (Rametsteiner and Simula 2003).

In addition to the certification of forest management, there is also the second type of forest certification. Certification of the chain of custody (CoC certification) verifies that certified wood material product has been taken from forests and managed according to responsible forest management standards (Buliga and Nichiforel 2019), and that it has been kept separate from noncertified material through the production process. This chapter is focused on certification of forest management.

The advantage of the instrument of certification is its voluntary nature and provision of benefits, not only to forests owners, managers, entrepreneurs, or timber companies. As a final result, certification helps to achieve social and environmental goals. This is facilitated by the set of criteria and standards. The FSC forest management standards include 10 principles and related criteria to confirm that the forests are managed economically and also in a way that preserves natural ecosystems and the social benefits of local communities and workers (Buliga and Nichiforel 2019). In order to receive FSC certification, all of them must be met. They include, for example, compliance with laws (including national laws, international agreements, etc.), contribution to maintaining, conservation and/or restoring of ecosystem services and environmental values, obligation to maintain and/or enhance the high conservation values through applying the precautionary approach (FSC 2015).

Among the main motivations for participating in forest certification are the following: landowner interest, knowledge and awareness, alignment of certification aims with landowner's values, low cost, market access, and benefits (e.g., Boakye-Danquah and Reed 2019). Public image or feedback on management practices are mentioned by community-based forestry practitioners, while their expectations for economic benefits are low (Crow and Danks 2010). According to research from the USA, access to professional forest management advice or information is a significant factor in seeking forest certification (e.g., Ma et al. 2012).

The criticism of forest certification is mainly focused on its minimal impact on tropical forest deforestation. Another disadvantage can be seen in dependency on local law. According to their research based on comparative analysis of Poland and Belarus, Niedziałkowski and Shkaruaba (2018; 180) stated that *the effectiveness of FSC depends solely on the strong support of the government and can collapse if the support is withdrawn.* In both cases, the role of the state actors was crucial in the forest certification process. This is in all probability broadly valid in countries where the state is a dominant forest owner, for example, postsocialist countries. Furthermore, a Romanian case study (Buliga and Nichiforel 2019) evaluated how much of the FSC standard is contingent on legal rules in the country. They concluded that 69% of certification requirements are addressed in law. They added that the effect of certification is mainly about contributing to a better enforcement of the existing legal rules.

It can be stated that forest certification can help with the introduction of climate smart forestry and that forest certification is a usable instrument for achieving environmental goals. It seems that especially small owners welcome the information and advice about SFM. In combination with information campaigns, forest certification can be seen as an effective additional instrument for fostering climate smart forestry, together with legal rules, and command and control instruments. However, it should be supplemented by raising the awareness of forest owners, or by demonstrating the economic benefit. Benefits are mostly affected by the demand for certified products, which in turn is related to general economic indicators (FAO 2016). Nevertheless, since the common reason for participating in forest certification schemes is public image, it should be effective also to raise public awareness. Disadvantages or problems of such instruments could be the necessity to have the independent third party, which supports and guarantees the certification program.

14.5   Private-Based Philanthropy

Most instruments require the involvement of different public administration levels. Specific legislation or other official documents with rules need to be issued and particular bureaus need to be empowered, so an instrument starts functioning. Instead, private-based philanthropy deals with the independent action of property owners that from different reasons pursue common interests, being its biodiversity enhancement, water retention, or CSF. While doing so, they (voluntary) sacrifice short-term economic benefits from using the land.

Private-based philanthropy may involve the action of an individual landowner or a nongovernmental organization. Its most spread form is so-called land trusts, private protected areas with long tradition particularly in the USA and Australia (Bennett et al. 2018). Within Europe, some forms of land trusts exist in the majority of Western European countries, especially the UK. Usually, their main goal is to acquire the land and to conserve it for biodiversity enhancement purposes. In order to gain money to acquire the land and to manage it, land trusts receive private donations. In the USA, the functioning of land trusts is supported by the existence of conservation easements. These are voluntary legal agreements restricting some property rights of the owner – therefore, a land trust does not buy the land itself, it only buys part of the land use rights and monitors if the conservation occurs (Horton et al. 2017; Graves et al. 2019). Landowners participating in conservation easements may receive some financial benefits, including income, tax credits, or tax relief on their decreased property value. Bastian (2017) proves that this financial motivation together with pursuing common interests is attractive for landowners – apart from financial incentives, a desire to protect the land in perpetuity or to maintain a connection with land is an important impulse.

Except for typical nongovernmental organization (such as land trusts) engagement, individual landowners may also pursue common interests voluntarily even when receiving no financial compensation. They do it especially if: a) they believe that what they do is the right thing, b) their living is not fully dependent on the economic use of the land. This is, for example, the position of the "philanthropic farmer" in Czechia, as described by Slavíková and Raška (2019), who decided to build natural water retention measures on his field from his own resources despite the fact that he reduces the extent of the farming land and subsequent agricultural subsidies.

This evidence shows that private actions may support (not only undermine) public policy goals. Their focus does not need to be limited to biodiversity conservation or water retention as mentioned in examples above. SFM or even CSF – if the private land in focus is a forest – might be considered as well. In Czechia, some existing land trusts aim at forest renewals considering the specific species composition adapted to changing climate. Čmelák was established 1994 and represents newly established land trusts in the form of a nongovernmental organization, where members of the NGO share common interest in deciding about the type of forest activities as the Board, or during Membership meetings. In the early1990s, members were concerned with the poor condition of forests in the North of Bohemia in the Jizera Mountains, mainly caused by sulfur emissions and bark beetle. Acting together with volunteers, and raising money from public donations, Čmelák bought the land. They changed spruce monoculture into mixed forest called "New Virgin Forest" (Špaček et al. 2020). Ongoing management of the New Virgin Forest is mostly financed by subsidies (grants) and donations. The PES scheme represents an important part of the donation in Čmelák (see next Sect. 14.5). Čmelák implements their activities in a relatively small area. Long-term goal of Čmelák's activities is to show novel ways to approach and implement financially self-sufficient SFM focused on nontimber forest products.

Private-based philanthropy will rarely solve problems worldwide. The rationale to promote and even establish private initiatives is mostly rooted in their potential to effectively overcome land-use conflicts and institutional misfits on fragmented land.

14.6   Market-Based Instruments: PES and PES-Like Schemes

This section introduces examples of the implementation of payments for ecosystem (or environmental) services (PES) scheme, while more detailed description of PES as an incentive tool for behavioral change to climate-smart practices can be found in Chap. 12 of this book: Gežík et al. (2021).

According to the previously mentioned definition of the PES scheme, the Land Trust Association Čmelák (Czechia) can be seen as an example of a PES-like scheme, since it complies with characteristics defined by Wunder (2005, p. 2): "(1) a voluntary transaction, in which (2) a well-defined environmental service (or a land-use likely to secure it) (3) is bought by at least one buyer (4) from at least one provider (5) if and only if its provision can be secured (conditionality)."

The main activities of Čmelák are based upon buying land with spruce monocultures in the Jizera Mountains and changing them into mixed-forest, so-called New Virgin Forest. Their activities are geared toward nature protection and conservation. Besides revenues from services (ecological education, sale of seedlings, etc.), subsidies, and grants, Čmelák obtains significant resources by selling Certificates of Patronage. Local citizens, public, or companies voluntarily buy this certificate to help with financing Čmelák's activities and extension of New Virgin Forest. Buyers become a patron of New Virgin Forest and contribute toward the creation of nonmaintained forest providing nontimber forest services (educational, recreational, etc.). The main problem may be seen to lie with the revenue generated by selling certificates, since this amount is sufficient for buy-off of new land; however, additional resources are required for other activities (maintaining New Virgin Forest).

Given that SFM in mountain forests can provide important regulation services (carbon sequestration and storage, water regulation and protection of soil, people, and infrastructure against hydrogeological hazards, etc.), groups of stakeholders may be willing to pay and compensate forest owners from losses occurring due to SFM and reducing timber harvesting as a means of lowering the risk of damage to buildings and people (Grilli et al. 2020).

The quantification of the monetary value of services is always important for performing PES. Olschewski et al. (2012) and Grilli, et al. (2020) describe monetary valued avalanche protection by forests in Swiss Alps using choice experiment and determination of willingness to pay for the protection. They conclude that willingness to pay is at a level of per household costs of alternative measures, and furthermore substantially higher than the costs of measures needed in forests to maintain protection based on engineering and silvicultural knowledge. The study shows examples where PES schemes should be an efficient solution for (in this case) avalanche protection in mountainous regions.

Studies indicate that the monetary valuation and setting the amount of payment is the most challenging factor when implementing PES schemes. Nevertheless, it is considered as an effective and already used instrument for changing forest management to SFM. PES scheme may provide important sources and compensations for forest owners for losses occurring for reducing timber harvesting.

14.7 Collaborative Innovation Networks

To ensure that instruments for CSF have impact, it is vital to check they are in accordance with present contexts. Standardized solutions and instruments should not simply be "put into use." Rather they must undergo a process where their appearance and usage become adapted, adjusted, and implemented in a manner that ensures correspondence with the expectations, motivations, and values of those affected by, or wielding, the tools. To ensure accordance between chosen instruments and society. Collaborative innovation networks (CoIN) may seem a promising method. Still, in their traditional format, CoINs are probably not by themselves sufficient to achieve the anticipated alignment between instrument and present challenges.

As a concept, CoIN was introduced to describe how multiple partners, through collaborative action, possess the power to successfully instigate products, services, policies, processes, business solutions etc. CoINs are mostly perceived as self-organized and motivated groups, consisting of actors like enterprises, universities, research institutes, and governments that, by means of collaboration, elaborate a collective vision. Such networks are highly plastic and adaptive, and able to adjust to local or temporal situations, both concerning time and space (Gloor 2007). Such networks can therefore easily be understood as viable networks, possessing great potential for CSF.

Several examples of CoIN in forestry and rural development can be located. The two presented below are representative for those found when trawling the Internet.

The *"Innovation Networks of Cork, Resins and Edibles in the Mediterranean basin"* aims to foster communication and knowledge transfer to highlight how enhanced fluxes and multi-actor networks can help to seed real and effective innovation of NWFPs and contribute to business discovery, social innovation and the co-design of locally adapted innovative value chains (https://www.incredibleforest.net). Still, from a CoIN-perspective the network lacks representation from the societal side, except through proxies, like NGOs.

The *"PIRIC-network"* of New Zealand (https://www.piric.org) aims for a sustainable future for New Zealand through accelerating the adoption of digital technologies into their Primary Industries and Regional Economies, realizing they can no longer act as individuals or organizations working alone. With industries and communities entering a time of significant disruption, rapid increase in consumer acceptability, and changes to the international trading environment. Even with a number of founding members from the societal sector, there still is a lack of real involvement into the core areas of activity by the relevant societies.

The common factor of these, as well as of the other examples found, was their focus on societal involvement, while still displaying a top-down structure. Nowadays, this may not be sufficient, since it is increasingly evident that it is both ethically and practically doubtful to let innovation processes be governed solely by the traditional actors, logics, and forces. This stems from the experience that, despite research and innovation improving our lives in almost every area, one also sees the transformative forces of innovation creating new risks and dilemmas.

To promote real involvement and reduce the risks of negative impacts from expert-dominated innovation, the concept of Responsible Research and Innovation (RRI) has been developed and implemented. To reduce negative "side effects" from research and innovation, RRI demands the involvement of citizens and communities in an upstream fashion, to ensure outcomes being aligned with the values of society (RRI-Tools 2020; Stokken and Børsen 2020). Thus, important funding frameworks, like Horizon 2020 and COST as well as several national funding schemes, require the implementation of RRI in all projects and activities.

RRI rests upon a normative framework, comprising the six policy keys: 1) ethics, 2) gender equality, 3) governance, 4) open access, 5) public engagement, and 6) science education. In our CSF context, the six keys denote that a CoIN process instigated to assure accordance between instrument and context, ought to be governed by civil servants that adhere to the NPG regime (Sørensen 2020) and the "be a partner to society" approach (Pahl-Wostl 2019). In particular, the governance and public engagement keys demand comprehension and the use of networks to be integrated parts of development and innovation processes.

To engage in networks in a manner where RRI is taken seriously, civil servants must employ a novel position, where both citizens and other relevant actors are provided with the possibility of being involved from the very start. Only then can the "current demand" for governed innovations, where people have a right to be heard, and where the inclusion of citizens' knowledge, resources, and motivation actively contributes to successful implementation, is met. To achieve this, citizen involvement must take the form of real debates (Force free discourses – in the terms of Habermas (1995)), where the say of the citizens surpasses tokenism (Arnstein 1969). An example of such an arrangement is how Ireland established citizens forums to foster dialogue around abortion to prevent domestic polarization. Another example is from Denmark, where a municipality that established ad hoc political committees at the core of their policymaking experienced that citizens' involvement transformed from demanding to actively seeking and promoting solutions (Sørensen 2020).

When problems, like those in Ireland and Denmark, transgress being a matter of finding the most bureaucratic sound or economically most viable solution, to being a matter of democracy, governance, ethics, and engagement on equal terms for all citizens, a problem often transforms from being a difficult question to a so-called wicked problem. The complexity and constant change that characterizes societal development has given rise to this concept (Churchman 1967, Rittel and Webber 2007), reflecting a reality in which societal problems in general tend to be governed by incomplete, contradictory, and/or changing requirements that are difficult or impossible to solve in a "right" way.

In our context, the core lesson in the development of instruments for CSF, therefore, relates to the need to always be looking for the feasible solution, understanding that the traditional positivistic science often fails to describe the societal processes encompassing policy development and implementation conditions. This calls both for a process-oriented approach, and a willingness to adapt and seek new solutions.

To get the best out of CoINs, it is therefore important not to "marry" one specific method or solution, but to keep an open view, looking for "solutions that can work" in this particular setting on the basis of a focus on the process rather than on the output. In this matter, being aware that collaboration possesses the power to contradict the traditional linear, deterministic models, while on the other hand generating results that are more depending on the conditions under which they are developed, than the actual resources and the factual actors involved. Thus, two parallel processes, involving the same set of actors, but separated in space, or sharing the same space but separated in time, may well end up in different solutions. And even more – that the choice of and inclusion of whom to consider as stakeholders largely will define the potential power of the outcome.

Approaches can, as argued, be utilized in a variety of modes of collaboration, for example, in partnership models, triple- or quadruple-helix models, communicative planning models, or actor-network models. The essence and their commonality are the ability to open up for broad and inclusive cooperation, where new solutions may be found and thus induce improved conditions for implementing. These may be reached through addressing three crucial aspects:

- *Perspectives*. More "heads" involved may give rise to a wider variety of solutions.
- *Inclusion*. Working together fosters trust.
- *Feasibility*. Trust eases the ground for easier and more successful implementation.

In market situations, this approach opens opportunities for partners/collaborators to increase their profits. In nonmarket situations, for example, regarding environmental services regulation and provisioning, giving rise to novel approaches, and in particular to what could be easier, more rapid and improved conditions for implementation.

CoIN further bridges with models of interacting spheres, like the classic three-fold sustainability model (see Fig. 14.2) or John Friedmann's 4-Life-worlds, in that they acknowledge the need for pluri-sectoral, or cross-sectoral cooperation. And, as well, representing sectors/realms outside of direct human control. Like natural-systems-based design, where one tries to mimic nature's solutions and to recreate them into modern day, human designed environmental solutions.

Thus, alliances with and among the stakeholders are needed, surpassing both the traditional bureaucratic approaches where the civil servants were the masters, and

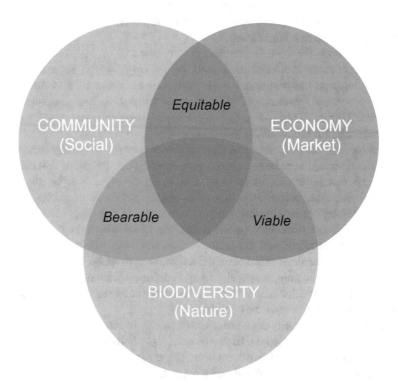

Fig. 14.2 The three spheres model of sustainable development (with overlaps labeled). (based on Thatcher 2015)

the approach denoted by New Public Management where the civil servants have the serving role. Rather, the servants must embrace the collaborative role, where the goal is to empower citizens as individuals and groups, and to build implementable and feasible solutions together. CoIN processes, in accordance with the Pan-European policy requirements for RRI, involving all relevant stakeholders and assuring that the solutions chosen serves the greater good, are therefore vital to achieve sustainability over time. Thus, CoIN possesses the potential to be understood as a tool to address wicked problems of great complexity within the field of CSF, at least if all relevant stakeholders are heard, instead of only the actors in control.

14.8   Corporate Social Responsibility

Corporate Social Responsibility (CSR) has no known accepted formal definition. CSR refers to the contributions of companies to the society as a whole or to particular communities, often directly related to companies' operations. It goes beyond legal regulations of the activity of companies resulting from volunteer decisions of companies to balance internal profitability and other corporate objectives with positive impacts in the society.

Although the concept can be traced back to the 1950s, the use of CSR in practice is relatively recent (Hajdúchová et al. 2019). The World Business Council for Sustainable Development defined CSR as "the ethical behaviour of a company towards society," or "management acting responsibly in its relationships with other stakeholders who have a legitimate interest in the business – not just the shareholders" (WBCSD 2000). In the same report, the WBCSD states that CSR "is the continuing commitment by business to behave ethically and contribute to economic development while improving the quality of life of the workforce and their families, as well as of the local community and society at large". This includes environmental responsibility, together with financial and social.

Starting in the 1990s, the global adoption of the concept of sustainable development in international and national level policy in making and regulatory frameworks has strongly contributed to integrate social and environmental dimensions into the responsibilities of businesses. This has contributed to make CSR a synonym of sustainability (Panwar and Hansen 2008). The integration of sustainable development into CSR has contributed to making CSR an important mechanism related to the prevention of undesirable impacts of companies to society (avoiding pollution or protecting species habitats) and/or supporting local communities around the world through donation of resources, time, or money, also from forest companies or companies operating in forest environments. Sustainable development is still today one of the major components of CRS together with social, environmental, and financial responsibility and company accountability (Lempiäinen 2011). With the general acceptance of the UN Sustainable Development Goals of the 2030 Agenda by the society at large, it is likely that sustainability will increase further its importance in CSR in the near future.

The growth of social responsibility within companies has evolved strongly in recent decades also due to changes taking place in societies in terms of the dominant values, rights, ideologies, and political and governance models, which influenced the expectations of the society from companies. Several other factors are also related to the increasing relevance of CSR, such as the contrast between the socioeconomic development context among sites in which companies operate and emergent drivers such as globalization, climate change, the biodiversity crisis, environmental catastrophes, forest fires and many others, increasing social and environmental awareness and, at the same time, creating opportunities for companies to become more socially responsible. Increasing connectivity and information exchange in societies have also increased the relevance of CSR, given the visibility of the work of companies as well as the impact of the public opinion and preferences on corporation's attitudes.

Social responsibility has become part of the business model of companies. Socially responsible entrepreneurship, socially responsible business, socially responsible marketing, and many other similar terms emerged in the businesses and corporation's arenas. Companies are today ranked according to their level of social responsibility (e.g., the Reputation Institute's World's Most Reputable Companies for Corporate Responsibility 2019), strongly impacting the public image of corporations.

ISO, the International Organization for Standardization, published in 2010 the standard ISO 26000 to address social responsibility, following years of a multi-stakeholder discussion process (ISO 2010). Based on the premise that "the objective of social responsibility is to contribute to sustainable development," ISO 26000 provides guidance on a series of essential topics, such as governance, human rights, labor practices, the environment, fair practices, consumer issues and community involvement and development, as well as practical guidance to integrate social responsibility in organizations, providing also examples of initiatives and tools for social responsibilities.

The social and environmental performance of companies has an effect on their investment efficiency (Guan et al. 2019). In the forest sector, companies benefit from CSR by improving relationships with stakeholders, maintain their legitimacy, address challenges, and achieve sustainable development as well as competitive advantages (Li and Gao 2019). Environment protection and promotion of employment seem to be the most relevant responsibilities disclosed by forest companies (Li and Gao 2019). SFM and its certification has been followed as a major CSR practice by the forest industry in general. In the Congo Basin, for example, certification is the aspect of CSR most valued by timber companies (Colaço and Simão 2018). Environmental themes related to forest operations, education, and health are priority aspects for these companies based on their disclosing. SFM and accountability are the two most frequently reported by large top forest companies in all continents (Vidal and Kozak 2008).

In practice, CSR can be regarded as economic instruments with a positive impact on increasing the supply of ecosystem services through monetary donations and other investments by companies. One example is the "Cash crops used to protect Costa Rican forest" by a subsidiary of NUEVA and a subsidiary of Schmidheiny to prevent degradation of forest in slopes due to slash and burn practices through planting macadamia trees in a buffer zone between farms and cloud forests on the slopes of a Costa Rican volcano (WBCSD 2000). Another example is the Farmer Training Center (FTC), established by Rio Tinto near the Kelian mine in Indonesia, to promote sustainable livelihood after the mine closure (WBCSD 2000). Examples of these kinds of actions have been compiled for countries of the Mediterranean region including Italy (establishment of protected area in the Brenta River Forest), Greece (local forest fire volunteer groups), Morocco ("Partnership for Moroccan Forests"), and Tunisia ("Pact for a Green Tunisia") (Górriz-Mifsud et al. 2018). Another type of CSR initiatives in the forest sector includes investments in offsetting the carbon footprint of companies through voluntary forest projects in Italy and Spain (Górriz-Mifsud et al. 2018).

Particularly after the signature of the Paris Agreement in 2015 (COP 21), climate change has become a major concern of societies and governmental bodies. The pressure of citizens (and consumers) has increased so much that for companies, committing to the great challenges of the Paris Agreement is a necessity (Allen and Craig 2016). This is even a stronger pressure for companies than it is for governments. The trend is mostly toward reducing the carbon footprint of companies, which is an important component of corporation disclosures. Therefore, climate change creates a new opportunity for CSR to support climate action initiatives and policies, which can include CSF, in mountain, and other types of forests.

14.9 Forest Concessions

A Forest concession "is a contract between a forest owner and another party permitting the harvesting and/or managing of specified resources from a given forest area" (Gray 2002). It is a frequent form of forest tenure and forest management in many countries and bioregions, both developing and developed (Landell-Mills and Ford 1999; Gray 2002). Concessions have been, however, much more frequently used in tropical countries in West and Central Africa, Southeast Asia, and Latin America (Tegegne et al. 2018). In other regions, concessions have been used mainly in Canada, in the USA, and in Latvia, Russia, Montenegro and Slovenia, in Europe (Landell-Mills and Ford 1999; Gray 2002; van Hensbergen 2016).

Forest concessions is a general mechanism often used to manage state or community land, but it can refer to several types of contracts and forest land tenure forms that might include a large range of possibilities that vary according to management objective, but that also vary from country to country. In Indonesia, for example, Natural Forest Timber Concessions, Industrial Plantation Forest Concessions, Ecosystem Restoration Concessions, Non Forest Product Concessions are commonly used concessions for private companies in public forestland and Community Forest Plantations, Community Forests and Village Forests in communal forests (Buergin 2016).

Forest concessions contracts are established between public or private forest landowners and a private, public, community, aboriginal group, or cooperative partner (Gray 2002), according to a system of established national forest policies at the country level (Søreide 2017). These contracts can be of two kinds: forest utilization, allowing harvesting and/or use rights (nonwood forest products, game, conservation, and tourism) in public forests, and forest management services (Gray 2002). In the latter case, contracts can be directed to conservation or protection of forest areas, including the management of protected areas. Forest concessions often require measures of environmental restoration and/or protection, such as afforestation, even for forest utilization (Gray 2002). The monitoring of the concession system is usually the responsibility of governments, which can, however, be given to NGOs or private companies (Søreide 2007).

International forestry remains aligned with the framework where certification is the core process/outcome, although additional novel social and political frameworks have emerged, such as the Ecosystem Services concept and the UN Sustainable Development Goals. Forest concessions are seen today from the point of view of such approaches to deal more with conservation, or as mechanisms to halt deforestation and forest degradation in tropical countries (Tegegne et al. 2018).

Forest concessions are a type of tenure with the potential for supporting CSF, including mountain forests, through either logging contracts based on sustainable forestry principles, or through management contracts directed to conservation of forests and provision of ecosystem services and the adoption of adaptation measures. However, the negative results of forest concessions as a tool for sustainable forestry can be questioned given the proneness of concession systems to corruption, illegal logging, and other practices that not just deprive countries of financial benefits, but increase deforestation and degradation of forest systems (Gray 2002; Søreide 2007).

In European countries, forest concessions are today rare, since governments are able to profitably take control of all forest activities and operations in public forestland (van Hensbergen 2016). In Slovenia, the forest concession system, was replaced recently in state forests that are now directly managed by the state-owned forest company.

14.10 Conclusions

There are many policy instruments for forest management used with varying degrees of success. A wide range of instruments respond to direct regulation, which has been seen as insufficient to deal with natural hazards and calamities (see Sect. 14.1). An example with the potential to overcome potential land-use conflicts, and effectively contribute to CSF using private lands and private initiatives, can be seen in private-based philanthropy (see Sect. 14.4). As mentioned previously, property owners (voluntarily) sacrifice short-term economic benefits from using the land.

PES and PES-like schemes (see Sect. 14.5) can make a significant contribution to the introduction of CSF, while many provide important sources and compensations for forest owners for losses (or economic benefits from using the land), which occur due to reductions in timber harvesting.

Instruments only work if they are implemented effectively. Thus, to ensure optimal implementation (See Sect. 14.6), the state must govern through alliances with the stakeholders, surpassing both the traditional bureaucratic approach where the civil servant knows the most, and the approach denoted from New Public Management where the civil servant has the serving role. Instead, the servant must take on the collaborative role where the goal is to empower citizens as individuals and groups to build implementable and feasible solutions together. Collaborative innovation processes (CoIN), in accordance with the Pan-Eurpoean policy requirement of RRI, both involving all relevant stakeholders and assuring that the solutions chosen serve the greater good, are therefore vital to achieve sustainability over time.

Another kind is corporate social responsibility (CSR; see Sect. 14.7). This concept connects elements of voluntary behavior, mandatory responsibilities, and societal expectations (Carroll 1999; Brammer et al. 2012). Furthermore, Forest Owners' Associations (FOAs; see Sect. 14.2) (traditionally also called "private forestry" or "community forestry") are governed under the mode of collective governance, but because of the strong position of industrial partners, the big FOAs, such as those forming the Federation, are governed according to the rules of CSR. For that reason the forest owners' associations in the Nordic countries use the word "family forestry" to denote private ownership at family level (Ackzell 2010).

One of the most advanced initiatives undertaken by the local and bigger regional FOAs, that delivered a new, voluntary governance instrument, is the PEFC developed alongside the FSC (see Sect. 14.3). PEFC is often considered as one of the most important private initiatives, due to the inclusion of stakeholder groups, such as environmental NGOs and social groups, for example, indigenous peoples, labor organizations, and forest owners in the schemes. PEFC is an example of customer-driven processes to foster SFM that regulates supply chains and markets of wood-based products, and is integral to environmental forest governance (Albrecht 2012). However, insufficient knowledge, low priority of biodiversity protection and mainstreaming nature's contributions to people, lack of transparency, as well as limited planning tools and resources, may reduce the effectiveness of PEFC. Despite their shortcomings, the PEFC schemes have proved particularly important for indigenous Sami people, who hold use rights to herd reindeer on about 30–50% of forestland in Norway, Finland, and Sweden (Blicharska et al. 2011).

A further example could also be a forest concession (see Sect. 14.8): contracts can be directed to conservation or protection of forest areas, including the management of protected areas. Forest concessions often require measures of environmental restoration and/or protection, such as afforestation, even if for forest utilization (Gray 2002). The monitoring of the concession system is of the responsibility of governments, which can, however, be given to an NGO or private companies (Søreide 2007).

In the implementation of CSF, the public sector plays, and will continue to play, a central role. Thus, it is important to take into account that the public sector is a political sector where citizens, politicians, and civil servants have a legitimate role as stakeholders in the governing of the state.

References

Ackzell L (2010.) Federation of Swedish Family Forest Owners Associations https://www.slideshare.net/CIFOR/federation-of-swedish-family-forest-owners-associations. Accessed 8 Jan 2021

Albrecht M (2012) Governance processes of sustainable forest management: from core markets to resource peripheries. In: M. Avdibegović, G. Buttoud, B. Marić, M. Shannon (Eds.). Assessing Forest Governance in a Context of Change. IUFRO Division 9: Forest Policy and Economics, Research Group 9.05.00 – Forest Policy and Governance. Proceedings of Abstracts from the IUFRO Seminar. 65–66, Sarajevo, Bosnia-Herzegovina, 2012, May, 9–13

Allen MW, Craig CA (2016) Rethinking corporate social responsibility in the age of climate change: a communication perspective. Int J Corp Soc Responsib 1:1

Amdam R (2020) Innovation in public planning. Calculat Commun Innov:33–52. https://doi.org/10.1007/978-3-030-46136-2_3

Arnstein SR (1969) A ladder of citizen participation. JAIP 35:216–224

Armitage DR (2005) Adaptive capacity and community-based natural resource management. Environ Manag 35(6):703–715. https://doi.org/10.1007/s00267-004-0076-z

Aurenhammer PK, Ščap Š, Triplat M, Krajnc N, Breznikar A (2017) Actors' potential for change in Slovenian Forest owner associations. Small-scale Forestry. https://doi.org/10.1007/s11842-017-9381-2

Baldwin R, Cave M, Lodge M (2011) Understanding regulation: theory, strategy and practice, 2nd edn. Oxford University Press, Oxford

Bastian CT, Keske CMH, McLeod DM, Hoag DL (2017) Landowner and land trust agent preferences for conservation easements: implications for sustainable land uses and landscapes. Landscape and Urban Paknning 157:1–13. https://doi.org/10.1016/j.landurbplan.2016.05.030

Battisti A, Larsson SA, Niemelä P (2015) Climate change and insect pest distribution range. Biology:1–15

Beland Lindahl K, Sandström C, Sténs A (2017) Alternative pathways to sustainability? Comparing forest governance models. Forest Policy Econ 77:69–78. https://doi.org/10.1016/J.FORPOL.2016.10.008

Bennet NJ, Whitty TS, Finkbeiner E, Pittman J, Bassett H, Gelcich S, Allison EH (2018) Environmental stewardship: a conceptual review and analytical framework. Environ Manag 611:597–614

Besseau P, Dansou K, Johnson F (2002) The International Model Forest Network (IMFN): elements of success. For Chron 78(5):648–654

Blicharska M, Angelstam P, Antonson H, Elbakidze M, Axelsson R (2011) Road, forestry and regional planners' work for biodiversity conservation and public participation: a case study in Poland's hotspot regions. J Environ Plan Manag 54(10):1373–1395

Boakye-Danquah J, Reed MG (2019) The participation of non-industrial private forest owners in forest certification programs: the role and effectiveness of intermediary organisations. Forest Policy Econ 100:154–163. https://doi.org/10.1016/j.forpol.2018.12.006

Böocher M (2012) A theoretical framework for explaining the choice of instruments in environmental policy. Forest Policy Econ 16:14–22

Brammer S, Jackson G, Matten D (2012) Corporate social responsibility and institutional theory: new perspectives on private governance. Soc Econ Rev 10:3–28

Braune A, Wittstock B (2011) Measuring environmental sustainability: the use of LCA based building performance indicators. LCM2011: Tools for Green and Sustainable Buildings

Bryson JM, Crosby BC, Bloomberg L (2014) Public value governance: moving beyond traditional public administration and the new public management. Public Adm Rev 74:445–456. https://doi.org/10.1111/puar.12238

Buergin R (2016) Ecosystem restoration concessions in Indonesia: conflicts and discourses. Crit Asian Stud 48(2):278–301

Buliga B, Nichiforel L (2019) Voluntary forest certification vs. stringent legal frameworks: Romania as a case study. J Clean Prod 207:329–342

Busenberg GJ (2004) Wildfire Management in the United States: the evolution of a policy failure. Rev Policy Res 21(2):145–156

Carroll AB (1999) Corporate social responsibility: evolution of a definitional construct. Bus Soc 38:268–295

Cashore B, Gale F, Meidinger E, Newsom D (2006) Introduction: forest certification in analytical and historical perspective. In: Cashore G, Meidinger N (eds) Confronting sustainability: forest certification in developing and transitioning countries. Yale Publishing Services Centre, Yale, pp 7–24

Chauvin CH (2012) Systemic tools and actors empowerment: the hold up of territory charters by the forest owning communes in France. In: M. Avdibegović, G. Buttoud, B. Marić, M. Shannon (Eds.). Assessing Forest Governance in a Context of Change. IUFRO Division 9: Forest Policy and Economics, Research Group 9.05.00 – Forest Policy and Governance. Proceedings of Abstracts from the IUFRO Seminar. 45–46, Sarajevo, Bosnia-Herzegovina, 2012, May, 9–13

Churchman CW (1967) Wicked problems. Manag Sci 14(4):B-141–B-142

Colaço R, Simão J (2018) Disclosure of corporate social responsibility in the forestry sector of the Congo Basin. Forest Policy Econ. 2018/07/01 92:136–147

Crow S, Danks C (2010) Why certify? Motivations, outcomes and the importance of facilitating organizations in certification of community-based forestry initiatives. Small-scale Forestry 9:295–211

Elliot C, Schlaepfer R (2001) Understanding forest certification using the advocacy coalition framework. Forest Policy Econ:257–266

FAO (2016) The Global Forest Resources Assessment 2015, Rom, 2nd edition, ISBN 978–92–5-109283-5

FAO (2018) Climate change for forest policy-makers – An approach for integrating climate change into national forest policy in support of sustainable forest management – Version 2.0. FAO Forestry Paper no.181. Rome, 68 pp.

Federal Ministry of Food, Agriculture and Consumer Protection (2011) Forest strategy 2020. Sustainable forest management – an opportunity and challenge for society. BMELV, Germany, Bonn

Fernandes PM, Guiomar N, Mateus P, Oliveira T (2017) On the reactive nature of forest fire-related legislation in Portugal: a comment on Mouraio and Martinho (2016). Land Use Policy 60:12–15

Fraser EDG, Dougill AD, Mabee WE, McAlpine P (2006) Bottom up and top down: analysis of participatory process for sustainability indicator identification as a pathway to community empowerment and sustainable environment management. J Environ Manage 78:114–127

FSC (2015). FSC Principles and criteria for forest stewardship. FSC-STD-01-001 V5–2 EN

Gežík V, Brnkaľáková S, Baštáková V, Kluvánková T (2021) Economic and social perspective of climate- smart forestry: incentives for behavioral change to climate- smart practices in the long-term. In: Managing Forest Ecosystems, Vol. 40, Tognetti R, Smith M, Panzacchi P (eds). Climate-Smart Forestry in Mountain Regions. Springer Nature, Switzerland, AG

Gloor PA (2007) Collaborative Innovation Newworks – How to Mint Your COINs? The 2007 International Symposium on Collaborative Technologies and Systems, May 21–25. 2007, Orlando, Florida, USA

Górriz-Mifsud E, Bugalho M, Corradini G, Valbuena P (2018) Chapter 14 – Financial incentives and tools for Mediterranean forests. In: Bourlion N, Garavaglia V, Picard N (eds) State of Mediterranean Forests 2018. Food and Agriculture Organization of the United Nations, Rome and Plan Bleu, Marseille, pp 229–242

Graves RA, Williamson MA, Belote RT, Brandt JS (2019) Quantifying the contribution of conservation easements to large-landscape conservation. Biol Conserv 232:83–96. https://doi.org/10.1016/j.biocon.2019.01.024

Gray JA (2002) Forest concession policies and revenue systems - country experience and policy changes for sustainable tropical forestry (English). World Bank Technical paper no. WTP 522. Forestry series. The World Bank, Washington, D.C. 126 pp.

Grilli G, Fratini R, Marone E, Sacchelli S (2020) A spatial-based tool for the analysis of payments for forest ecosystem services related to hydrogeological protection. Forest Policy Econ 111(102039):1–14

Guan X, Tian G, Tian G (2019) The mediating effect of corporate social responsibility on governance structure and investment efficiency based on PLS-SEM. Ekoloji 28(107):3579–3591

Habermas J (1995) Between facts and norms: contributions to a discourse theory of law and democracy. Polity Press, Cambridge

Hajdúchová I, Mikler C, Giertliová B (2019) Corporate social responsibility in forestry. J For Sci 65(11):423–427

Hall JE, Bonnell B (2004) Social and collaborative forestry: Canadian model forest experience. Canadian Forest Service, Ottawa

Higgins V, Dibden J, Potter C, Moon K, Cocklin C (2014) Payments for ecosystem services, neoliberalisation, and the hybrid governance of land management in Australia. J Rural Stud 36:463–474

Ho VM, Bonnell B, Kushalappa CG, Mooney C, Sarasin G, Svensson J, Verbisky R (2014) Governance solutions from the international model forest network. News 56:26–34

Hollberg A, Lützkendorf T, Habert G (2019) Top-down or bottom-up? – how environmental benchmark can support the design process. Build Environ 153:148–157

Holzinger K, Knill C (2004) Marktorientierte Umweltpolitik —ökonomischer Anspruch und politische Wirklichkeit. In: Reinhard, Zintl, Roland, Czada (Eds.), Politik und Markt (=PVS — Politische Vierteljahresschrift Sonderheft 34), pp. 232–255

Horton K, Knight H, Galvin KA, Goldstein KA, Herrington J (2017) An evaluation of landowners' conservation easements on their livelihoods and well-being. Biol Conserv 209:62–67. https://doi.org/10.1016/j.biocon.2017.02.016

IMNF (International Model Forest Network) (2019) 30 years of convening power. Model Forests, Multi-stakeholder Partnerships, and Sustainable Development

IMNF (International Model Forest Network) (2020) Landscapes, partnerships, Sustainability. IMNF Spring Newsletter https://imfn.net/. Accessed 8 Jan 2021

International Organization for Standardization (2010) Guidance on social responsibility. ISO Standard No. 26000. Retrieved from https://www.iso.org/obp/ui/#iso:std:iso:26000:ed-1:v1:en

Jensen SE, McPherson GR (2008) Living with fire: fire ecology and policy for the twenty-first century. University of California Press, Berkeley and Los Angeles

Kotecký V (2015) Contribution of afforestation subsidies policy to climate change adaptation in the Czech Republic. Land Use Policy 47:112–120. https://doi.org/10.1016/j.landusepol.2015.03.014

Kronholm T (2015). Forest owners' associations in a changing society. Doctoral thesis. Swedish University of Agricultural Sciences, Umeå. Acta Universitatis Agriculturae Sueciae, 102, 65 pp. https://pub.epsilon.slu.se/12729/1/kronholm_t_151022.pdf. Accessed 8 Jan 2021

Landell-Mills N, Ford J (1999) Privatising sustainable forestry: a global review of trends and challenges. Instruments for sustainable private sector forestry series. International Institute for Environment and Development, London, 112 pp

LaPierre L (2002) Canada's model forest program. Forest Chronicle 78(5):613–617

Lemos MC, Agrawal A (2006) Environmental governance. Annu Rev Env Resour 31:297–325

Lempiäinen A (2011) CSR in forest industry – case study of reporting and implementation of social responsibility in three international companies. MSc Thesis, University of Helsinki, 86 pp

Li Y, Gao L (2019) Corporate social responsibility of forestry companies in China: an analysis of contents, levels, strategies, and determinants. Sustainability 11:4379

Lockwood M (2010) Good governance for terrestrial protected areas: a framework, principles and performance outcomes. J Environ Manage 91(3):754–766

Lovrić M, Martinić I, Lovrić N, Landekić M, Šporčić M (2012) Role of NGO's in the implementation of the habitats directive. In: Avdibegović M, Buttoud G, Marić B, Shannon M (eds) Assessing Forest Governance in a Context of Change. IUFRO Division 9: Forest Policy and Economics, Research Group 9.05.00 – Forest Policy and Governance. Proceedings of Abstracts from the IUFRO Seminar. 61. Sarajevo, Bosnia-Herzegovina, pp 9–13

Ma Z, Butler BJ, Kittredge DB, Catanzaro P (2012) Factors associated with landowner involvement in forest conservation programs in the U.S.: implications for policy design and outreach. Land Use Policy 29:53–61

Makundi WR (1997) Global climate change and sustainable forest management – the challenge of monitoring and verification. Mitig Adapt Strat Glob Chang 2:133–155

Maryudi A, Devkota RR, Schusser C, Yufanyi C, Salla M, Aurenhammer H, Rotchanaphatharawit R, Krott M (2012) Back to basics: considerations in evaluating the outcomes of community forestry. Forest Policy Econ 14:1–5

McCarthy S, Matthews A, Riordan B (2003) Economic determinants of private afforestation in the Republic of Ireland. Land Use Policy 20:51–59

McManus P (2009) Environmental regulation. Elsevier Ltd., Sidneat

Mourao PR, Martinho VD (2016) Discussing structural breaks in the Portuguese regulation on forest fires – an economic approach. Land Use Policy 54:160–478

Mourao PR, Martinho VD (2019) Forest fire legislation: reactive or proactive? Ecol Indic 104:137–114

Muradian R, Rival L (2012) Between markets and hierarchies: the challenge of governing ecosystem services. Ecosyst Serv 1:93–100

Niedziałkowski K, Paavola J, Jędrzejewska B (2012) Participation and protected areas governance: the impact of changing influence of local authorities on the conservation of the Białowieża Primeval Forest, Poland. Ecol Soc 17(1):2. https://doi.org/10.5751/ES-04461-170102

Niedziałkowski K, Pietrzyk-Kaszyńska A, Pietruczuk M, Grodzińska-Jurczak M (2015) Assessing participatory and multilevel characteristics of biodiversity and landscape protection legislation: the case of Poland. J Environ Plan Manag 59(10):1891–1911. https://doi.org/10.1080/0964056 8.2015.1100982

Niedziałkowski K, Shkaruaba A (2018) Governance and legitimacy of the forest stewardship council certification in the national contexts – a comparative study of Belarus and Poland. Forest Policy Econ 97:180–188

Olschewski R, Bebi P, Teich M, Wissen Hayek U, Grêt-Regamey A (2012) Avalanche protection by forests — a choice experiment in the Swiss Alps. Forest Policy Econ:19–24. https://doi. org/10.1016/j.forpol.2012.02.016

Osborne SP (2006) The new public governance? Public Manag Rev 8:377–387. https://doi. org/10.1080/14719030600853022

Osborne SP (2010) Delivering public services: time for a new theory? Public Manag Rev 12:1–10. https://doi.org/10.1080/14719030903495232

Osman-Elasha B, Parrotta J, Adger N, Brockhaus M, Colfer CJP, Sohngen B, Dafalla T, Joyce LA, Nkem J, Robledo C (2009) Future socio-economic impacts and vulnerabilities. Chapter 4 in: Seppälä, R., Buck, A. and Katila, P. (Eds.) Adaptation of forests and people to climate change: a global assessment report, 101–122 https://researchgate.net/publication/233951303. Accessed 8 Jan 2021

Ostrom E (1990) Governing the commons: the evolution of institutions for collective action. Cambridge University Press, Cambridge, 280 pp

Paavola J, Hubacek K (2013) Ecosystem services, governance, and stakeholder participation: an introduction. Ecol Soc 18(4):42. https://doi.org/10.5751/ES-06019-180442

Pahl-Wostl C (2019) The role of governance modes and meta-governance in the transformation towards sustainable water governance. Environ Sci Policy 91:6–16

Paloniemi R, Apostolopoulou E, Cent J, Bormpoudakis D, Scott A, Grodzińska-Jurczak M, Tzanopoulos J, Koivulehto M, Pietrzyk-Kaszyńska A, Pantis JD (2015) Public participation and environmental justice in biodiversity governance in Finland, Greece, Poland and the UK. Environ Policy Gov 25:330–342. https://doi.org/10.1002/eet.1672

Panwar R, Hansen E (2008) Corporate social responsibility in forestry. Unasylva 230 59:45–48

Parkins JR, Dunn M, Reed MG, Sinclair AJ (2016) Forest governance as neoliberal strategy: a comparative case study of the Model Forest Program in Canada. J Rural Studies 45:270–278

Primmer E, Jokinen P, Blicharska M, Barton DN, Bugter R, Potschin M (2015) Governance of ecosystem services: a framework for empirical analysis. Ecosyst Serv 16:158–166

Prokofieva I, Gorriz E (2013) Institutional analysis of incentives for the provision of forest goods and services: an assessment of incentive schemes in Catalonia (North-East Spain). Forest Policy Econ 37:104–114. https://doi.org/10.1016/j.forpol.2013.09.005

Rametsteiner E (2009) Governance concepts and their application in forest policy initiatives from global to local levels. Small-scale Forestry 8:143–158. https://doi.org/10.1007/s11842-009-9078-2

Rametsteiner E, Simula M (2003) Forest certification-an instrument to promote sustainable forest management? J Environ Manage 67:87–98

Reed MS (2008) Stakeholder participation for environmental management: a literature review. Biol Conserv 141:2417–2431

Rittel H, Webber M (2007) Dilemmas in a General Theory of Planning (PDF). Archived from the original (PDF) on 2007-09-30. Retrieved 2020-12-14

Rosenkranz L, Seintsch B, Wippel B, Dieter M (2014) Income losses due to the implementation of the habitats directive in forests — conclusions from a case study in Germany. Forest Policy Econ 38:207–218. https://doi.org/10.1016/j.forpol.2013.10.005

RRI-Tools (2020) RRI-Tools. https://rri-tools.eu. Accessed 14 Dec 2020

Saarikoski H, Åkerman M, Primmer E (2012) The challenge of governance in regional forest planning: an analysis of participatory forest program processes in Finland. Soc Nat Resour 25:667–682

Sandström C (2009) Institutional dimensions of co-management: participation, power, and process. Soc Nat Resour 22(3):230–244

Schmithüsen F, Hirsch F (2010) Geneva timber and forest study paper 26. Private forest ownership in Europe. Geneva, Switzerland: UNECE http://www.unece.org/fleadmin/DAM/timber/publications/SP-26.pdf

Secco L, Pettenella D (2006) Participatory processes in forest management: the Italian experience in defining and implementing forest certification schemes. Schweiz Z Forstwes 157(10):445–452

Secco L, Pettenella D, Gatto P (2011) Forestry governance and collective learning process in Italy: likelihood or utopia? Forest Policy Econ 13:104–112

Seidl R, Schelhaas MJ, Rammer W, Verkerk PJ (2014) Increasing forest disturbances in Europe and their impact on carbon storage. Nat Clim Chang 4(9):806–810. https://doi.org/10.1038/nclimate2318

Shingo S (2017) Corporate Social Responsibility Kyoto-Style: Kyoto Model Forest's Approach to Collaborative Forest Management. 14 pp. https://imfn.net/wp-content/uploads/2019/01/CSR_Kyoto_Eng_final.pdf. Accessed 8 May 2020

Slavíková L, Raška P (2019) This is my land! Privately funded natural water retention measures in the Czech Republic. In: Hartmann T, Slavíková L, McCarthy S (eds) Nature-based flood risk management on private land. Springer. https://doi.org/10.1007/978-3-030-23842-1

Sippel A (2007) Forstliche Nutzung in FFH-Gebieten. Situationsanalyse und Perspektiven. Fachstudie erstellt durch die Forstliche Versuchs- und Forschungsanstalt Baden

Stavins R (1998) Market-based environmental policies. BCSIA Discussion Paper 98-02, ENRP Discussion Paper E-98-02, Kennedy School of Government, Harvard University

Stokken R, Børsen T (2020) Scientific literacy in a digital world. In: Halvorsen LJ, Stokken R, Rogne WM, Erdal IJ (eds) Digital samhandling - Fjordantologien 2020. pp. 15–39

Stupak I, Lattimore B, Titus BD, Smith CT (2011) Criteria and indicators for sustainable forest fuel production and harvesting: a review of current standards for sustainable forest management. Biomass Bioenergy 35:3287–3308

Søreide T (2007) Forest concessions and corruption. U4 ISSUE 3:2007 U4 Anti-Corruption Resource Centre Bergen, Norway, 26 pp

Sørensen E (2020) Interactive political leadership. 41–58. https://doi.org/10.1093/oso/9780198777953.003.0004

Špaček M, Kluvánková T, Louda J, Dubová L (2020) Documentation on strategic workshops (CINA) – collective forest management in Czechia and Slovakia, D4.2 subreport, project InnoForESt GA no. 763899. 43 pp. https://innoforest.eu/repository/d4-2-overview/d4-2-cina-report-innovation-region-slovakia-czech-republik/ Accessed 8 May 2020

Tegegne YT, Van Brusselen J, Cramm M, Linhares-Juvenal T, Pacheco P, Sabogal C, Tuomasjukka D (2018) Making forest concessions in the tropics work to achieve the 2030 agenda: voluntary guidelines, FAO and EFI. FAO Forestry Paper No. 180, Rome, 128pp

Thatcher A (2015) Defining Human Factors for Sustainable Development. https://www.research-gate.net/publication/273965629. Accessed 10 Dec 2020

Thompson GF (2003) Between hierarchies and markets: the logic and limits of network forms of organization. Oxford University Press

Tietenberg TH, Lewis L (2018) Environmental and natural resource economics, 11th edn. Routledge, New York. ISBN: 978-1-138-63229-5

Torfing J (2018) Collaborative innovation in the public sector: the argument. Public Manag Rev 00:1–11. https://doi.org/10.1080/14719037.2018.1430248

UNECE FAO (2018) Forests and water: valuation and payments for forest ecosystem services. United Nations, Geneva

van der Heijden J (2018) Understanding voluntary program performance: introducing the diffusion network perspective. Regulation & Governance: 44–62

van Hensbergen B (2016) Forest concessions-past present and future? Food and Agriculture Organization of the United Nations, Rome, 76 pp

Vatn A (2010) An institutional analysis of payments for environmental services. Ecol Econ 69:1245–1252

Vidal NG, Kozak RA (2008) Corporate responsibility practices in the forestry sector: definitions and the role of context. J Corporate Citizenship 31:59–75

Vizzari M, Lasserre B, Di Martino P, Marchetti M (2012) Promoting community-based natural resources management in Central Italy. In: M. Avdibegović, G. Buttoud, B. Marić, M. Shannon (Eds.). Assessing Forest Governance in a Context of Change. IUFRO Division 9: Forest Policy and Economics, Research Group 9.05.00 – Forest Policy and Governance. Proceedings of Abstracts from the IUFRO Seminar. 62–63, Sarajevo, Bosnia-Herzegovina, 2012, May, 9–13

WBCSD (2000) Corporate social responsibilities – meeting changing expectations. Geneva, 36 pp

Weatherall A, Nabuurs G-J, Velikova V, et al (2021) Defining Climate-Smart Forestry. In: Managing Forest Ecosystems, Vol. 40, Tognetti R, Smith M, Panzacchi P (eds). Climate-Smart Forestry in Mountain Regions. Springer Nature, Switzerland, AG

Weiss G, Slee RW, Ollonqvist P, Petenella D (eds) (2011) Innovation in forestry – territorial and value chain relationships. CABI. ISBN: 978-1-84593-689-1

Winkel G, Kaphengst T, Herbert S, Robaey Z, Rosenkranz L, Sotirov M (2009). EU Policy Options for the Protection of European forests Against Harmful Impacts. Albert-Ludwigs-University Institute of Forest and Environmental Policy, Freiburg & Ecologic Institute, Berlin (146 pp.)

Wu Z, Shao Y, Feng L (2019) Dynamic evolution model of a collaborative innovation network from the resource perspective and an application considering different government behaviors. Information 10(138):1–16

Wunder S (2005) Payments for environmental services: some nuts and bolts. Center for International Forestry Research, Jakarta

Chapter 15
The Role of Forests in Climate Change Mitigation: The EU Context

Matteo Vizzarri, Roberto Pilli, Anu Korosuo, Ludovico Frate, and Giacomo Grassi

Abstract The European Union (EU) aims at reaching carbon neutrality by 2050. Within the land use, land-use change, and forestry (LULUCF) sector, forestry will contribute to this target with CO_2 sink, harvested wood products (HWP), and use of wood for material or energy substitution. Despite the fact that the forest sink currently offsets about 9% of the total EU GHG emissions, evaluating its future mitigation potential is challenging because of the complex interactions between human and natural impacts on forest growth and carbon accumulation. The Regulation (EU) 2018/841 has improved robustness, accuracy, and credibility of the accounting of GHG emissions and removals in the LULUCF sector. For the forest sector, the accounting is based on the Forest Reference Level (FRL), i.e., a projected country-specific value of GHG emissions and removals against which the actual GHG emissions and removals will be compared. The resulting difference will count toward the EU GHG target for the period 2021–2030. Here, we provide an overview of the contribution of forests and HWP to the EU carbon sink for the period 2021–2025 (proposed FRLs) and focus on the contribution of mountain forests to the EU carbon sink, through exploring co-benefits and adverse side effects between climate regulation and other ecosystem services.

M. Vizzarri (✉) · R. Pilli · A. Korosuo · G. Grassi
European Commission, Joint Research Centre, Varese, Italy
e-mail: matteo.vizzarri@ec.europa.eu; roberto.pilli@ext.ec.europa.eu;
anu.korosuo@ec.europa.eu; giacomo.grassi@ec.europa.eu

L. Frate
Envix-Lab, Dipartimento di Bioscienze e Territorio, Università degli Studi del Molise,
Pesche, Italy

© The Author(s) 2022   507
R. Tognetti et al. (eds.), *Climate-Smart Forestry in Mountain Regions*, Managing
Forest Ecosystems 40, https://doi.org/10.1007/978-3-030-80767-2_15

15.1 Mitigation from EU Forests: Policy Context

The EU is at the forefront in implementing climate policies aiming at mitigation over the medium and long term. In order to contribute to maintaining the global temperature rise well below 2 °C above preindustrial levels (Paris Agreement[1]), in 2015, the EU committed to reduce GHG emissions by at least 40% by 2030 compared to 1990.[2] Since then, the EU has set out several climate policies and strategies to meet this target and is now preparing to increase its climate ambition, i.e., through further reducing emissions in 2030 (by −50%/−55% compared to 1990[3]) and moving toward reaching climate neutrality (i.e., a net balance between GHG emissions and removals) by the middle of this century. In particular, the European Green Deal (COM(2019) 640[4]) and the proposal for an EU Climate Law (COM(2020) 80[5]) are setting the basis for a comprehensive new climate legislative framework.

The LULUCF sector[6] – including mainly GHG fluxes from forests and wetlands and CO_2 fluxes from cropland and grasslands – plays a key role in mitigation. In particular, EU forests and harvested wood products (HWP) contribute to climate change mitigation by removing about 9% of the total GHG emissions from the atmosphere, which mainly originate from energy, transport, and agriculture sectors (EEA 2020). The mitigation potential of EU forests – which cover about 36% of European land area – refers to both their capacity to accumulate/release carbon during forest stands' development and the use of HWP for bioenergy purposes or material substitution. This not only is a result of recent mitigation actions but also derives from the legacy effects of historical management activities and external environmental changes (Grassi et al. 2019). Moreover, the effectiveness of these actions also depends on the uncertain effects from other human and natural perturbations, such as land-use changes, forest degradation, natural disturbances (Seidl et al. 2014), and climate change (Kirilenko and Sedjo 2007).

Given the difficulty in disentangling the natural and the anthropogenic contribution to the net GHG fluxes in the LULUCF sector – and especially on forests – a distinction had been applied between the *reporting* of current net emissions and the *accounting* of the impact of human activities on net emissions. The *reporting* refers to the implementation of standardized methodologies and approaches to describe all GHG exchanges between managed lands and the atmosphere (IPCC 2019). The

---

[1] https://unfccc.int/files/essential_background/convention/application/pdf/english_paris_agreement.pdf

[2] https://ec.europa.eu/clima/policies/strategies/2030_en#tab-0-0

[3] https://www.euractiv.com/section/climate-environment/news/environment-agenda-eu-climate-talks-enter-decisive-phase/

[4] https://eur-lex.europa.eu/legal-content/EN/TXT/?uri=CELEX%3A52019DC0640

[5] https://eur-lex.europa.eu/legal-content/EN/TXT/?uri=CELEX:52020PC0080

[6] The LULUCF sector includes the following categories: forest land, cropland, grassland, wetlands, settlements, other lands, and HWP (IPCC 2019).

accounting concerns the implementation of specific policy-agreed rules to evaluate the impact of anthropogenic activities for the climate mitigation targets (Krug 2018).

The Regulation (EU) 2018/841[7] (hereafter "LULUCF Regulation"), adopted in May 2018, aims to make the accounting in the LULUCF sector more credible and comparable to other sectors. To this aim, any change in policy/management compared to a base period (e.g., for cropland and grassland) or to an agreed benchmark (for forests) will be reflected in the accounting of future GHG emissions and removals (Grassi et al. 2018). For managed forest land, the LULUCF Regulation requires EU Member States (MS) to define national Forest Reference Levels (FRLs), i.e., the benchmark against which future forest emissions and removals will be compared to for accounting purposes. The FRL represents the projected average net GHG emissions in the periods 2021–2025 and 2026–2030 (i.e., first and second compliance period, respectively) based on the impact of management practices as documented in the period 2000–2009 (i.e., reference period, RP) on the evolution of age-related forest characteristics. This approach is based on (i) a transparent documentation of historical country-specific forest management practices (e.g., harvest intensity), (ii) the simulation of forest stands' development based on continued management practices and future evolution of age structure, and (iii) the exclusion of assumed policies and other influences from the market (Grassi et al. 2018). Based on this, MS will generate credits – if net emissions are lower than the proposed FRL – or debits, if net emissions are higher than the proposed FRL. The credits generated by the LULUCF sector can then contribute to reaching the targets in other sectors, up to a limit of 280 Mt CO_2e at EU scale for the period 2021–2030, while any LULUCF debits shall be compensated by extra emission reductions in other sectors. Credits generated on managed forest land (without considering dead wood and HWP) are capped at 3.5% of the total GHG emissions in the base year from all sectors excluding LULUCF. The LULUCF Regulation also introduces an internal flexibility mechanism, through which any removals exceeding this threshold in the first compliance period (2021–2025) can be transferred to the second compliance period (2026–2030) or to other MS that report debits, without affecting the overall target set at EU level.

In the next sections, we discuss the contribution of forests and HWP to the EU carbon sink in the period 2021–2025, with a particular focus on the mitigation potential of mountain forests.

15.2   Forest Reference Levels in EU

Article 8 of the LULUCF Regulation (and its Annex IV) lays down the accounting rules for managed forest land and includes principles and criteria to be followed in the determination of the FRL as well as the elements to be included in the National

---

[7] https://eur-lex.europa.eu/legal-content/EN/TXT/?uri=uriserv:OJ.L_.2018.156.01.0001.01.ENG

Forestry Accounting Plans (NFAPs). The NFAP is a report submitted to the European Commission (EC) and made public by individual MS, which contains the FRL value including the considered pools and gases, a description of the approaches and methods used for its determination, qualitative and quantitative information on forest management practices and intensity, adopted national policies, and historical and future harvesting rates.

MS submitted the draft NFAPs including the FRLs to the EC in December 2018. Based on the proposed FRLs as in the draft NFAPs and the feedbacks from an expert group,[8] the EC issued technical recommendations to MS (SWD (2019) 213[9]). Based on recommendations, as well as on the outcomes of several bilateral meetings and an additional expert group meeting, MS submitted the revised NFAPs including the FRLs to the EC between the end of 2019 and the beginning of 2020. The EC performed an in-depth assessment of the revised NFAPs and proposed FRLs between February and June 2020. In October 2020, the EC adopted the delegated act (C(2020) 7316[10]) reporting the FRLs to be applied by MS in the period 2021–2025 (see Table 15.1).

The projected EU forest sink (sum of individual MS' FRLs) in the period 2021–2025 is about 337 million tons CO_2e year^{-1} including the contribution from HWP, which corresponds to about 13% of the total.

MS made a considerable effort to implement robust and in some cases complex modelling approaches to simulate the evolution of the forest sink with regard to age-related dynamics, especially considering the living biomass pool (Korosuo et al. 2021; Vizzarri et al. 2021). This means that MS incorporated the dynamics of age-related forest characteristics (area, standing volume, increment) in the FRL simulation, excluding any additional influence from policy or market mechanisms. This is the main difference with respect to the Forest Management Reference Level developed under the Kyoto Protocol.[11] To prove that the FRLs have been built on data and information that accurately reflect the national circumstances, MS also provided additional documentation of their management practices – including harvest intensity – as implemented in the period 2000–2009, as well as a demonstration of their continuation within the compliance period. According to the

---

[8] The LULUCF Expert Group was established on 30 October 2018 and consisted of individuals appointed in their personal capacity, members from research institutions and non-governmental organizations, MS representatives, members from other public entities and third countries' authorities, and observers. Activities and other information about the LULUCF Expert Group are available at https://ec.europa.eu/transparency/regexpert/index.cfm?do=groupDetail.groupDetail& groupID=3638.

[9] https://eur-lex.europa.eu/legal-content/EN/TXT/?uri=CELEX%3A52019SC0213

[10] EC (European Commission). Commission Delegated Regulation (EU) … of 28.10.2020 amending Annex IV to Regulation (EU) 2018/841 of the European Parliament and of the Council as regards the forest reference levels to be applied by the Member States for the period 2021–2025. [C(2020)7316] [Internet]. Available from: https://ec.europa.eu/clima/sites/clima/files/forests/docs/c_2020_7316_delegated_regulation_en.pdf

[11] https://unfccc.int/topics/land-use/workstreams/land-use--land-use-change-and-forestry-lulucf/forest-management-reference-levels

Table 15.1 Forest Reference Levels including HWP for the period 2021–2025

Member State	Forest Reference Levels including HWP (2021–2025) [t $CO_2e\ yr^{-1}$]
Belgium	−1,369,009
Bulgaria	−5,105,986
Czech Republic	−6,137,189
Denmark	354,000
Germany	−34,366,906
Estonia	−1,750,000
Ireland	112,670
Greece	−2,337,640
Spain	−32,833,000
France	−55,399,290
Croatia	−4,368,000
Italy	−19,656,100
Cyprus	−155,779
Latvia	−1,709,000
Lithuania	−5,164,640
Luxembourg	−426,000
Hungary	−48,000
Malta	−38
Netherlands	−1,531,397
Austria	−4,533,000
Poland	−28,400,000
Portugal	−11,165,000
Romania	−24,068,200
Slovenia	−3,270,200
Slovakia	−4,827,630
Finland	−29,386,695
Sweden	−38,721,000
United Kingdom	−20,701,550
EU-28	−336,964,579

Source: Annex to C(2020) 7316
https://ec.europa.eu/clima/sites/clima/files/forests/docs/c_2020_7316_annex_en.pdf

LULUCF Regulation, MS should also ensure the consistency between modelled estimates and estimates reported in their GHG inventories or other historical information (e.g., from National Forest Inventories or statistics). In some cases, however, ensuring the consistency between the model outcomes and the GHG inventory for the period 2000–2009 was challenging. Indeed, this exercise required an additional model run that shows that the model can reproduce GHG inventory data for the period 2000–2009.

The overall forest sink at EU level (including HWP) as set within the FRL offers an overview of the mitigation potential of EU forests in the near future (Fig. 15.1).

If compared to GHG inventory estimates for the period 2000–2009 (submission 2020), the FRL estimates a 22% reduction of the forest C sink. The reduction of the

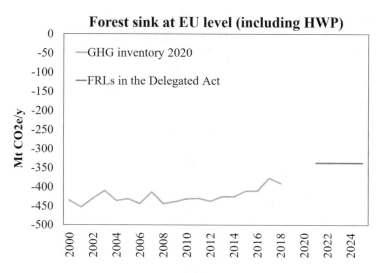

Fig. 15.1 Evolution of the EU forest sink (including HWP) in the period 2000–2025 based on different data sources and projections: EU GHG inventory (submission 2020) for the period 2000–2018 (solid blue line); FRLs in the Delegated Act (C(2020) 7316) (see Table 15.1) (solid red line). Values are in million tons $CO_2e\ yr^{-1}$

C sink largely depends on the effects of age-related dynamics and forest management (see also Grassi et al. 2018). The robustness of the FRLs is supported by independent analysis by Forsell et al. (2019), who, using similar assumptions as those underlying the MS FRLs, reached an estimate of similar magnitude.

15.3 Trade-Offs Between Climate and Other Forest Services: The Case of EU Mountain Forests

The mountain forests cover about 40% of the total forest area in the EU+27 UK (see Fig. 15.2 and detailed calculation in Box 15.1). The mountain forests in EU 27+UK provide a number of benefits to people, which encompass the provision of wood and non-wood products, regulation of climate, mediation of extreme events, conservation of biodiversity and habitats, and preservation of cultural features and aesthetics (Egan and Price 2017). Despite their multifunctional role, productivity and resilience of mountain forests are increasingly undermined by several ecological and socioeconomic factors, ranging from susceptibility to extreme climate events (e.g., avalanches, floods, landslides, windstorms, drought) to social segregation and low attractiveness for economic investments (Vizzarri et al. 2017). It is, therefore, extremely important to explore the relationships between mountain forests' services in terms of synergies and trade-offs (e.g., Rodríguez et al. 2006), as well as to understand the main drivers altering the ecosystem services' bundle, especially in mountain environments (e.g., Briner et al. 2013).

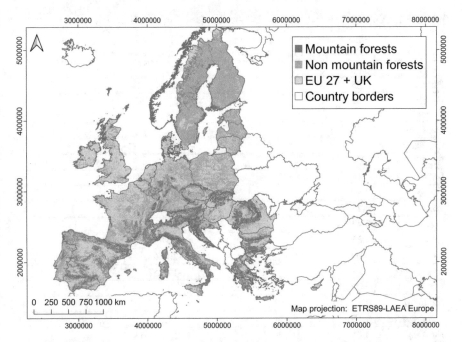

Fig. 15.2 Mountain/non-mountain forests in EU 27 + UK. See Box 15.1 for detailed explanation of mountain forests mapping

The mountain forests in the EU 27+UK contribute to remove about 127 million tons CO_2 from the atmosphere – forming about 35% of the total forest carbon sink (2015 value) (see Table 15.2 and detailed calculation in Box 15.1). If only forests located in high and scattered high mountains were considered (37% of the total mountain coverage; Sayre et al. 2018), CO_2 removals would be slightly more than 47 million tons (or about 13% of the total carbon sink; 2015 value). Hence, the major contribution to the total C sink comes from forests located in low and scattered low mountains, i.e., between 300 and 900 m a.s.l. with a slope > 50% (Sayre et al. 2018). The elevation range combined with a relatively ample forest extent at national scale may explain a larger contribution to the total C sink from mountain forests located in countries overlapping the most important massifs in Europe, such as the Alps and Apennines (more than 37% of the sink from Italy, Germany, and France), Pyrenees (15% of the sink from Spain), and Balkans (8% of the sink from Romania).

Within the European context, mountain forests are probably the best example of the complex, and sometimes competitive, interactions between different forest functions, spatial scales, and human activities (Fig. 15.3).

The economic revenues from forest resources are in most cases the key function attributed to forest land, above all in mountain areas. Here, this function is clearly "scaled" at local level, since forest management is based on activities and planning strategies which have a small scale and direct impact on local communities, through

Table 15.2 CO_2 emissions and removals in mountain forests in the EU 27+UK

Member State	Share of mountain forest cover [percentage of total forest cover]	CO_2 emissions (+)/removals (−) in mountain forests (t CO_2)
Belgium	0.09	−80,711
Bulgaria	0.83	−4,293,943
Czech Republic	0.51	−2,152,393
Denmark	0.00	
Germany	0.31	−17,557,365
Estonia	0.00	
Ireland	0.48	−504,541
Greece	0.97	−1,385,956
Spain	0.76	−22,055,007
France	0.42	−16,716,235
Croatia	0.65	−2,370,456
Italy	0.89	−20,565,547
Cyprus	0.91	−126,578
Latvia	0.00	
Lithuania	0.00	
Luxembourg	0.10	−34,079
Hungary	0.28	−1,137,033
Malta	0.00	
Netherlands	0.00	
Austria	0.86	−2,209,965
Poland	0.10	−2,283,392
Portugal	0.50	−4,323,900
Romania	0.79	−12,222,014
Slovenia	0.93	+823,697
Slovakia	0.91	−3,356,676
Finland	0.01	−434,988
Sweden	0.14	−6,480,554
United Kingdom	0.40	−7,290,015
EU 27+UK	**0.40**	**−126,757,652**

See Box 15.1 for methodology, data, and limitations

the economic and social benefits (i.e., on employment) provided by forest products (Pilli and Pase 2018). Within this context, carbon sequestration and timber provision have both synergies and trade-offs, depending on the management strategies (Gusti et al. 2020).

An increase in timber provision (i.e., harvesting rates) induces a temporary reduction of the standing volume at least at local level and in turn a relative reduction of the carbon stock and sink in aboveground living biomass. For this reason, several authors found that carbon sequestration rates in the short term (e.g., few decades) are higher under unmanaged scenarios (e.g., Mina et al. 2017). At the same

FOREST FUNCTIONS: SPATIAL SCALE AND INTERACTIONS

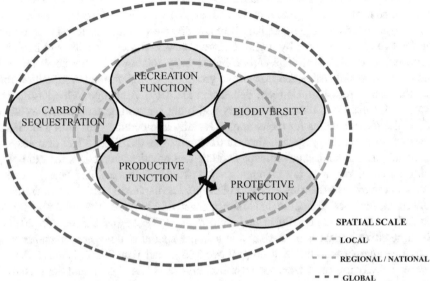

Fig. 15.3 Spatial scale (local, regional (or national), and global) and overall interactions between the main functions provided by forests (see the text for further information)

time, a shift in developmental stage toward younger forests (<140 years) due to changes in management regimes or land use may increase the carbon sink in living biomass in the medium to long term (Pugh et al. 2019). If properly balanced at regional or national level, however, an appropriate mixture of adaptive forest management strategies can be a means to both ensure sustainable timber provision and maintain or even enhance the forest carbon sink (Gusti et al. 2020). The positive effects of timber provision on carbon sequestration (in the medium term) are amplified if all pools (including their interactions) are considered. Indeed, timber also contributes to store carbon through the HWP pool, especially if in long-lived products. This adds to the mitigation impact arising from using wood products for substituting energy-intensive materials (e.g., steel, cement) and fossil fuel energy (Jonsson et al. 2021). For example, the management of coppice forests in mountain regions ensures on the one hand economic and social benefits, including the use of wood for energy purposes, and on the other hand is an alternative and partially in competition with carbon sequestration in HWP and the subsequent benefits for material substitution (Pilli and Pase 2018). Despite all these activities being carried out at local level, their effect on the carbon balance is generally accounted at national level, and, through the atmosphere, they have a direct impact on the overall carbon balance at global level.

Carbon sequestration and other services, such as biodiversity, progressively reduce while increasing the management intensity, i.e., from non-intervention to business as usual (Mina et al. 2017). Indeed, carbon sequestration positively correlates with biodiversity in mountain areas with low accessibility and less-intense management regimes (e.g., Lecina-Diaz et al. 2018). Isolated forests, often mountainous, are characterized by high density and diverse structures, and large amounts of deadwood, and in turn by higher carbon stocks and more species and habitats (Burrascano et al. 2013). The carbon-biodiversity relationship may be weak and vary across spatial scales and taxonomic groups, e.g., higher plant competition for light in denser forest stands (e.g., Sabatini et al. 2019). A change in the landscape pattern, for example, through clearcuts and other harvesting activities, may also affect the forest benefits for recreation, especially in mountain sites with a touristic vocation or in biodiversity-rich sites. In the latter case, the direct effect of management activities at landscape scale may add up to a broader interaction, for example, with the home range of big mammals, such as wolves, lynxes, and bears, mostly living in mountain territories, at least in Europe. The harvesting activities can reduce tree cover, limit soil protection, and in turn negatively influence the protective functions from forests, especially in mountain regions (Thees and Olschewski 2017). Such negative impact is exacerbated at a larger spatial scale, where management activities interact with the hydrological cycle (Sun and Vose 2016). Natural disturbances heavily impact carbon sequestration in forests (e.g., Thom and Seidl 2016). Storms, avalanches, insect outbreaks, and fires (in terms of their frequency, severity, and unpredictability) threaten the resilience and stability of mountain forest landscapes and inherently affect their carbon sequestration potential (e.g., Kulakowski et al. 2017). In particular, natural disturbances have negative effects on carbon uptake through reducing forest cover and subsequently living biomass, changing species composition, altering the carbon flows among pools and between the pools and the atmosphere (Pilli et al. 2016). Nevertheless, the effects of climate on ecosystem services (in particular carbon storage and biodiversity) in European mountain forests fluctuate over time and largely depend on site-specific characteristics, such as species composition and susceptibility to climate-related extreme events (e.g., drought), elevation, and topography (Mina et al. 2017).

15.4 Final Remarks

The EU is committed to reach climate neutrality by the middle of this century. Beyond expected contributions from the energy and transport sectors, the forest sector can also contribute to this target through carbon uptake in living biomass, dead organic matter, forest soils, and harvested wood products. However, the forest mitigation potential is driven by coupling age-related dynamics (growth rates, competition, and developmental stage) with management intensity, land-use change (land abandonment and forest regrowth), and severity and frequency of natural disturbances, whose effect may be exacerbated by climate in the long run. Therefore, the mitigation actions need to incorporate the expected response of forest carbon sink to external perturbations. On the policy side, the LULUCF Regulation poses the

Box 15.1: Calculation of CO_2 Emissions and Removals in EU Mountain Forests

Determination of the mountain forest area. The coverage of mountain forests in EU 27 was determined in QGIS 3.14™ by clipping the Copernicus High Resolution Layer (HRL) "Forest Type" (year 2015)[a] with the high resolution map of mountain areas (year 2010) by Karagulle et al. (2017)[b]. The outcome of this process is a raster layer with 20 m spatial resolution of mountain forests for EU 27 + UK (MFA). The area of mountain forests [ha] was determined for each of the EU 27 MS + UK (hereafter country) by dividing the number of pixels by 10,000 $[m^2]/400$ $[m^2] = 25$.

Calculation of CO_2 emissions and removals. To determine the CO_2 emissions and removals in mountain forest areas for each MS, Eq. [15.1] was used:

$$ER_{(mountain,i)} = w_i \, ER_{(FL-FL.i)} \tag{15.1}$$

where: $ER(mountain, i)$ is the value of net CO_2 emissions and removals in mountain forest areas in the i-th country (tons CO_2), wi is the ratio between the mountain forest area in the i-th country (derived from MFA) and the total forest area in the i-th country (derived from HRL) (dimensionless), and $ER(FL-FL.i)$ is the value of net CO_2 emissions and removals related to forest land remaining forest land (FL-FL) as reported for i-th country in the Common Reporting Format (CRF) Table 4 submitted to UNFCCC in 2020 for the inventory year 2015 (tons CO_2)[c].

The CO_2 emissions and removals in mountain forest areas for each country were then corrected [range between −0.31 and 0.11] to reflect the area consistency with FAOSTAT[d].

Limitations. This calculation of CO_2 emissions and removals is proportional to the total area of forest land in each country, and therefore does not take into account forest stands (age structure, species composition) and site characteristics (soil moisture, temperature, precipitation), land use (forest management and land-use change) and other drivers (e.g. natural disturbances) that may determine a spatial differentiation of the forest carbon sink at a finer scale than national.

(a)   Available online at https://land.copernicus.eu/pan-european/high-resolution-layers/forests/forest-type-1/status-maps/2015. Conifers and broadleaves classes were combined.

(b)   Available online at https://rmgsc.cr.usgs.gov/outgoing/ecosystems/Global/. High and scattered high (>900 m a.s.l.), and low and scattered low (301–900 m a.s.l.) mountains classes were merged to obtain a unique thematic class.

(c)   Available online at https://unfccc.int/ghg-inventories-annex-i-parties/2020.

(d)   Available online at http://www.fao.org/faostat/en/#data/GF.

basis for a credible and robust accounting system to ensure that the mitigation efforts in the forest sector are comparable to those in other sectors. Within this framework, the FRLs for the EU 27+UK already ensure that mitigation efforts will be compared against a robust benchmark that is based on the projected evolution of the forest sink, resulting from the impact of historical management on the future development of age-related forest characteristics. Such approach allows to not penalize countries, which will have a reduced sink purely due to age-related effects but will instead account for any additional impact from changes in policies (Grassi et al. 2018). On the management side, the forest mitigation potential in the EU can be enhanced by fostering resilience and adaptation, especially in fragile mountain environments, which increasingly suffer from land abandonment and susceptibility to extreme climate events, and properly balancing alternative management strategies (Gusti et al. 2020). For example, improved forest management, reforestation and forest restoration, and climate-smart forestry have large positive potentials for both mitigation and adaptation from forest ecosystems (Nabuurs et al. 2017; Bowditch et al. 2020; Smith et al. 2020). However, it is important to consider that the forest mitigation potential strongly varies across spatial and temporal scales. Forest management practices that are well-tailored to local circumstances and requirements are preferable in order to maximize the win-win combinations of forest ecosystem services, beyond carbon sequestration (e.g., Vizzarri et al. 2015). Advanced modelling tools (e.g., Shifley et al. 2017) and remote sensing techniques (e.g., Ceccherini et al. 2020) may allow policy-makers, forest managers, public administrators, and other stakeholders to provide evidence-based support to national policies and planning strategies toward mitigation targets.

Disclaimer The views expressed are purely those of the writers and may not in any circumstances be regarded as stating an official position of the European Commission or any other Government Agency.

References

Bowditch E, Santopuoli G, Binder F et al (2020) What is Climate-Smart Forestry? A definition from a multinational collaborative process focused on mountain regions of Europe. Ecosyst Serv 43:101113. https://doi.org/10.1016/j.ecoser.2020.101113

Briner S, Elkin C, Huber R (2013) Evaluating the relative impact of climate and economic changes on forest and agricultural ecosystem services in mountain regions. J Environ Manag 129:414–422. https://doi.org/10.1016/j.jenvman.2013.07.018

Burrascano S, Keeton WS, Sabatini FM, Blasi C (2013) Commonality and variability in the structural attributes of moist temperate old-growth forests: A global review. For Ecol Manag 291:458–479. https://doi.org/10.1016/j.foreco.2012.11.020

Ceccherini G, Duveiller G, Grassi G et al (2020) Abrupt increase in harvested forest area over Europe after 2015. Nature 583:72–77. https://doi.org/10.1038/s41586-020-2438-y

EEA (2020) Annual European Union greenhouse gas inventory 1990–2018 and inventory report 2020. EEA, Copenhagen

Egan PA, Price MF (2017) Mountain ecosystem services and climate change. A global overview of potential threats and strategies for adaptation. UNESCO, Paris

Forsell N, Korosuo A, Gusti M et al (2019) Impact of modelling choices on setting the reference levels for the EU forest carbon sinks: how do different assumptions affect the country-specific forest reference levels? Carbon Balance Manag 14:10. https://doi.org/10.1186/s13021-019-0125-9

Grassi G, Pilli R, House J et al (2018) Science-based approach for credible accounting of mitigation in managed forests. Carbon Balance Manag 13:8. https://doi.org/10.1186/s13021-018-0096-2

Grassi G, Cescatti A, Matthews R et al (2019) On the realistic contribution of European forests to reach climate objectives. Carbon Balance Manag 14:8. https://doi.org/10.1186/s13021-019-0123-y

Gusti M, Di Fulvio F, Biber P et al (2020) The effect of alternative forest management models on the forest harvest and emissions as compared to the forest reference level. Forests 11:794. https://doi.org/10.3390/f11080794

IPCC (2019) Refinement to the 2006 IPCC guidelines for national greenhouse gas inventories. Volume 4 – Agriculture, forestry and other land use. IPCC, Geneva

Jonsson R, Rinaldi F, Pilli R et al (2021) Boosting the EU forest-based bioeconomy: market, climate, and employment impacts. Technol Forecast Soc Change 163:120478. https://doi.org/10.1016/j.techfore.2020.120478

Karagulle D, Frye C, Sayre R et al (2017) Modeling global Hammond landform regions from 250-m elevation data. Trans GIS 21:1040–1060. https://doi.org/10.1111/tgis.12265

Kirilenko AP, Sedjo RA (2007) Climate change impacts on forestry. Proc Natl Acad Sci 104:19697–19702. https://doi.org/10.1073/pnas.0701424104

Korosuo A, Vizzarri M, Pilli R, Fiorese G, Colditz R, Abad Viñas R, Rossi S, Grassi G (2021) Forest reference levels under Regulation (EU) 2018/841 for the period 2021–2025, EUR 30403 EN, Publications Office of the European Union, Luxembourg, https://doi.org/10.2760/0521

Krug JHA (2018) Accounting of GHG emissions and removals from forest management: a long road from Kyoto to Paris. Carbon Balance Manag 13:1. https://doi.org/10.1186/s13021-017-0089-6

Kulakowski D, Seidl R, Holeksa J et al (2017) A walk on the wild side: disturbance dynamics and the conservation and management of European mountain forest ecosystems. For Ecol Manag 388:120–131. https://doi.org/10.1016/j.foreco.2016.07.037

Lecina-Diaz J, Alvarez A, Regos A et al (2018) The positive carbon stocks–biodiversity relationship in forests: co-occurrence and drivers across five subclimates. Ecol Appl 28:1481–1493. https://doi.org/10.1002/eap.1749

Mina M, Bugmann H, Cordonnier T et al (2017) Future ecosystem services from European mountain forests under climate change. J Appl Ecol 54:389–401. https://doi.org/10.1111/1365-2664.12772

Nabuurs GJ, Delacote P, Ellison D et al (2017) By 2050 the mitigation effects of EU forests could nearly double through climate smart forestry. Forests 8:484. https://doi.org/10.3390/f8120484

Pilli R, Pase A (2018) Forest functions and space: A geohistorical perspective of European forests. IForest 11:79–89. https://doi.org/10.3832/ifor2316-010

Pilli R, Grassi G, Kurz WA et al (2016) Modelling forest carbon stock changes as affected by harvest and natural disturbances. II. EU-level analysis. Carbon Balance Manag 11:20. https://doi.org/10.1186/s13021-016-0059-4

Pugh TAM, Lindeskog M, Smith B et al (2019) Role of forest regrowth in global carbon sink dynamics. Proc Natl Acad Sci 116:4382–4387. https://doi.org/10.1073/pnas.1810512116

Rodríguez JP, Beard TD Jr, Bennett EM et al (2006) Trade-offs across space, time, and ecosystem services. Ecol Soc 11:art28. https://doi.org/10.5751/ES-01667-110128

Sabatini FM, de Andrade RB, Paillet Y et al (2019) Trade-offs between carbon stocks and biodiversity in European temperate forests. Glob Chang Biol 25:536–548. https://doi.org/10.1111/gcb.14503

Sayre R, Frye C, Karagulle D et al (2018) A new high-resolution map of world mountains and an online tool for visualizing and comparing characterizations of global mountain distributions. Mt Res Dev 38:240–249. https://doi.org/10.1659/MRD-JOURNAL-D-17-00107.1

Seidl R, Schelhaas M-J, Rammer W, Verkerk PJ (2014) Increasing forest disturbances in Europe and their impact on carbon storage. Nat Clim Chang 4:806–810. https://doi.org/10.1038/nclimate2318

Shifley SR, He HS, Lischke H et al (2017) The past and future of modeling forest dynamics: from growth and yield curves to forest landscape models. Landsc Ecol 32:1307–1325. https://doi.org/10.1007/s10980-017-0540-9

Smith P, Calvin K, Nkem J et al (2020) Which practices co-deliver food security, climate change mitigation and adaptation, and combat land degradation and desertification? Glob Chang Biol 26:1532–1575. https://doi.org/10.1111/gcb.14878

Sun G, Vose J (2016) Forest management challenges for sustaining water resources in the anthropocene. Forests 7:68. https://doi.org/10.3390/f7030068

Thees O, Olschewski R (2017) Physical soil protection in forests – insights from production-, industrial- and institutional economics. For Policy Econ 80:99–106. https://doi.org/10.1016/j.forpol.2017.01.024

Thom D, Seidl R (2016) Natural disturbance impacts on ecosystem services and biodiversity in temperate and boreal forests. Biol Rev 91:760–781. https://doi.org/10.1111/brv.12193

Vizzarri M, Tognetti R, Marchetti M (2015) Forest ecosystem services: issues and challenges for biodiversity, conservation, and management in Italy. Forests 6:1810–1838. https://doi.org/10.3390/f6061810

Vizzarri M, di Cristofaro M, Sallustio L et al (2017) Strengthening integrated forest management in European mountain watersheds: insights from science to policy. In: Tognetti R, Scarascia-Mugnozza G, Hofer T (eds) Mountain watersheds and ecosystem services: balancing multiple demands of forest management in head-watersheds, EFI Technical report. European Forest Institute, Sarjanr, pp 63–76

Vizzarri M, Pilli R, Korosuo A et al (2021) Setting the forest reference levels in the European Union: overview and challenges. Carbon Balance Manage 16, 23. https://doi.org/10.1186/s13021-021-00185-4

Chapter 16
Smartforests Canada: A Network of Monitoring Plots for Forest Management Under Environmental Change

Christoforos Pappas, Nicolas Bélanger, Yves Bergeron, Olivier Blarquez, Han Y. H. Chen, Philip G. Comeau, Louis De Grandpré, Sylvain Delagrange, Annie DesRochers, Amanda Diochon, Loïc D'Orangeville, Pierre Drapeau, Louis Duchesne, Elise Filotas, Fabio Gennaretti, Daniel Houle, Benoit Lafleur, David Langor, Simon Lebel Desrosiers, Francois Lorenzetti, Rongzhou Man, Christian Messier, Miguel Montoro Girona, Charles Nock, Barb R. Thomas, Timothy Work, and Daniel Kneeshaw

Abstract Monitoring of forest response to gradual environmental changes or abrupt disturbances provides insights into how forested ecosystems operate and allows for quantification of forest health. In this chapter, we provide an overview of *Smartforests* Canada, a national-scale research network consisting of regional investigators who support a wealth of existing and new monitoring sites. The objectives of *Smartforests* are threefold: (1) establish and coordinate a network of high-precision monitoring plots across a 4400 km gradient of environmental and forest conditions, (2) synthesize the collected multivariate observations to examine the effects of global changes on complex above- and belowground forest dynamics

C. Pappas (✉) · N. Bélanger · E. Filotas · S. Lebel Desrosiers
Centre d'étude de la forêt, Université du Québec à Montréal, Montréal, QC, Canada

Département Science et Technologie, Téluq, Université du Québec, Montréal, QC, Canada
e-mail: christoforos.pappas@teluq.ca

Y. Bergeron · A. DesRochers
Centre d'étude de la forêt, Université du Québec à Montréal, Montréal, QC, Canada

Forest Research Institute, Université du Québec en Abitibi-Témiscamingue,
Amos, QC, Canada

O. Blarquez
Département de Géographie, Université de Montréal, Montréal, QC, Canada

H. Y. H. Chen
Faculty of Natural Resources Management, Lakehead University, Thunder Bay, ON, Canada

P. G. Comeau · C. Nock · B. R. Thomas
Department of Renewable Resources, University of Alberta, Edmonton, AB, Canada

© The Author(s) 2022
R. Tognetti et al. (eds.), *Climate-Smart Forestry in Mountain Regions*, Managing
Forest Ecosystems 40, https://doi.org/10.1007/978-3-030-80767-2_16

and resilience, and (3) analyze the collected data to guide the development of the next-generation forest growth models and inform policy-makers on best forest management and adaptation strategies. We present the methodological framework implemented in *Smartforests* to fulfill the aforementioned objectives. We then use an example from a temperate hardwood *Smartforests* site in Quebec to illustrate our approach for climate-smart forestry. We conclude by discussing how information from the *Smartforests* network can be integrated with existing data streams, from within Canada and abroad, guiding forest management and the development of climate change adaptation strategies.

L. De Grandpré
Natural Resources Canada, Canadian Forest Service, Laurentian Forestry Centre,
Quebec City, QC, Canada

S. Delagrange · F. Lorenzetti · C. Messier
Centre d'étude de la forêt, Université du Québec à Montréal, Montréal, QC, Canada

Institut des Sciences de la Forêt Tempérée (ISFORT), Université du Québec en Outaouais
(UQO), Ripon, QC, Canada

A. Diochon
Department of Geology, Lakehead University, Thunder Bay, ON, Canada

L. D'Orangeville
Faculty of Forestry and Environmental Management, University of New Brunswick,
Fredericton, NB, Canada

P. Drapeau · T. Work · D. Kneeshaw
Centre d'étude de la forêt, Université du Québec à Montréal, Montréal, QC, Canada

Département des sciences biologiques, Université du Québec à Montréal,
Montréal, QC, Canada

L. Duchesne
Direction de la Recherche Forestière, Ministère des Forêts, de la Faune et des Parcs du
Québec, Quebec City, QC, Canada

F. Gennaretti · B. Lafleur
Forest Research Institute, Université du Québec en Abitibi-Témiscamingue,
Amos, QC, Canada

D. Houle
Science and Technology Branch, Environment and Climate Change Canada,
Montreal, QC, Canada

D. Langor
Natural Resources Canada, Canadian Forest Service, Edmonton, AB, Canada

R. Man
Ontario Forest Research Institute, Ontario Ministry of Natural Resources and Forestry,
Sault Ste. Marie, ON, Canada

M. M. Girona
Forest Research Institute, Université du Québec en Abitibi-Témiscamingue,
Amos, QC, Canada

Restoration Ecology Group, Department of Wildlife, Fish and Environmental Studies,
Swedish University of Agricultural Sciences, Umeå, Sweden

16.1 Introduction

Canada is the third most forested country in the world with 347 million ha of forest land (The State of Canada's Forests 2020). This vast forest provides habitat for flora and fauna as well as crucial ecological, social, and economic services. Canadian forests contribute over $25 billion to Canada's gross domestic product and directly employ ca. 210,000 people in the forest industry (The State of Canada's Forests 2020). In addition to these direct economic benefits, forests provide critical ecological, social, and spiritual services. Furthermore, Canadian forests play a key role in the global carbon balance and thus affect Canada's international commitments regarding net carbon emissions (Luyssaert et al. 2008; Pan et al. 2011; Le Quéré et al. 2018; Baldocchi and Penuelas 2019).

However, the sustainability and resilience of forests are increasingly threatened by climate change as well as natural and anthropogenic disturbances, especially in high-latitude forests (ACIA 2005; Soja et al. 2007; Brandt et al. 2013; Gauthier et al. 2015; Trumbore et al. 2015; Brecka et al. 2020; DeSoto et al. 2020). Climate change will cause gradual long-term changes as well as increased frequency and severity of extreme events. These changes will contribute to increased uncertainty about future forest conditions that threaten the long-term viability of the forest sector and of human well-being (IPCC 2013; Brecka et al. 2020). Warming climate, drought stress, increasing frequency and severity of wildfires, and unprecedented outbreaks of insects and diseases are expected to reduce forest productivity and dramatically change forest composition, with concomitant impacts on biodiversity and ecosystem function, including the net carbon balance (Seidl et al. 2017; Navarro et al. 2018; Pugh et al. 2019).

While Canadian forest landscapes have always been dynamic due to the influence of a wide variety of natural biotic and abiotic disturbances, recent global changes are likely to alter the frequency and severity of these disturbances and lead to new disturbances not previously encountered. Increases in mean annual air temperature of 2.0 °C have been reported in western Canada during the period 1950–2003, compared with increases of only 0.5 °C in eastern Canada (Price et al. 2013). Similar discrepancies in precipitation have also been observed between eastern and western regions of Canada. In some areas, drought stress will reduce forest productivity and could threaten many ecosystem services. Given the vast extent of Canadian forests, effects of climate change will vary with geographic location, topography, forest composition, and local conditions. For example, western boreal forests are already drier than eastern boreal forests. Thus, slight increases in temperature or small decreases in precipitation may cause drought stress, reduced growth, increased tree mortality, and shifts in tree species composition, particularly in species-poor forests in western regions (Hogg et al. 2008; Michaelian et al. 2011; Peng et al. 2011; Chen et al. 2017; Cortini et al. 2017; Hisano et al. 2017; Searle and Chen 2017; Pappas et al. 2018). Another example is novel insect outbreaks, such as the invasion of the western boreal forest by the mountain pine beetle (Safranyik et al. 2010). However, in eastern boreal forests, warmer and drier conditions may increase tree growth (D'Orangeville et al. 2016, 2018). Several studies point toward a positive effect of the increased concentration of atmospheric CO_2 on tree growth (e.g., Tagesson et al. 2020), but reverse patterns have also been reported (Girardin et al. 2016). Translating how a carbon source (i.e., photosynthesis) is converted to a tree carbon sink (i.e., growth) remains challenging (Fatichi et al. 2019; Walker et al. 2020). Regional variation in forest responses to air temperature or precipitation change is also expected along a North-South gradient (Huang et al. 2010; Hogg et al. 2013; Chen et al. 2017). Projected increases in atmospheric CO_2 concentration, air temperature, evaporative demand, and surface net radiation are expected to intensify climate extremes, such as atmospheric and soil droughts (Held and Soden 2006; Dai 2012; Cook et al. 2014), with pronounced commensurate impacts on forest composition, structure, and function (Allen et al. 2015; Novick et al. 2016). Site-specific conditions will shape the impacts of climate change on forest function. Management options should be tailored accordingly by implementing case-specific solutions. Changes in soil conditions (nutrient concentrations, organic layer, permafrost) could scale up to long-term losses in productivity. Similarly, shifts in phenology can alter the synchrony between tree hosts and insect pests, potentially leading to increased damage to trees (Pureswaran et al. 2015). As a result, small-scale changes can have large consequences for forests, particularly when conditions are close to critical thresholds (Allen et al. 2015; Reyer et al. 2015; Trumbore et al. 2015).

Climate-smart forestry in the era of rapidly changing environmental conditions should provide tailored solutions for sustainable forest management based on a mechanistic understanding of forest function and of the influence of environmental stressors (Bowditch et al. 2020; Verkerk et al. 2020). This can be achieved by collecting and analyzing multivariate and multiscale observations of forest function (e.g., from the cell to the organism and to the ecosystem level) together with advanced understanding and numerical modelling of processes. Forest monitoring thus plays a central role in providing data to: (i) build improved knowledge on forest function at multiple spatiotemporal scales, as well as forest health and resilience to environmental change, and (ii) design and validate predictive modelling through numerical experiments. Precise temporal data, which are representative of the dynamic responses of forests in real time, are crucial for understanding and predicting the effects of global change (Kayler et al. 2015; Sass-Klaassen et al. 2016; Steppe et al. 2016). Accurate forest modelling, with state-of-the-art process-based simulation tools, allows for hypothesis testing and evaluating risk and uncertainty by conducting numerical experiments with hypothetical, yet realistic, scenarios of future climate conditions and/or forest stand composition (Fatichi et al. 2016; Mencuccini et al. 2019; Mastrotheodoros et al. 2020). To ensure the long-term viability of the forest sector, we also need well-coordinated research efforts that span a wide range of forest ecosystems and climatic conditions. While there have been many local-scale studies of changes associated with climate, we are lacking a comprehensive understanding of ecological responses and how these vary across the major forest types in Canada. We urgently need a Canada-wide concerted effort to document effects of climate variability, to experiment with different species mixes, and to model forest responses across large climatic gradients and forest types.

Smartforests Canada (https://smartforest.uqam.ca/) is a national project designed to address this gap through the establishment of a Canada-wide network of monitoring sites covering a wide spectrum of forest biomes and environmental conditions. More specifically, *Smartforests* aims to provide an improved understanding of how ecosystems, species, populations, and individual trees are influenced by changes in both physical (climate, soil) and biotic (competition, facilitation) environmental factors as well as interactions between these factors. The objectives of *Smartforests* are threefold: (1) establish a network of high-precision forest monitoring plots across a gradient of forest types and environmental conditions to examine the effects of global changes on complex above- and belowground forest dynamics, (2) synthesize multivariate data collected across the monitored forest stands to assess ecosystem functioning and resilience, and (3) assimilate the data and understanding of processes to guide the development of the next generation of forest growth simulation models and inform policy-making toward the best management and adaptation strategies for our forests.

In this chapter, we provide an overview of the *Smartforests* methodological approach and network of sites, and illustrate this multivariate and cross-scale *Smartforests* framework with an example of a temperate hardwood forest site in Quebec, Canada. We conclude with an outlook on how the *Smartforests* toolbox, based on state-of-the-art technology with automated and campaign-based measurements, can be deployed to quantify the multifaceted aspects of forest functioning and resilience under climate change. This holistic approach is firmly based on a balanced experimental design, which includes belowground, understory, and overstory forest components as well as biotic and abiotic factors affecting forest function. Information from the multivariate, multilevel *Smartforests* data streams can be integrated with global observation networks, where Canadian forests are currently underrepresented, and can be used for advanced forest growth modelling to guide forest management and the development of climate change adaptation strategies.

16.2 Methodological Framework

16.2.1 High-Precision Monitoring Plots

The backbone of the *Smartforests* approach is based on a Canada-wide network of high-precision forest monitoring plots (Fig. 16.1). These plots are designed to provide intensive and detailed spatiotemporal data on meteorological and soil conditions as well as various forest functions that are necessary to understand the response of forests to environmental stressors (Fig. 16.2). A tree-centered approach underlies the experimental design and instrumentation (Sass-Klaassen et al. 2016). Detailed observations are collected at different levels of spatial organization, spanning from the cell (e.g., plant tissue) to the organism (e.g., tree), the forest stand, and the landscape, and include campaign-based but also automated observations, coupled with

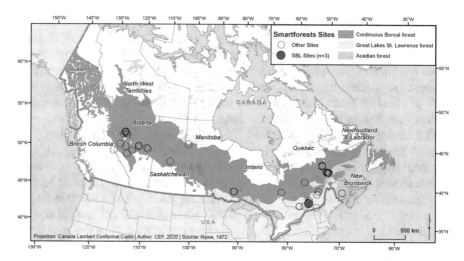

Fig. 16.1 An overview of the *Smartforests* network of sites. The strategically selected sites cover a wide range of environmental conditions as well as distinct forest types and vegetation characteristics. Point clusters occur in certain areas since, within each region, several monitoring plots are established to cover local-scale climatic gradients and environmental conditions. The case study used here to illustrate the implemented *Smartforests* approach is highlighted in red (Station de biologie des Laurentides; SBL)

remote sensing data (Fig. 16.2). It is important to underline that instrumentation and observed variables are not necessarily homogeneous across the network. Site-level priorities and specific research questions have resulted in tailored experimental designs to address the needs of specific research groups. Measured variables include meteorological (e.g., precipitation, air temperature, radiation, wind speed and direction, relative humidity) and soil conditions (temperature, water and nutrient availability), in addition to information on tree growth, reproduction, mortality, phenological changes in organisms, community turnover rates, net primary productivity, and trophic interactions. These high-precision monitoring plots allow us to: (1) collect long-term biological, ecological, and environmental data to document and understand changes in forest functioning with climate variability, (2) develop a network promoting ecological research and stimulating collaborations at national and international levels, and (3) provide a unique setting not only for research but also for educational activities, such as hosting teaching seminars, field classes, and facilitating student engagement through exposure to the scientific method.

The methodological design and the deployed instruments are tailored to cover a wide range of relevant ecophysiological and ecological processes and to quantify forest function including belowground as well as the understory and overstory components (Fig. 16.2). In the overstory, for example, we focus on measuring key biogeochemical processes describing the exchange of carbon, water, and energy in the soil-plant-atmosphere continuum, including measurements at the leaf level (e.g., leaf gas exchanges) and tree level (e.g., stem water fluxes and growth) but also at the landscape level (e.g., airborne thermal imaging), focusing on species interactions

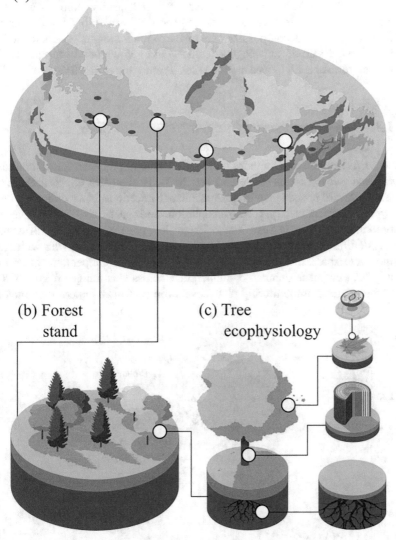

Fig. 16.2 The tree-centered approach implemented in the *Smartforests* network for tackling pressing environmental change questions in Canadian forests. Detailed tree ecophysiological observations are collected at the cell and whole plant level (subplot c). Plant ecophysiological insights and tree-level process understanding are upscaled to the forest stand level using information on stand demography and airborne imaging (subplot b). Findings at the forest stand level are synthesized across the *Smartforests* network of sites to better understand and model forest structure and function, health, and resilience to environmental change and provide Canada-wide guidance for forest management and policy-making (subplot a)

(Fig. 16.2). Understory vegetation is also monitored to quantify species demography and growth dynamics (Landuyt et al. 2019). Finally, soil conditions and belowground processes (e.g., temperature, water and nutrient availability, fine root growth, soil respiration, litter decomposition) are also explicitly monitored, acknowledging the fundamental role of soil biogeochemistry and belowground processes to tree growth and forest health (Vicca et al. 2012; Clemmensen et al. 2013).

16.2.2 The Smartforests Canada Network of Sites

The *Smartforests* network includes more than 100 high-precision forest monitoring plots spread across Canada (Figs. 16.1 and 16.3). The research efforts at these sites are geared toward pressing environmental change questions, in accordance with the *Smartforests* objectives, yet instrumentation and specific research questions explored at each site may vary. These forest plots cover a wide range of environmental conditions and include major North American forest types with widespread common tree species. The established network spans a temperature gradient of about 8 °C, i.e., mean annual air temperature across sites ranges from −3 °C to 5 °C. The gradient of monitoring plots covers forests from southern shade-tolerant

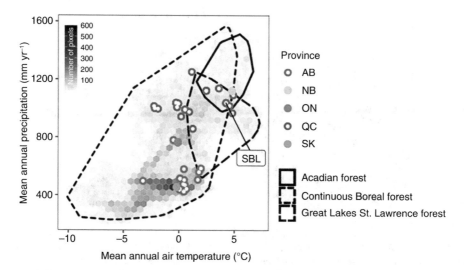

Fig. 16.3 Climate envelope of the forest types covered by the *Smartforests* network (color-coded dots denote the province, namely, Alberta, AB; New Brunswick, NB; Ontario, ON; Quebec, QC; Saskatchewan, SK; see also Fig. 16.1; for clarity, filled and open circles are used). Hexagons, color-coded with the density of 10′ resolution pixels, define the climate space of mean annual air temperature and total precipitation, as quantified using the WorldClim dataset (Fick and Hijmans 2017). Polygons show the specific forest types, as classified by Rowe 1972, which overlap the *Smartforests* network of sites. The case study used here to illustrate the implemented *Smartforests* approach is also highlighted (Station de biologie des Laurentides; SBL)

hardwoods to the boreal region along an East-West moisture gradient across the country, with mean annual total precipitation ranging from 380 mm in the West to 1252 mm in the East (Fig. 16.3). The latitudinal North-South gradient includes temperate shade-tolerant hardwood forests in southern Quebec, mixed temperate and temperate-boreal transition zone forests, as well as mixedwood and black spruce boreal forests in the Abitibi region. The East-West moisture gradient extends from the Acadian forest in New Brunswick to boreal forest plots on the North Shore of Quebec and to a series of boreal plots across Ontario, Manitoba, Saskatchewan, and Alberta. The wide range of environmental conditions and forest types covered by this network allows us to quantify forest dynamics, health, and resilience at the forest stand level and assess climate-change impacts by combining insights from the temperate, temperate/boreal transition, and boreal forest zones. The *Smartforests* network includes long-term monitoring sites, with more than a decade of detailed forest monitoring, as well as recently established research sites. The spatial and temporal gradient being covered by this network allows us to detect long-term effects of small changes in productivity or community relationships, including the influence of extreme events that may only affect a small number of sites, as a result of inter- and intraspecific differences in tree species, as well as adaptations of tree functioning to prevailing environmental conditions, e.g., species-specific responses of tree water use in humid forest stands of Eastern Canada (Oogathoo et al. 2020) vs. responses of the same species in drier sites in Central Canada (Pappas et al. 2018). Description of the design and efficacy of a network of observational plots across European mountain regions is presented in Chap. 5 of this book (Pretzsch et al. 2021).

16.3 Climate-Smart Forestry with High-Precision Monitoring Plots

16.3.1 From Forest Function to Forest Health and Resilience

Robust quantification of forest function including below- and aboveground components with processes occurring at cell, organism, and ecosystem levels provides the basis for assessing forest health and resilience to ongoing environmental change (Reyer et al. 2015). For example, tree growth is an indicator of tree age and vitality that is influenced by ontogeny, local competition, and climate (Dobbertin 2005). Temporal data on tree growth (e.g., annual tree ring widths, seasonal dendrometer-derived growth signals) can be used to characterize tree performance, vulnerability, and resilience to environmental changes over time (Lloret et al. 2011; Rogers et al. 2018; Pappas et al. 2020b). Moreover, tree water use and storage are indicators of drought-induced tree mortality risk (Martinez-Vilalta et al. 2018), and when combined with tree growth measurements, interspecific differences in species resilience could be quantified (Pappas et al. 2020b). Combining tree-level ecophysiological observations with remote sensing products covering forest landscapes provides a large-scale perspective of forest health that is useful in deriving early warning signals of critical transitions in forested ecosystems, e.g., drought-induced tree mortality (Camarero et al. 2015; Rogers et al. 2018; Cailleret et al. 2019). Discussion on how to implement tree-based monitoring platforms and large-scale forest observations is presented in Chaps. 10 and 11 of this book (respectively, Tognetti et al. 2021; Torresan et al. 2021). Thus, the *Smartforests* toolbox is geared to develop a quantitative and process-based understanding linking forest function under recent past and present environmental conditions with forest health and resilience to climate change using mechanistic understanding and modelling tools to predict future forest responses to environmental stressors.

16.3.2   Modelling and Adapting Forests to Climate Change

A comprehensive understanding of the interactions between forest dynamics, climate change, and management requires the use of simulation models optimized on robust data. Current tools employed by managers are still strongly influenced by the idea that forest ecosystems are at equilibrium with environmental conditions and tend to ignore the effects of changing environmental factors and vegetation acclimation and adaptation. The data collected under *Smartforests* are tailored to provide the necessary information to parameterize and validate state-of-the-art process-based models of forest growth and vegetation functioning (Fatichi et al. 2016; Prentice et al. 2015; Gennaretti et al. 2017). Process-based models facilitate prognostic simulations with a scenario analysis approach to evaluate, both at stand and landscape level, the benefits, compromises, and uncertainties associated with different management strategies and climate change scenarios. Simulation models constitute a relevant approach to investigate the dynamics, function, and structure of forests at scales that may be difficult to capture via experimental field research alone.

In addition, models are powerful tools for evaluating the long-term consequences of variation in baseline conditions and management strategies on forests at different spatial scales (Elkin et al. 2013; Bugmann et al. 2019). This approach is becoming more relevant with the acknowledged need to develop management strategies that will ensure that forest ecosystems maintain the variety of services on which our society depends, despite uncertainties associated with rapidly changing ecological, climatic, and economic conditions (Albrich et al. 2020). Managers and decision-makers already rely on simulation models, for example, to determine annual allowable cut levels. In many regions of Canada, forests are now being managed as ecosystems rather than just for fiber. While this approach is a large step forward in that it considers multiple values, it still often fails to consider changing forest conditions and natural disturbance regimes. As such, the effects of climate change are not considered in simulation models used for determining the annual allowable cut, which may overestimate timber supply. There is, therefore, an urgent need to adapt current simulation models and develop new models that better integrate uncertainty in resource availability, risk management in planning processes, and the multiple spatial and temporal scales over which climate change influences forests (Boucher et al. 2018; Boulanger et al. 2019; Gauthier et al. 2015; Mina et al. 2020).

Models that represent key ecosystem processes and are parameterized, calibrated, and validated with multivariate observations from the *Smartforests* network can be used to test and compare scenarios for developing adaptive forest management strategies. The data collected with *Smartforests* will contribute to the improvement of parameter estimates and to the accuracy of the processes being modelled, such as tree water use and growth, tree mortality, plant phenology, plant nutrition, natural disturbance dynamics, plant succession, plant inter- and intraspecific interactions, plant-microbe interactions, as well as the effect of climate change on these processes. Simulation results from these models provide virtual numerical experiments which can be used to inform managers and decision-makers on the best strategies to use in order to mitigate the negative impacts of global change and to adapt to and exploit opportunities linked with future environmental conditions. The models will be crucial to evaluating the consequences of alternative management approaches under different climate scenarios and thus providing managers with means of incorporating newly acquired data into decision-making. A review of mechanistic and empirical models, which are currently available to predict forest growth, is presented in Chap. 7 of this book (Bosela et al. 2021).

16.4   A *Smartforests* Case Study

16.4.1   Site Description

To illustrate the *Smartforests* methodological approach, we used the established high-precision monitoring plots at the Station de biologie des Laurentides (SBL; 45.987 N, 74.005 E), which is located within the temperate hardwood forest of the Great Lakes – St. Laurent region of Quebec (Figs. 16.1 and 16.4). The SBL site is a 16.4 km² research and teaching forest operated by the Université de Montréal (https://sbl.umontreal.ca/) and is situated in a transitional mixedwood forest within the southern Laurentians region (Fig. 16.4). The area is characterized by a continental climate with a mean annual air temperature of 5.5 °C and total precipitation of 1050 mm (long-term 1980–2010 averages from meteorological observations in St. Jerome; Environment and Climate Change Canada). Overstory vegetation includes common tree species in North America's temperate, boreal, and temperate/boreal transition zone (Table 16.1). Due to its geographic position, topography, and disturbance history, the area has developed into a mosaic of tree species, with dominant species being sugar and red maple (*Acer saccharum* Marsh. and *Acer rubrum* L.) mixed with American beech (*Fagus grandifolia* Ehrh.), white and yellow birch (*Betula papyrifera* Marsh. and *Betula alleghaniensis* Britt.), balsam fir (*Abies balsamea* (L.) Mill.), and bigtooth aspen (*Populus grandidentata* Michx.). Patches of forest stands dominated with red oak (*Quercus rubra* L.) occur at the southeast edge of the SBL region, corresponding to the northern species distribution range (Fig. 16.4). Understory vegetation consists primarily of striped maple (*Acer*

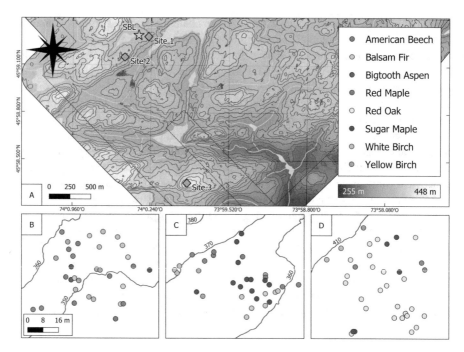

Fig. 16.4 Location of the three forest plots at the Station de biologie des Laurentides (SBL) together with the exact locations of the trees selected for continuous monitoring of stem growth and water use, color-coded according to species

pensylvanicum L.), hobblebush (*Viburnum lantanoides* Michx.), wood fern (*Dryopteris* spp.), and lycopods (*Lycopodium* spp.). The forest floor is a moder humus form, and most of the soils, which are thin, are well-drained Orthic Ferro-Humic or Humo-Ferric Podzols (Soil Classification Working Group 1998). To better capture local-scale topographic gradients and heterogeneity related to vegetation composition, three distinct forest plots were established within the SBL region for automated monitoring and campaign-based surveys (Fig. 16.4, Table 16.1). Average stand age, density, and basal area of overstory tree species (i.e., stem diameter ≥ 5 cm) in these plots are around 80 year, 1060 stems ha^{-1} and 33 m^2 ha^{-1}, respectively (Table 16.1).

16.4.2   *Automated Measurements of Forest Functioning*

Tree-level ecophysiological measurements are collected with automated equipment at the three forest plots at SBL and, together with concurrently recorded environmental variables (e.g., meteorological and soil conditions), are used to characterize tree and forest growth dynamics and water use (Fig. 16.2). Meteorological variables include hourly recorded rainfall, snow depth, albedo, relative humidity, air

Table 16.1 An overview of the tree species at Station de biologie des Laurentides (SBL) selected for continuous monitoring of stem growth and water use, together with their leaf type, taxon group, wood anatomy, and stand demography (stem density and basal area for overstory trees with DBH ≥ 5.0 cm)

Species name	Common name	Leaf type	Taxon group	Wood anatomy	Density [stems ha^{-1}]; Basal area [m^2 ha^{-1}]
Abies balsamea	Balsam fir	Needle-leaved	Gymnosperm	Tracheids	91; 0.9
Acer rubrum	Red maple	Broad-leaved	Angiosperm	Diffuse-porous	152; 2.7
Acer saccharum	Sugar maple	Broad-leaved	Angiosperm	Diffuse-porous	315; 4.5
Betula alleghaniensis	Yellow birch	Broad-leaved	Angiosperm	Diffuse-porous	24.3; 1.2
Betula papyrifera	White birch	Broad-leaved	Angiosperm	Diffuse-porous	67; 5.6
Fagus grandifolia	American beech	Broad-leaved	Angiosperm	Diffuse-porous	30; 0.4
Quercus rubra	Red oak	Broad-leaved	Angiosperm	Ring-porous	224; 16.2

temperature (above and below the canopy), wind speed and direction, and solar radiation, while soil conditions are characterized by hourly measurements of soil volumetric water content (VWC), water potential, and temperature recorded at a depth of approximately 10 cm (Fig. 16.5; Spectrum Technologies, Aurora, IL, US). Across the study region, 48 micro-stations are deployed. For each micro-station, two replicates of two soil variables are included (i.e., soil temperature and VWC or soil temperature and water potential) for a robust characterization of local-scale environmental heterogeneity (Fig. 16.5).

Tree ecophysiological monitoring includes: (1) sap flow measurements with both custom-made thermal dissipation sensors (20 mm long stainless-steel probes, with a 2 mm diameter; Granier 1987; Lu et al. 2004; Pappas et al. 2018) and commercial sensors (3-N, East 30, Pullman, WA, USA) measuring sap flow with the heat ratio method (Burgess et al. 2001) and (2) stem radius change measurements with two types of high-frequency and precision stem dendrometers (DC3, Ecomatik, Munich, Germany; DRL26C, Environmental Measuring Systems, Brno, Czech Republic). More than 100 trees are instrumented with sap flow and stem dendrometers in the three study plots, and all dominant tree species in the area are represented in the measurements being conducted (Fig. 16.4). This large sample size allows for detailed quantification of temporal dynamics in tree growth and water use and their inter- and intraspecific differences as well as for upscaling estimates to the forest stand level (Fig. 16.2). Measuring sap flow with two different methods allows us to infer sap flow at different sapwood depths and to derive species-specific radial profiles of sap ascent in tree stems (Fig. 16.6). Such information not only is useful for

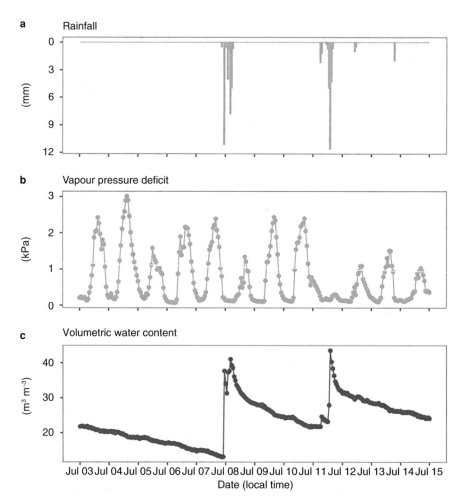

Fig. 16.5 Hourly time series for key meteorological and soil variables at the Station de biologie des Laurentides (SBL, Site 2; Fig. 16.4) for the period July 3 to July 15, 2020. Rainfall events (subplot a) resulted in drops in daytime vapor pressure deficit (subplot b) and concurrently increased soil water content (subplot c)

pinpointing interspecific differences in stem hydraulics and water use but also allows for robust transpiration estimates at the forest stand level (Berdanier et al. 2016). Further, combining constant heat (thermal dissipation method) and pulse-based (heat ratio method) sap flow measuring techniques allows us to minimize uncertainties related to method-specific assumptions and limitations (Steppe et al. 2010; Rabbel et al. 2016; Peters et al. 2018, 2020; Flo et al. 2019). By doing so, tree water use and transpiration rates can be assessed in terms of both temporal dynamics and absolute rates. Moreover, when combining tree-level data using different measuring techniques with forest stand characteristics (e.g., stem density) and

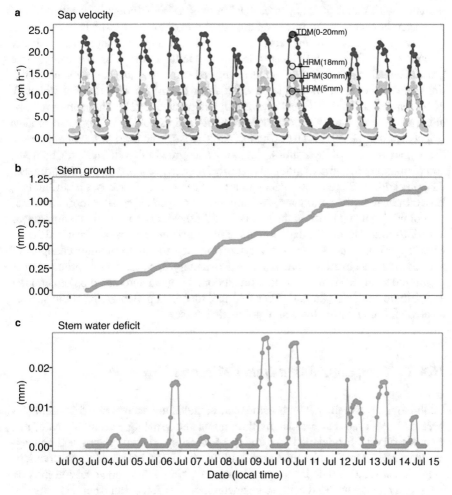

Fig. 16.6 Hourly time series of concurrently recorded ecophysiological variables using sap flow and dendrometer sensors affixed to the stem of an American beech (*Fagus grandifolia*) with DBH = 15.1 cm and sapwood depth, S_d = 3.9 cm, at the Station de biologie des Laurentides (SBL, Site 2; Fig. 16.4) for the period July 3 to July 15, 2020. Continuous ecophysiological monitoring includes sap velocity (subplot a) measured with two approaches, namely, the thermal dissipation method (TDM; 20 mm long probes; data processed with the *TREX* R Package; Peters et al. 2020) and the heat ratio method (HRM; at 5 mm, 18 mm, and 30 mm sapwood depths), and stem radius changes, used to derive stem growth (subplot b) and stem water status (stem water deficit; subplot c) computed with the *treenetproc* R package (Haeni et al. 2020; Knüsel et al. 2021)

species-specific allometry (e.g., sapwood area), transpiration estimates can be derived at the forest stand level (Oishi et al. 2008; Link et al. 2014; Matheny et al. 2014; Renner et al. 2016; Hassler et al. 2018). Two different measuring techniques are also deployed for continuously monitoring stem radius changes, namely, dendrometers mounted with tangential pulling force (DRL26C) and dendrometers mounted with radial pressing force (DC3), thus reducing potential weaknesses of each specific technique to infer tree growth and hydraulics at a subdaily resolution (Fig. 16.6). Processing of the recorded stem radius changes is useful to empirically disentangle: (i) irreversible changes (expansion) in the stem radius as a result of tree growth and (ii) reversible stem radius changes due to water use, i.e., subdaily variations and seasonal variation in stem water storage (King et al. 2013; Zweifel et al. 2016; Pappas et al. 2018; Haeni et al. 2020; Knüsel et al. 2021). These observations, along with concurrent sap flow measurements and established theoretical models (Mencuccini et al. 2013, 2016; Chan et al. 2016), can be used to provide process-based partitioning of hydraulic-, osmotic- and growth-driven stem fluctuations, complementing the aforementioned empirical approach for processing dendrometer data. In addition, 30 motion-sensing time-lapse cameras (Wingscapes BirdCam Pro; Ebsco Industries, Birghminham, AL) are also deployed at the sites to monitor leaf phenology from bud break in the spring to leaf senescence in the fall, taking images at various times during the day throughout the year. The cameras are installed 30 cm from the soil surface at an angle varying from 45° to 70° to monitor canopies of various heights covering understory and overstory species. Tree reproduction is monitored at the scale of the whole forest domain by an automated pollen counter and particle analyzer (RapidE, Plair SA) that identifies pollen grains at the species level in real time during the whole pollination season.

16.4.3   Campaign-Based Data Collection

Continuous measurements with automated methods are accompanied by field surveys and observational campaigns throughout the growing season. To date, this includes surveys for detailed monitoring of forest stand demography and aboveground tree biomass. To upscale tissue- and tree-level ecophysiological observations to the forest stand, which is the spatial scale at which policy- and decision-making is typically made, we need detailed characterization of forest stand demography and species-specific allometric characteristics (Waring and Landsberg 2011). Thus, at SBL, several circular subplots were established, and tree species and their stem diameter at breast height (1.3 m above the ground surface; DBH [cm]) were recorded (for trees with DBH $\geq$ 5.0 cm) and used to estimate site- and species-level stem density and basal area (Table 16.1).

In addition, during the summer 2020 field campaign, more than 350 tree cores were collected with a "biomass-oriented" design (Babst et al. 2014; Pappas et al. 2020a). The analysis of these tree cores will permit: (1) detailed characterization of species-specific sapwood allometry, a key parameter for quantitative estimates of tree- and forest-level water use and transpiration, and (2) reconstruction of forest stand's aboveground biomass increments to assess temporal variability in tree growth and the strength of the forest carbon sink (Babst et al. 2014; Pappas et al. 2020a). Moreover, when annual tree growth and biomass allocation patterns are combined with sub-annual (i.e., seasonal, daily, hourly) information on tree growth from the stem dendrometers, then the temporal spectrum of variability in growth can be characterized from hourly to decadal time scales. Such cross-scale characterization of species-specific growth patterns can provide novel insights into species responses to environmental change and, ultimately, offer a quantitative understanding of species-specific resilience (Pappas et al. 2020b).

16.5 Outlook

The resulting knowledge from this pan-Canadian *Smartforests* network, which is designed to encompass a broad range of forests and climates in Canada, will pave the way for the development of innovative adaptation strategies to ensure sustainable forest management and will enhance our understanding of ecosystem functioning. Within the context of sustainable forest management, identifying how these forests change with and respond to climate fluctuations will support development of strategies to preserve economic and non-economic ecosystem services. The *Smartforests* network and the integration of site-specific measurements with larger-scale remote sensing products and model outputs will serve to identify early warning signals of forest responses to either subtle changes in climate or extreme events. In doing so, forest managers will be able to react quickly to develop adaptation strategies. Moreover, research across the *Smartforests* network will provide quantitative insights into the terrestrial biogeochemical cycles that will lead to constrained estimates of the strength of the forest carbon sink across Canada. Such constraints are necessary for Canada-wide estimates of forest carbon budgets and for guiding policy-making on climate change mitigation strategies. The collected observations across the *Smartforests* network will also contribute substantially to existing global networks of forest monitoring plots, including sap flow observations (SAPFLUXNET; Poyatos et al. 2020) and plant functional traits (TRY; Kattge et al. 2020), where Canadian forests are currently underrepresented.

References

ACIA (2005) Impacts of a warming arctic: arctic climate impact assessment. Cambridge University Press, Cambridge

Albrich K, Rammer W, Turner MG, Ratajczak Z, Braziunas KH, Hansen WD, Seidl R (2020, September) Simulating forest resilience: a review. Glob Ecol Biogeogr 29:2082–2096. https://doi.org/10.1111/geb.13197

Allen CD, Breshears DD, McDowell NG (2015) On underestimation of global vulnerability to tree mortality and forest die-off from hotter drought in the Anthropocene. Ecosphere 6(8):art129. https://doi.org/10.1890/ES15-00203.1

Babst F, Bouriaud O, Papale D, Gielen B, Janssens IA, Nikinmaa E et al (2014) Above-ground woody carbon sequestration measured from tree rings is coherent with net ecosystem productivity at five eddy-covariance sites. New Phytol 201:1289–1303. https://doi.org/10.1111/nph.12589

Baldocchi D, Penuelas J (2019) The physics and ecology of mining carbon dioxide from the atmosphere by ecosystems. Glob Chang Biol 25(4):1191–1197. https://doi.org/10.1111/gcb.14559

Berdanier AB, Miniat CF, Clark JS (2016) Predictive models for radial sap flux variation in coniferous, diffuse-porous and ring-porous temperate trees. Tree Physiol 36:932–941. https://doi.org/10.1093/treephys/tpw027

Bosela M, Merganičová K, Torresan C, et al (2021) Modelling future growth of mountain forests under changing environments. In: Managing Forest Ecosystems, Vol. 40, Tognetti R, Smith M, Panzacchi P (Eds): Climate-Smart Forestry in Mountain Regions. Springer Nature, Switzerland, AG

Boucher D, Boulanger Y, Aubin I, Bernier PY, Beaudoin A, Guindon L, Gauthier S (2018) Current and projected cumulative impacts of fire, drought, and insects on timber volumes across Canada. Ecol Appl 28(5):1245–1259. https://doi.org/10.1002/eap.1724

Boulanger Y, Arseneault D, Boucher Y, Gauthier S, Cyr D, Taylor AR et al (2019) Climate change will affect the ability of forest management to reduce gaps between current and pre-settlement forest composition in southeastern Canada. Landsc Ecol 34(1):159–174. https://doi.org/10.1007/s10980-018-0761-6

Bowditch E, Santopuoli G, Binder F, del Río M, La Porta N, Kluvankova T et al (2020) What is climate-smart forestry? A definition from a multinational collaborative process focused on mountain regions of Europe. Ecosyst Serv 43(7):101113. https://doi.org/10.1016/j.ecoser.2020.101113

Brandt JP, Flannigan MD, Maynard DG, Thompson ID, Volney WJA (2013) An introduction to Canada's boreal zone: ecosystem processes, health, sustainability, and environmental issues. Environ Rev 21(4):207–226. https://doi.org/10.1139/er-2013-0040

Brecka AFJ, Boulanger Y, Searle EB, Taylor AR, Price DT, Zhu Y et al (2020) Sustainability of Canada's forestry sector may be compromised by impending climate change. For Ecol Manag 474(July):118352. https://doi.org/10.1016/j.foreco.2020.118352

Bugmann H, Seidl R, Hartig F, Bohn F, Brůna J, Cailleret M et al (2019) Tree mortality submodels drive simulated long-term forest dynamics: assessing 15 models from the stand to global scale. Ecosphere 10(2). https://doi.org/10.1002/ecs2.2616

Burgess SSO, Adams MA, Turner NC, Beverly CR, Ong CK, Khan AAH, Bleby TM (2001) An improved heat pulse method to measure low and reverse rates of sap flow in woody plants. Tree Physiol 21(15):1157

Cailleret M, Dakos V, Jansen S, Robert EMR, Aakala T, Amoroso MM et al (2019) Early-warning signals of individual tree mortality based on annual radial growth. Front Plant Sci 9(1):1–14. https://doi.org/10.3389/fpls.2018.01964

Camarero JJ, Gazol A, Sangüesa-Barreda G, Oliva J, Vicente-Serrano SM (2015) To die or not to die: early warnings of tree dieback in response to a severe drought. J Ecol 103(1):44–57. https://doi.org/10.1111/1365-2745.12295

Chan T, Hölttä T, Berninger F, Mäkinen H, Nöjd P, Mencuccini M, Nikinmaa E (2016) Separating water-potential induced swelling and shrinking from measured radial stem variations reveals a cambial growth and osmotic concentration signal. Plant Cell Environ 39(2):233–244. https://doi.org/10.1111/pce.12541

Chen L, Huang J-G, Alam SA, Zhai L, Dawson A, Stadt KJ, Comeau PG (2017) Drought causes reduced growth of trembling aspen in western Canada. Glob Chang Biol:1–15. https://doi.org/10.1111/gcb.13595

Clemmensen KE, Bahr A, Ovaskainen O, Dahlberg A, Ekblad A, Wallander H et al (2013) Roots and associated fungi drive long-term carbon sequestration in boreal forest. Science 339(3):1615–1618. https://doi.org/10.1016/b978-0-408-01434-2.50020-6

Cook BI, Smerdon JE, Seager R, Coats S (2014) Global warming and 21st century drying. Clim Dyn:1–21. https://doi.org/10.1007/s00382-014-2075-y

Cortini F, Comeau PG, Strimbu VC, Hogg EH(T), Bokalo M, Huang S (2017) Survival functions for boreal tree species in northwestern North America. For Ecol Manag 402:177–185. https://doi.org/10.1016/j.foreco.2017.06.036

D'Orangeville L, Duchesne L, Houle D, Kneeshaw D, Côté B, Pederson N (2016) Northeastern North America as a potential refugium for boreal forests in awarming climate. Science 352(6292):1452–1455. https://doi.org/10.1017/CBO9781107415324.004

D'Orangeville L, Houle D, Duchesne L, Phillips RP, Bergeron Y, Kneeshaw D (2018) Beneficial effects of climate warming on boreal tree growth may be transitory. Nat Commun 9(1):1–10. https://doi.org/10.1038/s41467-018-05705-4

Dai A (2012) Increasing drought under global warming in observations and models. Nat Clim Chang 2(8):1–7. https://doi.org/10.1038/nclimate1633

DeSoto L, Cailleret M, Sterck F, Jansen S, Kramer K, Robert EMR et al (2020) Low growth resilience to drought is related to future mortality risk in trees. Nat Commun 11(1):1–9. https://doi.org/10.1038/s41467-020-14300-5

Dobbertin M (2005) Tree growth as indicator of tree vitality and of tree reaction to environmental stress: a review. Eur J For Res 124(4):319–333. https://doi.org/10.1007/s10342-005-0085-3

Elkin C, Gutiérrez AG, Leuzinger S, Manusch C, Temperli C, Rasche L, Bugmann H (2013) A 2 °C warmer world is not safe for ecosystem services in the European Alps. Glob Chang Biol:1–14. https://doi.org/10.1111/gcb.12156

Fatichi S, Pappas C, Ivanov VY (2016) Modeling plant-water interactions: an ecohydrological overview from the cell to the global scale. Wiley Interdiscip Rev Water 3(3):327–368. https://doi.org/10.1002/wat2.1125

Fatichi S, Pappas C, Zscheischler J, Leuzinger S (2019) Modelling carbon sources and sinks in terrestrial vegetation. New Phytol. https://doi.org/10.1111/nph.15451

Fick SE, Hijmans RJ (2017) WorldClim 2 : new 1-km spatial resolution climate surfaces for global land areas. Int J Climatol. https://doi.org/10.1002/joc.5086

Flo V, Martinez-Vilalta J, Steppe K, Schuldt B, Poyatos R (2019) A synthesis of bias and uncertainty in sap flow methods. Agric For Meteorol 271(3):362–374. https://doi.org/10.1016/j.agrformet.2019.03.012

Gauthier S, Bernier P, Kuuluvainen T, Shvidenko AZ, Schepaschenko DG (2015) Boreal forest health and global change. Science 349(6250):819–822. https://doi.org/10.1126/science.aaa9092

Gennaretti F, Gea-Izquierdo G, Boucher E, Berninger F, Arseneault D, Guiot J (2017) Ecophysiological modeling of photosynthesis and carbon allocation to the tree stem in the boreal forest. Biogeosciences 14(21):4851–4866. https://doi.org/10.5194/bg-14-4851-2017

Girardin MP, Hogg EH, Bernier PY, Kurz WA, Guo XJ, Cyr G (2016) Negative impacts of high temperatures on growth of black spruce forests intensify with the anticipated climate warming. Glob Chang Biol 22(2):627–643. https://doi.org/10.1111/gcb.13072

Granier A (1987) Evaluation of transpiration in a Douglas-fir stand by means of sap flow measurements. Tree Physiol 3(4):309–320. https://doi.org/10.1093/treephys/3.4.309

Haeni M, Wilhelm M, Peters RL, Zweifel R (2020) Treenetproc – clean, process and visualise dendrometer data. R package version 0.1.4. Github repository: https://github.com/treenet/treenetproc

Hassler SK, Weiler M, Blume T (2018) Tree-, stand- and site-specific controls on landscape-scale patterns of transpiration. Hydrol Earth Syst Sci 22(1):13–30. https://doi.org/10.5194/hess-22-13-2018

Held IM, Soden BJ (2006) Robust responses of the hydrologic cycle to global warming. J Clim 19:5686–5699. https://doi.org/10.1175/JCLI3990.1

Hogg EH, Brandt JP, Michaelian M (2008) Impacts of a regional drought on the productivity, dieback, and biomass of western Canadian aspen forests. Can J For Res 38(6):1373–1384. https://doi.org/10.1139/X08-001

Hogg EH, Barr AG, Black TA (2013) A simple soil moisture index for representing multi-year drought impacts on aspen productivity in the western Canadian interior. Agric For Meteorol 178–179:173–182. https://doi.org/10.1016/j.agrformet.2013.04.025

Huang JA, Tardif JC, Bergeron Y, Denneler B, Berninger F, Girardin MP (2010) Radial growth response of four dominant boreal tree species to climate along a latitudinal gradient in the eastern Canadian boreal forest. Glob Chang Biol 16(2):711–731. https://doi.org/10.1111/j.1365-2486.2009.01990.x

IPCC (2013) In: Stocker TF, Qin D, Plattner G-K, Tignor M, Allen SK, Boschung J et al (eds) Climate change 2013: the physical science basis. Contribution of working group I to the fifth assessment report of the intergovernmental panel on climate change. Cambridge University Press, Cambridge/New York

Kattge J, Bönisch G, Díaz S, Lavorel S, Prentice IC, Leadley P et al (2020) TRY plant trait database – enhanced coverage and open access. Glob Chang Biol 26(1):119–188. https://doi.org/10.1111/gcb.14904

Kayler ZE, De Boeck HJ, Fatichi S, Grünzweig JM, Merbold L, Beier C et al (2015) Experiments to confront the environmental extremes of climate change. Front Ecol Environ 13(5):219–225. https://doi.org/10.1890/140174

King G, Fonti P, Nievergelt D, Büntgen U, Frank D (2013) Climatic drivers of hourly to yearly tree radius variations along a 6°C natural warming gradient. Agric For Meteorol 168:36–46. https://doi.org/10.1016/j.agrformet.2012.08.002

Knüsel S, Peters RL, Haeni M, Wilhelm M, Zweifel R (2021) Processing and extraction of seasonal tree physiological parameters from stem radius time series. Forests 12(6):1–14. https://doi.org/10.3390/f12060765

Landuyt D, De Lombaerde E, Perring MP, Hertzog LR, Ampoorter E, Maes SL et al (2019) The functional role of temperate forest understorey vegetation in a changing world. Glob Chang Biol 25(11):3625–3641. https://doi.org/10.1111/gcb.14756

Le Quéré C, Barbero L, Hauck J, Andrew RM, Canadell JG, Sitch S, Korsbakken JI (2018) Global carbon budget 2018. Earth Syst Sci Data 10(4):2141–2194

Link P, Simonin K, Maness H, Oshun J, Dawson T, Fung I (2014) Species differences in the seasonality of evergreen tree transpiration in a Mediterranean climate: analysis of multiyear, half-hourly sap flow observations. Water Resour Res 50:1–26. https://doi.org/10.1002/2013WR014023. Received

Lloret F, Keeling EG, Sala A (2011) Components of tree resilience: effects of successive low-growth episodes in old ponderosa pine forests. Oikos 120(12):1909–1920. https://doi.org/10.1111/j.1600-0706.2011.19372.x

Lu P, Urban L, Zhao P (2004) Granier's thermal dissipation probe (TDP) method for measuring sap flow in trees: theory and practice. Acta Bot Sin 46(6):631–646. https://doi.org/316-646

Luyssaert S, Schulze E-D, Börner A, Knohl A, Hessenmöller D, Law BE et al (2008) Old-growth forests as global carbon sinks. Nature 455(7210):213–215. https://doi.org/10.1038/nature07276

Martinez-Vilalta J, Anderegg WRL, Sapes G, Sala A (2018) Greater focus on water pools may improve our ability to understand and anticipate drought-induced mortality in plants. New Phytol 12:0–3. https://doi.org/10.1111/nph.15644

Mastrotheodoros T, Pappas C, Molnar P, Burlando P, Manoli G, Parajka J et al (2020) More green and less blue water in the Alps during warmer summers. Nat Clim Chang 10(2):155–161. https://doi.org/10.1038/s41558-019-0676-5

Matheny A, Bohrer G, Vogel CS, Morin T, He L, de Frasson RPM et al (2014) Species-specific transpiration responses to intermediate disturbance in a northern hardwood forest. J Geophys Res Biogeo 119(12):2292–2311. https://doi.org/10.1002/2014JG002804.Received

Mencuccini M, Hölttä T, Sevanto S, Nikinmaa E (2013) Concurrent measurements of change in the bark and xylem diameters of trees reveal a phloem-generated turgor signal. New Phytol 198(4):1143–1154. https://doi.org/10.1111/nph.12224

Mencuccini M, Salmon Y, Mitchell P, Hölttä T, Choat B, Meir P et al (2016) An empirical method that separates irreversible stem radial growth from bark water content changes in trees: theory and case studies. Plant Cell Environ:290–303. https://doi.org/10.1111/pce.12863

Mencuccini M, Manzoni S, Christoffersen B (2019) Modelling water fluxes in plants: from tissues to biosphere. New Phytol 222(3):1207–1222. https://doi.org/10.1111/nph.15681

Michaelian M, Hogg EH, Hall RJ, Arsenault E (2011) Massive mortality of aspen following severe drought along the southern edge of the Canadian boreal forest. Glob Chang Biol 17(6):2084–2094. https://doi.org/10.1111/j.1365-2486.2010.02357.x

Mina M, Messier C, Duveneck M, Fortin M-J, Aquilué N (2020) Network analysis can guide resilience-based management in forest landscapes under global change. Ecol Appl 00(00):e02221. https://doi.org/10.1002/eap.2221

Navarro L, Morin H, Bergeron Y, Girona MM (2018) Changes in spatiotemporal patterns of 20th century spruce budworm outbreaks in eastern Canadian boreal forests. Front Plant Sci 9(December):1–15. https://doi.org/10.3389/fpls.2018.01905

Novick KA, Ficklin DL, Stoy PC, Williams CA, Bohrer G, Oishi AC et al (2016) The increasing importance of atmospheric demand for ecosystem water and carbon fluxes. Nat Clim Chang 1(9):1–5. https://doi.org/10.1038/nclimate3114

Oishi AC, Oren R, Stoy PC (2008) Estimating components of forest evapotranspiration: a footprint approach for scaling sap flux measurements. Agric For Meteorol 148(11):1719–1732. https://doi.org/10.1016/j.agrformet.2008.06.013

Oogathoo S, Houle D, Duchesne L, Kneeshaw D (2020) Vapour pressure deficit and solar radiation are the major drivers of transpiration of balsam fir and black spruce tree species in humid boreal regions, even during a short-term drought. Agric For Meteorol 291(10):108063. https://doi.org/10.1016/j.agrformet.2020.108063

Pan Y, Birdsey RA, Fang J, Houghton R, Kauppi PE, Kurz WA et al (2011) A large and persistent carbon sink in the world's forests. Science 333(6045):988–993. https://doi.org/10.1126/science.1201609

Pappas C, Matheny AM, Baltzer JL, Barr A, Black TA, Bohrer G et al (2018) Boreal tree hydrodynamics: asynchronous, diverging, yet complementary. Tree Physiol 38(7):953–964. https://doi.org/10.1093/treephys/tpy043

Pappas C, Maillet J, Rakowski S, Baltzer JL, Barr AG, Black TA et al (2020a) Aboveground tree growth is a minor and decoupled fraction of boreal forest carbon input. Agric For Meteorol 290(9):0–1. https://doi.org/10.1016/j.agrformet.2020.108030

Pappas C, Peters RL, Fonti P (2020b) Linking variability of tree water use and growth with species resilience to environmental changes. Ecography 43(9):1386–1399. https://doi.org/10.1111/ecog.04968

Peng C, Ma Z, Lei X, Zhu Q, Chen H, Wang W et al (2011) A drought-induced pervasive increase in tree mortality across Canada' s boreal forests. Nat Clim Chang 1(12):467–471. https://doi.org/10.1038/nclimate1293

Peters RL, Fonti P, Frank DC, Poyatos R, Pappas C, Kahmen A et al (2018) Quantification of uncertainties in conifer sap flow measured with the thermal dissipation method. New Phytol. https://doi.org/10.1111/nph.15241

Peters RL, Pappas C, Hurley AG, Poyatos R, Flo V, Zweifel R et al (2020) Assimilate, process, and analyse thermal dissipation sap flow data using the TREX R package. Methods Ecol Evol:1–9. https://doi.org/10.1111/2041-210x.13524

Poyatos R, Granda V, Flo V, Adams MA, Adorján B, Aguadé D et al (2020) Global transpiration data from sap flow measurements: the SAPFLUXNET database. Earth Syst Sci Data Discuss 2020(10):1–57. https://doi.org/10.5194/essd-2020-227

Prentice IC, Liang X, Medlyn BE, Wang YP (2015) Reliable, robust and realistic: the three R's of next-generation land-surface modelling. Atmos Chem Phys 15(10):5987–6005. https://doi.org/10.5194/acp-15-5987-2015

Pretzsch H, Hilmers T, Uhl E, et al (2021) Efficacy of trans-geographic observational network design for revelation of growth pattern in mountain forests across Europe. In: Managing Forest Ecosystems, Vol. 40, Tognetti R, Smith M, Panzacchi P (Eds): Climate-Smart Forestry in Mountain Regions. Springer Nature, Switzerland, AG

Price DT, Alfaro RI, Brown KJ, Flannigan MD, Fleming RA, Hogg EH et al (2013) Anticipating the consequences of climate change for Canada' s boreal forest ecosystems. Environ Rev 365(12):322–365

Pugh TAM, Arneth A, Kautz M, Poulter B, Smith B (2019) Important role of forest disturbances in the global biomass turnover and carbon sinks. Nat Geosci 12(9):730–735. https://doi.org/10.1038/s41561-019-0427-2

Pureswaran DS, De Grandpré LD, Paré D, Taylor A, Barrette M, Morin H et al (2015) Climate-induced changes in host tree-insect phenology may drive ecological state-shift in boreal forests. Ecology 96(6):1480–1491. https://doi.org/10.1890/13-2366.1

Rabbel I, Diekkrüger B, Voigt H, Neuwirth B (2016) Comparing ΔTmax determination approaches for Granier-based sapflow estimations. Sensors 16(12):1–16. https://doi.org/10.3390/s16122042

Renner M, Hassler SK, Blume T, Weiler M, Hildebrandt A, Guderle M et al (2016) Dominant controls of transpiration along a hillslope transect inferred from ecohydrological measurements and thermodynamic limits. Hydrol Earth Syst Sci 20:2063–2083. https://doi.org/10.5194/hess-2015-535

Reyer CPO, Brouwers N, Rammig A, Brook BW, Epila J, Grant RF et al (2015) Forest resilience and tipping points at different spatio-temporal scales: approaches and challenges. J Ecol 103(1):5–15. https://doi.org/10.1111/1365-2745.12337

Rogers BM, Solvik K, Hogg EH, Ju J, Masek JG, Michaelian M et al (2018) Detecting early warning signals of tree mortality in boreal North America using multiscale satellite data. Glob Chang Biol 24(6):2284–2304. https://doi.org/10.1111/gcb.14107

Rowe JS (1972) Forest regions of Canada. Canadian Forest Service Publications, Ottawa

Safranyik L, Carroll AL, Régnière J, Langor DW, Riel WG, Shore TL et al (2010) Potential for range expansion of mountain pine beetle into the boreal forest of North America. Can Entomol 142(5):415–442. https://doi.org/10.4039/n08-CPA01

Sass-Klaassen U, Fonti P, Cherubini P, Gričar J, Robert EMR, Steppe K, Bräuning A (2016) A tree-centered approach to assess impacts of extreme climatic events on forests. Front Plant Sci 7(7):1–6. https://doi.org/10.3389/fpls.2016.01069

Searle EB, Chen HYH (2017) Persistent and pervasive compositional shifts of western boreal forest plots in Canada. Glob Chang Biol 23(2):857–866. https://doi.org/10.1111/gcb.13420

Seidl R, Thom D, Kautz M, Martin-Benito D, Peltoniemi M, Vacchiano G et al (2017) Forest disturbances under climate change. Nat Clim Chang 7(6):395–402. https://doi.org/10.1038/nclimate3303

Soil Classification Working Group (1998) The Canadian system of soil classification. NRC Research Press, Ottawa

Soja AJ, Tchebakova NM, French NHF, Flannigan MD, Shugart HH, Stocks BJ et al (2007) Climate-induced boreal forest change: predictions versus current observations. Glob Planet Chang 56:274–296. https://doi.org/10.1016/j.gloplacha.2006.07.028

Steppe K, De Pauw DJW, Doody TM, Teskey RO (2010) A comparison of sap flux density using thermal dissipation, heat pulse velocity and heat field deformation methods. Agric For Meteorol 150(7–8):1046–1056. https://doi.org/10.1016/j.agrformet.2010.04.004

Steppe K, von der Crone JS, De Pauw DJW (2016) TreeWatch.net: a water and carbon monitoring and modeling network to assess instant tree hydraulics and carbon status. Front Plant Sci 7(7):1–8. https://doi.org/10.3389/fpls.2016.00993

Tagesson T, Schurgers G, Horion S, Ciais P, Tian F, Brandt M et al (2020) Recent divergence in the contributions of tropical and boreal forests to the terrestrial carbon sink. Nat Ecol Evol 4(2):202–209. https://doi.org/10.1038/s41559-019-1090-0

The State of Canada's Forests. Annual Report 2020 (2020) Natural Resources Canada. Canadian Forest Service, Ottawa, 88 p. https://cfs.nrcan.gc.ca/publications?id=40219&lang=en_CA

Tognetti R, Valentini R, Belelli Marchesini L, Gianelle D, Panzacchi P, Marshall JD (2021) Continuous monitoring of tree responses to climate change for smart forestry – a cybernetic web of trees. In: Managing Forest Ecosystems, Vol. 40, Tognetti R, Smith M, Panzacchi P (Eds): Climate-Smart Forestry in Mountain Regions. Springer Nature, Switzerland, AG

Torresan C, Luyssaert S, Filippa G, Imangholiloo M, Gaulton R (2021) Remote sensing technologies for assessing climate-smart criteria in mountain forests. In: Managing Forest Ecosystems,

Vol. 40, Tognetti R, Smith M, Panzacchi P (Eds): Climate-Smart Forestry in Mountain Regions. Springer Nature, Switzerland, AG

Trumbore S, Brando P, Hartmann H (2015) Forest health and global change. Science 349(6250):814–818. https://doi.org/10.1126/science.aac6759

Verkerk PJ, Costanza R, Hetemäki L, Kubiszewski I, Leskinen P, Nabuurs GJ et al (2020) Climate-smart forestry: the missing link. Forest Policy Econ 115(4). https://doi.org/10.1016/j.forpol.2020.102164

Vicca S, Luyssaert S, Peñuelas J, Campioli M, Chapin FS, Ciais P et al (2012) Fertile forests produce biomass more efficiently. Ecol Lett 15(6):520–526. https://doi.org/10.1111/j.1461-0248.2012.01775.x

Walker AP, De Kauwe MG, Bastos A, Belmecheri S, Georgiou K, Keeling RF et al (2020) Integrating the evidence for a terrestrial carbon sink caused by increasing atmospheric CO2. New Phytol. https://doi.org/10.1111/nph.16866

Waring RH, Landsberg JJ (2011) Generalizing plant-water relations to landscapes. J Plant Ecol 4(1–2):101–113. https://doi.org/10.1093/jpe/rtq041

Zweifel R, Haeni M, Buchmann N, Eugster W (2016) Are trees able to grow in periods of stem shrinkage? New Phytol 211(3):839–849. https://doi.org/10.1111/nph.13995

Chapter 17
Climate-Smart Forestry in Brazil

Marcos Giongo, Micael Moreira Santos, Damiana Beatriz da Silva,
Jader Nunes Cachoeira, and Giovanni Santopuoli

Abstract Brazil is the second largest forested country in the world with a high level of naturalness and biodiversity richness, playing a significant role in the adoption of mitigation and adaptation strategies to climate change. Although the Brazilian federal government is mainly responsible for the protection of natural ecosystems, the decentralization process, which demands competences of the states and municipalities, allowed the establishment of several agencies and institutions dealing with monitoring, assessment, and management of forest ecosystems through a complex and interrelated number of forest policies. Nevertheless, the deforestation rate, with a consequent loss of biodiversity and ecosystem services, represents critical challenges, attracting worldwide attention. The variety of mitigation and adaptation measures adopted over the years represents viable tools to face climate change and to promote climate-smart forestry in Brazil. Notwithstanding the positive effects achieved in the last decade, a better coordination and practical implementation of climate-smart forestry strategies is required to reach nationally and internationally agreed objectives.

This chapter aims to depict the Brazilian forestry sector, highlighting the management strategies adopted overtime to counteract climate change.

M. Giongo (✉) · M. M. Santos · D. B. da Silva · J. N. Cachoeira
Environmental Monitoring and Fire Management Center, University of Tocantins, Gurupi, Brazil
e-mail: giongo@uft.edu.br; damisb@uft.edu.br; jadernunes@mail.uft.edu.br

G. Santopuoli
Dipartimento Agricoltura, Ambiente e Alimenti, Università degli Studi del Molise, Campobasso, Italy

Centro di Ricerca per le Aree Interne e gli Appennini (ArIA), Università degli Studi del Molise, Campobasso, Italy
e-mail: giovanni.santopuoli@unimol.it

R. Tognetti et al. (eds.), *Climate-Smart Forestry in Mountain Regions*, Managing Forest Ecosystems 40, https://doi.org/10.1007/978-3-030-80767-2_17

17.1 Introduction

Climate change currently represents one of the main themes in global political agendas and presents crucial challenges for the world, requiring coordinated actions at all scales and contexts. Brazil presents a high value of natural and biodiversity richness, including the world's largest tropical forest. Balancing urbanization exploitation, continuous demand for ecosystem services by society, and promoting sustainable use of natural resources is very challenging, particularly in the developing countries. The implicit factors that determine vulnerability to the impacts of climate change are complex and closely related to the level of development of a given community (Parikh 2000). For this reason, Brazil, as a developing country, is more vulnerable to climate change because of less capacity for adaptation, technological enhancement, and financial and institutional structures to further development (Silva 2010).

Mitigation and adaptation measures are required to reduce the country's vulnerability to climate change. These measures aim to reduce (mitigation) and tackle (adaptation) the impacts of climate change. In other words, mitigation actions aim to avoid the uncontrollable, while adaptation actions aim to manage the inevitable (Laukkonen et al. 2009). As recently highlighted by Verkerk et al. (2020), a successful mitigation strategy must take adaptation measures into account to ensure the resilience of forest ecosystems. This assumption is the core of the emerging concept of climate-smart forestry (CSF) that deals with the enhancement of forests' resilience and the delivering of ecosystem services while connecting mitigation, adaptation, and societal demands (Verkerk et al., 2020; Bowditch et al., 2020). In Brazil, mitigation actions have historically been prioritized compared to the adaptation actions, which have been included more effectively in the Brazilian agenda to face climate change in the last decades (Rodrigues Filho et al. 2016).

The United Nations Framework Convention on Climate Change (UNFCCC) recognizes adaptation strategies as an important measure to respond to the adverse effects of climate change and to prepare for future impacts. According to the Intergovernmental Panel on Climate Change (IPCC), adaptation is the process of adjusting to the current and expected climate and to impacts (IPCC 2014). In human systems, adaptation actions aim to reduce or avoid damages to natural resources, exploring beneficial opportunities, through accurate interventions by facilitating adjustment to the expected impacts of climate change (Scarano and Ceotto 2015).

The need for mitigation and adaptation measures in Brazil is crucial, with consequences at a global scale due to the extensive forest cover, the high value of forest biodiversity, the significant exploitation of water resources, and the huge pool of carbon stored in the forest ecosystems and other natural resources. For these reasons, efforts to develop new actions and strategies to tackle climate change are strongly recommended and encouraged by the current socioeconomic development conditions.

Brazil is the seventh largest emitter of greenhouse gases (GHG) in the world, representing 3.4% of global emissions, following China, the United States, the

European Union, India, Indonesia, and Russia. According to the National Inventory of Greenhouse Gas Emissions, "energy," "industrial processes," "agriculture and livestock," "land-use changes and forestry," and "waste treatment" are the most important sources of GHG emissions in Brazil. Though most of the developing countries highlight a general increment of emissions caused by the energy sector, Brazil's emission trend is strongly dependent on deforestation activities, increasing and decreasing with the deforestation rate (Claudio Angelo and Rittl 2019). The contribution to the GHG emissions from the Brazilian energy sector (i.e., burning fuels and fugitive emissions) is lower than other developing countries (e.g., Russian Federation, India, and China), representing 27% of the gross emissions. On the contrary, land-use change and forestry is the most impactful sector with 38% of the total CO_2 emissions in 2015 (Pao and Tsai 2011). Agriculture represents 25% of gross CO_2 emissions, caused mainly by cattle enteric fermentation and by the use of both animal manure and synthetic fertilizers. Of the industrial processes, the production of steel and cement contributes the most, registering about 6% of the CO_2 emissions in 2015. Disposal of solid waste and treatment of domestic and industrial sewage are the main impacting factors of waste treatment, with 4% of CO_2 emissions (MCTIC 2017).

Regarding land-use change, the conversion from forest area to agricultural and livestock lands is the primary source of GHG in terms of gross emissions. The sector had a significant peak of emissions in 1995, related to the intense conversion of forest areas to pasture areas in the Amazon biome. In the period 1995 and 2004, the conversion rate increased significantly, requiring the implementation of the Action Plan for Prevention and Control of the Legal Deforestation in the Amazon (PPCDAm), which resulted in an evident reduction of deforestation in the Amazon biome. In addition, looking at the net emissions, which consider the carbon stored due to the growing stock within forests and natural resources, the land-use change contributed to more than 60% of total emissions in the period 1990–2009, representing the main source of GHG in Brazil. However, it decreased to 27% for the period 2005–2010, following the agriculture and energy sectors with 32% and 30% of emissions, respectively (MCTIC 2017).

Looking at the biome level, according to the inventory carried out in 2015, the emissions caused by land-use change in the Amazon biome represent about 47% of the total CO_2 gross emissions, followed by Atlantic Forest and Cerrado biomes with 25% and 21%, respectively. Nevertheless, it is important to highlight that in the previous inventories, particularly for the period 1990–2002, the Cerrado biome was the most exploited biome and the highest source of CO_2, while, subsequently, the deforestation focused on the Atlantic forest biome. The emissions caused by land-use change of Caatinga, Pampas, and Pantanal biomes were slightly lower, with 4%, 3%, and 0.6%, respectively (MCTIC 2017).

Considering the important role that Brazilian's forest ecosystems play in counteracting global climate change, this chapter aims to describe the climate change policies and actions adopted by Brazilian governments over time through five sections, including the present introduction. Section 17.2 will introduce the Brazilian forest ecosystems, highlighting the main aspects of forest degradation and the forestry sector, including timber market and commercialization. Section 17.3 will focus on the forest monitoring programs, particularly focused on deforestation and forest fire, adopted by federal governments and municipalities and their interconnections. Section 17.4 will describe the mitigation and adaptation strategies adopted in Brazilian's forest ecosystems to face climate change, while Sect. 17.5 will conclude with final considerations.

17.2 Forest Ecosystems and Forestry

17.2.1 Forest Area, Changes, and Trend

Brazil is the second largest forest country (the first one is the Russian Federation), with about 500 million hectares of both natural forests and forest plantations, representing 59% of its territory (FAO and UNEP 2020; MAPA 2019). If correctly managed, these forests can strongly support climate change mitigation, reducing the negative impacts caused by society. Nevertheless, deforestation is the leading cause of loss of forest cover in Brazil and represents a critical threat to forest biodiversity, with a consequent loss of 7.6% of forest species and 10% of the native vegetation species between 2000 and 2018. According to the IBGE (2020) report, between 2016 and 2018, the replacement of forest areas with agriculture, pasture, and urban sprawl consumed about 1% of Brazilian territory. Similarly, between 1985 and 2017, native vegetation decreased by 9%, while agricultural lands increased by 37% (Souza et al. 2020).

Brazilian's forest ecosystems present a complex forest structure defined in six biomes: *Amazon*, *Caatinga*, *Cerrado*, *Atlantic Forest*, *Pampa*, and *Pantanal*, with distinctive characteristics and types of vegetation and fauna. The Amazon biome is particularly important, being the largest biome in Brazil and South America, covering eight countries beyond Brazil for a total extension of about 6.4 million km^2 (Lentini et al. 2005). In Brazil, it covers approximately 4.2 million km^2, equal to 63% of Amazon biome and 49% of country area (MMA 2006a). The loss of natural vegetation within the Amazon biome in Brazil was c.440,000 km^2 (11.1%) between 1985 and 2019, mostly due to the conversion into agricultural and livestock lands (Souza et al. 2020). However, in 2019, native vegetation (forest and non-forest) covered approximately 3.5 million km^2 (83% of the total area of the Amazon biome).

More precisely, 1995 and 2004 were the years with the highest deforestation rate of Amazon biome of the last 30 years, with a loss of 29,100 km^2 and 27,800 km^2, respectively, and an average value of about 20,629 km^2 year^{-1}. The annual rate of deforestation was lower in the period 2005–2012, with values of 4600 km^2 in 2012, 5900 km^2 in 2013, and 10,100 km^2 in 2019, which resulted in the highest value in the last 10 years (Assis et al. 2019). Moreover, forest fires, which are very frequent in Brazil (Santopuoli et al., 2016a; Assis et al. 2019), continuously contribute to the deforestation rate (Fig. 17.1).

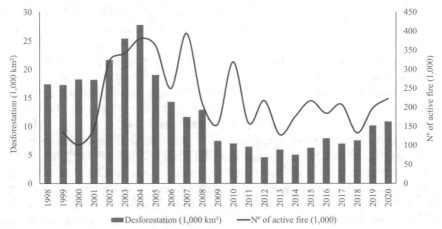

Fig. 17.1 Annual deforestation rates and outbreaks of active fire between 1998 and 2020. (Source Assis et al. 2019)

The Cerrado is the second largest biome in Brazil covering about 25% of the national territory and presenting a high level of endemism, with native vegetation covering 1 million km², more than 50% of the biome in 2010 (MMA and IBAMA 2011) and even more in 2002 which was equal to 60.4% (FAGRO 2007). According to the data carried out by TerraClass Cerrado Project, in 2013, the natural vegetation area was about 54% of the total Cerrado. The different values are associated with the methods adopted, such as the different scales and the minimum mapping size rather than differences in the biome area (MMA 2015). The Cerrado biome reached higher values of deforestation rate between 2001 and 2004 with a loss of about 29,000 km² of native vegetation areas per year, as reported in the PRODES system. Deforestation in the Cerrado has been gradually decreasing, in 2019 reaching the lowest value in terms of deforested areas, with 6483 km² deforested. The deforestation rate decreased by 33% compared to 2010, corresponding with the year in which the federal government adopted the Action Plan for Preservation and Control of Deforestation and Burnings in the Cerrado (PPCerrado).

The Atlantic Forest is extended in the south, southeast, midwest, and northeast regions, covering 15 Brazilian states. It represents the biome with the smallest natural vegetation area and with the highest human population, about 70% of the Brazilian inhabitants. From 2002 to 2009, about 2990 km² (0.27% of the biome) of natural vegetation were lost.

Though the Pantanal is a floodplain recognized by the United Nations Educational, Scientific and Cultural Organization (UNESCO) as a Natural Heritage and World Biosphere Reserve, the annual rate of deforestation for the Pantanal biome was 638 km², with a total loss of natural vegetation equal to 4467 km² (2.9% of the biome) between 2002 and 2009.

The Caatinga, an exclusively Brazilian biome, lost 16,035 km² (1.9% of the biome) between 2002 and 2009. Slightly lower, about 2514 km² (1.4% of the biome) was the loss of natural vegetation within the Pampa biome (MMA and IBAMA 2011).

The drivers of land-use change in the Amazon rainforest are mainly related to economic opportunities. In some regions, particularly those with indigenous communities, the need for fertile soils and agricultural lands is still significant, often representing a traditional custom of such communities (Santopuoli et al. 2016a; Ometto et al. 2011). The paths of change in land use and land cover in the Amazon, in space and time, are shaped by different actors and institutional arrangements, which in different socioeconomic, biophysical, and political contexts characterize the patterns of deforestation and land use (Ometto et al. 2011).

The heterogeneity among Brazilian regions, not only in the Amazon, characterizes the different patterns of land use and land cover and, consequently, impacts on the deforestation rate in the whole country. Among some factors that promote deforestation, the most impacting are the interaction between agricultural expansion, timber trade, population growth, and the construction of roads and public governance, all of which can interact in different ways, depending on the temporal and spatial dynamics of each region (Arraes et al. 2012).

Half a century ago, occupation and, consequently, deforestation, mainly in the Amazon region, were the result of a developmental and integrationist model based on occupation policies founded on geopolitics (Alencar et al. 2004). Such a model can be summarized as integrationist aimed at the rapid opening and expansion of agricultural frontier areas and road construction, such as Transamazônica and Belém-Brasília, which connects the southern part with the northern part of Brazil (Martins et al. 2009). The great challenge that has been faced by Brazil, mainly in the Amazon region, is the maintenance of ecosystem services offered by the forest and its complex ecological processes, with the evident need for the growth and development of populations and communities (Davidson and Artaxo 2004).

17.2.2 Forestry Sector: Products and Market

In Brazil, forestry-based industries comprise several segments, such as cellulose and paper, corrugated cardboard, charcoal, furniture, and mechanically processed wood (i.e., sawn wood, reconstituted, plywood, and laminated panels) and higher-value aggregate products, in addition to several non-timber products (SBS 2008).

According to the national report of forest products (IBGE 2019), the products related to silviculture (i.e., forest exploitation of forest plantations) and timber extractivism (i.e., the harvesting of timber from natural resources) between 2016 and 2019 provided an average value of about R$ 20.6 billion. The silvicultural activities only registered a value of R$ 15.5 billion in 2019. The income derived from cellulose production was the highest among the silvicultural products, with 29.3% of total silvicultural income, occupying the fourth place in the ranking of the country's total exports. Wood for other purposes (e.g., furniture industry, shipbuilding, civil construction, manufacture of pallets, wooden panels, laminate floors, posts) represented 28.9% of the total forestry sector, being in the second position regarding the generation of value of the sector. The production of charcoal is the third largest generator of value in silviculture, representing 25.2%, followed by firewood with 13.9% and non-wood products with 2.7% (IBGE 2019). The income provided by timber extractivism reached a value of R $ 4.4 billion in 2019 and was mainly focused to store roundwood (IBGE 2019).

Though Brazil was the world's second largest cellulose producer in 2018, following the United States, it was the first in cellulose and timber exports. The export to China and Europe was 55% of the total exports (IBÁ 2019). Regarding paper production, Brazil ranks eighth in the world, with 10.4 million tons in 2018 (IBÁ 2019). At the global level, Brazil is eighth regarding the production of wood panels and sawn wood with 0.2 million m^3 and 9.1 million m^3, respectively. Regarding the production of charcoal, Brazil is the world leader, accounting for 11% of all charcoal produced globally (IBÁ 2019).

17.3   Forest Monitoring Programs

17.3.1   Monitoring of Deforestation and Forest Fires

Since the 1970s, through the establishment and strengthening of strategic partnerships, the National Institute for Space Research (INPE), the Brazilian Agricultural Research Corporation (Embrapa), and the Brazilian Institute of Geography and Statistics (IBGE), and all federal agencies of the indirect administration, have been developing technologies and methodologies for monitoring Brazilian natural resources. These assist inspection and monitoring in areas threatened by deforestation, as well as actions to prevent and fight fire (MMA 2017a). In the last decades, several efforts were made to develop and define new methodologies and techniques for mapping and monitoring Brazilian biomes to support policy- and decision-makers, providing timely and reliable information.

In an attempt to reverse the dependence on obtaining satellite images provided by equipment from other nations, on July 6, 1988, the governments of Brazil and China signed a partnership agreement involving INPE and CAST (Chinese Academy of Space Technology) for the development of a program to build two advanced remote sensing satellites, called CBERS Program (cbers.inpe.br). This program made Brazil the pioneer in providing free images from medium spatial resolution sensors, thus becoming a global example of the scope of Earth observation, by making remote sensing an easily accessible tool. The CBERS Program satellites are essential for major strategic national projects, such as PRODES, for assessing deforestation in the Amazon, and DETER, for assessing deforestation in real time, among other systems. Currently, the CBERS Program presents its sixth satellite launched into orbit, CBERS 04A.

Currently, there are four monitoring systems (i.e., PRODES, DETER, TerraClass, and Queimadas) to assess deforestation and forest fire in the Amazon biome. Moreover, INPE developed a further system called DEGRAD, which was replaced by DETER in December 2016. INPE has been monitoring the rate of clear-cut deforestation in the Amazon since 1988 through the Brazilian Satellite Forest Monitoring Project (PRODES), providing one of the most consistent maps of deforestation in tropical forest regions in the world (Ometto et al. 2011). PRODES system focuses on monitoring clear-cut deforestation in the Legal Amazon, providing crucial information for assessing the yearly regional deforestation rate and supporting the government to develop forest policies (INPE 2020). PRODES provides high-quality data supporting the assessment of the GHG emissions of the forestry sector, which are necessary to obtain funding according to the UNFCCC, based on the reduced deforestation rate.

Since 2004, DETER (Real-Time Deforestation Detection system) deals with the real-time detection of land-use changes in the Amazon biome. More precisely, DETER is an alert system aimed to support IBAMA (Brazilian Institute of the Environment and Renewable Natural Resources) in the detection and control of deforestation and forest degradation, as well as other national and local governments about the use of natural resources. TerraClass system was developed through the collaboration between INPE and Embrapa, with the aims to provide maps of land use and land cover of deforested areas belonging to the Legal Amazon in 2004, 2008, 2010, 2012, and 2014 (Almeida et al. 2016). Furthermore, in 2008, the Institute of Man and Environment of the Amazon (Imazon) developed a further monitoring system, the Deforestation Alert System (SAD), to assess deforestation in Amazon. Imazon is a nonprofit Brazilian research organization, which, among other tasks, reports monthly the rates of deforestation and forest degradation of the Amazon biome. The integration of SAD and DETER products allows an accurate temporal and spatial evaluation of the deforestation rate in Amazon because they use different monitoring methods (Fig. 17.2).

Caatinga, Cerrado, Atlantic Forest, Pampa, and Pantanal are part of the monitoring programs started in 2002, through the Project for Conservation and Sustainable Use of Biological Diversity (PROBIO). Moreover, in 2008, the Ministry of the Environment (MMA) and the Brazilian Institute for the Environment and Renewable Resources (IBAMA) signed an agreement to carry out the Project for Monitoring Deforestation in Brazilian Biomes by Satellite (PMDBBS), with the support of the United Nations Development Programme (UNDP). These monitoring activities aimed to assess the loss of native vegetation of the abovementioned biomes to counteract illegal deforestation actions integrating data from different projects between 2002 and 2011. In 2013, the Ministry of the Environment (MMA) promoted the union of different public institutions to develop the first version of Mapping of Use and Vegetation Coverage of the Cerrado: TerraClass Cerrado, based on the methods defined through the TerraClass Amazon project (MMA 2013).

Recognizing the importance of having periodic information about the forest cover changes, the Ministry of the Environment established the permanent Environmental Monitoring Program for Brazilian Biomes (PMABB), through the Ordinance No. 365 on November 27, 2015. The program involved different agencies of the federal government, dealing with monitoring activities through remote sensing to promote the harmonization of the monitoring products. This step was

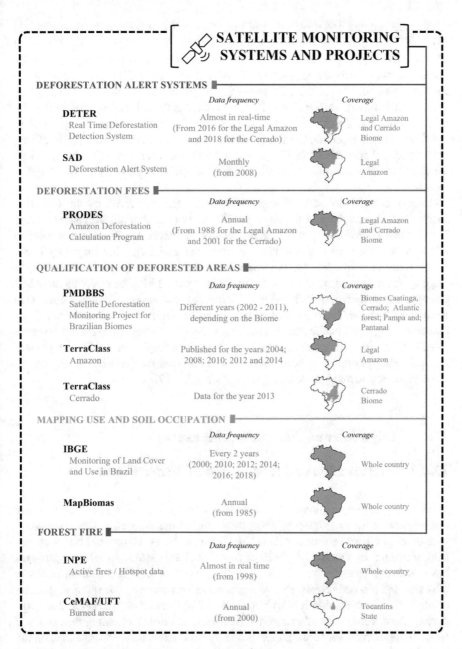

Fig. 17.2 Land-use cover monitoring systems in Brazil

very important, allowing comparison of maps from different programs and with different spatial and temporal scales developing official monitoring data on different forest biomes (MMA 2017a). The program implementation followed three steps: (1) implementation and monitoring of the Amazon and Cerrado (period 2016–2017),

(2) implementation and monitoring of the Atlantic Forest (period 2016–2017), and (3) implementation and monitoring of the Caatinga, Pampa, and Pantanal (2017–2018). However, there is no record of the progress of the Program, nor the achievement of the expected results outlined in the second edition of the *Strategy for the Environmental Monitoring Program for Brazilian Biomes* published in 2017.

A further important step of monitoring systems is the detection of active fires by satellite imagery. The monitoring activities aimed to detect the active spot within the forest fire, to calculate the fire risk, and to map burnt area scars. The platform, developed and operated by INPE, provides data from several satellites in quasi-real time, where outbreaks of at least 30 m long and 1 m wide are detected (burned.dgi. inpe.br).

Land-use and land cover changes in Brazil were monitored by the Brazilian Institute of Geography and Statistics (IBGE) and by the Annual Mapping of Land Cover and Use in Brazil project (MapBiomas). However, INGE aims to spatialize and quantify the land use and land cover of the entire Brazilian territory every 2 years and presents data for the years 2000, 2010, 2012, 2014, 2016, and 2018 (ibge.gov.br). Moreover, in 2015, a new project called MapBiomas was launched through an initiative of SEEG/OC (System of Estimates of Greenhouse Gas Emissions from the Climate Observatory), in collaboration with universities, NGOs, and technology companies, to produce annual maps of land use and land cover in Brazil since 1985 (mapbiomas.org). The project is implemented by the Google Earth Engine platform, which offers a wide processing capacity in the cloud, through extensive machine learning algorithms (Fig. 17.2).

17.4 Mitigation and Adaptation Measures

17.4.1 Preservation and Restoration of Native Forests

In order to promote sustainable forest development by reconciling the use of natural resources with the protection of ecosystems and to make forest policy compatible with other public government policies, the National Forest Program (PNF) was created, instituted by Decree No. 3420 of April 20, 2000 (MMA 2000). The program has broad objectives that range, for example, from encouraging the sustainable use of native and planted forests, recovering permanently preserved forests, supporting economic and social initiatives by populations that live and depend on forests, supporting the development of grassroots industries, and even expanding the domestic and foreign markets for forest products and by-products. The NPF was based on the participative and integrated processes involving federal, state, district, and municipal governments, as well as organized civil society (Brazil 2020). For this reason, the National Plan for the Recovery of Native Vegetation (PLANAVEG) was created as the main instrument for applying the National Policy for the Recovery of Native Vegetation (PROVEG), instituted by Decree No. 8972 of January 23, 2017. As reported in PLANAVEG, implementation, monitoring, and evaluation are carried out by the National Commission for the Recovery of Native Vegetation (CONAVEG) and aim to expand and strengthen public policies, financial incentives, markets, recovery technologies, good agricultural practices, and other measures necessary for the recovery of native vegetation of at least 12 million hectares (Mha) by the year 2030, mainly in permanent preservation areas (APPs) and legal reserve areas (RL), as well as in degraded areas with low agricultural productivity. PLANAVEG is based on eight strategic initiatives: (i) sensitization, (ii) seeds and seedlings, (iii) markets, (iv) institutions, (v) financial mechanisms, (vi) rural extension, (vii) spatial planning and monitoring, and (viii) research and development, designed to motivate, facilitate, and implement the recovery of native vegetation (MMA 2017b). Most of the 12 million hectares mentioned in the PLANAVEG are concentrated in the Amazon and Atlantic Forest biomes with about 76%, while the Cerrado biome represents 17% and the Caatinga, Pantanal, and Pampa biomes together represent 5% of the total goal. Funding to support PLANAVEG came from different sources, such as the government's budget, national and multilateral financial institutions, funds such as the Global Environment Facility (GEF), bilateral government agreements, donations, the private sector, and foundations (MMA 2017b). Established in 2017, CONAVEG was responsible for implementing the plan through two thematic advisory groups: one for the practical implementation of PLANAVEG and one for monitoring vegetation recovery. However, after a few meetings, the groups were deactivated due to the process of implementing the new structure of the Ministry of the Environment (MMA) and the process of extinguishing the federal collegiate bodies. Subsequently, CONAVEG was ended, hindering the implementation of PLANAVEG, due to the loss of the forum for discussion and coordination of actions and standards related to the different strategies adopted by the plan (Crouzeilles et al. 2019).

17.4.2 Financial Incentive Programs for Forest Conservation in Brazil

17.4.2.1 REDD+ in Brazil

The REDD+ (Reducing Emissions from Deforestation and Forest Degradation) program is the most important funding source for promoting the development of strategies for mitigating climate change and protecting tropical forests. In Brazil, the inter-ministerial working group on REDD + works to negotiate and build a national strategy based on discussions on four main points: financial architecture, technical aspects, governance and investment arrangements, and positive economic incentives (Toni 2011).

The Forest Code (Law No. 12.654/2012) became the main tool for implementing REDD+ in Brazil (Euler 2016), as it provides (i) the mandatory Rural Environmental Registry (CAR) for all rural properties in order to monitor the conservation status of forests; (ii) the institution of the Environmental Reserve Quota (CRA), a mechanism for offsetting the mandatory maintenance of forest cover required by law (art. 44, Law No. 12,651/2012); and (iii) the possibility of payment or incentives for environmental service, in order to present additionally for national and international greenhouse gas reduction markets (Art. 41, II, § 4 and § 5, Law No. 12,651/2012).

In Brazil, the national REDD+ strategy (ENREDD +) was developed between 2010 and 2016 to accomplish the rules established in the Warsaw Framework (MMA 2016a). The strategy should have met its objectives within 2020, and thereafter it will be reassessed for the next period of implementation. The general objective is to contribute to the mitigation of climate change through the elimination of illegal deforestation, conservation and the recovery of forest ecosystems, and the development of a sustainable low-carbon forest economy, to generate economic, social, and environmental benefits. The implementation of ENREDD+ should support the decentralization of funding derived by the payments for REDD+ results, promoting the development of a national REDD+ system that acts in an integrated manner at the federal and state levels (May et al. 2011). Moreover, the National Commission for REDD+ (CONAREDD+) was established by Decree No. 10,144 of November 28, 2019, to support coordination and monitoring, to improve the effectiveness of the REDD+ National Strategy, and to prepare the requirements for accessing to the payments for the achievement of REDD+ results. Resolution No. 6, of July 6, 2017, of CONAFREDD+, defines the allocation of payments for emission reductions within the Amazon biome, highlighting that most of the payments (60%) were assigned to the states of Acre, Amapá, Amazonas, Maranhão, Mato Grosso, Pará, Rondônia, Roraima, and the Tocantins. However, two criteria were considered for funding allocation: (i) the area of native forests, conservation units, and indigenous lands and (ii) the deforestation reduction rate. The remaining 40% of payments were for the federal government, for its efforts to conserve native forests in conservation units and indigenous lands and the reduction of deforestation. Funding can be directly allocated to the Amazonian states, or through the Amazon fund of the federal government. For direct funding, each state can define its own REDD+ results-based payment initiatives, beyond the REDD+ resources that they already receive through projects supported by the Amazon Fund. The Amazon Fund, launched in 2009, supports governmental and non-governmental projects in a diversified manner and aims to promote a sharing of the benefits of REDD+ received by the federal government at different scales and to promote the continuous reduction of deforestation (May et al. 2011). The management of the Amazon Fund is carried out by the National Bank of Economic and Social Development (BNDES), which also raises funds in coordination with the Ministry of the Environment, as well as contracting and monitoring the projects and sustained actions. Most of the funding came from Norway (the largest contribution) and Germany. Though to a lesser extent, Petrobrás (Petróleo Brasileiro S.A.) contributes to implementing projects financed by the Amazon Fund. From the beginning to 2017, the amount of funding received exceeded 1.2 billion dollars (MMA 2019).

For the first time in 2012, the Forest Code highlighted the importance of payments for ecosystem services (PES) to implement REDD+ in Brazil. Several federation units approved specific rules, to date, and any progresses about their implementation were reported at the federal level. However, in January 2021, Law 14119 established the National Policy for PES, aimed at contributing to climate regulation and the reduction of emissions from deforestation and forest degradation. Besides, it will support and promote the conservation and recovery of native vegetation, wildlife, and the natural environment in rural areas, forests, and forests located in urban areas, as well as water resources mainly located in hydrographic basins with critical plant coverage.

17.4.2.2 Financing Climate Change Adaptation

The Forest Code contains principles that detail incentives to promote the protection and the preservation of native vegetation, which are essential to the restoration and conservation of the environment (Costa 2016). The National Plan for the Recovery of Native Vegetation (PLANAVEG) aims to strengthen public policies and financial incentives and presents, among its eight strategic initiatives, the development of financial mechanisms to encourage the recovery of native vegetation (MMA 2017b).

One of the main instruments for balancing the sustainable use of forest resources and the mitigation of climate change is credit for financing forest activities. In Brazil, to meet the demands of companies, cooperatives, communities, family farmers, peoples, and traditional communities, financial instruments should support forest management, recovery of native vegetation in permanently preserved areas (APP) and legal reserves, forest plantations for industrial use, forest products, marketing, and working capital (MMA 2016b).

According to the information contained in the Forest Financing Guide concerning the lines in force in the second half of 2016, about 32 lines were available for funding among programs and subprograms, such as the National Program for the Strengthening of Family Farming (PRONAF), the National Program for Supporting the Medium Rural Producer (PRONAMP), the Program for Reducing Greenhouse Gas Emissions in Agriculture (ABC Program), the financing programs offered by the National Development Bank (BNDES), and the Constitutional Funds. Different financial instruments are available to cover wide situations and objectives, depending on the size and type of interested organizations or companies.

17.4.3   Forestry Production

17.4.3.1   Forestry and Production Incentive Policies

According to Antonangelo and Bacha (1998), the development of Brazilian silviculture can be split into three periods: (i) 1500–1965 before the definition of reforestation incentives, (ii) 1966–1988 testing and validity of reforestation incentives, and (iii) 1989 to nowadays use of reforestation incentives. Forestry activity in Brazil began shortly after its discovery, through the exploitation of pau-brasil, which became the main economic activity carried out in the country (Siqueira 1990). At the end of the first period, numerous efforts for planting and restoring forest ecosystems started to support forestry activities, even if these efforts were insignificant compared to the damages that occurred due to the deforestation rate. The few plantations aimed to supply the demand for sleepers and energy for rail transport demand rubber for the pneumatic industry and tannins for tanneries (Valverde et al. 2012). Despite this, in the period 1500–1965, there were mainly pioneering efforts to introduce plantations of eucalyptus and pine species (Antonangelo and Bacha 1998; Santopuoli et al. 2016b).

In the second period, there was a significant development of forestry due to the growth of forest science, the increasing number of forest professionals, the expansion of forest plantations, and the increased interest in the forestry business (Antonangelo and Bacha 1998). In 1966, Law No. 5106 of 1966 was established to support forestry companies, deducting up to 50% of the income tax amount, as they proved investments in afforestation or reforestation (minimum 10,000 trees per year), previously approved by the Ministry of Agriculture. The forest plantations area increased rapidly at the beginning of the second period (1966–1969) with about 310,000 hectares of forest plantations (Hora 2015), which further increased until 1979. Thereafter, there was a slight decrease in the annual average forest plantation area, as incentives were focused on non-timber species with a consistent reduction in tree species (Bacha 2008).

The incentive policies were the starting point and one of the main factors for the expansion of reforestation in Brazil. In the beginning, tax incentives were aimed at the development of reforestation to supply the already existing industries, consumers of paper and cellulose, and the charcoal steel industry, since, close to these units, the natural forest reserves threatened the future supply of the industries (Bacha 1991). The policy incentives represented the starting point for the development of forest plantations in Brazil. They were one of the main factors fostering the expansion of reforestation in Brazil. However, Bacha (1991) highlights that in the long term, incentives can reduce the companies' investments with negative impacts on the forestry and forest ecosystems, and thus it can't be an isolated factor for economic and industrial policies.

As Hora (2015) points out, the establishment of the Forest Code of 1965 (Law 4771, repealed by Law No. 12,651), the Tax Incentives Law (Law No. 5106 of September 2, 1966), the creation of the Brazilian Institute of Forestry Development

(IBDF) in 1967, and the creation of higher education courses focused on forestry in the 1960s opened a new vision of the Brazilian forest policy.

Currently, the forestry sector in Brazil drives the national economy with a sectorial gross domestic product (GDP) of R$ 86.6 billion, which represents 1.3% of Brazilian GDP and 6.9% of industrial GDP. These values grew by 13.1% in 2018 when compared to 2017, while the national average represented an increase in the national GDP of 1.1%, of agriculture and livestock involving only 0.1%, the service sector 1.3%, and the industry in general by 0.6%. In addition, since 2012, exports of forest products have grown by 12.3%, and in 2018, the Brazilian forestry sector was responsible for generating R$ 12.8 billion in federal taxes, corresponding to 0.9% of the entire collection of Brazil (IBÁ 2019).

The most planted forest genera in Brazil are *Pinus* and *Eucalyptus*, being the best adapted to the edaphoclimatic characteristics, having the highest levels of productivity. In the case of eucalyptus, the first studies were carried out at the beginning of the twentieth century, when Edmundo Navarro de Andrade started comparative tests between the genus *Eucalyptus* and native species (Valverde et al. 2012). In recent years, forest masses of this kind have been used to produce charcoal for the steel industry and for the production of cellulose, paper, panels, cleaning products, flavorings, and medicines, in addition to the use of eucalyptus lumber to grow each day (Valverde et al. 2012).

Between 2000 and 2018, there was an expansion (about 70%) of productive forests (IBGE 2020). In 2018, the total area of planted forests in Brazil reached 7.83 million hectares, of which 5.7 million hectares were occupied by eucalyptus (72.8% of the total planted areas), 1.6 million hectares (20.4%) by pine, while other species, including rubber, acacia, teak, and paricá, covered approximately 590 thousand hectares (7.5%). In the period 2009–2018, the surface of the eucalypt plantations increased by 1,013,507 hectares, going from 4.7 to 5.7 million hectares, while pine plantations decreased by 225,415 hectares, from 1.8 million hectares to 1.6 million hectares (IBÁ 2019). In terms of productivity, Brazil has the most productive eucalyptus and pine forests in the world, with an average of 35 m^3 ha year^{-1} and 29 m^3 ha year^{-1} for eucalyptus and pine, respectively. In addition to productivity, Brazil has one of the shortest rotations in the world, when it comes to the time between planting and harvesting trees, both for eucalyptus and pine (FAO 2001). Figure 17.3 shows a graph of Brazil's average productivity and rotation in relation to other important players in the world.

According to the report of the Brazilian Tree Industry (IBÁ 2019), in 2018, there were 7.83 million hectares of planted forests in Brazil, within which 36% for the production of pulp and paper, 29% for round wood, 12% for charcoal steel segment, 10% for financial investments, 6% for wood panels and laminate flooring, and the remaining for other timber products.

The expectations of the forestry sector are to increase the production of planted forests to supply the forest products and services market. In addition to the increase in pulp exports, making Brazil the largest player in the world market, new wood-based products have emerged, such as immunized wood artifacts, chips, and new panels (MDF, MDP, OSB) (Valverde et al. 2012).

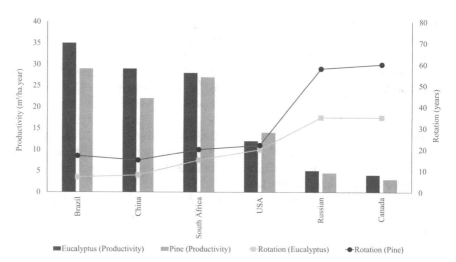

Fig. 17.3 Productivity and average rotation of different pine and eucalyptus producing countries (Source FAO 2001)

The Brazilian Tree Industry (IBÁ) in 2015 carried out a study with scenarios and trends for the planted forest sector (disregarding some factors), presented in the National Plan for the Development of Planted Forests (MAPA 2018), with the following projections until the year 2025: (i) the planted area should have an annual increment rate of 1.2%; (ii) the volume of wood produced would grow at an annual rate of 3.9%. The National Plan for the Development of Planted Forests presents the goal of more than two million hectares planted with commercial forests by the year 2030.

Despite the good expectations of the forestry sector, there are still several hindering factors to the expansion of forestry in Brazil: such as the empirical criticism without technical-scientific foundations of forest plantations; model of land-based forest production, concentrating and under extensive monoculture resulting from the Policy of Tax Incentives for Reforestation (in force from 1965 to 1988) that still persists; environmental management policies with complex legislation that are difficult to apply; the land policy that inhibits foreign investment in Brazilian land; precarious and deficient basic infrastructures for the production; legal regarding constitutional guarantees of property rights and free enterprise; and the scarce resources and policies aimed at research and development, among other obstacles (Valverde et al. 2012; Mendes et al. 2016).

17.4.3.2 Agroforestry Systems

In 2012, the Ministry of Agriculture and Supply launched the Low-Carbon Agriculture Program (Plano ABC) to promote the reduction of GHG in agriculture, to improve the efficiency in the use of natural resources, and to improve the resilience of productive systems and rural communities, including the agricultural sector in the practices of adaptation to climate change. Among other specific objectives, the ABC Plan reports the expansion of technologies: recovery of degraded pastures, crop-livestock-forest integration (iLPF) and agroforestry systems (SAFs), no-till system, biological nitrogen fixation, and planted forests (MAPA 2012).

In the last years, both the agroforestry systems (SAFs) and the crop-livestock-forest integration (iLPF) reached great evidence for their efforts to develop scientific research and projects to promote local development and environmental conservation. Despite SAFs being one of the oldest land-use techniques, currently it represents a recent frontier in the advancement of research and agriculture (Brant 2015). SAFs are gaining more space every day among agricultural enterprises, in view of their economic and environmental heterogeneity and for allowing the exploitation of finite resources in the long term (Steenbock and Vezzani 2013).

The abovementioned Forest Code established general guidelines for the restoration and exploration of legal reserve (RL) areas through SAFs, without distinguishing different SAF types and achievable objectives. In this case, the competent environmental agency is responsible for establishing acceptable criteria and standards for the restoration, exploration, and management of the RL areas (Martins and Ranieri 2014). In this context, Miccolis et al. (2019) highlighted that these knowledge and policy gaps, therefore, leave a wide margin for interpretation, leading to many uncertainties that discourage technicians from making recommendations and farmers to adopt SAFs in these areas. It is important to highlight that the Forest Code promotes the use of SAFs for the restoration of permanent preservation areas (APPs) using exotic species up to 50%, ensuring the maintenance of the ecological functions of native species.

In 2006, the National Forestry Plan with Native Species and Agroforestry Systems (PENSAF) was presented as part of the priorities of the National Forest Program. It established the basic conditions for the development of silviculture with native species and SAFs, directly providing financial income for rural owners and generating economic, social, and environmental benefits for Brazil (MMA 2006b). PENSAF was valid for 10 years with a budget estimated at approximately R$ 90 million distributed among information systems, science and technology, inputs, technical assistance and rural extension, credit, market and trade in forest products, legislation, monitoring, and control. Despite the existence of a program, presented as the first federal public policy for SAFs, its practical implementation is less known, with a lack of documents about the use of financial resources and achieved results, also because of the changes in the technical staff of government agencies.

17.4.4 National Policy on Climate Change

The National Policy on Climate Change (PNMC), instituted by Law No. 12187/2009, appears to formalize Brazil's voluntary commitment to the UNFCCC to reduce GHG emissions between 36.1% and 38.9% of projected emissions by 2020, ensuring that economic and social development contributes to the protection of the global climate. Considering the Brazilian GDP growth (5% or year until 2020) and the additional renewable energy demand, the emission of 3236 million tons CO_2-eq within 2020 was estimated.

Subsequently, Decree No. 9578/2018, which regulates Law No. 12187/2009, fixed the target of the emissions between 1168 and 1259 million tons CO_2-eq within 2020. Moreover, the PNMC established sectoral targets for meeting the aggregate target, the most relevant of which is the 80% reduction in emissions from deforestation in the Amazon, compared to the average verified between 1996 and 2005. At the time of its approval, in 2009, the PNMC Law represented an enormous role for Brazil, since few countries had legal instruments to establish their strategies to face the problem of climate change. In the same year, the National Fund for Climate Change (FNMC) was created, presenting itself as an unprecedented initiative for a developing country, idealized to become one of the most important instruments of politics (Senate Federal 2019).

As one of the instruments of the National Policy on Climate Change, in 2016, the Federal Government's National Plan for Adaptation to Climate Change was proposed to support initiatives for the management and reduction of climate risk in the long term, as established in the Ministerial Ordinance No. 150 of May 10, 2016. The plan was prepared between 2013 and 2016 by the Executive Group of the Interministerial Committee on Climate Change (GEx-CIM), as highlighted in the National Policy on Climate Change (Law No. 12,187/09) and its regulatory decree (Decree No. 7390/10).

Furthermore, Brazilian governments fixed a new target for the year 2025 according to the Paris Agreement signed at the COP21, established in the Intended Nationally Determined Contribution (iNDC). At the end of 2015, Brazil submitted its iNDC proposal and was unique among developing countries in presenting an absolute target (Brandão Jr. et al. 2018), highlighting reductions in GHG emissions by 33% within 2025 and by 43% within 2030, compared with emissions in 2005. Such contributions consist of achieving emission levels of 1.3 GtCO2-eq in 2025 and 1.2 GtCO2-eq in 2030. Between 2004 and 2012, Brazil's GDP increased by 32%, while emissions decreased by 52% (GWP-100; IPCC AR5), breaking the trend between economic growth and increased emissions during this period, reducing the per capita emissions from 14.4 tCO2-eq (GWP-100; IPCC AR5) in 2004 to 6.5 tCO2-eq (GWP-100; IPCC AR5) in 2012. The efforts to reduce emissions were visible with per capita values comparable to those that some developed countries have considered equitable and ambitious for their average emissions per capita in 2030. However, the per capita values should further decrease to 6.2 tCO2-eq (GWP-100; IPCC AR5) in 2025 and 5.4 tCO2-eq (GWP100; IPCC AR5) in 2030.

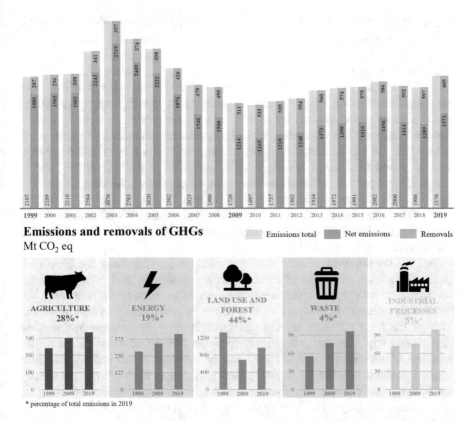

Emissions and removals of GHGs
Mt CO$_2$ eq

Fig. 17.4 Net GHG emissions from Brazil in MtCO$_2$-eq between 1999 and 2019. (Source SEEG 2020)

Figure 17.4 shows the graph with data on sectoral emissions of greenhouse gases in Brazil between 1999 and 2019 (in millions of tons of CO$_2$ equivalent, MtCO$_2$-eq), referring to the analysis of Brazilian emissions of Greenhouse Gases and their implications for Brazil's climate goals (SEEG 2020).

17.4.5   Forestry Practices for Adapting to Climate Change

Actions and strategies for climate change adaptation can be effectively implemented if poverty is reduced. According to the definition of IPCC, vulnerability to climate change is the propensity or predisposition of a given system to be adversely affected by climate change, including climate variability, extremes, and dangers. To reduce forest vulnerability and minimize the negative impacts caused by climate change, adaptation strategies are necessary. To increase resilience and reduce vulnerability, poverty must be reduced, and nature must be protected and restored (Scarano and

Ceotto 2015). While adaptation policies primarily address vulnerability and risks, sustainable development policies aim to reduce poverty through economic growth, address inequality through the redistribution of wealth, and prevent environmental degradation using resources sustainably (Agrawal and Carmen We read 2015).

In recent years, ecosystem-based adaptation (EbA), which is based on the use of ecosystem services to reduce human vulnerability to climate change, has gained space between managers and researchers as a new approach to tackling climate change. The benefits of EbA strategies include reducing vulnerability to gradual and extreme events, maintaining the ecological integrity of ecosystems, carbon sequestration, greater food security, sustainable management of water resources, and an integrated approach to territorial management, all of which generate multiple economic, social, environmental, and cultural benefits for society (MMA 2016c). In 2016, Brazil launched the National Adaptation Plan, where EbA is the fundamental part of the plan, even if no spatially explicit subnational priorities were defined in the plan (Kasecker et al. 2018).

Brazil's commitment to the iNDC is explicit in its propensity for EbA, as it states: "The implementation of climate change adaptation policies and measures contributes to building the resilience of populations, ecosystems, infrastructure and production systems, by reducing vulnerabilities or providing ecosystem services." Though there are still few experiences about the EbA implementation, the few applications implemented demonstrated the power of this tool due to the great richness and biological diversity of Brazil and the fact that Brazil has a tradition of community involvement (ICLEI 2014). EbA experiences in Brazil were funded by the International Climate Initiative (Internationale Klimaschutzinitiative, IKI), the World Bank, the Deutsche Gesellschaft für Internationale Zusammenarbeit (GIZ), the World Wildlife Fund (WWF), and FGV/GVces (ICLEI 2014).

Sustainable forest management of public Brazilian forests is based on mechanisms to promote forest management through (i) the creation and direct management of national, state, and district forests, (ii) the non-contributory allocation of forest management rights to local communities, and (iii) contributory forestry concessions in which the right to manage the forest is defined before the bidding process (SFB 2006).

17.4.6 Integrated Fire Management

Fire plays an important role as a tool for agricultural and landscape management and contributes significantly to the emission of GHG. Historically, fire was used as a tool for several traditional events in Brazil. Over the years, fire control activities have become extremely important on a global scale, and the increasing investments in research, as well as the intensification of fight actions, resulted, for a long time, in the prioritization of interventions aimed at fire-use restrictions (Toni and Pereira 2015). In Brazil, such as at the global level, for a long time, fire is conceived as a threat to the human population and natural resources. For this reason, most of the fire policies aimed to avoid the use of fire and to promote inspection, suppression, and prevention actions through restrictive laws (Falleiro et al. 2016). Nevertheless, the so-called "zero fire" policies commonly led to extensive forest fires, with duration ranging from a few hours to several days, particularly in the protected areas of Cerrado biome (Mistry et al. 2019).

By contrast, in the last decade, the use of Integrated Fire Management (IFM) gained great attention and application in Brazil. Officially, in 2014, the Brazilian government focused to adopt the concept of IFM, mainly in conservation units and indigenous lands, to reintroduce fire as a management tool in the Cerrado biome (Eloy et al. 2019). In particular, to explore and maintain the traditional knowledge about the use of fire, the first experiences with MIF were implemented in the state of Mato Grosso, in 2007, focused on the ecological aspects, such as the effects of fire on animals and fruit plants (Falleiro 2011). The IFM principles consider the use of fire by local communities to promote, as one of the tools, controlled burning in the beginning of the dry season for productive and conservation purposes in fire-resistant vegetation, to create mosaics with different burning periods, and to protect fire-sensitive vegetation from forest fires (Schmidt and Eloy 2020). The IFM consists of training residents of local communities as fire management agents to carry out controlled burning and incorporating ecological knowledge and practices within fire management (Falleiro et al. 2016). Controlled burning in the protected areas can be implemented according to the fire management plan, and it is also regulated by the national forest code. The National Integrated Fire Management Policy (PNMIF) was developed to reduce the occurrence of forest fires and damages caused by fire and to fulfill the need to establish a national policy as highlighted by the Forest Code. PNMIF provides a series of management measures to gradually replace the use of fire in rural areas, promoting the use of fire in a controlled manner, especially among traditional and indigenous communities, and increasing their capacity to cope with forest fires. Nevertheless, the process for PNMIF adoption is still at the initial stages. In the context of the IFM, tools and methodologies have been developed to assist management actions in conservation units and indigenous lands.

The tools that have made a considerable contribution come from remote sensing data, through methodologies that use spectral mixture analysis (SMA) to map vegetation conditions (green and dry vegetation) and soil detection. This methodology is very useful in planning IFM actions, as it provides data that is easy to interpret and apply, which can be assessed by indigenous and local inhabitants through smartphones and generates information on priority areas for the realization of FIM. Despite recent advances in fire management policies and practices in protected areas and indigenous lands in the Cerrado and the consequent results in the reduction of large forest fires, IFM programs are still in an initial stage, therefore requiring new studies and experiences for the improvement of the IFM in Brazil.

In the Brazilian context, IFM plays a crucial role and is strongly recommended to integrate, into the management actions, the local know-how for improving the positive effects of MIF on conservation and resilience of natural ecosystems (Schmidt et al. 2018).

17.5   Final Considerations

Despite the numerous environmental legal institutions, with detailed and complex legislation, the various policies to encourage reforestation and restoration of degraded areas, the extensive protected areas, the technologies of monitoring systems, and the large database of data obtained over decades, among other mechanisms, related to mitigation and adaptation to climate change, more practical actions are strongly necessary to adopt strategies and techniques for the climate-smart forestry. Despite all the positive factors listed above, especially when referring to Brazilian environmental legislation, there is still a great difficulty for public authorities to put into practice the actions to reach the policy objectives. These challenges are exacerbated by complex political frameworks and by continuous changes of structure and competences of environmental agencies belonging to the federal jurisdiction (IBAMA and ICMBio).

As a consequence, a decentralized setting of responsibility was developed, within which the federal government distributed the responsibilities to the states and municipalities regarding monitoring, conservation, and restoration of natural ecosystems and degraded areas through decentralized policies and tools. However, the federal governments remain mainly responsible for the management of natural resources, particularly for public lands, which are very large.

In addition, non-governmental organizations (NGOs) operating in Brazil act in a complementary manner to the actions of the federal government and have thus been playing an important role in mitigating and adapting to climate change in the country. The actions carried out by NGOs put pressure on the federal public authorities to improve their transparency mechanisms and improve the quality of results obtained.

Despite the conflicts that still exist, Brazil is a protagonist and one of the world's pioneers in signing international commitments to reduce GHG emissions, such as the Paris Agreement and the iNDC. This aspect highlights the interest of Brazil to improve the development and future implementation of mitigation and adaptation measures. Moreover, Brazil plays an important role at the global level to face climate change, due to the large forest area, the complex forest structure, and richness of forest biodiversity, across the different biomes, as well as the amount of forest carbon stock.

The abundance of natural resources, and the interest in more effective management of natural resources, including the restoration efforts of degraded natural resources, requires the integration of policies, actions, and tools that support timely monitoring on a large scale. Promoting the use of remote sensing can represent a viable strategy to support researchers and policy- and decision-makers in limiting forest damages, improving resilience, mitigation, and adaptation actions. The harmonization of a large quantity of information, derived from the different databases and platforms, is a crucial point to address in the future to support the development of climate-smart forestry.

References

Agrawal A, Carmen Lemos M (2015) Adaptive development. Nat Clim Chang 5:185–187. https://doi.org/10.1038/nclimate2501

Alencar A, Nepstad D, McGrath D, Moutinho P, Pacheco P, Diaz MDCV, Soares Filho B (2004) Desmatamento na Amazônia: indo além da "emergência crônica", vol 90. Ipam, Belém

Angelo C, Rittl C (2019) Análise das Emissões Brasileiras de Gases de Efeito Estufa e suas implicações para as metas do Brasil, 1970 - 2018. Sist. Estim. Emiss. Gases Efeito Estufa 1–33

Antonangelo A, Bacha CJC (1998) As fases da silvicultura no Brasil. Rev Bras Econ 52:207–238

Assis LFFG, Ferreira KR, Vinhas L et al (2019) TerraBrasilis: a spatial data analytics infrastructure for large-scale thematic mapping. ISPRS Int J Geo-Inf 8:513. https://doi.org/10.3390/ijgi8110513

Bacha CJC (1991) A expansão da silvicultura no Brasil. Rev Bras Econ 45:145–168

Bacha JCC (2008) Análise da evolução do reflorestamento no Brasil. Rev Econ Agric 55:5–24

Bowditch E, Santopuoli G, Binder F, del Río M, La Porta N, Kluvankova T et al (2020) What is Climate-Smart Forestry? A definition from a multinational collaborative process focused on mountain regions of Europe. Ecosyst Serv 43:101113. https://doi.org/10.1016/j.ecoser.2020.101113

Brandão Jr. A, Barreto P, Lenti F, et al (2018) Emissões do setor mudança do uso da terra

Brant HSC (2015) Os Sistemas Agroflorestais com funções ecológicas ressaltadas em áreas de conservação no Brasil. Cadernos da Disciplina Sistemas Agroflorestais 7

Brasil (2016) Pretendida contribuição nacionalmente determinada. 9:10

Brasil (2020) Decreto no 3.420, de 20 de abril de 2000

Costa MM (2016) Financiamento para a restauração ecológica no Brasil. In: da Silva APM, Marques HR, Sambuichi RHR (eds) Mudanças no Código Florestal Brasileiro: desafios para a implementação da nova lei. IPEA, Rio de Janeiro, p 359

Crouzeilles R, Rodrigues RR, Strassburg BBN et al (2019) Relatório Temático sobre Restauração de Paisagens e Ecossistemas, São Carlos

da Hora AB (2015) Análise da formação da base florestal plantada para fins industriais no Brasil sob uma perspectiva histórica

Davidson EA, Artaxo P (2004) Globally significant changes in biological processes of the Amazon Basin: results of the large-scale biosphere-atmosphere experiment. Glob Chang Biol 10:519–529. https://doi.org/10.1111/j.1529-8817.2003.00779.x

de Almeida CA, Coutinho AC, Esquerdo JC, Dalla M et al (2016) High spatial resolution land use and land cover mapping of the Brazilian legal Amazon in 2008 using Landsat-5/TM and MODIS data. Acta Amaz 46:291–302. https://doi.org/10.1590/1809-4392201505504

de Arraes RA, Mariano FZ, Simonassi AG (2012) Causas do desmatamento no Brasil e seu ordenamento no contexto mundial. Rev Econ e Sociol Rural 50:119–140. https://doi.org/10.1590/S0103-20032012000100007

de Falleiro RM, Santana MT, Berni CR (2016) As contribuições do Manejo Integrado do Fogo para o controle dos incêndios florestais nas Terras Indígenas do Brasil. Biodivers Bras – BioBrasil 6:88–105

Eloy LA, Bilbao B, Mistry J, Schmidt IB (2019) From fire suppression to fire management: advances and resistances to changes in fire policy in the savannas of Brazil and Venezuela. Geogr J 185:10–22. https://doi.org/10.1111/geoj.12245

Euler AMC (2016) O acordo de Paris e o futuro do redd+ no Brasil. Cad Adenauer XVII 2:85–104

FAGRO (2007) Mapeamento de Cobertura Vegetal do Bioma Cerrado, Brasília

Falleiro RDM (2011) Resgate do Manejo Tradicional do Cerrado com Fogo para Proteção das Terras Indígenas do Oeste do Mato Grosso: um Estudo de Caso. Biodivers Bras 2:86–96

FAO (2001) Mean annual volume increment of selected industrial forest plantation species. 1–27

FAO and UNEP (2020) The state of the world's forests 2020: forests, biodiversity and people

IBÁ (2019) Relatório 2019 Report 2019, Brasília

IBGE (2019) Produção da extração vegetal e da silvicultura 2019. Produção da extração Veg e da Silvic 8

IBGE (2020) Monitoramento da Cobertura e Uso da Terra 2016–2018, Rio de Janeiro

ICLEI (2014) Adaptação Baseada em Ecossistemas: Oportunidades para políticas públicas em mudanças climáticas. Fundação Grupo Boticário, Curitiba

INPE (2020) Monitoramento do Desmatamento da Floresta Amazônica Brasileira por Satélite

IPCC (2014) Summary for policy makers. In: Field CB, Barros VR, Dokken DJ, Mach KJ, Mastrandrea MD, Bilir TE, Chatterjee M, Ebi KL, Estrada YO, Genova RC, Girma B, Kissel ES, Levy AN, MacCracken S, Mastrandrea PR, White LL (eds) Climate Change 2014: impacts, adaptation, and vulnerability. Part A: global and sectoral aspects. Contribution of working group II to the fifth assessment report of the intergovernmental panel on climate change. Cambridge University Press, Cambridge/New York, pp 1–32

Kasecker TP, Ramos-Neto MB, da Silva JMC, Scarano FR (2018) Ecosystem-based adaptation to climate change: defining hotspot municipalities for policy design and implementation in Brazil. Mitig Adapt Strateg Glob Chang 23:981–993. https://doi.org/10.1007/s11027-017-9768-6

Laukkonen J, Blanco PK, Lenhart J et al (2009) Combining climate change adaptation and mitigation measures at the local level. Habitat Int 33:287–292. https://doi.org/10.1016/j.habitatint.2008.10.003

Lentini M, Pereira D, Celentano D, Pereira R (2005) Fatos Florestais da Amazônia 2005. Instituto do Homem e Meio Ambiente da Amazônia, Belém

MAPA (2012) Plano Setorial de Mitigação e Adaptação às Mudanças Climáticas para Consolidação da Economia de Baixa Emissão de Carbono na Agricultura – PLANO ABC, 1st edn, Brasília

MAPA (2018) Plano nacional de desenvolvimento de florestas plantadas, Brasília

MAPA (2019) Florestas do Brasil em resumo 2019, Brasília

Martins TP, Ranieri VEL (2014) Agroforestry as an alternative to legal reserves. Ambient Soc 17:79–96. https://doi.org/10.1590/S1414-753X2014000300006

Martins OS, Rettmann R, Pinto EDPP et al (2009) Redd – summary of the Brazil case, Brasília

May PH, Millikan B, Gebara M (2011) The context of REDD+ in Brazil: drivers, agents and institutions, 2nd edn. CIFOR, Bogor

MCTIC (2017) Estimativas Anuais de Emissões de Gases de Efeito Estufa no Brasil, 3o. Ministèrio de Ciência, Tecnologia, Inovações e Comunicações

MCTIC (2019) Estimativas Anuais de Emissões de Gases de Efeito Estufa no Brasil. Ministério da Ciência, Tecnologia, Inovações e Comunicações. Secretaria de Políticas para a Formação e Ações Estratégicas. Coordenação-Geral do Clima, Brasília

Mendes L, Treichel M, Beling RR (2016) Anuário Brasileiro da Silvicultura Brazilian forestry and timber yearbook 2016, Santa Cruz do Sul

Miccolis A, Peneireiro FM, Marques HR (2019) Restoration through agroforestry in Brazil: options for reconciling livelihoods with conservation. In: van Noordwijk M (ed) Sustainable development through trees on farms: agroforestry in its fifth decade. World Agroforestry (ICRAF) Southeast Asia Regional Program, Bogor, pp 205–227

Mistry J, Schmidt IB, Eloy L, Bilbao B (2019) New perspectives in fire management in South American savannas: the importance of intercultural governance. Ambio 48:172–179. https://doi.org/10.1007/s13280-018-1054-7

MMA (2000) Programa Nacional de Florestas, Brasília

MMA (2006a) Uso e Cobertura da Terra na Floresta Amazônica, Brasília

MMA (2006b) Plano Nacional de Silvicultura com Espécies Nativas e Sistemas Agroflorestais – PENSAF, Brasília

MMA (2013) Projeto TERRACLASS Cerrado, Brasília

MMA (2015) Mapeamento do uso e cobertura do Cerrado, Brasília

MMA (2016a) ENREDD+: estratégia nacional para redução das emissões provenientes do desmatamento e da degradação florestal, conservação dos estoques de carbono florestal, manejo sustentável de florestas e aumento de estoques de carbono florestal. Ministério do Meio

Ambiente. Secretaria de Mudanças Climáticas e Qualidade Ambiental. Departamento de Políticas de Combate ao Desmatamento, Brasília

MMA (2016b) Guia de Financiamento Florestal: 2016. Ministério do Meio Ambiente, Brasília

MMA (2016c) National adaptation plan. Ministry of Environment, Brasília

MMA (2017a) Estratégia do Programa Nacional de Monitoramento Ambiental dos Biomas Brasileiros, Brasília

MMA (2017b) Plano Nacional de Recuperação da Vegetação Nativa, Brasília

MMA (2019) Fundo Amazônia: Relatório de atividades 2019, Brasília

MMA, IBAMA (2011) Monitoramento do desmatamento nos biomas brasileiros por satélite, Brasília

Ometto JP, Aguiar APD, Martinelli LA (2011) Amazon deforestation in Brazil: effects, drivers and challenges. Carbon Manag 2:575–585. https://doi.org/10.4155/cmt.11.48

Pao HT, Tsai CM (2011) Modeling and forecasting the CO2 emissions, energy consumption, and economic growth in Brazil. Energy 36:2450–2458. https://doi.org/10.1016/j.energy.2011.01.032

Parikh J (2000) Inequity: a root cause of climate change. IHDP Updat Mag Int Hum Dimens Progr Glob Environ Change 3:2–3

Rodrigues Filho S, Lindoso DP, Bursztyn M, Nascimento CG (2016) Climate in trance: mitigation and adaptation policies in Brazil. Rev Bras Climatol 19:74–90. https://doi.org/10.5380/abclima.v19i0.48874

Santopuoli G, Cachoeira JN, Marchetti M, Viola MR, Giongo M (2016a) Network analysis to support environmental resources management. A case study in the Cerrado, Brazil. Land Use Policy 59:217–226

Santopuoli G, Marchetti M, Giongo M (2016b) Supporting policy decision makers in the establishment of forest plantations, using SWOT analysis and AHPs analysis. A case study in Tocantins (Brazil). Land Use Policy 54:549–558. https://doi.org/10.1016/j.landusepol.2016.03.013

SBS (2008) Fatos e números do Brasil florestal, São Paulo

Scarano FR, Ceotto P (2015) Brazilian Atlantic forest: impact, vulnerability, and adaptation to climate change. Biodivers Conserv 24:2319–2331. https://doi.org/10.1007/s10531-015-0972-y

Schmidt IB, Eloy L (2020) Fire regime in the Brazilian Savanna: recent changes, policy and management. Flora Morphol Distrib Funct Ecol Plants 268:151613. https://doi.org/10.1016/j.flora.2020.151613

Schmidt IB, Moura LC, Ferreira MC et al (2018) Fire management in the Brazilian savanna: first steps and the way forward. J Appl Ecol 55:2094–2101. https://doi.org/10.1111/1365-2664.13118

Senado Federal (2019) Avaliação da Política Nacional sobre Mudança do Clima, Brasília

SEEG (2020) Seeg 8 - Análise das emissões Brasileiras de gases de efeito estufa e suas implicações para as metas de clima do Brasil 1970-2019. Renew. Sustain. Energy Rev. 3:1–41

SFB (2006) Forest management for sustainable production of goods and services in Brazil, Brasília

Silva CHRT (2010) Cop16, Metas Voluntárias E Reforma Do Código Florestal: O Desmatamento No Brasil E a Mitigação Da Mudança Global Do Clima, Brasília

Siqueira JDP (1990) A atividade florestal como um dos instrumentos de desenvolvimento do Brasil. Congr Florest Bras 6:8–15

Souza CM, Shimbo JZ, Rosa MR et al (2020) Reconstructing three decades of land use and land cover changes in brazilian biomes with landsat archive and earth engine. Remote Sens 12. https://doi.org/10.3390/RS12172735

Steenbock W, Vezzani FM (2013) Agrofloresta – Aprendendo a Produção com a Natureza

Toni F (2011) Decentralization and REDD+ in Brazil. Forests 2:66–85. https://doi.org/10.3390/f2010066

Toni F, Pereira LEC (2015) Fogo nas savanas tropicais da América Latina: entre a criminalização e o manejo. 1–5

Valverde SR, Mafra JWA, da Miranda MA et al (2012) Silvicultura Brasileira – Oportunidades e desafios da economia verde. Coleção Estud sobre diretrizes para uma Econ verde no Bras 39

Verkerk PJ, Costanza R, Hetemäki L et al (2020) Climate-smart forestry: the missing link. For Policy Econ 115. https://doi.org/10.1016/j.forpol.2020.102164

Index

© The Author(s) 2022
R. Tognetti et al. (eds.), *Climate-Smart Forestry in Mountain Regions*, Managing
Forest Ecosystems 40, https://doi.org/10.1007/978-3-030-80767-2

Printed in the United States
by Baker & Taylor Publisher Services